Chemo-Enzymatic Cascade Reactions

Chemo-Enzymatic Cascade Reactions

Dunming Zhu

WILEY-VCH

Author

Prof. Dunming Zhu
Tianjin Institute of Industrial Biotechnology
Chinese Academy of Sciences
32 Xi Qi Dao
Tianjin Airport Economic Area
300308 Tianjin
China

Cover
Gear wheels © picoStudio/Shutterstock, Medicine abstract background © Zoezoe33/Shutterstock

Library of Congress Card No.:
applied for

British Library Cataloguing-in-Publication Data
A catalogue record for this book is available from the British Library.

Bibliographic information published by the Deutsche Nationalbibliothek
The Deutsche Nationalbibliothek lists this publication in the Deutsche Nationalbibliografie; detailed bibliographic data are available on the Internet at <http://dnb.d-nb.de>.

Print ISBN: 978-3-527-34451-2
ePDF ISBN: 978-3-527-81427-5
ePub ISBN: 978-3-527-81428-2
oBook ISBN: 978-3-527-81426-8

Typesetting SPi Global, Chennai, India
Printing and binding CPI Group (UK) Ltd, Croydon, CR0 4YY

Printed on acid-free paper

C083361_040321

Contents

Preface

After doing research in chemical catalysis for more than 10 years, I accidently entered the field of biocatalysis, that has trapped me until now. I began to realize that biocatalysis can offer solutions to some synthetic problems which the traditional chemistry may not be able to solve, and combination of chemical reaction with biotransformation provides unprecedented tools for the needs in fine chemical industry. I have witnessed the rapid expansion of this research field and their applications to solve real industrial problems. In our own endeavor, we recently developed some microbial catalysts that effectively transform phytosterols, a renewable material and by-product of oil industry, into a series of key intermediates for the production of steroidal medicines. They have been implemented at industrial scale to replace the previous synthetic routes from diosgenin which requires a large amount of field to grow the plant and waste-producing process to extract it from the plant. When integrated into the synthesis of steroidal drugs, these bioprocesses are leading to dramatic improvement in production efficiency and environmental impact for steroidal pharmaceutical industry.

While organic chemical reactions have made major contributions to the production of fine chemicals that meet our growing needs in materials, pharmaceuticals, food additives, cosmetics among others, they also have adverse effects on the environment. Advances in biotechnology offer new avenues for addressing these problems faced by chemical industry. In this context, integration of biotransformation with chemical reaction will lead to novel green synthetic processes for the production of the desired chemical products, because they provide complementary advantages in the molecular construction and functional group transformation.

When Dr. Lifen Yang at Wiley contacted me about writing a book, chemoenzymatic cascade reactions came to my mind. In spite of the rapid progress in biocatalysis, I often get questions from researchers in organic synthesis about what biocatalysis can offer and their concerns about the handling of enzymes and the scale-up of enzymatic reactions. On the other hand, the researchers in enzymology often cannot find valuable applications for the enzymes in their hands. Actually, enzyme is just a class of unique catalysts and can work well with other types of catalysts in the same reaction flask under proper operation to achieve unprecedented transformations. Therefore, I hope that this book not only presents the state-of-art of this intersection of biocatalysis and traditional chemical reaction, but also serves as an introduction to those who are interested in this growing field.

Along the long journey of writing, Dr. Yang, Ms. Shirly Samuel, and Ms. Katherine Wong have shown kind patience and offered immense assistance. I greatly appreciate their professional assistance that make the book possible. I am greatly indebted to my colleague Professor Peter C. K. Lau who has proofread the manuscript and given numerous constructive suggestions. I am also grateful to my colleagues and students who have made excellent contributions to our projects on enzyme discovery and engineering, biocatalytic reaction mechanism and integration of biotransformation into organic synthesis. I also thank my wife Dr. Ling Hua, my son James, and daughter Elizabeth for their support because I had to spend much more time on my computer than I should spend with them. I also enjoyed writing when our cat Ellie was lying by my computer.

Source: Dunming Zhu.

1

Introduction

Chemical processes are vital for the manufacturing of goods that meet the human's growing needs. Especially since early last century, organic chemical reactions have greatly contributed to the production of fine chemicals as materials, pharmaceuticals, food additives, cosmetics, etc., which are related to our daily lives. In the meantime, fine chemical industry has contributed to increasing air pollution and environmental contamination, which have adverse effects on the earth, our health, and the quality of our daily lives. The *E* (environmental) factor (mass of waste/mass of product, kgs/kg), which is often used to assess the environmental impact of a manufacturing process, for the production of fine chemicals is usually 5–50, or even higher for pharmaceuticals ($25 \geq 100$) [1]. Therefore, it is desirable to develop green organic chemical processes for the manufacturing of the desired chemical products, thus enabling sustainable development of the fine chemical industry [2].

On the other hand, Nature has created and evolved a diversity of enzymes that catalyze numerous kinds of reactions in live organisms and show advantages over traditional chemical reactions, such as high chemo-, regio- and stereoselectivities, and mild reaction conditions. Enzymes can catalyze various reactions that are difficult to be achieved by traditional chemical reactions. When being incorporated into organic synthesis, enzyme catalysis can reduce the number of reaction steps by eliminating the protection and deprotection steps or redesigning the synthetic route with enzymatic reactions to achieve greater efficiency or atom economy [3]. For example, 19-Nor-steroids are key intermediates for the production of contraceptives, such as norethindrone, mifepristone, and tibolone. By employing the biocatalytic hydroxylation at C-19 of steroids, the synthesis of 19-nor-steroids can be achieved in three chemoenzymatic steps [4]. However, the chemical demethylation of steroids usually requires many more steps as shown in Scheme 1.1 [5].

Retrosynthetic analysis involving both chemical and enzyme catalysis enables the design of novel synthetic sequences for the preparation of complex organic molecules such as active pharmaceutical ingredients [6]. Chemoenzymatic cascade reactions thus play an important role in developing green chemical processes by reducing the waste, energy consumption, and production cost. Great advances in this field have been achieved in the last two decades, and industries are paying increasing attention to enzyme application in the green production of chemicals used for pharmaceuticals, food additives, cosmetics, and so on [7].

Chemo-Enzymatic Cascade Reactions, First Edition. Dunming Zhu.

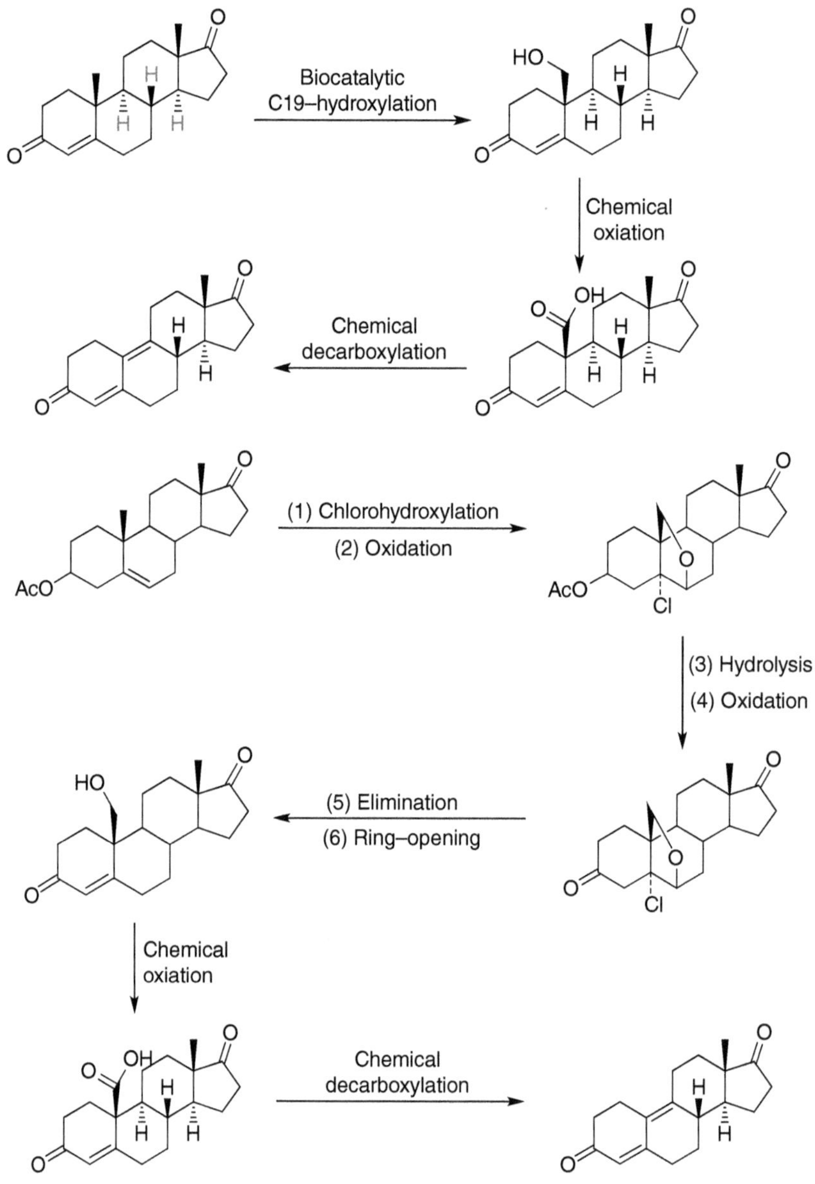

Scheme 1.1 Chemoenzymatic and chemical demethylation of steroids. Source: Based on Wang et al. [5].

In the first chapter of this book, the unique features of enzyme catalysis compared to traditional chemical reactions will be discussed. Next, different modes of chemoenzymatic transformations will be covered with some examples of the operating processes. In the remaining chapters, recent advances in this field, organized according to the modes of chemoenzymatic transformations, will be presented in more detail.

1.1 Advantages of Enzyme Catalysis

1.1.1 Chemoselectivity

Enzyme catalysis is usually highly chemoselective, and the specific transformation of one functional group can be achieved without affecting the other active functional groups in the same molecule, a feature that otherwise cannot be realized by traditional chemical reactions. For example, chemical hydrolysis of nitrile group to carboxylic acid requires strong basic or acidic conditions at elevated temperature, under which the ester group is also hydrolyzed. Thus, it is impossible to chemo-selectively hydrolyze the nitrile group indiscriminately in the presence of ester group in the same molecule. On the other hand, this challenge can be addressed by using nitrilases, which can catalyze the chemo-specific hydrolysis of nitriles under neutral conditions to give the corresponding carboxylic acids in the presence of other acid- or base-sensitive functional groups [8]. For example, ethyl (*R*)-4-cyano-3-hydroxybutyate was hydrolyzed by a recombinant nitrilase from *Aarabidopsis thaliana* (AtNIT2) to give ethyl (*R*)-3-hydroxyglutarate. The ester group in the molecule remained intact (Scheme 1.2) [9]. This key building block for the synthesis of the cholesterol-lowering drug, rosuvastatin, was produced with excellent biocatalyst productivity (55.6 g/g wet cells weight) and space-time productivity (625.5 g/l d).

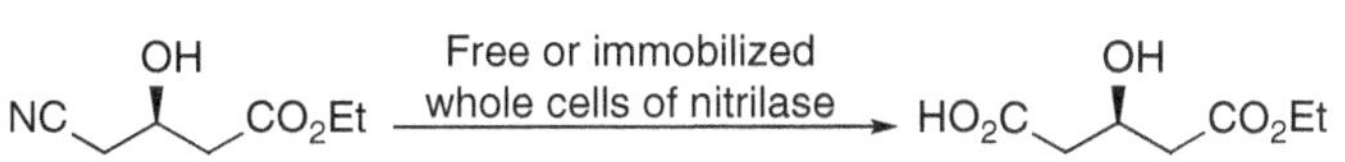

Scheme 1.2 Nitrilase-catalyzed chemospecific hydrolysis of ethyl (*R*)-4-cyano-3-hydroxybutyate. Source: From Yao et al. [9]. © 2014, John Wiley & Sons.

Reduction of aldehyde and ketone to give alcohols is an important transformation in organic synthesis. Traditional chemical carbonyl reduction methods often show low chemo-selectivity toward either aldehyde or ketone. To achieve selective reduction of aldehyde in the presence of keto group, and vice versa, careful selection of the reducing agent and control of the reaction conditions are usually required. On the other hand, Nature has evolved many aldehyde or ketone reductases, which can catalyze the chemo-specific reduction of aldehyde functional group in the presence of keto group, or vice versa, to give the corresponding alcohols. For example, an NADPH-dependent aldehyde reductase from *Oceanospirillum sp.* MED92 (OsAR) catalyzes the selective reduction of the aldehyde group in 4-acetylbenzaldehyde without reducing the keto group, affording 4-acetylbenzyl alcohol as the sole product (Scheme 1.3) [10].

Carboxylic acid functional group is difficult to be reduced and usually requires strong reducing agents, which in turn can often reduce C=O, C=N, and other functional groups. Thus, it is quite challenging to achieve chemo-specific reduction of carboxylic acid group in the presence of other reducible functional groups such as C=O and C=N groups in the same molecule. Furthermore, the reduction of carboxylic acid is difficult to stop at the aldehyde intermediate, since the latter can

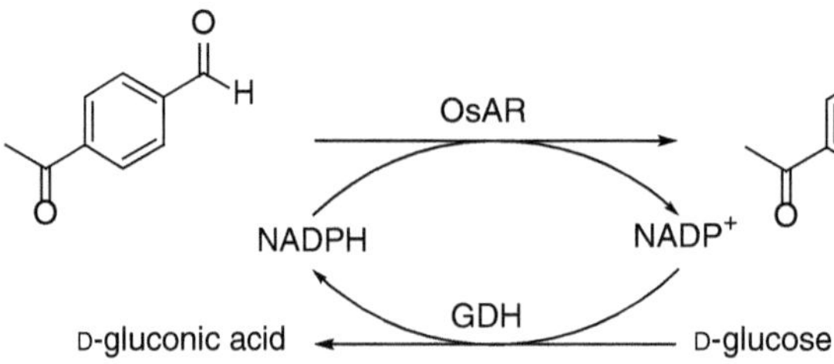

Scheme 1.3 The reduction of 4-acetylbenzaldehyde catalyzed by OsAR. Source: Based on Li et al. [10].

be further reduced to the corresponding alcohol by the reducing agents [11, 12]. However, by using carboxylic acid reductases (CAR, E.C.1.2.1.30) as biocatalyst these problems can be solved, because they catalyze the selective reduction of carboxylic acids into the corresponding aldehydes under mild conditions [13, 14]. This enzyme does not catalyze the reduction of other functional groups such as keto groups and C=N double bonds. Scheme 1.4 shows the enzymatic reduction of a representative keto acid (4-methyl-5-oxo-5-phenylpentanoic acid) to the corresponding keto aldehyde, leaving the keto group intact using a recombinant CAR from *Mycobacterium marinum* (MmCAR) [15].

Scheme 1.4 CAR-catalyzed reduction of keto acid to keto aldehyde.

Ene reductase [16], carbonyl reductase [17], and imine reductase [18] catalyze the specific reductions of C=C, C=O, or C=N functional group, respectively. Ene reductase catalyzes the reduction of C=C bond without affecting the C=O or C=N functional group in the molecule (Scheme 1.5) [19]. This is difficult to be realized by the metal-catalyzed hydrogenation reaction since the C=O or C=N functional group may also be hydrogenated during the reduction of the C=C bond.

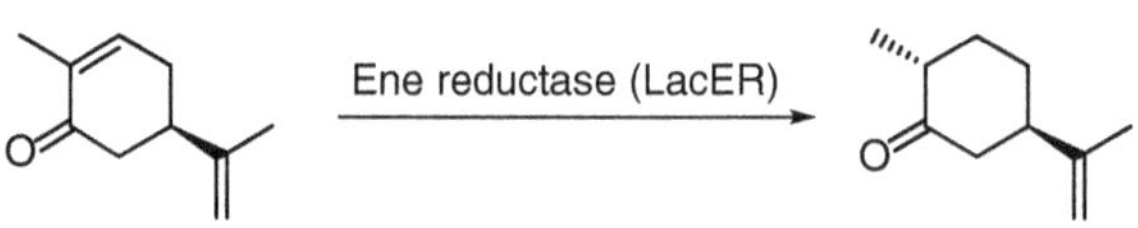

Scheme 1.5 Reduction of (*R*)-carvone by an ene reductase (LacER) from *Lactobacillus casei*. Source: Based on Chen et al. [19].

1.1.2 Regioselectivity

Enzymes can differentiate the same functional group at different positions in one molecule; thus, enzymatic reactions are usually highly regioselective. Incorporation of regiospecific enzymatic reaction into organic synthesis often simplifies the synthetic route of a target compound by eliminating the protecting and de-protecting steps. The regiospecific reduction of either the trifluoromethyl or the methyl keto group in the methyl/trifluoromethyl diketones was achieved by using some commercially available ketoreductases (KREDs) to give either the *R* or the *S* enantiomer with >98% enantiomeric excess (ee), as shown in Scheme 1.6 [20]. Among the three keto groups in the molecule of cortisone, the keto group at 11-position was highly regio- and stereospecifically reduced to give 11β-hydrocortisone by a mutant 11β-hydroxysteroid dehydrogenase (11β-HSDH) from guinea pig (Scheme 1.7) [21].

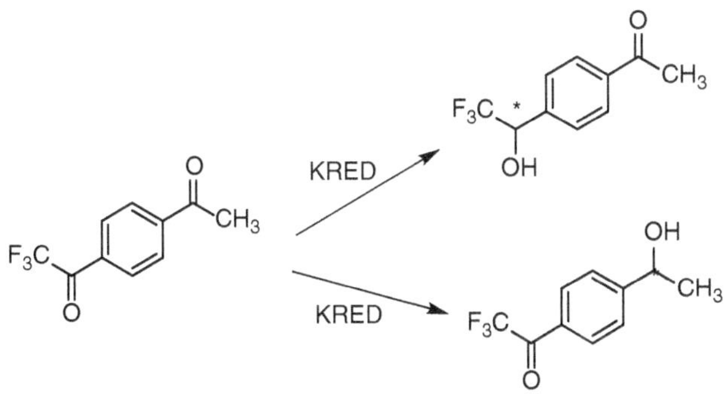

Scheme 1.6 Enzymatic regiospecific reduction of the methyl/trifluoromethyl diketone. Source: Based on Grau et al. [20].

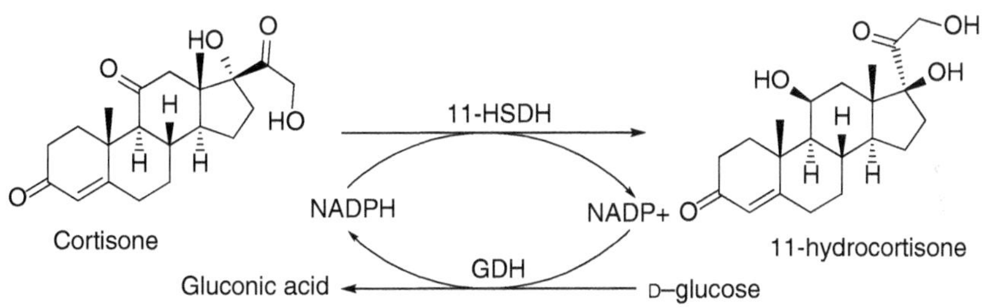

Scheme 1.7 Enzymatic regio- and stereospecific reduction of cortisone. Source: From Zhang et al. [21]. © 2014, Springer Nature.

Like alcohol dehydrogenase, other oxidoreductases also show excellent regioselectivity. For example, *meso*-diaminopimelate dehydrogenase (*meso*-DAPDH, EC 1.4.1.16) acts on the D-configuration amino group to generate L-2-amino-6-oxopimelate (Scheme 1.8) [22].

Hydrolases such as lipases and nitrilases often exhibit exquisite regioselectivity. An immobilized lipase from *Staphylococcus xylosus* catalyzed the acylation of one of the hydroxyl groups of ascorbic acid, leading to the lipophilic ascorbyl esters in good yield (Scheme 1.9) [23]. The optically active 3-alkylglutaric acid monoesters bearing various alkyl substituents were prepared by the selective hydrolysis of prochiral 3-alkylglutaric acid diesters using commercially available lipase *Candida antarctica*

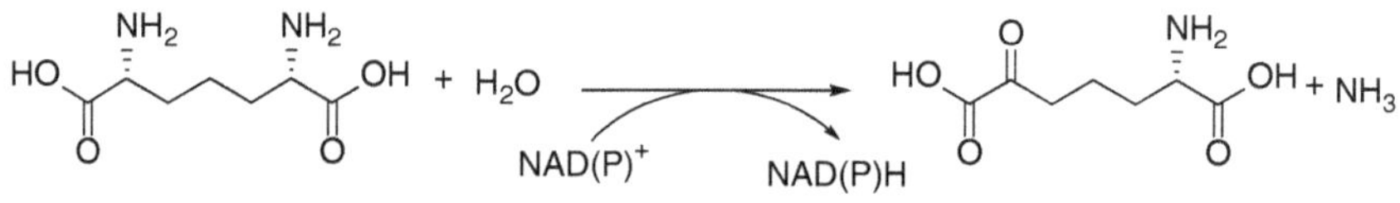

Scheme 1.8 Enzymatic regio- and stereospecific oxidative deamination of *meso*-diaminopimelate. Source: Adapted from Akita et al. [22].

lipase B (CAL-B). The unreacted ester group can then be converted to amino group, affording 3-substituted γ-aminobutyric acids, the important γ-aminobutyric acid (GABA) derivatives (Scheme 1.10) [24].

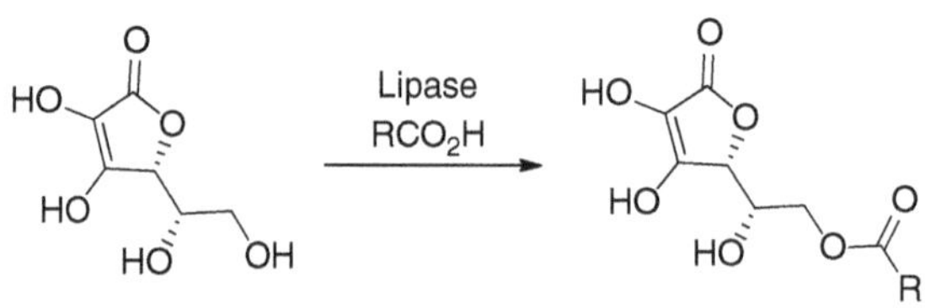

Scheme 1.9 Lipase-catalyzed regioselective acylation of hydroxyl groups. Source: Adapted from Kharrat et al. [23].

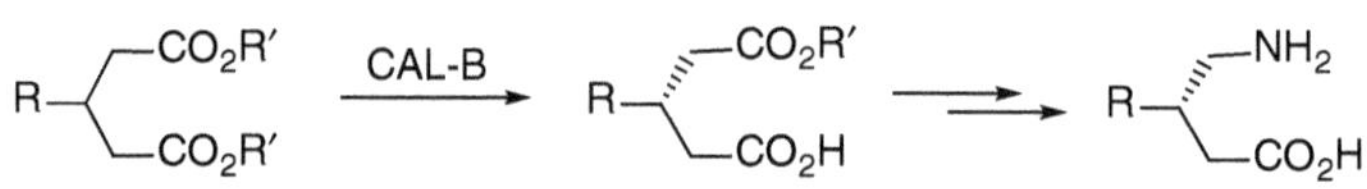

Scheme 1.10 Lipase-catalyzed selective hydrolysis of diesters. Source: Adapted from Jung et al. [24].

A nitrilase (bll6402) from *Bradyrhizobium japonicum* not only catalyzed the regiospecific hydrolysis of cyano group with less steric hindrance in cyclic dinitriles but also acted exclusively on one of the two exactly same CN groups in a molecule to produce the corresponding cyanocarboxylic acids, as shown in Scheme 1.11 [25, 26]. This transformation is not possible by traditional chemical hydrolysis. Furthermore, the desymmetric hydrolysis of prochiral 3-substituted glutaronitriles was achieved using several nitrilases of different origins and their mutant enzymes. The generated optically active 3-substituted-4-cyanobutanoic acids can be further transformed

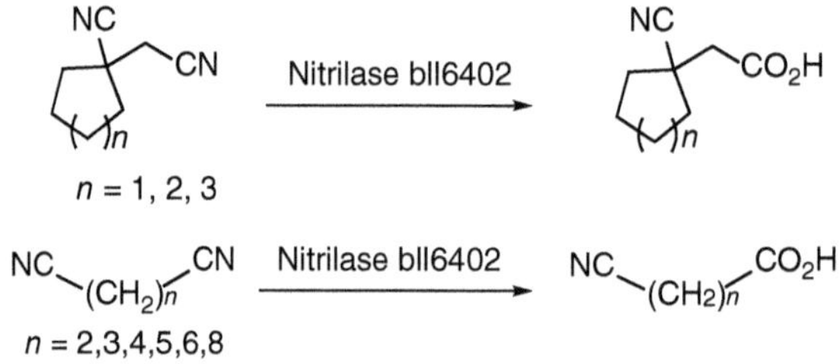

Scheme 1.11 Nitrilase-catalyzed the regiospecific hydrolysis of dinitriles. Source: Veselá et al. [25]; Zhu et al. [26].

into pharmaceutically important GABA derivatives such as the currently marketed drugs, (*S*)-Pregabalin and (*R*)-Baclofen (Scheme 1.12) [27, 28].

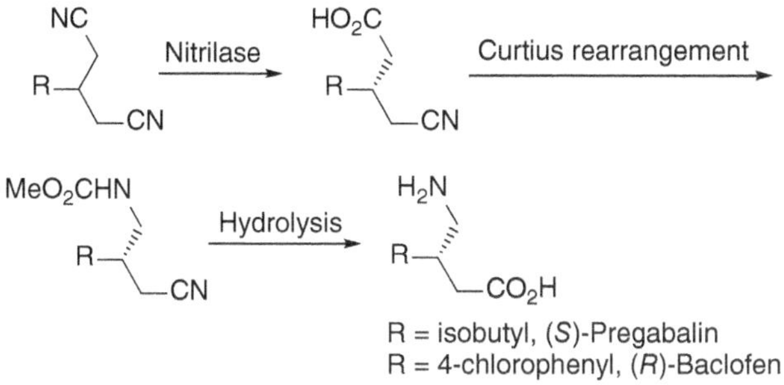

Scheme 1.12 Nitrilase-catalyzed desymmetric hydrolysis of prochiral 3-substituted glutaronitriles. Source: Duan et al. [27]; Yu et al. [28].

1.1.3 Stereoselectivity

The exquisite stereoselectivity of enzymes often outperforms the traditional chemical catalysis, making enzyme catalysis an alternative or complimentary toolbox for asymmetric transformations [29]. For chemical reduction of ketones, high enantioselectivity is achieved only when the two substituents flanking the carbonyl functional group are sterically and/or electronically very different. For example, the transition-metal-catalyzed hydrogenation of diaryl ketones usually requires an *ortho*-substituent on one of the aryl groups to realize high enantiocontrol in the product formation. The chemical reduction of diaryl ketones with only a *para*- or *meta*-substituent on one of the aryl groups affords the diarylmethanol in low enantiomeric purity (ee often being less than 50%) [30]. In contrast, high enantioselectivity has been achieved by enzymatic reduction. A carbonyl reductase from red yeast *Sporobolomyces salmonicolor* AKU4429 (SSCR) and its mutant enzymes effectively catalyzed the enantioselective reduction of diaryl ketones to give the corresponding chiral alcohols with up to 92% ee [31, 32]. Recently, a mutant alcohol dehydrogenase from *Ralstonia* sp. (RasADH) catalyzed the reduction of these ketones with ee values of 93, 95% for *p*-Cl, *p*-CH_3, respectively, or 97% for *p*-OCH_3 (Scheme 1.13).

(*R*)-Tetrahydrothiophene-3-ol is a key component in Sulopenem, a potent antibacterial prodrug with broad-spectrum antibacterial activity against most gram-positive and gram-negative bacteria. A straightforward synthesis of this alcohol is the reduction of tetrahydrothiophene-3-one; but the near spatial symmetry of the ketone results in low optical purity (23–82% ee) by chemical reduction. However, both high yield (81–88%) and >99% ee of (*R*)-tetrahydrothiophene-3-ol could be obtained by using a mutant Alcohol dehydrogenase from *Lactobacillus kefir* (LKADH) as biocatalyst when tetrahydrothiophene-3-one at a concentration of 100 g/l was used (Scheme 1.14). This biocatalytic process has successfully replaced an original multistep hazardous process starting from an achiral pool of

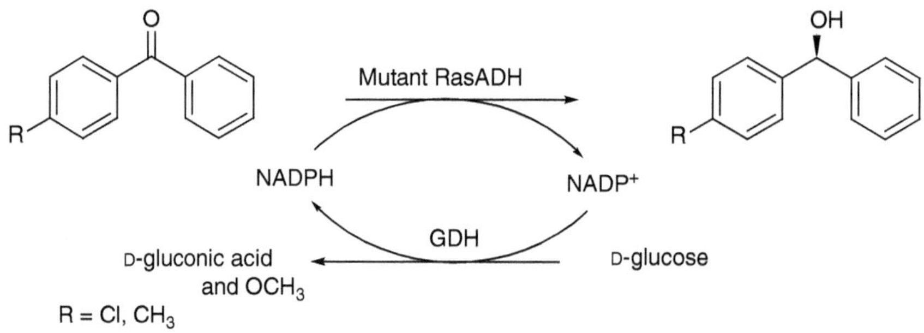

Scheme 1.13 Mutant RasADH-catalyzed reduction of diarylketones.

substrates [34]. Similarly, highly (*R*)- and (*S*)-selective variants of alcohol dehydrogenase from *Thermoethanolicus brockii* (TbSADH) catalyzed the reduction of tetrahydrofuran-3-one and other difficult-to-reduce ketones to both enantiomers of the corresponding chiral alcohols with high ee values (Scheme 1.14) [33].

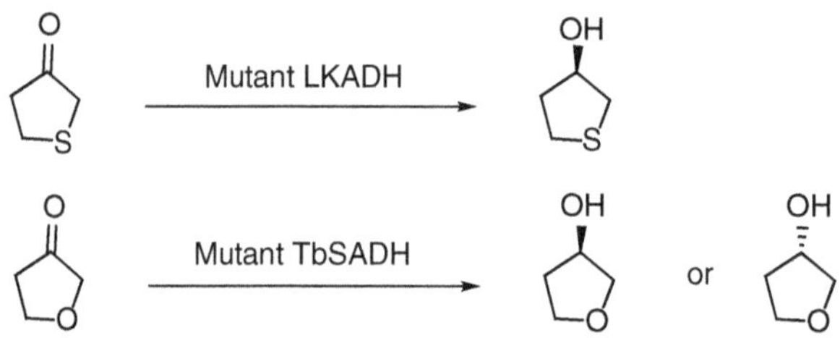

Scheme 1.14 Ketoreductase-catalyzed reduction of tetrahydrothiophene-3-one and tetrahydrofuran-3-one. Source: Based on Sun et al. [33].

(13*R*, 17*S*)-Ethyl secol is a key intermediate for the synthesis of steroidal medicines, such as gestodene and levonorgestrel. A straightforward approach to access (13*R*, 17*S*)-ethyl secol is the desymmetric reduction of ethyl secodione, which creates two chiral centers in one reaction step. The chemical reduction of ethyl secodione at low temperature led to the isolation of (13*R*, 17*R*)-ethyl secol in 25–30% yield and 75% ee [35]. Recently, by using an engineered carbonyl reductase (RasADH F12) from *Ralstonia* sp. as the biocatalyst, reductive desymmetrization of ethyl secodione and other 2,2-disubstituted cyclopentadiones led to the production of essentially one single stereoisomer, (13*R*, 17*S*)-ethyl secol, in up to 94% isolated yield (Scheme 1.15) [36].

1.1.4 Mild Reaction Conditions

In addition to high chemo-, regio-, and stereoselectivity, it is well known that enzymatic reactions are usually carried out in aqueous buffer under neutral or close to neutral pH and at room temperature. These mild conditions can keep the potentially labile functional groups in the molecule intact in the synthesis of a target compound without involving the protection/deprotection steps of a

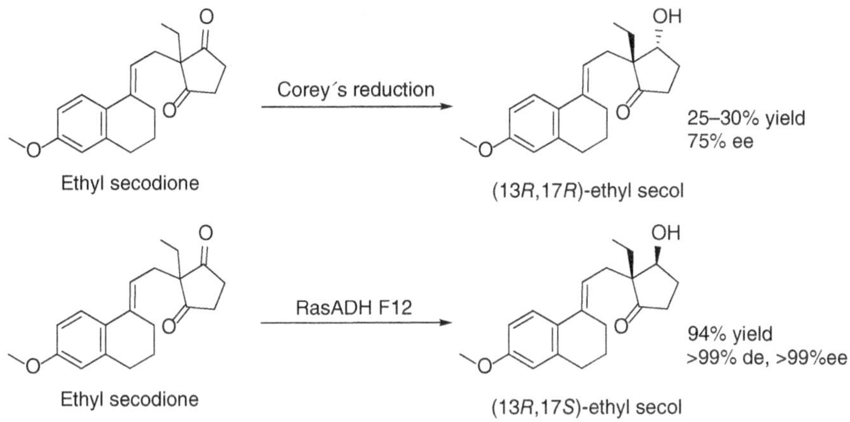

Scheme 1.15 Chemical and biocatalytic reduction of ethyl secodione. Source: Adapted from Chen et al. [36].

traditional chemical transformation. Additionally, the enzymatic reaction may proceed efficiently and cleanly, and the formation of unwanted by-products can be minimized. One example is the reduction of α-bromoacetophenones, an important transformation for the preparation of optically active β-bromo alcohols, epoxides, or diols. The reduction, using metal hydrides as the reducing agent, often causes loss of the bromo group, leading to low yield of α-bromohydrins and difficulty in product purification. In order to address this challenge [37], the isolated carbonyl reductase from *Candida magnolia* (CMCR) was employed in combination with a D-glucose dehydrogenase/D-glucose cofactor regeneration system for the reduction of α-bromoacetophenones. The reaction was performed in a two-phasic reaction medium and the concomitant loss of bromine was effectively prevented. The optically pure α-bromohydrins were prepared in 79–92% yields and >99% ee (Scheme 1.16) [38].

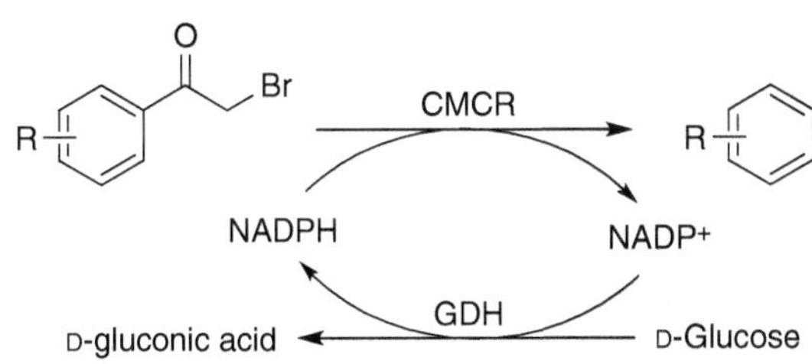

Scheme 1.16 Enzymatic reduction of α-bromoacetophenones. Source: From Ren et al. [38]. © 2012, Elsevier.

Chemical hydrolysis of nitriles requires strong basic or acidic conditions and elevated reaction temperature. Therefore, for the chemical hydrolysis of β-hydroxy nitriles to the β-hydroxy acid, it is difficult to avoid the undesirable elimination of OH group that results in the formation of the unsaturated by-products, because it

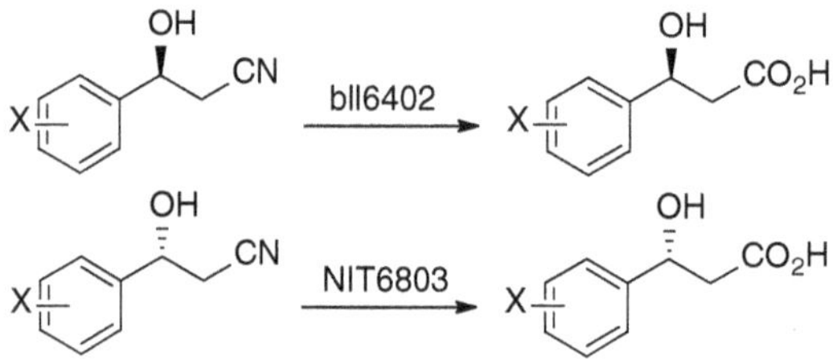

Scheme 1.17 Nitrilase-catalyzed hydrolysis of β-hydroxy nitriles. Source: From Ankati et al. [40]. © 2009, American Chemical Society.

cannot tolerate such harsh reaction conditions [39]. A couple of nitrilases (bll6402, NIT6803) from different sources have been shown to effectively catalyze the hydrolysis of β-hydroxy nitriles without affecting the β-hydroxy group. The corresponding β-hydroxy carboxylic acids were obtained in excellent yields (Scheme 1.17) [40].

1.2 Modes of Chemoenzymatic Transformations

Because of the unparalleled selectivity and mild reaction conditions, enzymes are expected to play an exciting role in "green chemistry." Chemoenzymatic cascade strategies offer unprecedented opportunities for developing efficient and sustainable synthetic technologies to address the issues of health, environment, energy, and security that we face today.

Chemoenzymatic transformations can be carried out in one of the three modes: separate-pot two-step, one-pot two-step, and one-pot one-step (Scheme 1.18). This is a simplified presentation because more than one enzymatic reaction can be combined with multiple chemical transformations to achieve a synthetic goal. Especially for the "separate-pot two-step" mode, infinite number of either enzymatic or chemical reactions can be incorporated into a synthetic route of a target compound as the case may be. For the other two modes, the number of reactions may be limited by the compatibility of chemical reactions and biotransformations.

1.2.1 "Separate-Pot Two-Step" Mode

For the "separate-pot two-step" mode, the sequential steps of chemical reaction and biotransformation are carried out in separate reactors, with the intermediate being isolated or partially purified. The chemical reaction and biotransformation can be performed under quite different reaction conditions, and thus have been widely used to achieve complex synthetic goals. An example is the chemoenzymatic process for the preparation of β-thymidine. β-Thymidine is a precursor for the production of anti-AIDS drugs stavudine (d4T) and zidovudine (AZT). The enzymatic transglycosylation of guanosine and thymine yielded 5-methyluridine (5-MU) and guanine. The resulting 5-MU was then converted into β-thymidine by bromination and hydrogenation, as shown in Scheme 1.19 [41].

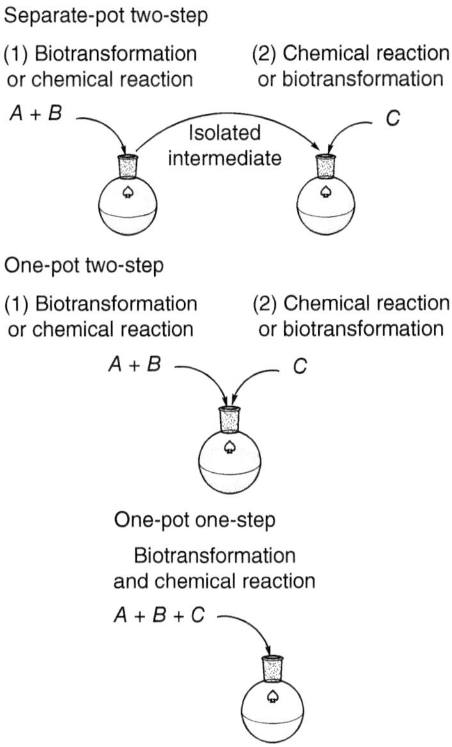

Scheme 1.18 Three simplified modes of chemoenzymatic cascade transformations.

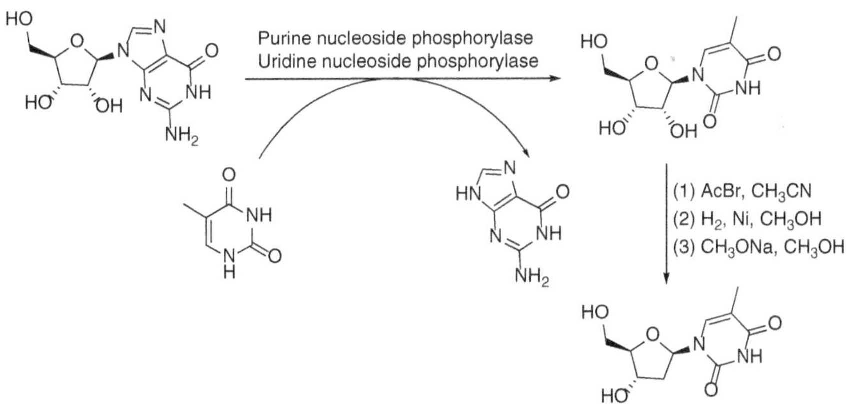

Scheme 1.19 Chemoenzymatic synthesis of β-thymidine by a separate-pot two-step process. Source: Adapted from Gordon et al. [41].

1.2.2 "One-Pot Two-Step" Mode

In the case of "one-pot two-step" mode, the chemical reaction and biotransformation are carried out sequentially in one pot without the isolation of the intermediates. This helps the simplification of process operation and reduction in solvent usage, waste generation, and operational cost. The enzymatic reduction of α-azidoacetophenone derivatives generates 2-azido-1-arylethanols with excellent optical purity, which can react with alkynes employing click chemistry to afford optically pure triazole-containing β-adrenergic receptor blocker analogs. The bioreduction of ketones and Cu-catalyzed "click" reaction was first carried out in the "separate-pot two-step" mode, in which the 2-azido-1-arylethanol intermediates were isolated (Scheme 1.20) [42]. This process was later performed in a "one-pot two-step" mode. The corresponding chiral 1,2,3-triazole-derived diols were prepared in high yields and excellent enantio- and diastereoselectivities under very mild conditions in aqueous medium (Scheme 1.21) [43].

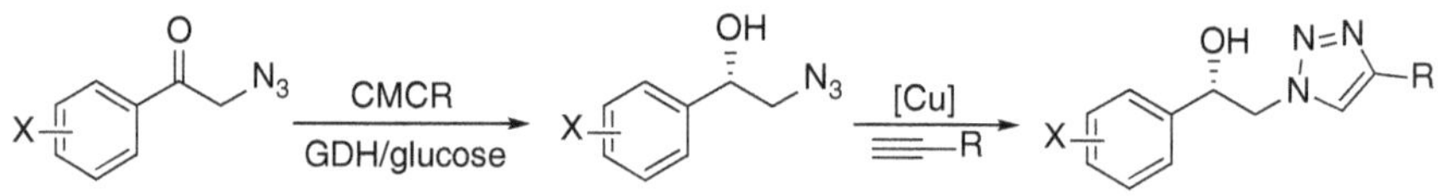

Scheme 1.20 Chemoenzymatic synthesis of 1,2,3-triazole-derived diols by a separate-pot two-step process. Source: Adapted from Ankati et al. [42].

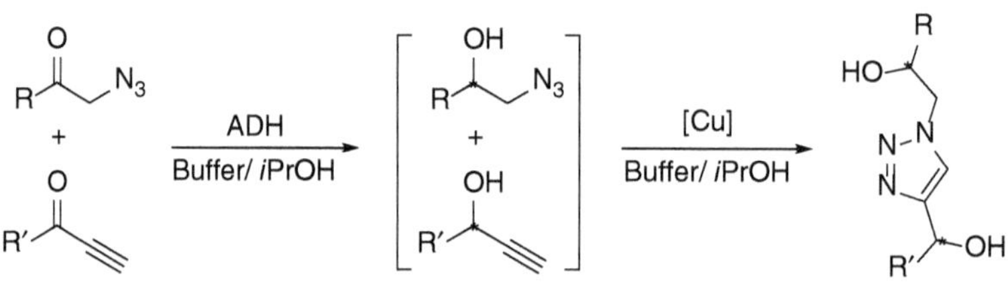

Scheme 1.21 Chemoenzymatic synthesis of 1,2,3-triazole-derived diols by a one-pot two-step process. Source: Adapted from Cuetos et al. [43].

1.2.3 "One-Pot One-Step" Mode

The third mode of chemoenzymatic transformation is "one-pot one-step," in which the chemical reaction and biotransformation proceed concurrently without high concentration accumulation of the intermediates. An example is our recent report that deracemization of 2-methyl-1,2,3,4-tetrahydroquinoline was achieved by using a mutant cyclohexylamine oxidase (CHAO) with the reducing agent $BH_3{\cdot}NH_3$. The bio-oxidation of (*S*)-2-methyl-1,2,3,4-tetrahydroquinoline led to the imine intermediate, which was immediately reduced by the chemical reducing agent, resulting in the production of (*R*)-2-methyl-1,2,3,4-tetrahydroquinoline with 76% isolated yield and 98% ee after cycles of enantioselective oxidation and nonselective reduction (Scheme 1.22) [44]. Tetrahydroquinoline is a "privileged" scaffold or substructure in many biologically active natural products and therapeutic agents used in cancer drug development.

Both chemical reaction and biotransformation possess advantageous features and shortcomings in terms of their capability for molecular construction and functional transformation. Combination of chemical reaction and biotransformation

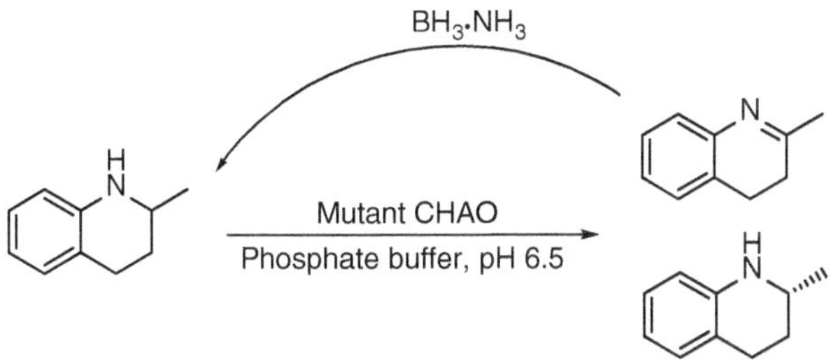

Scheme 1.22 Chemoenzymatic deracemization of 2-methyl-1,2,3,4-tetrahydroquinoline by a one-pot one-step process. Source: From Li et al. [44]. © 2014, American Chemical Society.

may give full play to their advantages, and thus offer tremendous opportunity for developing efficient and sustainable processes by designing novel synthetic routes from "starting materials to products" with retrosynthesis strategy [3, 6]. This can be relatively easily realized by the "separate-pot two-step" mode, because both biocatalytic and chemical transformations are individually treated as a synthetic step without considering the compatibility of the reaction conditions that are often quite different for chemical reactions and biotransformations. However, this "separate-pot" mode is less efficient than the two "one-pot" modes. The "one-pot two-step" and "one-pot one-step" modes are actually cascade reactions defined in organic chemistry [45]. A cascade reaction is a chemical process that comprises at least two consecutive reactions such that each subsequent reaction occurs by virtue of the chemical functionality formed in the previous step. In the strictest definition of cascade reaction, the reaction conditions do not change among the consecutive steps of a cascade and no new reagents are added after the initial step. This kind of sequential reactions is also called domino reaction. But a rather inclusive definition, which does not preclude the addition of new reagents or the change of conditions after the first reaction, is often adopted in the literature. In this book, the inclusive definition of cascade reaction is adopted, and includes one-pot sequential and concurrent transformations.

The undeniable benefits of cascade reactions include higher atom economy and production efficiency, and less resource consumption and waste generation. Especially, in the "one-pot one-step" mode, if one carries out several transformations in one synthetic operation, the involved savings can be considerable. However, for this module, there are many challenges, such as the reaction condition compatibility of the chemical reaction with biotransformation, the inhibition or inactivation of (bio)catalysts, and so on, which must be addressed before industrial application [46, 47]. In this context, scientists and engineers are designing chemical catalysts for use in aqueous reaction medium and evolving enzyme catalysts with tolerance of organic solvents, so that these catalysts can work under the same reaction conditions (Scheme 1.23). Process engineering involving various strategies such as reaction medium engineering [48] and reaction site separation [46] has also been carried out for developing efficient and economical chemoenzymatic cascade processes. When further advances in (bio)catalyst design and process engineering have been achieved in the future, we can expect that chemoenzymatic cascade reactions will definitely play an ever-increasing role in the transition of the currently non-sustainable fine chemical industry to a green and sustainable one.

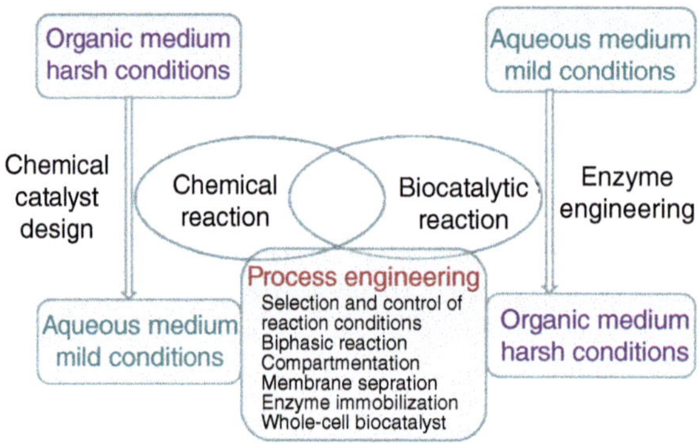

Scheme 1.23 Scientists and engineers are working together to develop efficient and economical chemoenzymatic cascade.

References

1 Sheldon, R.A. (2017). The E factor 25 years on: the rise of green chemistry and sustainability. *Green Chemistry* 19 (1): 18–43.

2 Zimmerman, J.B., Anastas, P.T., Erythropel, H.C., and Leitner, W. (2020). Designing for a green chemistry future. *Science* 367 (6476): 397–400.

3 Green, A.P. and Turner, N.J. (2016). Biocatalytic retrosynthesis: redesigning synthetic routes to high-value chemicals. *Perspectives in Science* 9: 42–48.

4 Lu, W., Chen, X., Feng, J. et al. (2018). A fungal P450 enzyme from *Thanatephorus cucumeris* with steroid hydroxylation capabilities. *Applied and Environmental Microbiology* 84 (13): e00503–e00518.

5 Wang, Y., Ju, W., Tian, H. et al. (2018). Scalable synthesis of cyclocitrinol. *Journal of the American Chemical Society* 140 (30): 9413–9416.

6 de Souza, R.O.M.A., Miranda, L.S.M., and Bornscheuer, U.T. (2017). A retrosynthesis approach for biocatalysis in organic synthesis. *Chemistry - A European Journal* 23 (50): 12040–12063.

7 Sheldon, R.A. and Woodley, J.M. (2018). Role of biocatalysis in sustainable chemistry. *Chemical Reviews* 118 (2): 801–838.

8 Wang, M.-X. (2015). Enantioselective biotransformations of nitriles in organic synthesis. *Accounts of Chemical Research* 48 (3): 602–611.

9 Yao, P., Li, J., Yuan, J. et al. (2015). Enzymatic synthesis of a key intermediate for rosuvastatin by nitrilase-catalyzed hydrolysis of ethyl (R)-4-cyano-3-hydroxybutyate at high substrate concentration. *ChemCatChem* 7 (2): 271–275.

10 Li, G., Ren, J., Wu, Q. et al. (2013). Identification of a marine NADPH-dependent aldehyde reductase for chemoselective reduction of aldehydes. *Journal of Molecular Catalysis B: Enzymatic* 90: 17–22.

11 Brewster, T.P., Miller, A.J.M., Heinekey, D.M., and Goldberg, K.I. (2013). Hydrogenation of carboxylic acids catalyzed by half-sandwich complexes of iridium and rhodium. *Journal of the American Chemical Society* 135 (43): 16022–16025.

12 Korstanje, T.J., Ivar van der Vlugt, J., Elsevier, C.J., and de Bruin, B. (2015). Hydrogenation of carboxylic acids with a homogeneous cobalt catalyst. *Science* 350 (6258): 298–302.

13 Winkler, M. (2018). Carboxylic acid reductase enzymes (CARs). *Current Opinion in Chemical Biology* 43: 23–29.

14 Strohmeier, G.A., Eiteljörg, I.C., Schwarz, A., and Winkler, M. (2019). Enzymatic one-step reduction of carboxylates to aldehydes with cell-free regeneration of ATP and NADPH. *Chemistry – A European Journal* 25 (24): 6119–6123.

15 France, S.P., Hussain, S., Hill, A.M. et al. (2016). One-pot cascade synthesis of mono- and disubstituted piperidines and pyrrolidines using carboxylic acid reductase (CAR), ω-transaminase (ω-TA), and imine reductase (IRED) biocatalysts. *ACS Catalysis* 6 (6): 3753–3759.

16 Toogood, H.S. and Scrutton, N.S. (2018). Discovery, characterization, engineering, and applications of ene-reductases for industrial biocatalysis. *ACS Catalysis* 8 (4): 3532–3549.

17 Moore, J.C., Pollard, D.J., Kosjek, B., and Devine, P.N. (2007). Advances in the enzymatic reduction of ketones. *Accounts of Chemical Research* 40 (12): 1412–1419.

18 Scheller, P.N., Fademrecht, S., Hofelzer, S. et al. (2014). Enzyme toolbox: novel enantiocomplementary imine reductases. *ChemBioChem* 15 (15): 2201–2204.

19 Chen, X., Gao, X., Wu, Q., and Zhu, D. (2012). Synthesis of optically active dihydrocarveol via a stepwise or one-pot enzymatic reduction of (R)- and (S)-carvone. *Tetrahedron: Asymmetry* 23 (10): 734–738.

20 Grau, B.T., Devine, P.N., DiMichele, L.N., and Kosjek, B. (2007). Chemo- and enantioselective routes to chiral fluorinated hydroxyketones using ketoreductases. *Organic Letters* 9 (24): 4951–4954.

21 Zhang, D., Zhang, R., Zhang, J. et al. (2014). Engineering a hydroxysteroid dehydrogenase to improve its soluble expression for the asymmetric reduction of cortisone to 11β-hydrocortisone. *Applied Microbiology and Biotechnology* 98 (21): 8879–8886.

22 Akita, H., Fujino, Y., Doi, K., and Ohshima, T. (2011). Highly stable meso-diaminopimelate dehydrogenase from an Ureibacillus thermosphaericus strain A1 isolated from a Japanese compost: purification, characterization and sequencing. *AMB Express* 1 (1): 43.

23 Kharrat, N., Aissa, I., Sghaier, M. et al. (2014). Lipophilization of ascorbic acid: a monolayer study and biological and antileishmanial activities. *Journal of Agricultural and Food Chemistry* 62 (37): 9118–9127.

24 Jung, J.-H., Yoon, D.-H., Kang, P. et al. (2013). CAL-B catalyzed desymmetrization of 3-alkylglutarate: "olefin effect" and asymmetric synthesis of pregabalin. *Organic & Biomolecular Chemistry* 11 (22): 3635–3641.

25 Veselá, A.B., Rucká, L., Kaplan, O. et al. (2016). Bringing nitrilase sequences from databases to life: the search for novel substrate specificities with a focus on dinitriles. *Applied Microbiology and Biotechnology* 100 (5): 2193–2202.

26 Zhu, D., Mukherjee, C., Biehl, E.R., and Hua, L. (2007). Nitrilase-catalyzed selective hydrolysis of dinitriles and green access to the cyanocarboxylic acids of pharmaceutical importance. *Advanced Synthesis & Catalysis* 349 (10): 1667–1670.

27 Duan, Y., Yao, P., Ren, J. et al. (2014). Biocatalytic desymmetrization of 3-substituted glutaronitriles by nitrilases. A convenient chemoenzymatic access to optically active (S)-Pregabalin and (R)-Baclofen. *SCIENCE CHINA Chemistry* 57 (8): 1164–1171.

28 Yu, S., Yao, P., Li, J. et al. (2019). Improving the catalytic efficiency and stereoselectivity of a nitrilase from *Synechocystis* sp. PCC6803 by semi-rational engineering en route to chiral γ-amino acids. *Catalysis Science & Technology* 9 (6): 1504–1510.

29 De Wildeman, S.M.A., Sonke, T., Schoemaker, H.E., and May, O. (2007). Biocatalytic reductions: from lab curiosity to "first choice". *Accounts of Chemical Research* 40 (12): 1260–1266.

30 Wu, J., Ji, J.-X., Guo, R. et al. (2003). Chiral [RuCl2(dipyridylphosphane)(1,2-diamine)] catalysts: applications in asymmetric hydrogenation of a wide range of simple ketones. *Chemistry – A European Journal* 9 (13): 2963–2968.

31 Truppo, M.D., Pollard, D., and Devine, P. (2007). Enzyme-catalyzed enantioselective diaryl ketone reductions. *Organic Letters* 9 (2): 335–338.

32 Li, H., Zhu, D., Hua, L., and Biehl, E.R. (2009). Enantioselective reduction of diaryl ketones catalyzed by a carbonyl reductase from *Sporobolomyces salmonicolor* and its mutant enzymes. *Advanced Synthesis & Catalysis* 351 (4): 583–588.

33 Sun, Z., Lonsdale, R., Ilie, A. et al. (2016). Catalytic asymmetric reduction of difficult-to-reduce ketones: triple-code saturation mutagenesis of an alcohol dehydrogenase. *ACS Catalysis* 6 (3): 1598–1605.

34 Liang, J., Mundorff, E., Voladri, R. et al. (2010). Highly enantioselective reduction of a small heterocyclic ketone: biocatalytic reduction of tetrahydrothiophene-3-one to the corresponding (R)-alcohol. *Organic Process Research & Development* 14 (1): 188–192.

35 Contente, M.L., Molinari, F., Serra, I. et al. (2016). Stereoselective enzymatic reduction of ethyl secodione: preparation of a key intermediate for the total synthesis of steroids. *European Journal of Organic Chemistry* 2016 (7): 1260–1263.

36 Chen, X., Zhang, H., Maria-Solano, M.A. et al. (2019). Efficient reductive desymmetrization of bulky 1,3-cyclodiketones enabled by structure-guided directed evolution of a carbonyl reductase. *Nature Catalysis* 2 (10): 931–941.

37 Rocha, L.C., Ferreira, H.V., Pimenta, E.F. et al. (2010). Biotransformation of α-bromoacetophenones by the marine fungus *Aspergillus sydowii*. *Marine Biotechnology* 12 (5): 552–557.

38 Ren, J., Dong, W., Yu, B. et al. (2012). Synthesis of optically active α-bromohydrins via reduction of α-bromoacetophenone analogues catalyzed by an isolated carbonyl reductase. *Tetrahedron: Asymmetry* 23 (6): 497–500.

39 Hann, E.C., Sigmund, A.E., Fager, S.K. et al. (2003). Biocatalytic hydrolysis of 3-hydroxyalkanenitriles to 3-hydroxyalkanoic acids. *Advanced Synthesis & Catalysis* 345 (6–7): 775–782.

40 Ankati, H., Zhu, D., Yang, Y. et al. (2009). Asymmetric synthesis of both antipodes of β-hydroxy nitriles and β-hydroxy carboxylic acids via enzymatic reduction or sequential reduction/hydrolysis. *The Journal of Organic Chemistry* 74 (4): 1658–1662.

41 Gordon, G.E.R., Bode, M.L., Visser, D.F. et al. (2011). An integrated chemo-enzymatic route for preparation of β-thymidine, a key intermediate in the preparation of antiretrovirals. *Organic Process Research & Development* 15 (1): 258–265.

42 Ankati, H., Yang, Y., Zhu, D. et al. (2008). Synthesis of optically pure 2-azido-1-arylethanols with isolated enzymes and conversion to triazole-containing β-blocker analogues employing click chemistry. *The Journal of Organic Chemistry* 73 (16): 6433–6436.

43 Cuetos, A., Bisogno, F.R., Lavandera, I., and Gotor, V. (2013). Coupling biocatalysis and click chemistry: one-pot two-step convergent synthesis of enantioenriched 1,2,3-triazole-derived diols. *Chemical Communications* 49 (26): 2625–2627.

44 Li, G., Ren, J., Yao, P. et al. (2014). Deracemization of 2-methyl-1,2,3,4-tetrahydroquinoline using mutant cyclohexylamine oxidase obtained by iterative saturation mutagenesis. *ACS Catalysis* 4 (3): 903–908.

45 Nicolaou, K.C., Edmonds, D.J., and Bulger, P.G. (2006). Cascade reactions in total synthesis. *Angewandte Chemie International Edition* 45 (43): 7134–7186.

46 Schmidt, S., Castiglione, K., and Kourist, R. (2018). Overcoming the incompatibility challenge in chemoenzymatic and multi-catalytic cascade reactions. *Chemistry – A European Journal* 24 (8): 1755–1768.

47 Gröger, H. and Hummel, W. (2014). Combining the 'two worlds' of chemocatalysis and biocatalysis towards multi-step one-pot processes in aqueous media. *Current Opinion in Chemical Biology* 19: 171–179.

48 Kourist, R. and González-Sabín, J. (2020). Non-conventional media as strategy to overcome the solvent dilemma in chemoenzymatic tandem catalysis. *ChemCatChem* 12 (7): 1903–1912.

2

"Separate-pot Two-step" Chemoenzymatic Transformation

The majority of biocatalysts have poor stability in organic solvents and at high temperatures. Biocatalytic reactions are usually carried out in aqueous solution and at ambient temperature. On the other hand, organic chemical reactions mainly occur in organic solvents and/or at elevated temperatures. As such, many reported chemoenzymatic processes so far were carried out in separate reaction vessels with chemical and biocatalytic reactions as distinct steps. The purification step is performed after each reaction to isolate the intermediate product because the unreacted substrates, by-products, and catalysts may prevent subsequent reaction. For this "none-one-pot" chemoenzymatic transformation, the chemical reaction and the biotransformation are carried out separately, and reaction condition compatibility of the chemical reaction with biotransformation is not required. Therefore, it is quite flexible to combine them into the synthetic route of a target compound. Hence, biotransformations have been increasingly incorporated into the synthesis of fine chemicals such as pharmaceuticals, food additives, cosmetics, and so on.

When biotransformation is integrated with chemical reactions in the synthesis of a target molecule, shorter synthetic routes with fewer reaction steps can be designed, and much efficient preparation of the target compound can be achieved. An early example is the production of cortisone, a well-known steroidal drug. This steroid active pharmaceutical ingredient (API) was first synthesized by 31 chemical reaction steps only to yield 1 g of cortisone from 615 g of deoxycholic acid (purified from beef bile) at an economic cost of \$200/g in 1949. Introduction of a biotransformation step via microbial 11 α-hydroxylation of progesterone using *Rhizopus arrhizus* ATCC 11145 or *Aspergillus niger* ATCC 9142 markedly reduced the required chemical steps to 11 as well as brought down the production costs to only \$1/g in 1979 [1]. Recently, biocatalysis has been applied to enable efficient and economical production of sitagliptin, an oral antihyperglycemic (antidiabetic drug) of the dipeptidyl peptidase-4 (DPP-4) inhibitor. Originally prepared by asymmetric hydrogenation of an enamine precursor at high pressure (250 psi) using a chiral rhodium catalyst, the process suffered from inadequate stereoselectivity and contamination of rhodium in the product. Besides, additional purification steps were needed to upgrade both optical and chemical purity, resulting in lower yield and higher production cost [2]. Implementation of the asymmetric amination of

Chemo-Enzymatic Cascade Reactions, First Edition. Dunming Zhu.

prositagliptin ketone by a transaminase produced sitagliptin with >99.95% ee. The biocatalytic process resulted in 10 to 13% increase in overall yield, 53% increase in productivity (kg/l per day), 19% reduction in total waste, elimination of all heavy metals, and reduction in total manufacturing cost [3]. These paradigms underscore the various benefits of implementation of biotransformation into the synthesis of target chemicals.

As our understanding of the enzymes' properties and reaction mechanisms have deepened and the technologies in molecular biology and fermentation engineering have been advanced, more and more useful enzymes have become available and applied in organic syntheses. This chapter, organized by the enzyme type, will describe the recent advances in the implementation of biotransformations into the synthesis of fine chemicals.

2.1 Lipases

Lipases (triacylglycerol ester hydrolases EC 3.1.1.3) are a class of hydrolases that catalyze the hydrolysis of triglycerides to glycerol and free fatty acids. Lipases catalyze the hydrolysis of carboxylic acid esters under aqueous conditions, and the reverse ester-forming reactions, such as esterification, interesterification, and transesterification, in nonaqueous media. Some lipases are readily available commercially in large quantity at low cost, and some of them are also stable even in organic solvents. Therefore, in addition to applications in numerous industrial processes including oils and fats, food and baking, dairy, detergents, leather and paper processing, cosmetics and perfume, biodiesel and bioremediation, lipases are also widely utilized as biocatalysts in synthetic organic chemistry.

The applications of lipases in organic synthesis are mainly focused on the kinetic resolution of chiral alcohols, amines, and carboxylic acids. (*S*)-α-cyano-3-phenoxybenzyl alcohol (*S*-CPBA) is an important intermediate for the synthesis of many pyrethroids, a class of synthetic pesticides similar to the natural pesticide pyrethrum. This chiral alcohol was prepared by a chemoenzymatic process, in which racemic α-cyano-3-phenoxybenzyl acetate (CPBAc) was prepared from *m*-phenoxybenzaldehyde (*m*-PBA), sodium cyanide, and acetyl anhydride in the presence of a phase-transfer catalyst, followed by the resolution of the racemic ester via a highly enantioselective lipase-catalyzed transesterification to *n*-butanol in diisopropyl ether (Scheme 2.1). The immobilized *Pseudomonas* sp. lipase catalyzed the transesterification reaction to give *S*-CPBA with high conversion (46% out of 50%) and >96% ee. The remaining (*R*)-ester was easily racemized and recovered with the use of triethylamine in diisopropyl ether or toluene [4].

The racemic cyanohydrin acetate can also be synthesized by a one-pot process, in which $SnBr_2$ mediated the addition of TMSCN onto 2-naphthaldehyde to give cyanohydrin trimethylsilyl intermediates, followed by the reaction with acetyl bromide. The resulting racemic α-acetoxy-2-naphthylacetonitrile was then kinetically resolved by the highly enantioselective hydrolysis of the (*S*)-enantiomer of the racemate catalyzed by a lipase from *Burkholderia cepacia*, affording the

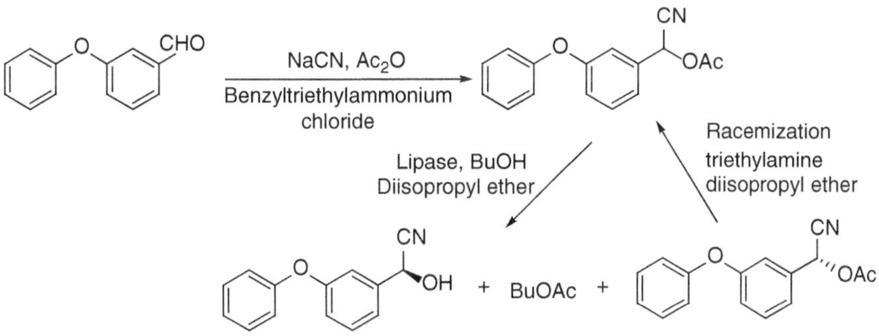

Scheme 2.1 Chemoenzymatic synthesis of (*S*)-α-cyano-3-phenoxybenzyl alcohol.

(*R*)-acetate with >99% ee. The formed (*S*)-cyanohydrin spontaneously decomposed into 2-naphthaldehyde, which could be recovered to serve as the starting material (Scheme 2.2) [5].

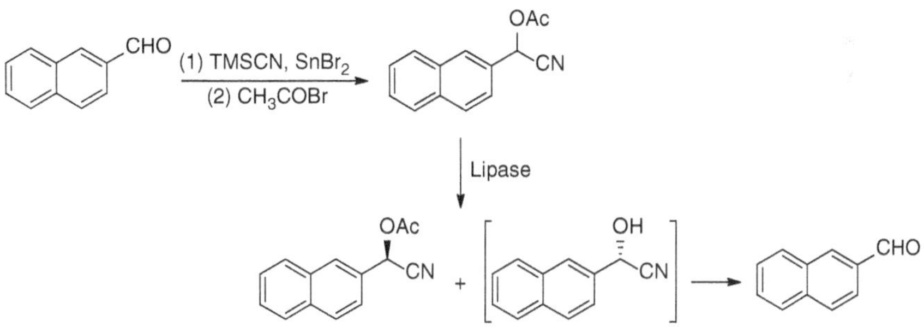

Scheme 2.2 Chemoenzymatic synthesis of (*R*)-α-acetoxy-2-naphthylacetonitrile. Source: Adapted from Kimura et al. [5].

Proxyphylline, (±)-3, 7-(2-hydroxypropyl)-1,3-dimethyl-3,7-dihydro-1H-purine-2,6-dione, has been widely used in clinical practice as a cardiac stimulant, vasodilator, and bronchodilator for over 60 years. The racemate of this compound could be synthesized by triethylamine-mediated regioselective ring opening of propylene oxide with theophylline (1, 1,3-dimethyl-7H-purine-2,6-dione) in boiling methanol with 74% isolated yield. Some commercially available lipases were tested for the kinetic resolution of racemic proxyphylline via O-acetylation with vinyl acetate, but low enantioselectivity was observed. As such, racemic proxyphylline was first converted into the corresponding acylated proxyphylline by treatment with acetic anhydride or with the appropriate acyl chloride in dry dichloromethane in the presence of triethylamine and a catalytic amount of 4-(*N*,*N*)-dimethylaminopyridine (DMAP). The kinetic resolution of acylated proxyphylline was then achieved by using immobilized *Candida antarctica* lipase B (CAL-B) as the biocatalyst for hydrolysis or methanolysis of its esters in acetonitrile (Scheme 2.3). (*S*)-(+)-Butanoate of proxyphylline and (*R*)-(−)-proxyphylline were prepared in 45 and 46% yields with 97 and 96% ee, respectively [6].

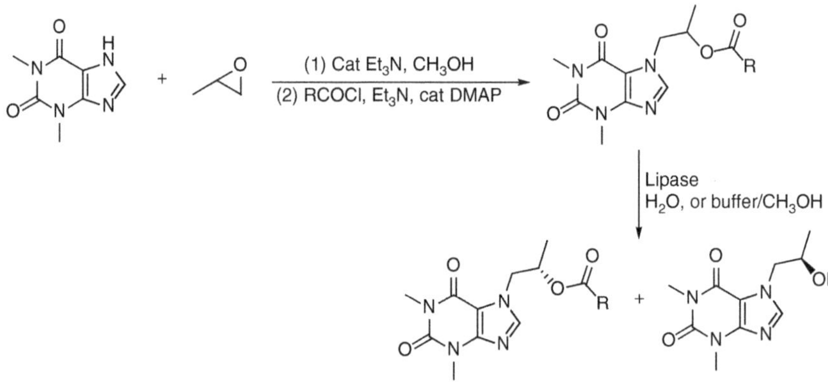

Scheme 2.3 Synthesis and lipase-catalyzed kinetic resolution of proxyphylline.

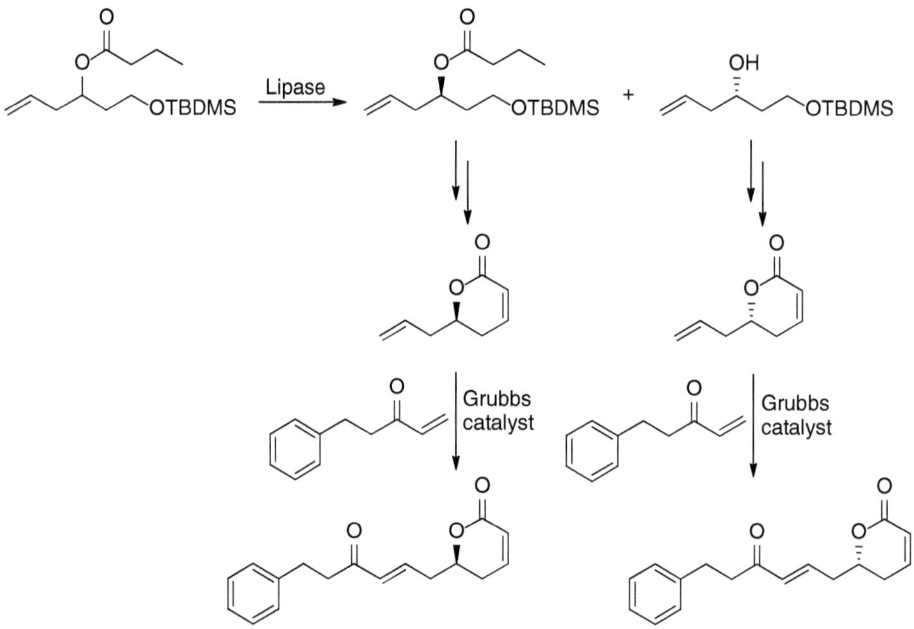

Scheme 2.4 Chemoenzymatic synthesis of (*R*)- and (*S*)-Rugulactone. Source: Adapted from Reddipalli et al. [7].

Rugulactone is a natural product with inhibitory activity against the nuclear factor κB (NF-κB) activation pathway. The key structure of Rugulactone is the chiral dihydro α-pyrone moiety, and the two enantiomers have quite different biological activity. (*R*)- or (*S*)-6-allyl-5,6-dihydro-pyran-2-one is thus used to construct (*R*)- or (*S*)-Rugulactone, respectively, by cross-metathesis reaction with 5-phenyl-pent-1-en-3-one in the presence of a Grubbs second-generation catalyst. (*R*)- and (*S*)-6-allyl-5,6-dihydro-pyran-2-one were prepared by lipase-catalyzed kinetic resolution of racemic 1-(*tert*-butyldimethylsilyloxy)hex-5-en-3-yl butyrate followed by chemical transformations as shown in Scheme 2.4 [7]. They can also be synthesized from methyl 3-oxohex-5-enoate by ketoreductase-catalyzed reduction, followed by two chemical reactions (Scheme 2.5) [8].

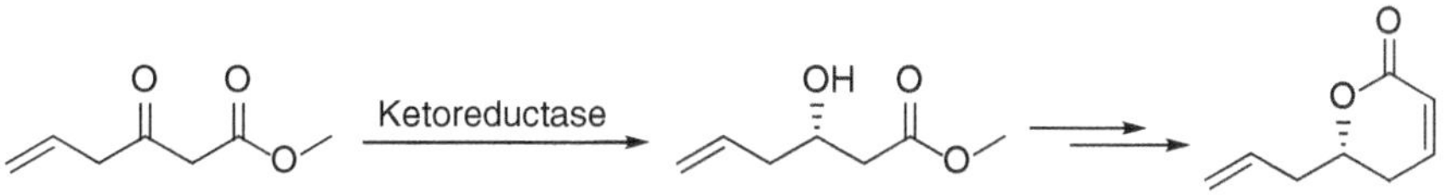

Scheme 2.5 Chemoenzymatic synthesis of (*S*)-6-allyl-5,6-dihydro-pyran-2-one. Source: Adapted from Tyrikos-Ergas et al. [8].

Recently, a more efficient approach has been reported for the preparation of optically pure 6-substituted 5,6-dihydropyran-2-one derivatives. This chemoenzymatic process involves lipase-catalyzed dynamic kinetic resolution (DKR) of racemic homoallylic alcohols and sequential ring-closing metathesis (RCM). By employing commercially available CAL-B as the biocatalyst, ruthenium complex $[Ru(CO)_2Cl(\eta^5\text{-}C_5Ph_5)]$ as the racemization catalyst, and vinyl crotonate as the acyl donor, a variety of racemic homoallylic alcohols were converted into the corresponding crotonates via DKR in high yields and excellent enantioselectivity. In order to avoid the complication in the subsequent RCM reaction, the product of enzymatic DKR was isolated, and used for the next step RCM reaction, generating the desired 5,6-dihydropyran-2-ones in 55–75% yields and 94–>99% ee (Scheme 2.6) [9].

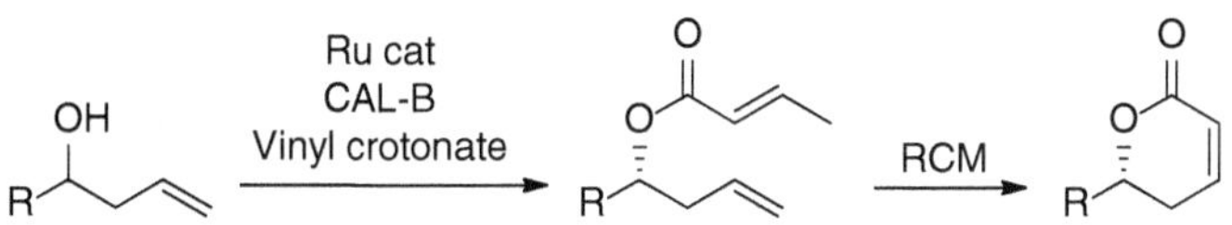

Scheme 2.6 Chemoenzymatic synthesis of optically pure 6-substituted 5,6-dihydro-pyran-2-one derivatives. Source: Adapted from Koszelewski et al. [9].

In addition to transesterification and hydrolysis of the corresponding esters, the racemic alcohols can also be kinetically resolved by lipase-catalyzed direct acylation with various acyl donors such as vinyl acetate. Michael addition of imidazole to cyclohex-2-en-1-one under microwave conditions produced racemic 3-(1H-imidazol-1-yl)cyclohexanone, which was reduced to give *cis*-3-(1H-imidazol-1-yl)cyclohexanol. The resulting racemic *cis*-alcohol was successfully resolved through the acetylation catalyzed by a lipase from *Pseudomonas cepacia* lipase (PSL-C) to furnish (1*R*,3*S*)-*cis*-3-(1*H*-Imidazol-1-yl)cyclohexyl acetate and (1*S*,3*R*)-*cis*-3-(1*H*-Imidazol-1-yl)cyclohexanol in 97 and 98% ee, respectively (Scheme 2.7). A family of novel imidazolium-based chiral ionic liquids were then prepared via quaternization with alkyl halides, followed by anion exchange with inorganic salts, and their properties as phase-transfer catalysts in the Michael addition of diethyl malonate to *trans*-chalcone were investigated [10].

Chiral 1-(2-hydroxycycloalkyl)imidazoles are useful starting materials for the preparation of enantiopure ionic liquids. The synthesis of these enantiopure imidazole derivatives can be achieved by lipase-catalyzed kinetic resolution of the

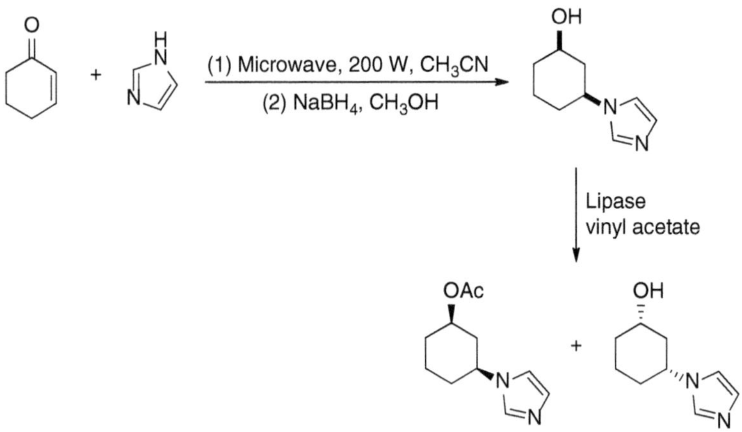

Scheme 2.7 Synthesis and lipase-catalyzed kinetic resolution of *cis*-3-(1*H*-Imidazol-1-yl)cyclohexanol.

racemic imidazole alcohols. For example, CAL-B or PSL-C catalyzed the acetylation of (*R*,*R*)-*trans*-2-(1*H*-imidazol-1-yl)cycloalkanols using vinyl acetate as acyl donor in organic solvents such as methyl *tert*-butyl ether (MTBE) or tetrahydrofuran (THF). The enzymatic acetylation was coupled with the ring-opening reaction of epoxides using imidazole as nucleophile, and a chemoenzymatic process was developed for the synthesis of enantiopure imidazole derivatives (Scheme 2.8). The chemical synthesis of *trans*-2-(1*H*-imidazol-1-yl)cycloalkanols and enzymatic kinetic resolution were performed in batch or continuous flow mode [11].

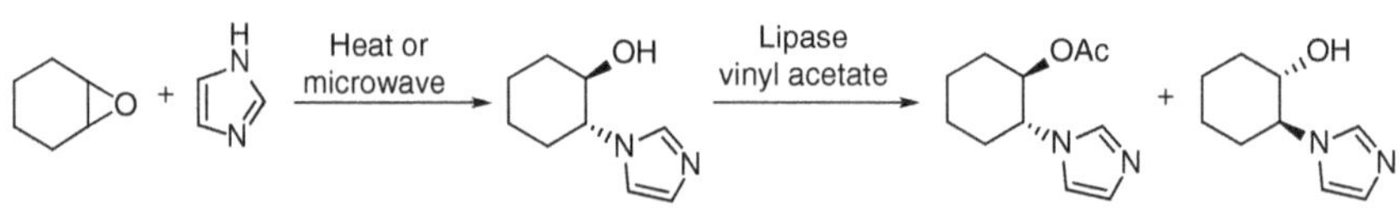

Scheme 2.8 Synthesis and lipase-catalyzed kinetic resolution of *trans*-2-(1*H*-imidazol-1-yl)cyclohexanol.

Chiral mandelic acid and its analogs are important intermediates for the synthesis of many pharmacologically relevant compounds such as semisynthetic antibiotics, anticholinergic, antiplatelet/antithrombotic, vasodilator, antitumor, and anti-obesity agents. A large number of asymmetric methods have been developed for the production of optically active mandelic acid derivatives in the past few decades. Recently, a chemoenzymatic method has been reported for the synthesis of substituted mandelic acid methyl esters starting from substituted benzaldehydes and trichloromethane or tribromomethane. The racemic mandelic acid methyl esters were synthesized by the reaction of benzaldehydes with trichloromethane or tribromomethane, followed by the esterification in methanol in the presence of $SOCl_2$. The kinetic resolution of these racemic methyl esters was then achieved by using lipases from *Pseudomonas fluorescens* or *B. cepacia* to give the corresponding methyl (*R*)-(−)-mandelates with up to >99 % ee (Scheme 2.9). Methyl

(*R*)-(−)-mandelate was used to prepare (*R*)-(+)-pemoline in 98% ee by reacting with guanidine hydrochloride under basic conditions. The optically active pemoline is a dopaminergic agent used in the treatment of attention-deficit hyperactivity disorder (ADHD) and narcolepsy [12].

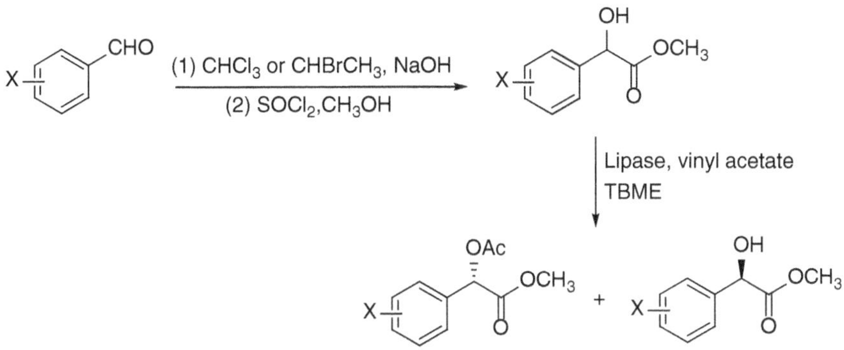

Scheme 2.9 Synthesis and lipase-catalyzed kinetic resolution of methyl mandelate.

The addition of monolithiumacetylide to aliphatic aldehydes afforded the corresponding propargyl alcohols in high yields. The commercially available lipase Novozym-435 catalyzed the kinetic resolution of the resulting racemic propargylic alcohols using vinyl acetate as acyl donor in *tert*-butyl methyl ether, yielding the (*S*)-acetates (Scheme 2.10). They were then used in the highly enantioselective synthesis of paecilocin A and 3-butyl-7-hydroxyphthalide [13].

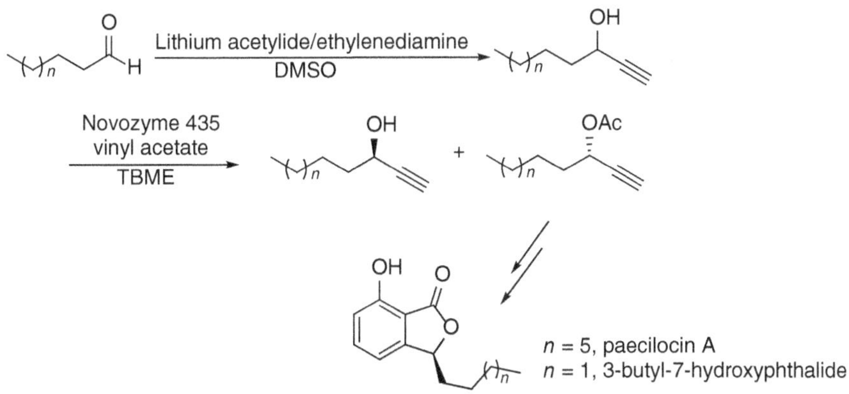

Scheme 2.10 Synthesis and lipase-catalyzed kinetic resolution of propargylic alcohols.

Ramatroban is a pharmacologically active compound marketed for the treatment of allergic rhinitis and asthma, and it also shows potential applications for the treatment of coronary artery disease. This compound has a chiral 2,3,4,9-tetrahydro-1*H*-carbazol-3-amine structure, and the (*R*)-enantiomer is 10 to 100 times more active than its enantiomeric counterpart. A chemoenzymatic method was developed to introduce chirality in the core structure of

Ramatroban by employing a lipase for the kinetic resolution. In this manner, 4,9-dihydro-1*H*-carbazol-3(2*H*)-one was synthesized and reduced by $NaBH_4$ to give (±)-2,3,4,9-tetrahydro-1*H*-carbazol-3-ol. The racemate was then resolved by using CAL-B as biocatalyst and vinyl acetate as acyl donor, affording (*R*)-2,3,4,9-tetrahydro-1*H*-carbazol-3-yl acetate and (*S*)-2,3,4,9-tetrahydro-1*H*-carbazol-3-ol. The (*S*)-alcohol was further converted into (*R*)-2,3,4,9-tetrahydro-1*H*-carbazol-3-amine in a one-pot chemical process and used for the synthesis of (*R*)-Ramatroban (Scheme 2.11). Excellent enantioselectivity was also attained for the kinetic resolution of racemic 2,3,4,9-tetrahydro-1*H*-carbazol-3-amine using lipase CAL-B and ethyl methoxyacetate as acyl donor, but low conversion prevented its application in the synthesis of (*R*)-Ramatroban [14].

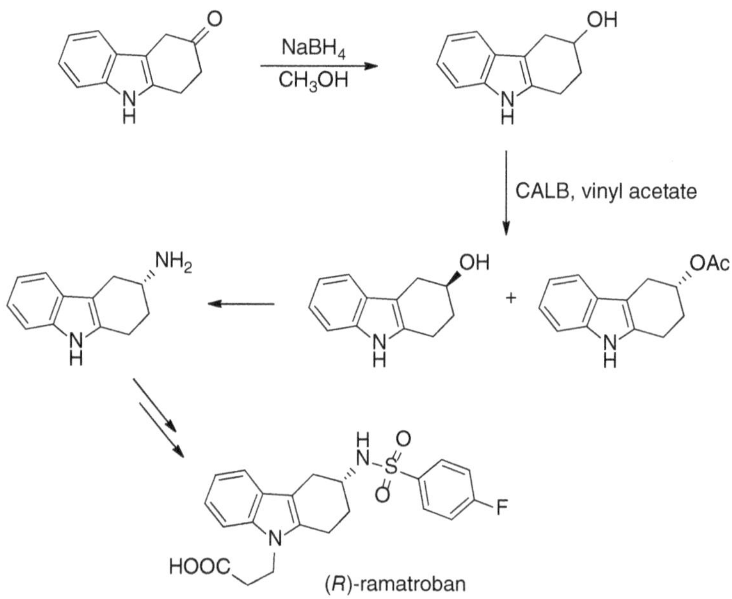

Scheme 2.11 Chemoenzymatic synthesis of (*R*)-Ramatroban.

Lipases have also been employed as efficient biocatalysts in the chemoenzymatic preparation of enantiomerically pure amines. A family of 3-substituted pentane-1,5-diamines can be synthesized by converting the corresponding diols in three reaction steps. The resulting prochiral diamines were desymmetrized by using PSL-C as the biocatalyst and ethyl methoxyacetate as acyl donor in 1,4-dioxane, generating (*S*)-monoamides in 33–59% isolated yield and 54–99% ee. After the stereoselective protection of one of the two amino groups, the (*S*)-monoamides were acylated with allyl chloroformate to produce optically active (*R*)-amido carbamates (Scheme 2.12) [15].

Diethylaminosulfur trifluoride (DAST) is a fluorinating reagent used for the conversion of alcohols, aldehydes, and unhindered ketones to the corresponding alkyl fluorides and geminal difluorides, respectively. Lipase-catalyzed kinetic

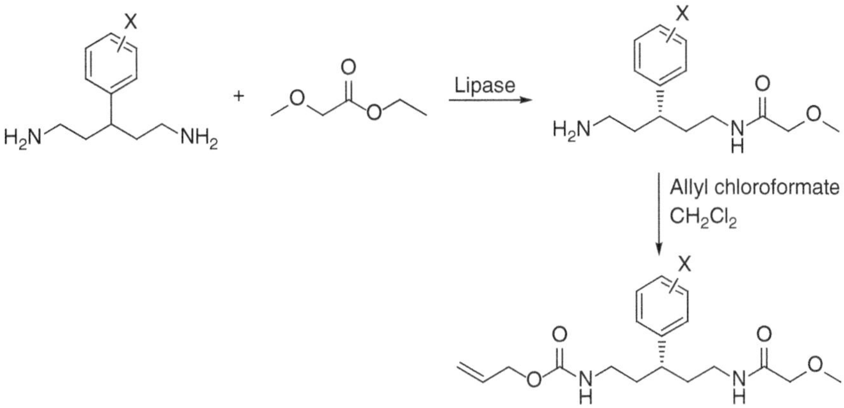

Scheme 2.12 Chemoenzymatic synthesis of optically active amines via lipase-catalyzed desymmetrization of prochiral diamines. Source: Adapted from Ríos-Lombardía et al. [15].

resolution of alcohols was combined with DAST fluorination for the synthesis of optically pure organofluorine compounds. The racemic *cis*-2-azido-1-indanols were kinetically resolved by employing commercial lipase Amano AK and vinyl acetate in diisopropyl ether to give the (*R*,*S*)-acetate and (*S*,*R*)-2-azido-1-indanol. The (*R*,*S*)-acetate was converted into (*R*,*S*)-2-azido-1-indanol by treatment with sodium methoxide in methanol. The resulting (*S*,*R*)- and (*R*,*S*)-2-azido-1-indanols were fluorinated with DAST, affording the four diastereoisomers of the corresponding 1-fluoro-2-azido-indane. The four diastereoisomers of 1-fluoro-2-amino-indane were then prepared by the reduction of azido group under Staudinger's conditions (triphenylphosphine, MeOH) or catalytic hydrogenation using platinum oxide as catalyst under hydrogen atmosphere (Scheme 2.13) [16].

Lipase-catalyzed hydrolysis of esters into the optically pure carboxylic acids has also been combined with chemical reactions for the efficient synthesis of precursors of β-substituted-γ-amino acids. A number of aliphatic and aromatic 3-substituted-3-cyano-2-(ethoxycarbonyl)propanoic acid ethyl esters were hydrolyzed by employing commercially available lipases to afford 3-substituted-3-cyano-2-(ethoxycarbonyl)propanoic acid in high yields and stereoselectivity. Thermal decarboxylation of 3-substituted-3-cyano-2-(ethoxycarbonyl)propanoic acid in aqueous solution produced 3-substituted-3-cyanopropanoic acid ethyl esters, which can be transformed to the corresponding optically active β-substituted-γ-amino acids by hydrolysis and the reduction of the nitrile group (Scheme 2.14) [17].

(1*S*,4*R*)-8-Hydroxy-1,2,3,4-tetrahydro-1,4-methanonaphthalen-5-yl propionate can serve as a key intermediate for the synthesis of Yimitasvir, a novel NS5A inhibitor in a phase III clinical trial for the treatment of hepatitis C virus (HCV) infection. A commercially available lipase AK was found to catalyze the desymmetric hydrolysis of the corresponding prochiral diphenol diester to give (1*S*,4*R*)-8-Hydroxy-1,2,3,4-tetrahydro-1,4-methanonaphthalen-5-yl propionate with 99.0%

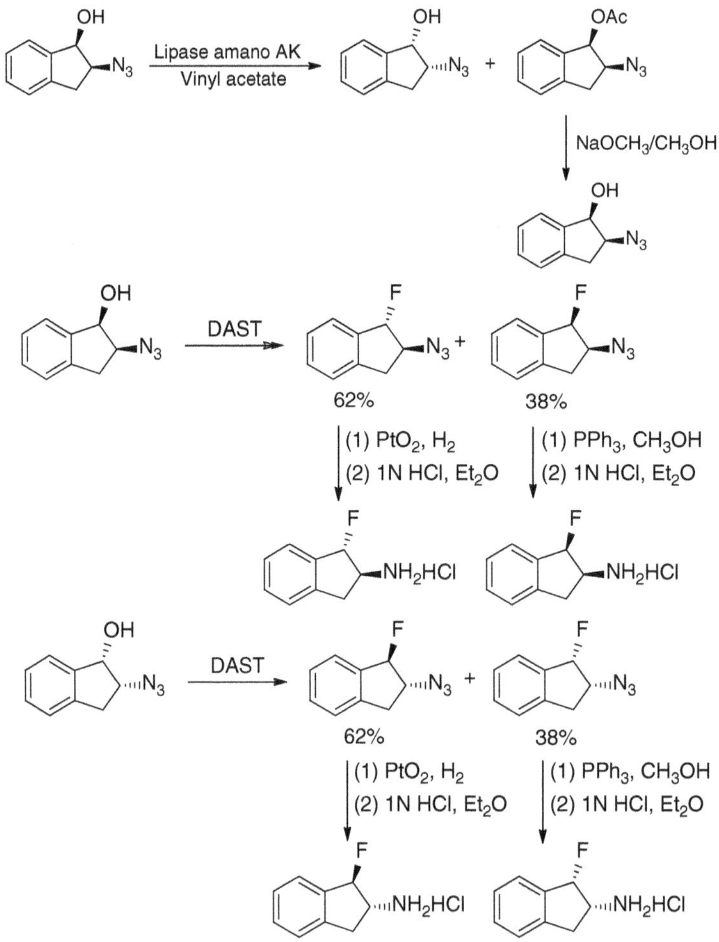

Scheme 2.13 Chemoenzymatic synthesis of optically active 1-fluoro-2-amino-indane. Source: Adapted from Iacazio and Réglier [16].

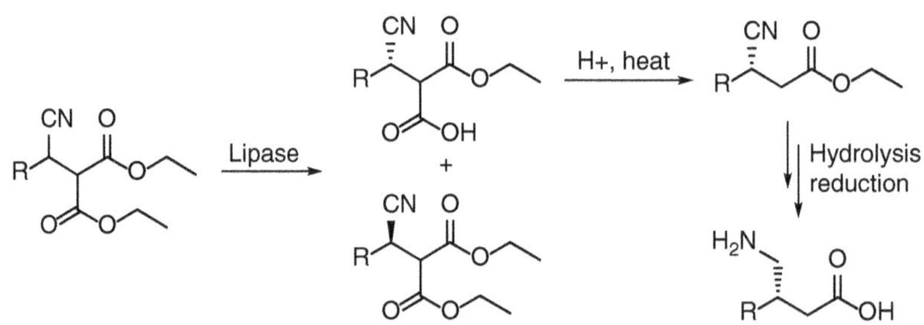

Scheme 2.14 Chemoenzymatic synthesis of optically active β-substituted-γ-amino acids. Source: Adapted from Mukherjee and Martinez [17].

purity and 97.8% ee in 81.6% overall yield (Scheme 2.15). The diester was prepared by reacting the diphenol compound with propionyl chloride in the presence of trimethylamine and catalytic amount of 4-DMAP. The chemoenzymatic process was performed at kilogram scale and offered a cost-effective strategy to replace the late-stage crystallization-induced resolution process for the synthesis of yimitasvir [18].

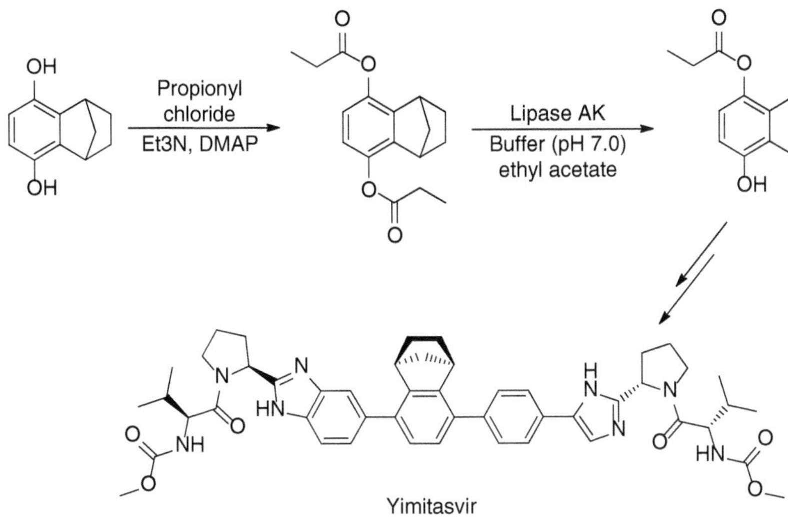

Scheme 2.15 Chemoenzymatic synthesis of Yimitasvir.

meso-1,2-Cyclopentanedimethanol can also be desymmetrized via lipase-catalyzed acetylation or hydrolysis of the diacetate. From this diol, (1*S*, 2*R*)-1,2-cyclopentanedimethanol monoacetate was prepared in 97% ee by acetylation using lipase Amano PS as biocatalyst and vinyl acetate as acyl donor, while (1*R*, 2*S*)-1,2-cyclopentanedimethanol monoacetate was obtained in 95% ee via the complementary monohydrolysis of the diacetate using the same lipase. (1*S*, 2*R*)-1,2-Cyclopentanedimethanol monoacetate was first converted to (1*R*, 2*S*)-azidoalcohol, followed by oxidation to give (1*R*, 2*S*)-azidoaldehyde, which was then transformed into the *cis*-fused bicyclic pyrrolidine via Passerini reaction with *n*-butyl isocyanide and propionic acid, and subsequent two-step process of azide reduction with concomitant acyl migration and SN_2 cyclization of the resulting secondary amide with the secondary alcohol (Scheme 2.16). (1*R*, 2*S*)-1,2-Cyclopentanedimethanol monoacetate could be transformed into the *trans*-fused bicyclic pyrrolidines by coupling Ugi reaction with a subsequent cyclization process (Scheme 2.17). The resulting *cis*-fused bicyclic pyrrolidines could serve as the key intermediate for the synthesis of the antiviral drug telaprevir (Scheme 2.18) [19].

Serine endopeptidase subtilisin A, which cleaves the amide bond in proteins and is commercially available as Alcalase, can be used as a practical biocatalyst for the formation of amide bond in the preparation of oligopeptides. However, the peptide's C-terminal carboxylic acid needs to be activated as ester and this often causes

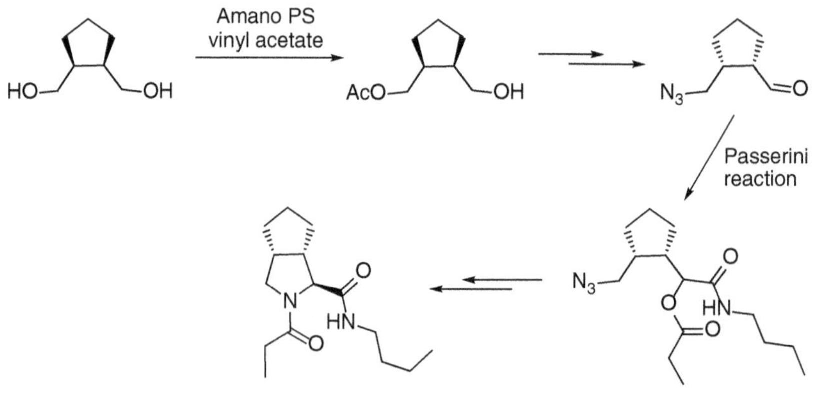

Scheme 2.16 Enzymatic desymmetrization of *meso*-1,2-cyclopentanedimethanol and subsequent transformation into *cis*-fused bicyclic pyrrolidine.

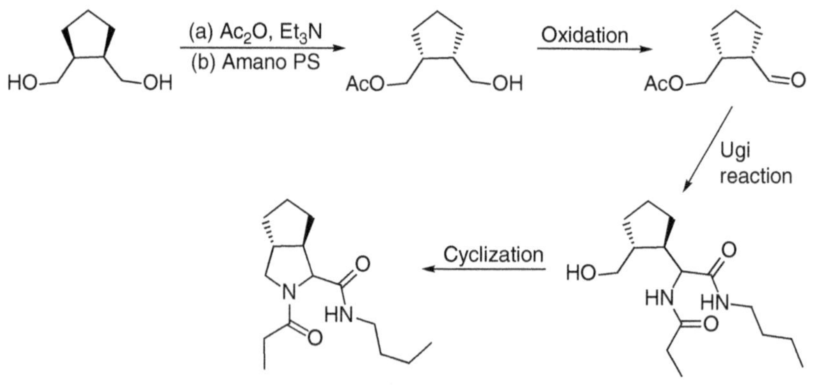

Scheme 2.17 Enzymatic desymmetrization of *meso*-1,2-cyclopentanedimethanol and subsequent transformation into *trans*-fused bicyclic pyrrolidine.

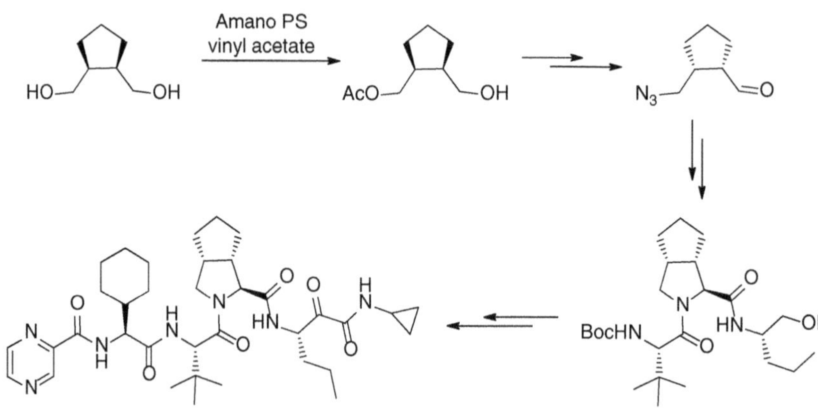

Telaprevir

Scheme 2.18 Chemoenzymatic synthesis of important API telaprevir. Source: Adapted from Moni et al. [19].

the racemization of the activated peptide esters, presenting a challenge for this approach. In this context, a transition-metal-catalytic method has been developed for the preparation of peptide enol esters without racemization. After screening a diversity of ligands and solvents, ruthenium was used to catalyze the alkyne addition in the presence of (+)-2,3-O-isopropylidene-2,3-dihydroxy-1,4-bis(diphenylphosphino)butane as ligand in dry THF. The peptide enol esters were synthesized without racemization and used as acyl donors in the Alcalase-catalyzed peptide synthesis with other peptide fragments in organic solvents (Scheme 2.19) [20].

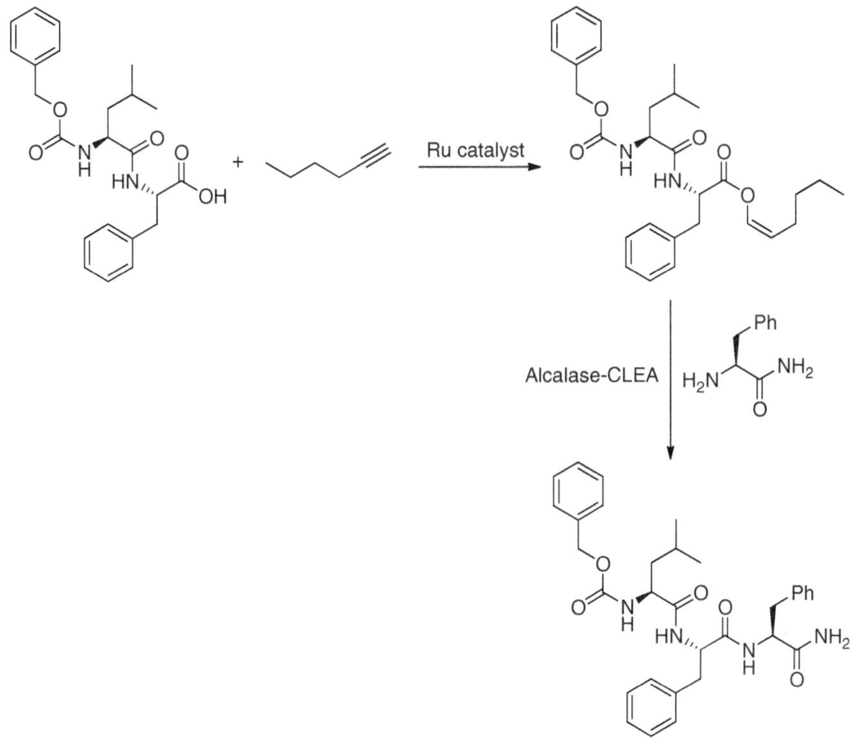

Scheme 2.19 Chemoenzymatic synthesis of peptides. Source: Adapted from Schröder et al. [20].

2.2 Nitrilases

Chemical hydrolysis of nitriles into carboxylic acids often requires drastic conditions of strong bases, acids, or elevated temperatures. On the other hand, nitrilases catalyze the hydrolysis of nitriles under very mild conditions with high efficiency and selectivities. As such, nitrilase-catalyzed reactions have been integrated into the synthesis of various important fine chemicals. Statins are a large class of drugs prescribed for the clinic treatment of hypercholesterolemia and dyslipidemia. Ethyl (*R*)-4-cyano-3-hydroxybutyrate is a useful intermediate in the synthesis of statin side chains. This has been prepared from inexpensive epichlorohydrin, first by converting it into 3-hydroxyglutaronitrile using two chemical steps. The resulting prochiral dinitrile was then desymmetrized by nitrilase in phosphate buffer. Finally, esterification with ethanol and sulfuric acid catalyst generated the desired ethyl (*R*)-4-cyano-3-hydroxybutyrate (Scheme 2.20) [21].

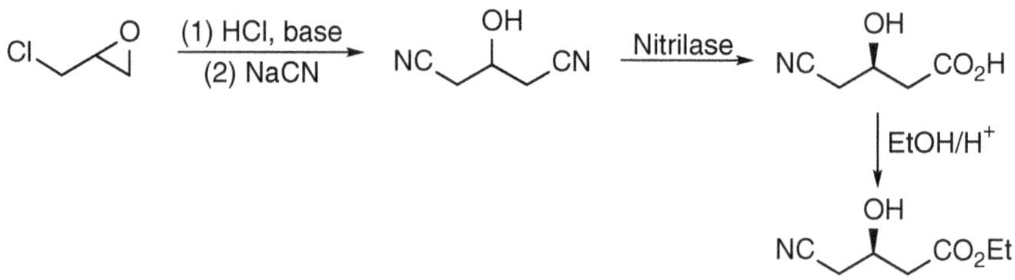

Scheme 2.20 Chemoenzymatic synthesis of ethyl (*R*)-4-cyano-3-hydroxybutyrate. Source: Based on Bergeron et al. [21].

A chemoenzymatic route involving nitrilase-catalyzed hydrolysis under mild reaction conditions has also been developed for the synthesis of enantiomerically pure 1,4-benzodioxane-2-carboxylic acid derivatives, which are useful precursors for pharmaceuticals and biologically active compounds. The condensation of catechol or 3,4-dihydroxybenzaldehyde with 2-chloroacrylonitrile in the presence of potassium carbonate gave racemic 2-cyano-1,4-benzodioxane and 2-cyano-6-formyl-1,4-benzodioxane, respectively. Nitrilase catalyzed the kinetic resolution of these cyano compounds to generate the optically active 1,4-benzodioxane-2-carboxylic acids (Scheme 2.21). Otherwise, compounds with formyl functional group are not stable under the reaction conditions of chemical nitrile hydrolysis for the preparation of these optically active 1,4-benzodioxane-2-carboxylic acids [22].

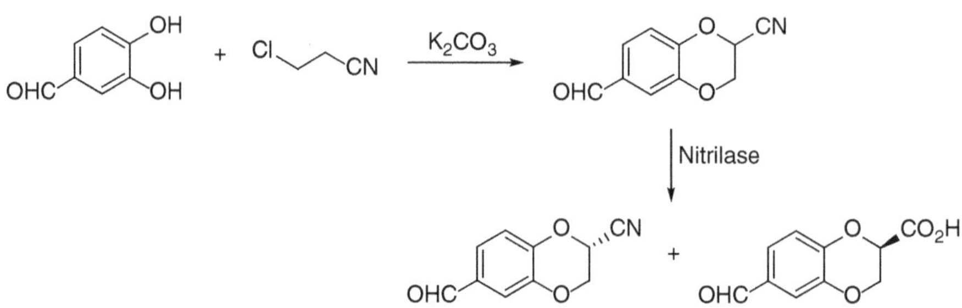

Scheme 2.21 Chemoenzymatic synthesis of (*R*)-6-formyl-1,4-benzodioxane-2-carboxylic acid.

Another chemoenzymatic process involving nitrilase-catalyzed hydrolysis has been the synthesis of (*S*)-Pregabalin, a blockbuster drug for the treatment of neuropathic pain and partial seizures. In this process, nitrilase catalyzed the regio- and stereospecific hydrolysis of racemic isobutylsuccinonitrile to give

(3*S*)-3-cyano-5-methylhexanoic acid, which was then converted to (*S*)-Pregabalin via hydrogenation of the nitrile group catalyzed by a heterogeneous nickel catalyst (Scheme 2.22). The recovered (3*R*)-isobutylsuccinonitrile was readily epimerized under basic conditions and reused [23].

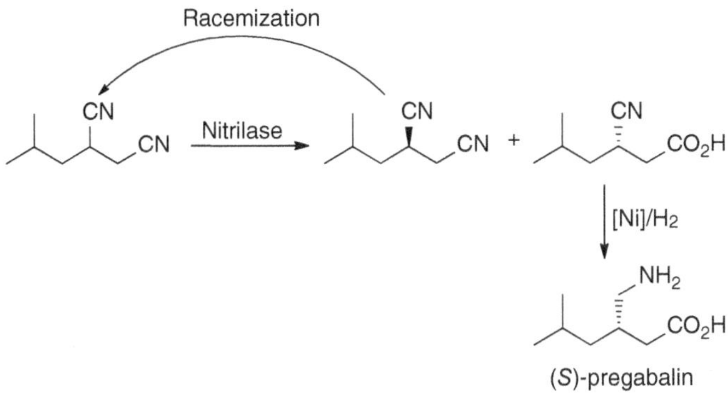

Scheme 2.22 Chemoenzymatic synthesis of (*S*)-Pregabalin via nitrilase-catalyzed kinetic resolution.

In the above method, the key step was the nitrilase-catalyzed kinetic resolution of a racemic dinitrile. A variation of the chemoenzymatic synthesis of (*S*)-Pregabalin was the desymmetrization of a prochiral dinitrile as a key step. Here, a nitrilase (BjNIT6402) from *Bradyrhizobium japonicum* catalyzed the desymmetric hydrolysis of 3-isobutylglutaronitrile, affording (*S*)-3-(cyanomethyl)-5-methylhexanoic acid in 93% yield and 90% ee. This was then transformed into (*S*)-Pregabalin via Curtius rearrangement followed by acidic hydrolysis. Similarly, (*R*)-Baclofen was prepared by enzymatic hydrolysis of 3-(4′-chlorophenyl)glutaronitrile with a nitrilase (HsNIT) from *Herbaspirillum* sp, and subsequent Curtius rearrangement and acidic hydrolysis (Scheme 2.23) [24]. The enzymatic hydrolysis efficiency was improved by using mutants of a nitrilase from *Synechocystis* sp. PCC6803, and a variety of 3-substituted glutaronitriles were desymmetrically hydrolyzed to optically active 3-substituted 4-cyanobutanoic acid. By combining with well-established chemical transformations nitrilase-catalyzed desymmetric hydrolysis of prochiral 3-substituted glutaronitriles offers an attractive approach to access chiral β-substituted γ-amino acids, a class of important γ-aminobutyric acid (GABA) derivatives [25].

The nitrilase-catalyzed resolution of *rac*-3-oxocyclohexane-1-carbonitrile could be integrated into the chemoenzymatic synthesis of various APIs, such as soluble epoxide inhibitor (Scheme 2.24) [26].

2.3 Carbonyl Reductases

Carbonyl reductases (also called ketoreductases or alcohol dehydrogenases) catalyze the asymmetric reduction of ketones to give the corresponding chiral alcohols.

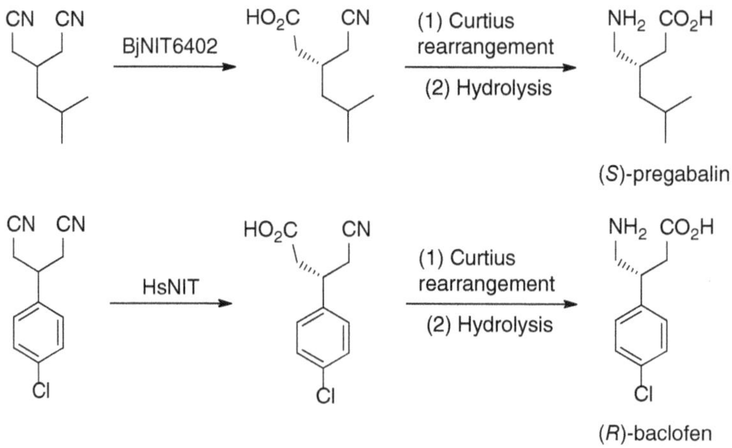

Scheme 2.23 Chemoenzymatic synthesis of (*S*)-Pregabalin and (*R*)-Baclofen via nitrilase-catalyzed desymmetric hydrolysis. Source: Based on Duan et al. [24].

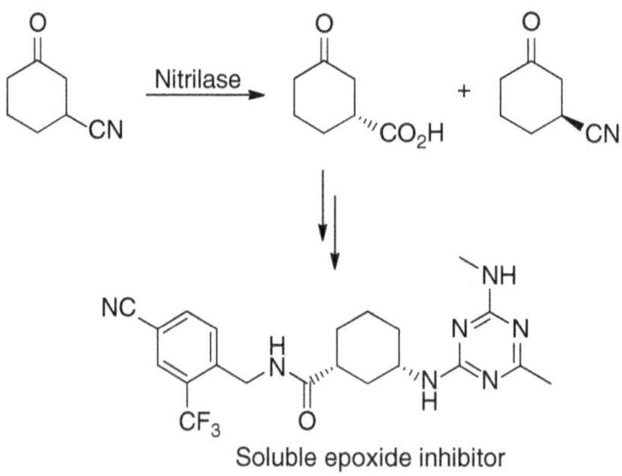

Scheme 2.24 Chemoenzymatic synthesis of soluble epoxide inhibitor via nitrilase-catalyzed kinetic resolution. Source: Based on Hadi et al. [26].

The ready availability, broad substrate range, exquisite chemo, regio, and stereoselectivities as well as the mild reaction conditions of these enzymes have made them the catalyst of choice for ketone reductions. Enzymatic reduction has often been integrated with chemical ketone formation and/or further transformation of the generated alcohols to achieve specific synthetic goals. To access optically active triazole-containing β-adrenergic receptor blocker analogs, a chemoenzymatic method was developed by employing enantioselective carbonyl reductases with "click chemistry." (*S*)-2-Azido-1-arylethanols were initially synthesized via enzymatic reduction of the corresponding α-azidoacetophenone derivatives effected by a recombinant carbonyl reductase from *Candida magnoliae* (CMCR), and then converted to optically pure β-hydroxy triazoles with high yields via copper(I)-catalyzed [3 + 2] Huisgen's cycloaddition (Scheme 2.25) [27].

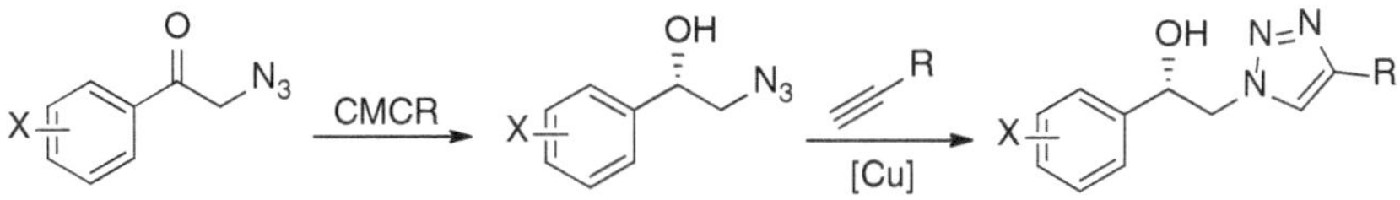

Scheme 2.25 Chemoenzymatic synthesis of optically pure β-hydroxy triazoles. Source: Adapted from Ankati et al. [27].

The biocatalytic ketone reduction has been combined with chemical transformations for the asymmetric synthesis of the active component (*R*)-enantiomer of denopamine and salmeterol, two marketed chiral drugs. The substituted α-bromo-acetophenones were reduced by *Rhodotorula rubra* whole cells in the presence of sodium lauryl sulfate, and the resulting (*R*)-alcohols were chemically transformed into β-adrenoreceptor agonists (*R*)-(−)-denopamine and (*R*)-(−)-salmeterol in good overall yields and high enantiomeric purity (Scheme 2.26) [28].

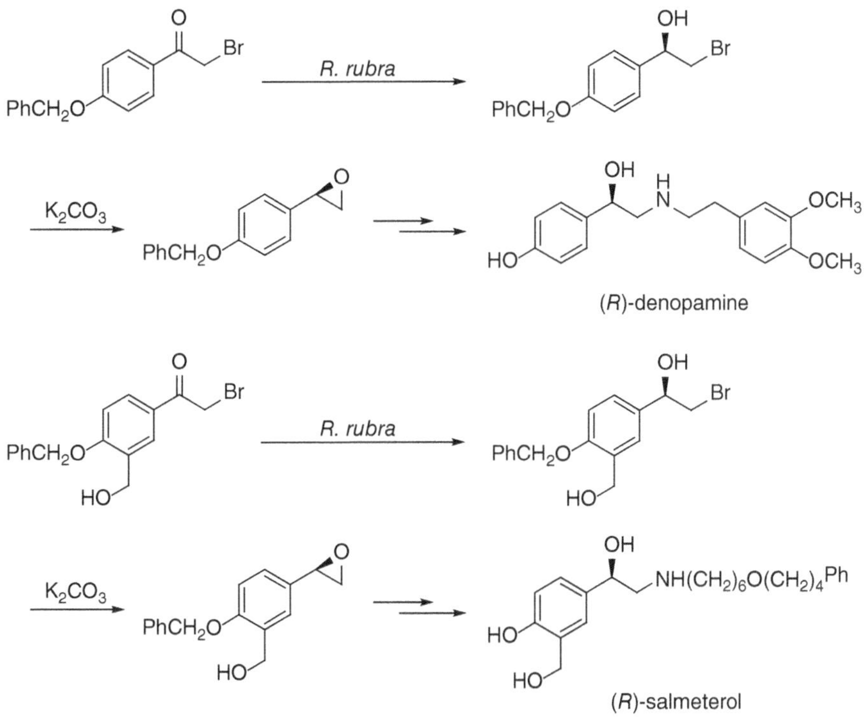

Scheme 2.26 Chemoenzymatic synthesis of (*R*)-(−)-denopamine and (*R*)-(−)-salmeterol. Source: Based on Goswami et al. [28].

The reductions of 1-arylacetones can be achieved by employing enantiocomplementary alcohol dehydrogenases, generating the (*S*)- and (*R*)-enantiomers of the corresponding alcohols with up to >99 % ee. These optically pure 1-aryl-2-propanols react with benzaldehydes via oxa-Pictet–Spengler reaction to form the 1-aryl-3-methylisochroman derivatives in 47–92% yields and syn/anti ratio of up to 99:1. The structural motif of 1-functionalized isochromans widely exists in synthetic and natural products with various biological activities. A diversity of enantiomerically pure compounds with this structural motif have been prepared by this two-step chemoenzymatic process (Scheme 2.27) [29].

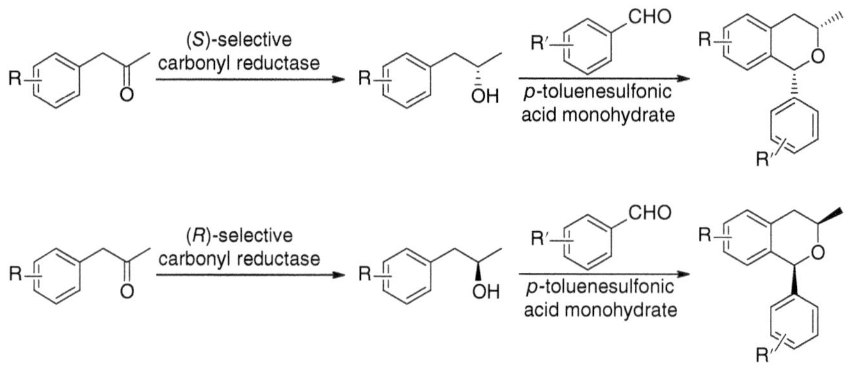

Scheme 2.27 Chemoenzymatic synthesis of 1-aryl-3-methylisochromans. Source: Based on Simon et al. [29].

Ethyl 3,3-dimethyl-2-oxobutanoate could be reduced by a ketoreductase to give the hydroxy ester, which was then hydrolyzed to (*R*)-3,3-dimethyl-2-hydroxybutyric acid in 82% isolated yield and >99.5% ee. Subsequent reaction of the enantiomerically pure hydroxyacid with L-proline methyl ester furnished the diastereomerically pure (*S*,*R*)-lactone after treatment of the reaction mixture with catalytic amount of TsOH in toluene. The enantiopure lactone is a key intermediate for the synthesis of a potent thrombin inhibitor (Scheme 2.28) [30].

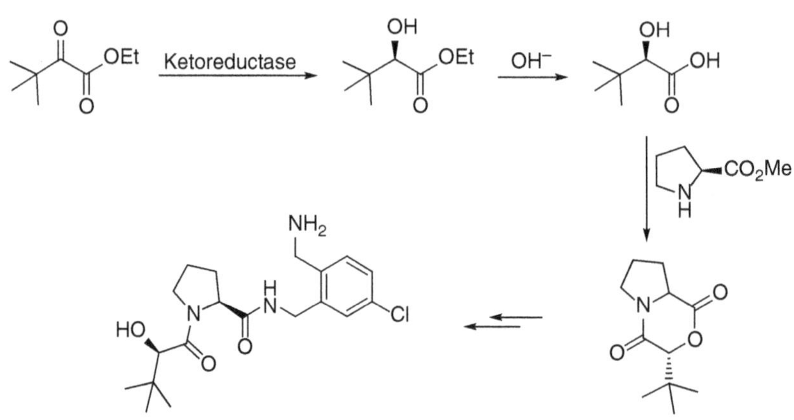

Scheme 2.28 Chemoenzymatic synthesis of a thrombin inhibitor. Source: Adapted from Nelson et al. [30].

Similarly, the α-keto group of ethyl 2,4-dioxo-4-phenylbutyrate was bioreduced by yeast strain *Pichia pastoris* CBS 704 to give ethyl (*R*)-2-hydroxy-4-oxo-4-phenylbutyrate, which was then converted by chemical reduction into ethyl (*R*)-2-hydroxy-4-phenylbutyrate (Scheme 2.29), an important intermediate in the manufacture of a variety of angiotensin-converting enzyme (ACE) inhibitors such as benazepril, cilazapril, and enalapril [31].

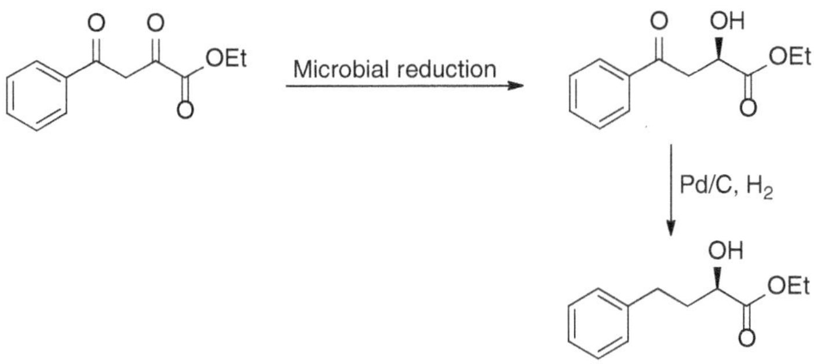

Scheme 2.29 Chemoenzymatic synthesis of ethyl (*R*)-2-hydroxy-4-phenylbutyrate.

The alcohol dehydrogenase from *Ralstonia* sp. (RasADH) catalyzed the reduction of methyl 4-(*p*-bromophenyl)-4-oxobutanoate to give methyl (*R*)-4-(*p*-bromophenyl)-4-hydroxybutanoate with >97% ee, which was converted into a lactone with an identical ee value and excellent yield by treatment with a HCl solution. The brominated lactone obtained underwent various Pd-catalyzed coupling reactions. $Pd(PPh_3)_2Cl_2$ catalyzed the Suzuki coupling of the brominated lactone with phenylboronic acid affording the (*R*)-product with 75% conversion. Sonogashira coupling with phenylacetylene proceeded smoothly using $Pd(PPh_3)_2Cl_2$ and CuI as catalysts at 100 °C, generating the (*R*)-configurated coupling product in 60% yield. Similarly, Heck transformation with styrene under the action of $Pd(OAc)_2$ at 120 °C produced the (R)-lactone derivative with 69% conversion (Scheme 2.30). It is worth noting that in all three Pd-catalyzed coupling reactions, the lactone derivatives were obtained without loss of enantioselectivity. This chemoenzymatic approach offers a great way to accessing functionalized chiral lactones with potential biological activities [32].

ω-Hydroxyl α-diazo esters such as benzyl 2-diazo-7-hydroxyoctanoate can undergo ring closure to form oxygen-containing heterocycles under the action of a transition metal catalyst, which promotes the stereoselective intramolecular insertion of the hydroxyl group into the *in situ* formed metal-carbenoid. The optically pure ω-hydroxyl α-diazo esters can be obtained from the corresponding ω-keto α-diazo esters using highly enantioselective carbonyl reductases. Integration of the enzymatic reduction with the metal-catalyzed ring-closure reaction should lead to the chiral O-containing heterocycles bearing two stereogenic centers. Indeed, the (*R*)-selective alcohol dehydrogenase (LbADH) and (*S*)-selective enzyme Gre2p catalyzed the reduction of benzyl 2-diazo-7-oxooctanoate, furnishing the corresponding (*R*)- and (*S*)-alcohol, respectively. By employing $Rh_2(OAc)_4$ as catalyst, enantiopure

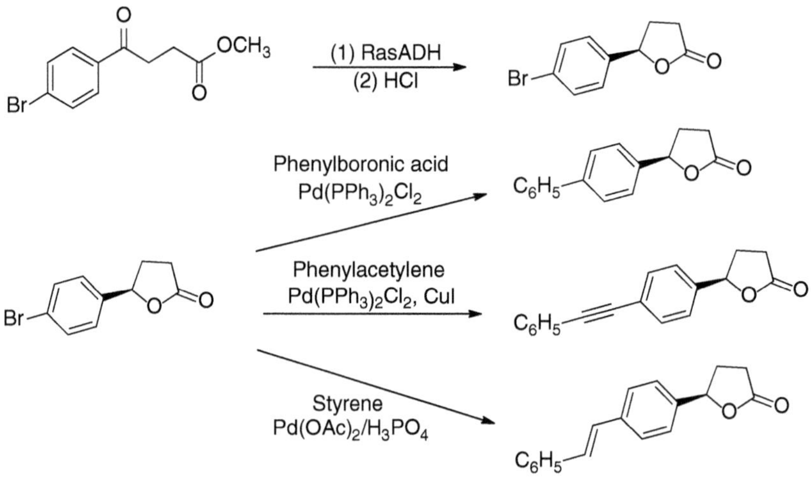

Scheme 2.30 Chemoenzymatic synthesis involving enzymatic ketone reduction and Pd-catalyzed coupling reactions.

benzyl (*R*)-2-diazo-7-hydroxyoctanoate was cyclized to give two diastereomers of benzyl 7-methyloxepane-2-carboxylate in 73:27 dr (diastereomeric ratio) and >98% ee with *trans*-configured oxepane as the major product (Scheme 2.31). Similarly, the enzymatic reduction and subsequent Rh-catalyzed transformation of benzyl 2-diazo-6-oxoheptanoate generated benzyl 6-methyloxane-2-carboxylate [33].

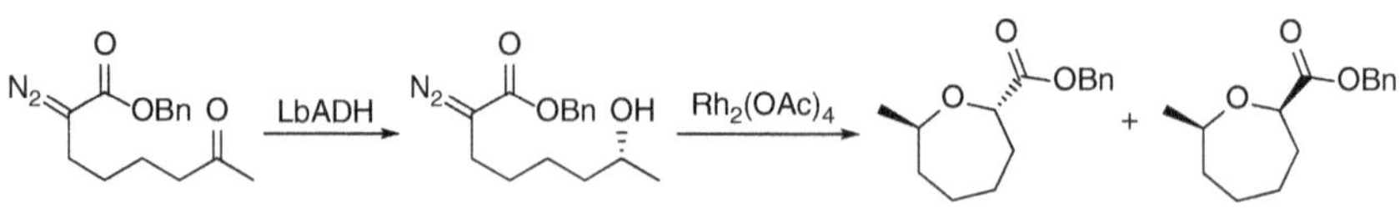

Scheme 2.31 Chemoenzymatic synthesis of benzyl 7-methyloxepane-2-carboxylate.

The biocatalytic reduction of β-ketosulfides using *Pseudomonas monteilii* ZMU-T04 generated the corresponding (*S*)-β-hydroxyl-sulfides. The Ti(OiPr)$_4$/(*S*)-BINOL complex catalyzed the asymmetric sulfoxidation of β-hydroxyl-sulfides to furnish the corresponding β-hydroxyl-sulfoxides with up to 99:1 dr and >99% ee. A wide range of aryl β-hydroxyl-sulfoxides were synthesized by this two-step enantioselective synthetic strategy (Scheme 2.32) [34].

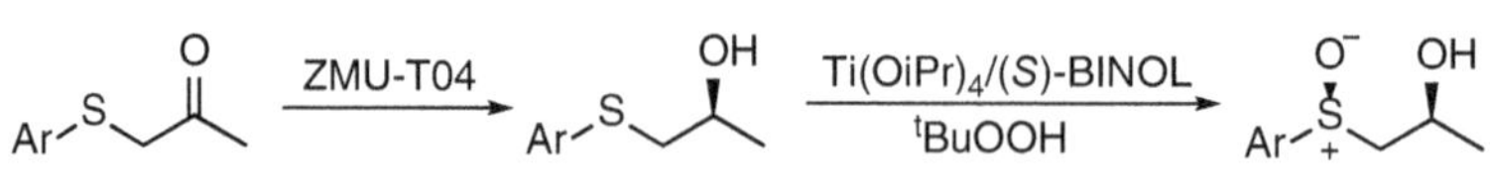

Scheme 2.32 Chemoenzymatic synthesis of optically active aryl β-hydroxyl-sulfoxides. Source: Based on Cui et al. [34].

The chemoenzymatic strategy has also been applied to the synthesis of (*S*)-duloxetine, a marketed pharmaceutical for the treatment of major depressive disorders. 3-(Dimethylamino)-1-(2-thienyl)-1-propanone was prepared from 2-acetylthiophene, and then reduced employing a carbonyl reductase from *Rhodosporidium toruloides*. (*S*)-3-(Dimethylamino)-1-(2-thienyl)-1-propanol was obtained with high yield and ee value using *Escherichia coli* cells harboring the carbonyl reductase and a glucose dehydrogenase, and further transformed into (*S*)-duloxetine (>98.5% ee) with 60% overall yield from 2-acethylthiophene (Scheme 2.33) [35].

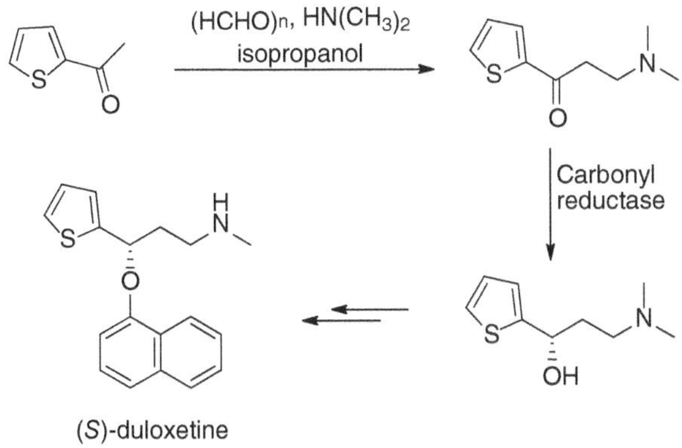

Scheme 2.33 Chemoenzymatic synthesis of (*S*)-duloxetine. Source: Based on Chen et al. [35].

2.4 Ene Reductases

Ene reductase (ER) catalyzes the reduction of alkenes, producing two chiral centers, under mild reaction conditions with high chemo, regio, and stereoselectivity. In contrast to the *cis*-hydrogenation by transition-metal-based homogeneous catalysts, ERs usually catalyze *trans*-hydrogenation, which represents a highly useful asymmetric transformation. A diversity of α, β-unsaturated aldehydes, ketones, carboxylic acids, and derivatives (esters, lactones, cyclic imides, cyclic acid anhydrides) and α,β-unsaturated nitroalkene can be reduced by using ERs, providing a promising sustainable tool for the asymmetric hydrogenation of C=C double bond. The ER-catalyzed reduction of activated C=C bonds has been integrated into multistep chemical reactions, serving as emerging methodologies for the synthesis of important compounds [36].

(*R*)-Flurbiprofen has been reported as a potential reducer of β-amyloid level, which causes plaque formation in Alzheimer's disease. An ER (YqjM) from *Bacillus subtilis* catalyzed the reduction of a variety of methyl 2-phenylacrylate derivatives, giving enantiomerically pure (*R*)-profen derivatives. (*R*)-Flurbiprofen methyl ester was then synthesized by coupling this C=C bond bioreduction with a Pd-catalyzed coupling reaction, as shown in Scheme 2.34 [37].

(*S*)-3-Aminomethyl-5-methylhexanoic acid (Pregabalin) is the active ingredient in Lyrica, a class of drugs known as GABA analogs. A chemoenzymatic process

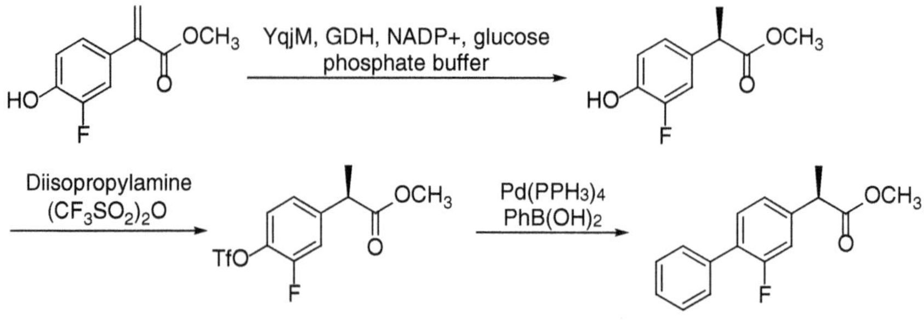

Scheme 2.34 Chemoenzymatic synthesis of (*R*)-flurbiprofen methyl ester. Source: Based on Pietruszka et al. [37].

for the manufacture of pregabalin involving lipase-catalyzed kinetic resolution of a β-cyanodiester was described in Section 2.1. Nitrilase-based chemoenzymatic approaches have also been reported to access β-substituted γ-amino acids. The ER-catalyzed asymmetric reduction of β-cyanoacrylate esters was combined with the chemical preparation of (*Z*)- and (*E*)-β-cyanoacrylate esters to produce the corresponding optically active β-cyano esters in high yield and enantiomerically pure form (Scheme 2.35), which could be converted into the β-substituted γ-amino acids by hydrogenation of the cyano group and hydrolysis of the ester. The *E*/*Z* configuration of alkenes exerted dramatic impact on the stereochemical outcome of the bioreduction of C=C bond; by substrate- and enzyme engineering both stereoisomers were accessible with high conversion and perfect stereoselectivity (Scheme 2.35) [38]. However, the ERs tested showed little activity toward (*E*)-5-methyl-3-cyanohex-2-enoic acid. Interestingly, a 12-oxophytodienoic acid reductase OPR1 from *Lycopersicon esculentum* and its variants effectively catalyzed the reduction of (*E*)-5-methyl-3-cyanohex-2,4-dienoic acid to give (*R*)-5-methyl-3-cyanohex-4-enoic acid in 68–96% conversion and >99% ee. It was then transformed into (*S*)-pregabalin via Raney Ni-catalyzed hydrogenation (Scheme 2.36) [39].

Guerbet alcohols are attractive intermediates in the production of plasticizers, lubricants, and surfactants. They are industrially produced by self-condensation reaction of primary aliphatic alcohols at elevated reaction temperature and high-pressure conditions. Such harsh reaction conditions result in low selectivity and formation of undesired by-products. A chemoenzymatic approach involving organocatalytic and enzymatic reactions has been developed to address these issues. In this new process, for example, 1-hexanol was oxidized using hypochlorite as oxidant under the catalysis of 2,2,6,6-tetramethylpiperidin-1-oxyl (TEMPO), followed by the lysine-catalyzed homoaldol condensation of the resulting aldehyde, affording 2-butyl-2-octenal. The 2-branched α,β-unsaturated aldehyde was then reduced by employing an ER from *Gluconobacter oxydans* and an alcohol dehydrogenase from *Rhodococcus* sp. in a one-pot format together with a GDH/glucose cofactor regeneration system, thus giving 2-butyl-1-octanol in 78% yield (Scheme 2.37).

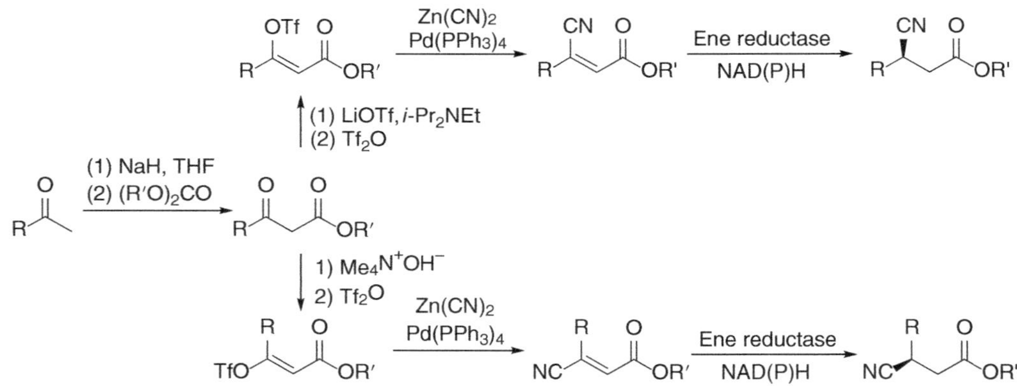

Scheme 2.35 Chemoenzymatic processes for the preparation of optically active β-cyano esters. Source: Aapted from Winkler et al. [38].

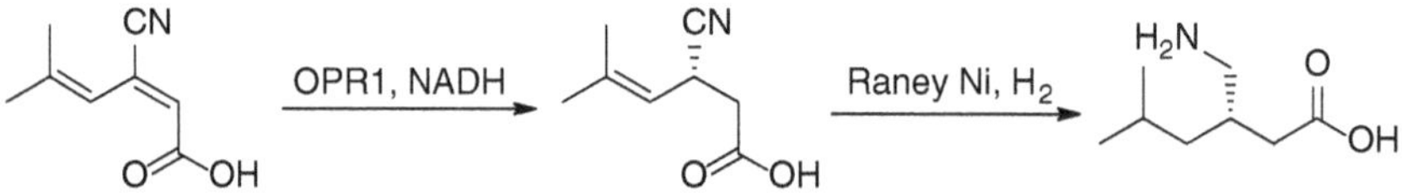

Scheme 2.36 Chemoenzymatic preparation of (*S*)-pregabalin. Source: Adapted from Winkler et al. [39].

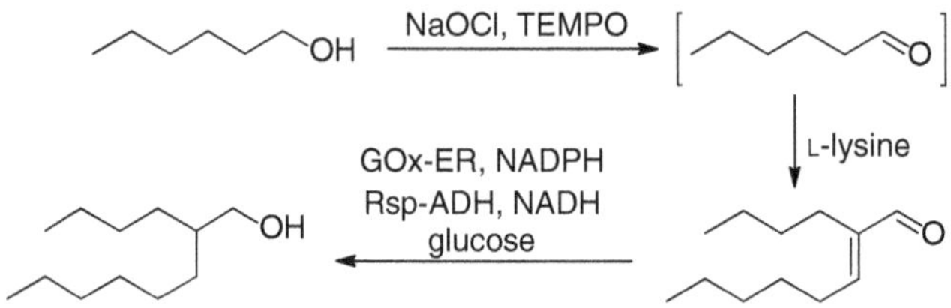

Scheme 2.37 Chemoenzymatic preparation of 2-butyl-1-octanol.

The whole chemoenzymatic process was performed at room temperature and the desired target product was obtained in 62% overall yield [40].

2.5 Transaminases

Transaminases (TA), a class of pyridoxal phosphate (PLP)-dependent enzymes, catalyze the amino transfer reactions between an amino donor and a carbonyl functional group. The overall process of transamination is composed of two half-reactions, the transfer of the amino group from amino donor to the PLP cofactor and the transfer of amino group from the PLP cofactor to the amino acceptor. This enzymatic amino transfer reaction offers an attractive method to synthesize chiral amines and has been applied in chemoenzymatic processes for the production of high-value fine chemicals with chiral amine motifs [41]. For example, the ω-transaminase from *Vibrio fluvialis* JS17 (Vf-TA) has been employed in the chemoenzymatic synthesis of (*S*)-rivastigmine, a potent drug for the treatment of Alzheimer's disease, starting from the acetophenone precursor (Scheme 2.38) [42].

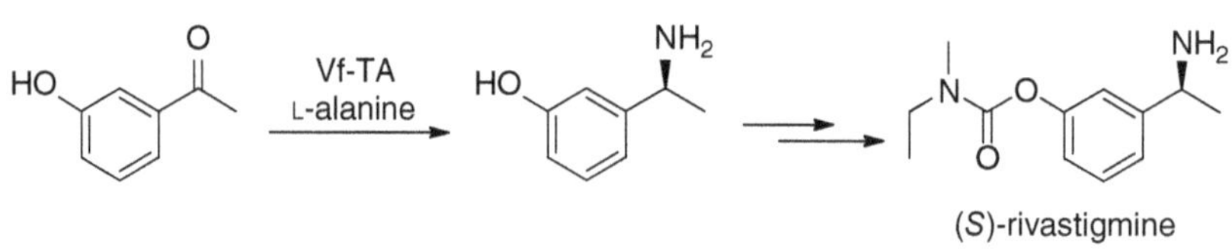

Scheme 2.38 Chemoenzymatic synthesis of (*S*)-Rivastigmine. Source: Adapted from Fuchs et al. [42].

Transaminase Vf-TA was also employed in the synthesis of a JAK2 kinase inhibitor AZD1480. After having screened 60 different TAs, Vf-TA was found to be the most promising with (*S*)-α-methylbenzylamine (*S*-MBA) as the amino

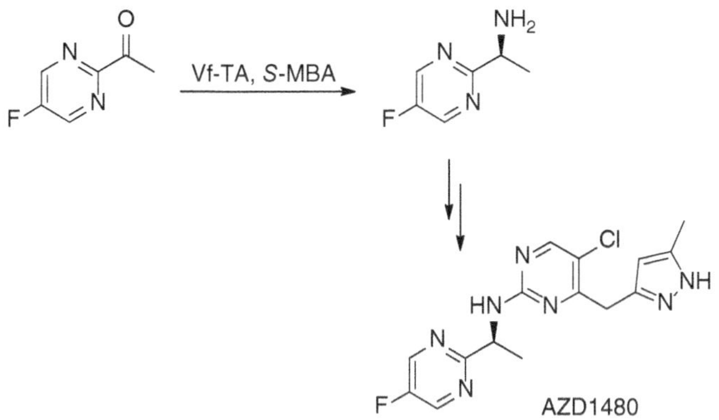

Scheme 2.39 Chemoenzymatic synthesis of AZD1480.

donor (Scheme 2.39). As such, by using commercial enzyme preparations of Vf-TA, (*S*)-1-(5-fluoropyrimidin-2-yl)ethylamine was prepared in 77% yield and 99.8% ee. When a recombinant *E. coli* whole-cell biocatalyst harboring the Vf-TA gene was used, this process afforded (*S*)-1-(5-fluoropyrimidin-2-yl)ethylamine with 66% yield and 97.3% ee [43]. Imagabalin ((3*S*,5*R*)-3-amino-5-methyloctanoic acid) acts as a ligand for the $\alpha 2\delta$ subunit of the voltage-dependent calcium channel and shows preclinical efficacy for the treatment of generalized anxiety disorder. A variant of Vf-TA was successfully applied to the transamination of (*R*)-ethyl 5-methyl 3-oxooctanoate with (*S*)-methylbenzylamine as the amino donor, giving (3*S*,5*R*)-ethyl 3-amino-5-methyloctanoate. Imagabalin was then produced by hydrolysis of the resulting β-amino ester (Scheme 2.40) [44].

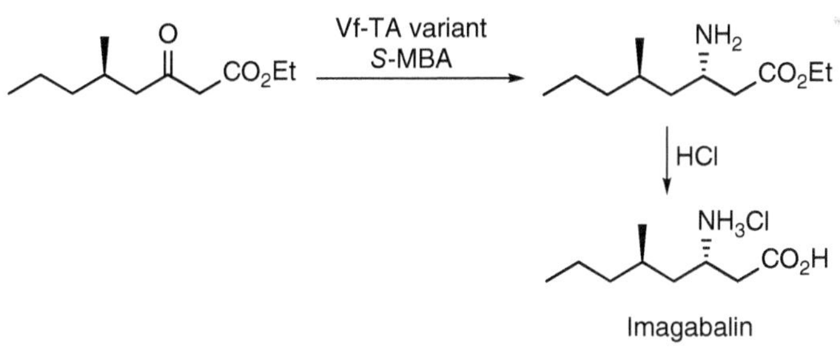

Scheme 2.40 Chemoenzymatic synthesis of Imagabalin. Source: Adapted from Midelfort et al. [44].

Chiral 1,4-diazepanes are important structural units in natural products and pharmaceutical molecules, such as Suvorexant, a selective, dual orexin receptor antagonist for the treatment of primary insomnia. The core chiral diazepane ring has been assembled using a TA-catalyzed asymmetric amino transfer to methanesulfonic acid 2-[(5-chloro-benzooxazol-2-yl)-(3-oxo-butyl)amino]ethyl ester, followed by a seven-membered ring annulation. 5-Chloro-2-((*R*)-5-methyl-[1,4]diazepan-1-yl)-benzooxazole hydrochloride was obtained in 62% yield and 99% ee using a mutant

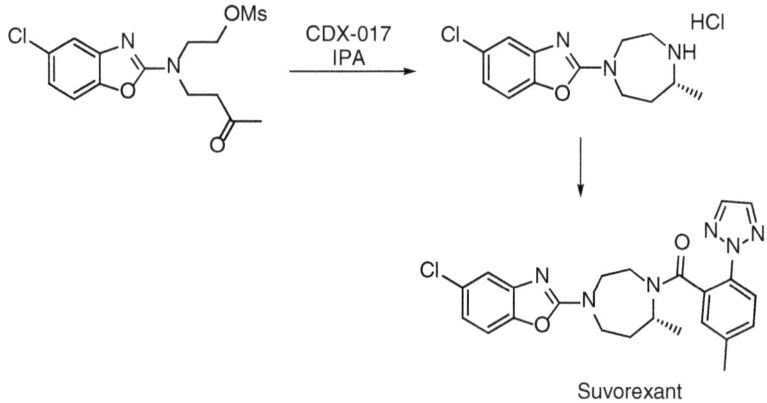

Scheme 2.41 Chemoenzymatic preparation of Suvorexant.

TA CDX-017 as the biocatalyst and isopropylamine (IPA) as amino donor under optimized reaction conditions (Scheme 2.41). Suvorexant was then synthesized by coupling 5-chloro-2-((*R*)-5-methyl-[1,4]diazepan-1-yl)-benzooxazole hydrochloride with 5-methyl-2-[1,2,3]triazol-2-yl-benzoic acid under Schotten–Baumann conditions [45]. Another dual orexin receptor antagonist MK-6096 was also synthesized by a chemoenzymatic process involving a TA-catalyzed reaction as the key step (Scheme 2.42) [46].

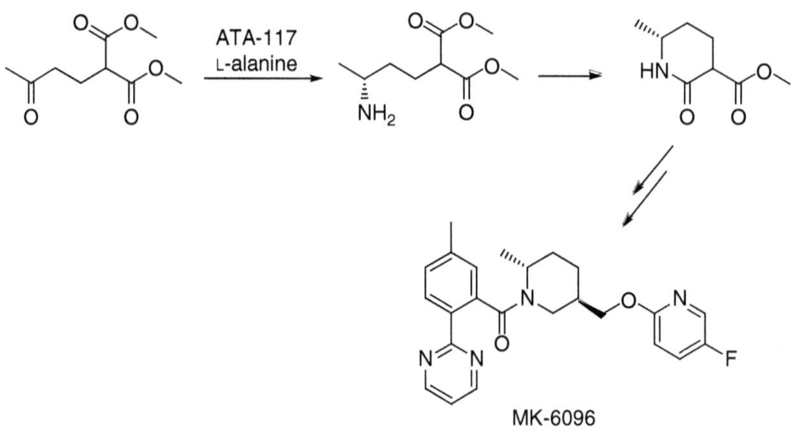

Scheme 2.42 Chemoenzymatic synthesis of MK6096. Source: Adapted from Girardin et al. [46].

The mutant TA CDX-017 also catalyzed the amino transfer reaction to ethyl (7-oxo-6,7,8,9-tetrahydropyrido[1,2-a]indol-10-yl)acetate, affording ethyl [(7*R*)-7-amino-6,7,8,9-tetrahydropyrido[1,2-a]indol-10-yl]acetate hydrochloride in 81% yield and 98–99% ee, which was further transformed into MK-7246 (Scheme 2.43). A cost-effective chemoenzymatic process has thus been developed for the kilogram-scale manufacturing of MK-7246, a potent and selective CRTH2 antagonist for the potential treatment of respiratory disease [47]. This process greatly improved the production efficiency and reduced the environmental impact compared to the previous methods.

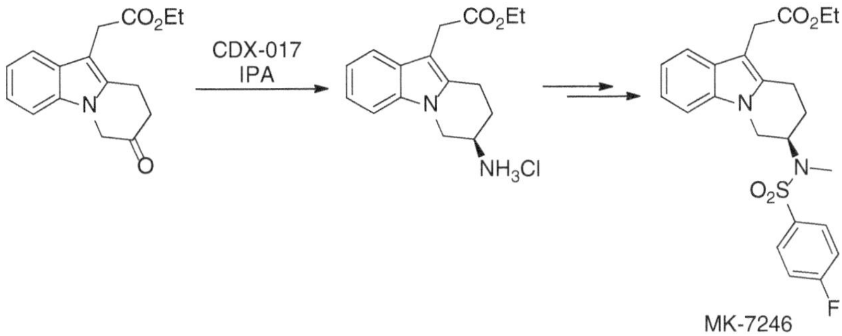

Scheme 2.43 Chemoenzymatic preparation of MK-7246.

Cinacalcet, N-[(1*R*)-1-(naphthyl)ethyl]-3-[3-(trifluoromethyl)-phenyl]propane-1-amine, is a calcimimetic approved drug for the treatment of secondary hyperparathyroidism. The core structure of Cinacalcet is the chiral amine, (*R*)-(+)-1-(1-naphthyl)ethylamine. This chiral amine can be prepared by the transamination of 1-acetonaphthone. After screening with a diversity of commercially available TAs, ArRmut11-M117F/G279A was selected to establish a chemoenzymatic process for the preparation of Cinacalcet. ArRmut11-M117F/G279A catalyzed the amino transfer reaction from isopropylamine (*i*PrNH$_2$) to 1-acetonaphthone, affording (*R*)-(+)-1-(1-naphthyl)ethylamine in 77% yield. The isolated amine was coupled with 3-[3-(trifluoromethyl)phenyl]propan-1-ol using the iridium catalyst [Cp*IrCl$_2$]$_2$ as the catalyst in toluene, giving Cinacalcet with an overall yield of 60% for the two-step reactions (Scheme 2.44). An attempt

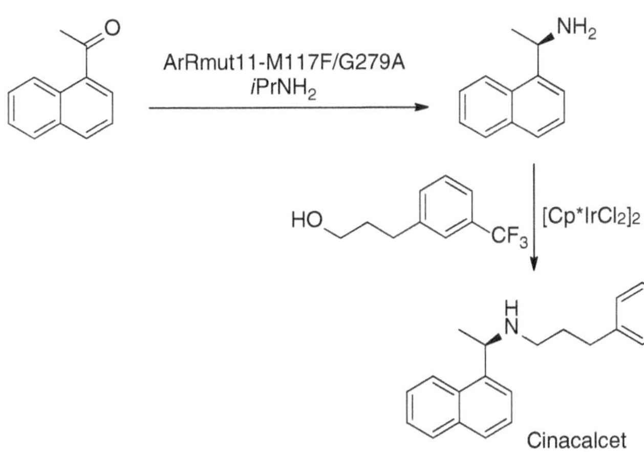

Scheme 2.44 Chemoenzymatic synthesis of Cinacalcet.

made to not remove the insolubles of the transamination reaction in the one-pot two-step setting was not successful [48].

2.6 Imine Reductases

Another biocatalytic strategy to construct chiral amine moiety is imine reductase (IRED)-catalyzed reduction, which has emerged as a valuable tool for the synthesis of *N*-heterocyclic molecules in recent years. By using (*R*)- and (*S*)-selective IREDs, stereocomplementary synthesis of pharmaceutically relevant 2-aryl-substituted pyrrolidines was achieved [49]. A chemoenzymatic route to chiral pharmaceutically important morphinan skeletons was developed by employing IREDs in the key step for the creation of chiral center. Various phenyl-substituted 1-benzyl-1,2,3,4,5,6,7,8-octahydroisoquinoline (1-benzyl-OHIQ) derivatives, the precursors to morphinan skeletons, were prepared with excellent enantiomeric purity from the corresponding 1-benzyl-3,4,5,6,7,8-hexahydroisoquinolines (1-benzyl-HHIQ), which were in turn synthesized by the reaction of 2-(1-cyclohexenyl)ethylamine with the corresponding aryl acetic acids, followed by intramolecular cyclization in the presence of $POCl_3$ in toluene (Scheme 2.45) [50].

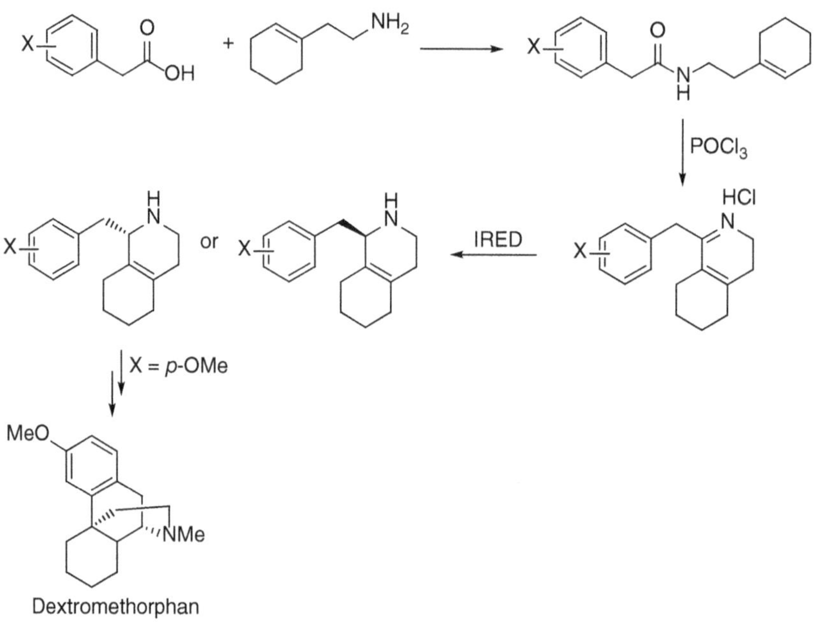

Scheme 2.45 Chemoenzymatic synthesis of phenyl-substituted 1-benzyl-1,2,3,4,5,6,7,8-octahydroisoquinolines. Source: Adapted from Yao et al. [50].

Dibenz[c,e]azepine is a privileged bridged biaryl architecture in bioactive compounds, chiral organocatalysts, and molecular switches. Both (*R*)- and (*S*)-5-methyl-6,7-dihydro-5*H*-dibenzo[c,e]azepine were prepared in excellent conversion and ee values by a chemoenzymatic route involving biocatalytic imine reduction in the last step to establish the chiral center (Scheme 2.46). The imine precursor 7-methyl-5*H*-dibenzo[c,e]azepine was synthesized via a Suzuki–Miyaura cross-coupling of 2-acetylphenylboronic acid and *N*-Boc-2-bromobenzylamine, followed by intramolecular imine formation by treatment with trifluoroacetic acid [51].

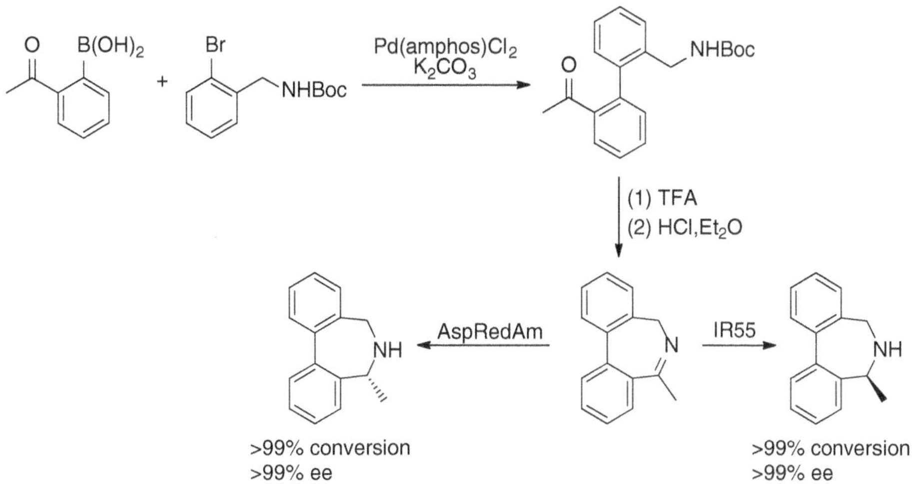

Scheme 2.46 Chemoenzymatic synthesis of (*R*)- and (*S*)-5-methyl-6,7-dihydro-5*H*-dibenzo[c,e]azepine.

Seven-membered *N*-heterocyclic azepane and benzazepine are structure cores in many bioactive compounds that find various applications in medicinal and pesticidal chemistry. IREDs were found to reduce the preformed 2-aryl 7-membered imines from the corresponding aminoketones. The enantioenriched 2-arylazepanes were obtained in up to >99% ee with good conversion by using suitable *R*- or *S*-selective IREDs. The subsequent stereospecific replacement of the α-C-H bond by an aryl or alkyl substituent through a sequential formation of *N*′-substituted aryl or allyl urea and stereospecific intramolecular aryl or vinyl migrations to the α-position within lithiated ureas produced the corresponding α-tertiary amine derivatives in 46–90% yields and 98:2 to >99:1 enantiomeric ratio (Scheme 2.47) [52].

Seven-membered 2-phenyl cyclic imine was observed to dissociate into the corresponding aminoketone in aqueous phosphate buffer in a wide pH range

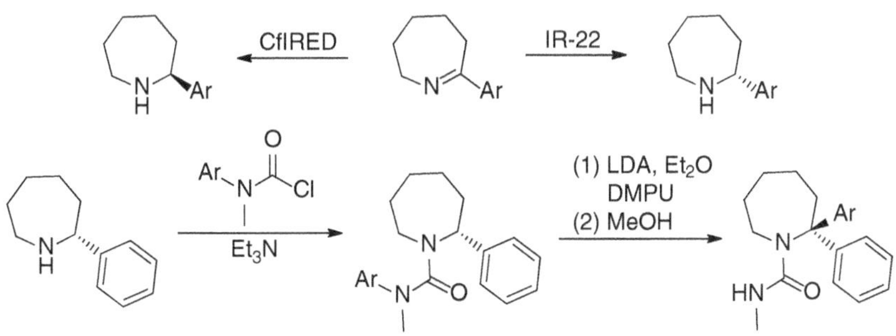

Scheme 2.47 Chemoenzymatic synthesis of α-tertiary amine derivatives. Source: Adapted from Zawodny et al. [52].

from 1 to 8. In the above imine reduction in Tris-HCl buffer, an equilibrium between the cyclic imine and aminoketone was proposed and it was possible to synthesize azepane from the corresponding aminoketones. In a recent study, the intramolecular asymmetric reductive amination of 4-((2-aminoethyl)(5-chlorobenzo[d]oxazol-2-yl)amino)butan-2-one was performed to yield 5-chloro-2-((*S*)-5-methyl-[1,4]diazepan-1-yl)benzooxazole or the (*R*)-enantiomer by using (*S*)- or (*R*)-selective IREDs (Scheme 2.48). 5-Chloro-2-((*R*)-5-methyl-[1,4]diazepan-1-yl)benzooxazole was the key structural component of Suvorexant, a selective dual orexin receptor antagonist for the treatment of primary insomnia. Some other optically pure 5-methyl-1-substituted 1,4-diazepanes were synthesized via this IRED-mediated intramolecular reductive amination and the aminoketone precursors were prepared from the corresponding aldehydes as shown in Scheme 2.49.

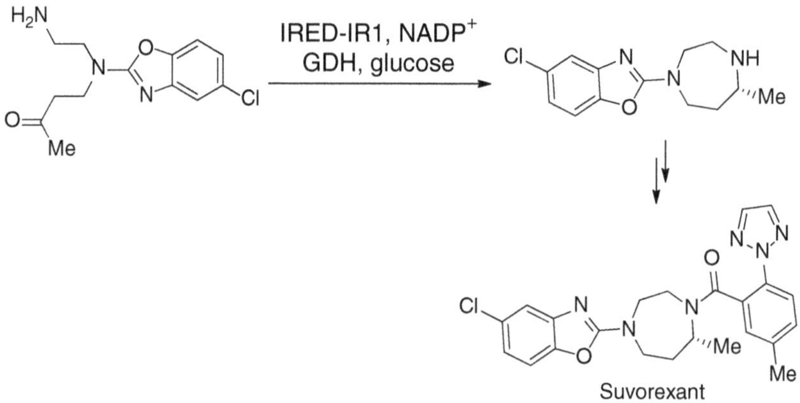

Scheme 2.48 Chemoenzymatic synthesis of Suvorexant.

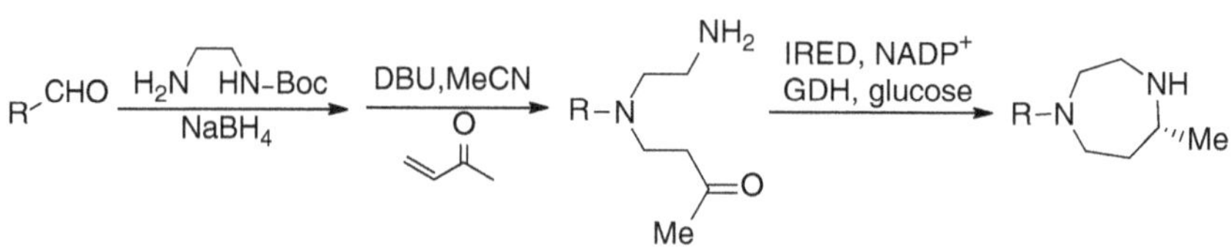

Scheme 2.49 Chemoenzymatic synthesis of optically pure 5-methyl-1-substituted 1,4-diazepane.

2.7 Cytochromes P450s

Cytochrome P450 monooxygenases have the exceptional ability of selectively inserting oxygen into nonreactive C=H bonds, yielding hydroxyl groups that can be transformed into other functional groups. In this context, chemoenzymatic fluorination of unactivated organic compounds has been achieved by coupling P450-catalyzed hydroxylation with nucleophilic substitution of the newly generated hydroxyl group by fluorine. The variants of P450BM3 (the bacterial long-chain fatty acid hydroxylase from *Bacillus megaterium*) were found to catalyze the regioselective hydroxylation of several cyclopentenone derivatives, and the resulting hydroxylated products were deoxofluorinated using DAST as the nucleophilic fluorinating agent to furnish the fluorinated products (Scheme 2.50). Similarly, the methylester of the anti-inflammatory drug ibuprofen was regioselectively fluorinated by this chemoenzymatic method, as shown in Scheme 2.51 [53].

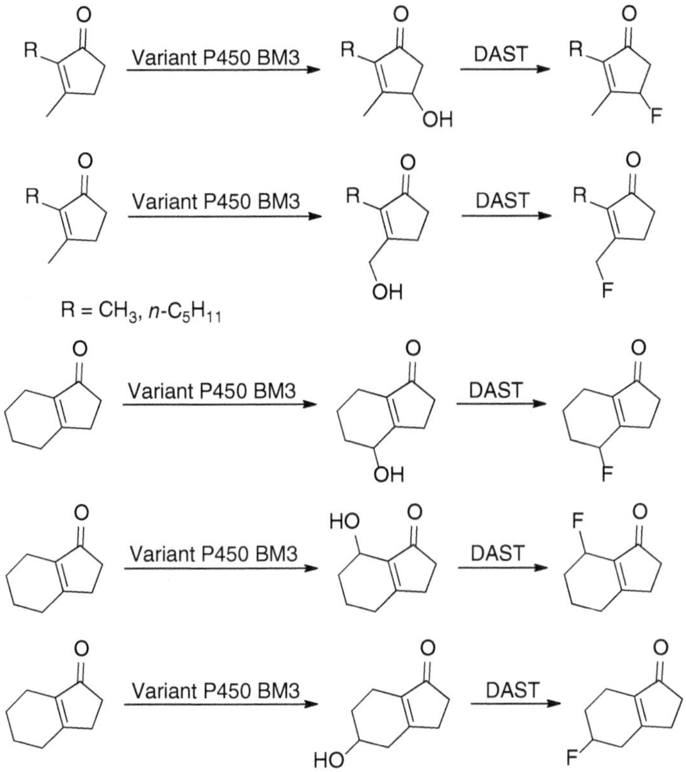

Scheme 2.50 Chemoenzymatic fluorination of cyclopentenone derivatives.

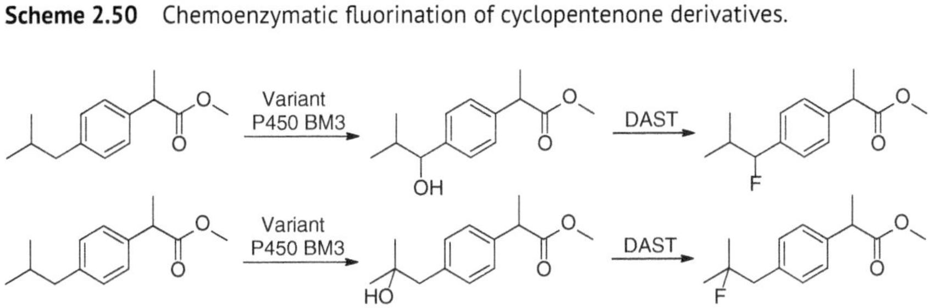

Scheme 2.51 Chemoenzymatic fluorination of methylester of ibuprofen. Source: Adapted from Rentmeister et al. [53].

Cytochrome P450 BM3 mutants also catalyzed the regio- and enantioselective hydroxylation of 6-iodotetralone at C4 to give (*R*)-4-hydroxy-6-iodotetralone. The

C-I functional group of the newly formed (*R*)-alcohol serves as the reactive site for further C—C bond-forming transformations. As such, three types of Pd-catalyzed cross-coupling reactions with (*R*)-4-hydroxy-6-iodotetralone had been carried out. Suzuki cross-coupling with $PhB(OH)_2$ generated (*R*)-4-hydroxy-6-phenyltetralone in 96% yield and 99.4% ee. Carbonylation with CO in methanol produced (*R*)-4-hydroxy-6-methoxycarbonyltetralone in 94% yield and 99.1% ee. Carbonylative Suzuki cross-coupling using $PhB(OH)_2$ under the CO atmosphere afforded (*R*)-4-hydroxy-6-benzoyltetralone in 85% yield and 99% ee (Scheme 2.52). In all cases, by employing $PdCl_2(PPh_3)_2$ as the precatalyst, a potential problem of the undesired Pd-catalyzed racemization at the benzylic alcohol position during the reaction was avoided [54].

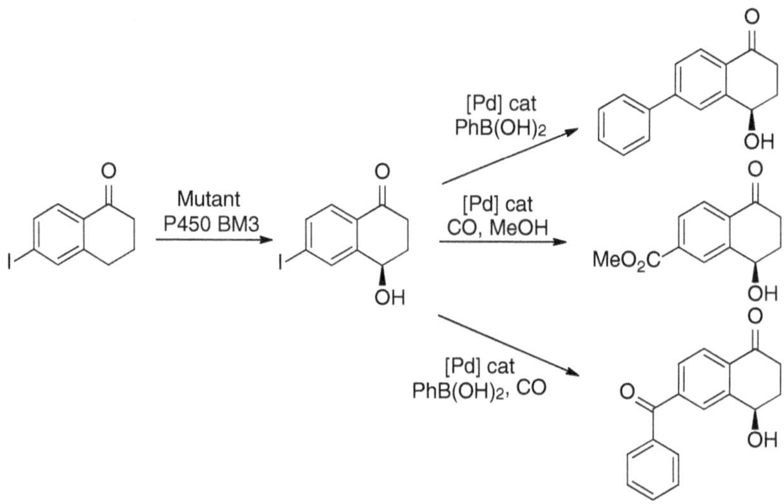

Scheme 2.52 Chemoenzymatic synthesis of compounds with the 4-hydroxy-1-tetralone skeleton.

The engineered P450BM3 enzymes also showed demethylase activity. The enzymatic demethylation converted methyl ethers into alcohols that can be deoxofluorinated to yield the desired fluorine-containing compounds. Polysaccharides play important roles in a wide range of biological processes, and subtle changes in their structures can dramatically alter their properties, thus requiring the use of homogeneous synthetic materials in their biological studies. Traditionally, oligosaccharide synthesis requires the installation and removal of protecting groups to differentiate each position of the monosaccharide moiety. This remains a challenging task. However, the mutant P450BM3 enzymes were found to catalyze the regioselective demethylation of a number of protected monosaccharide substrates. For example, 1,2,3,4,6-pentamethyl-α-mannopyranoside was selectively demethylated at six-position by a mutant P450BM3. The deprotected 6-OH was fluorinated using DAST as the fluorinating agent, and the 6-fluorinated sugar was then prepared by deprotection with BBr_3 in methylene dichloride (Scheme 2.48).

The 3-methoxy group of 1-β-benzoyl-2,3,4,6-tetramethylgalactose was selectively cleaved by another mutant P450 BM3 enzyme. The resulting 3-OH product was deoxygenated using Barton conditions, providing the 3-H galactose in good yield by following global deprotection, as shown in Scheme 2.53. This offered a highly efficient means to access a diversity of substituted monosaccharides and polysaccharide for studies in chemistry, biology, and medicine [55]. The chemoenzymatic substitution of a methoxy group for fluorine, a challenging transformation in traditional organic chemistry, had also been applied to the 5-phenyloxazoline derivative and Corey lactones, as shown in Scheme 2.54 [53].

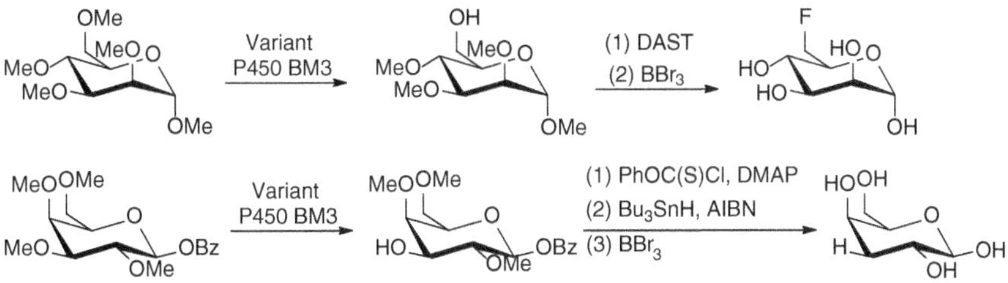

Scheme 2.53 Chemoenzymatic selective functionalization of monosaccharides.

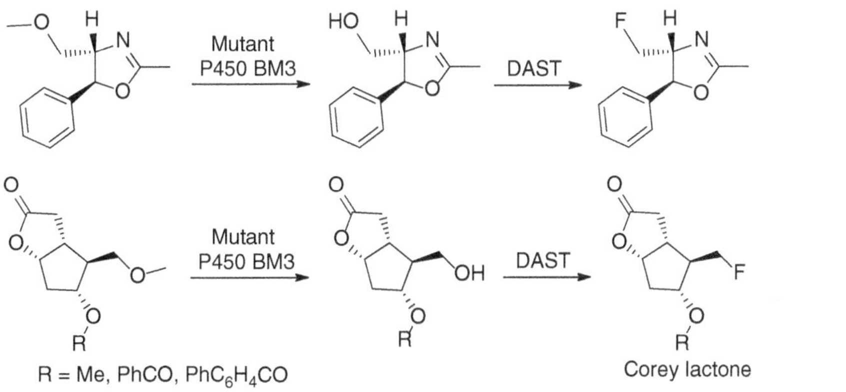

Scheme 2.54 Chemoenzymatic substitution of a methoxy group for fluorine. Source: From Rentmeister et al. [53]. © 2008, Springer Nature.

P450 enzymes mediate the selective C–H functionalization of complex molecules, offering an effective strategy for the late-stage manipulation of bioactive natural product scaffolds to produce a library of their derivatives. This enables the evaluation of the structure–activity relationships of these compounds and optimization of their pharmacological properties. For example, the evolved P450 BM3 variants catalyzed the highly regioselective hydroxylation of C9 and C14 in parthenolide carbocyclic backbone. A diversity of 9- and 14-substituted parthenolide derivatives were generated by the acylation [56], alkylation [57], metal-catalyzed carbene O–H insertion [57], or carbamoylation [58], following the hydroxylation (Scheme 2.55). The resulting C9- and C14-functionalized parthenolide analogs were evaluated for

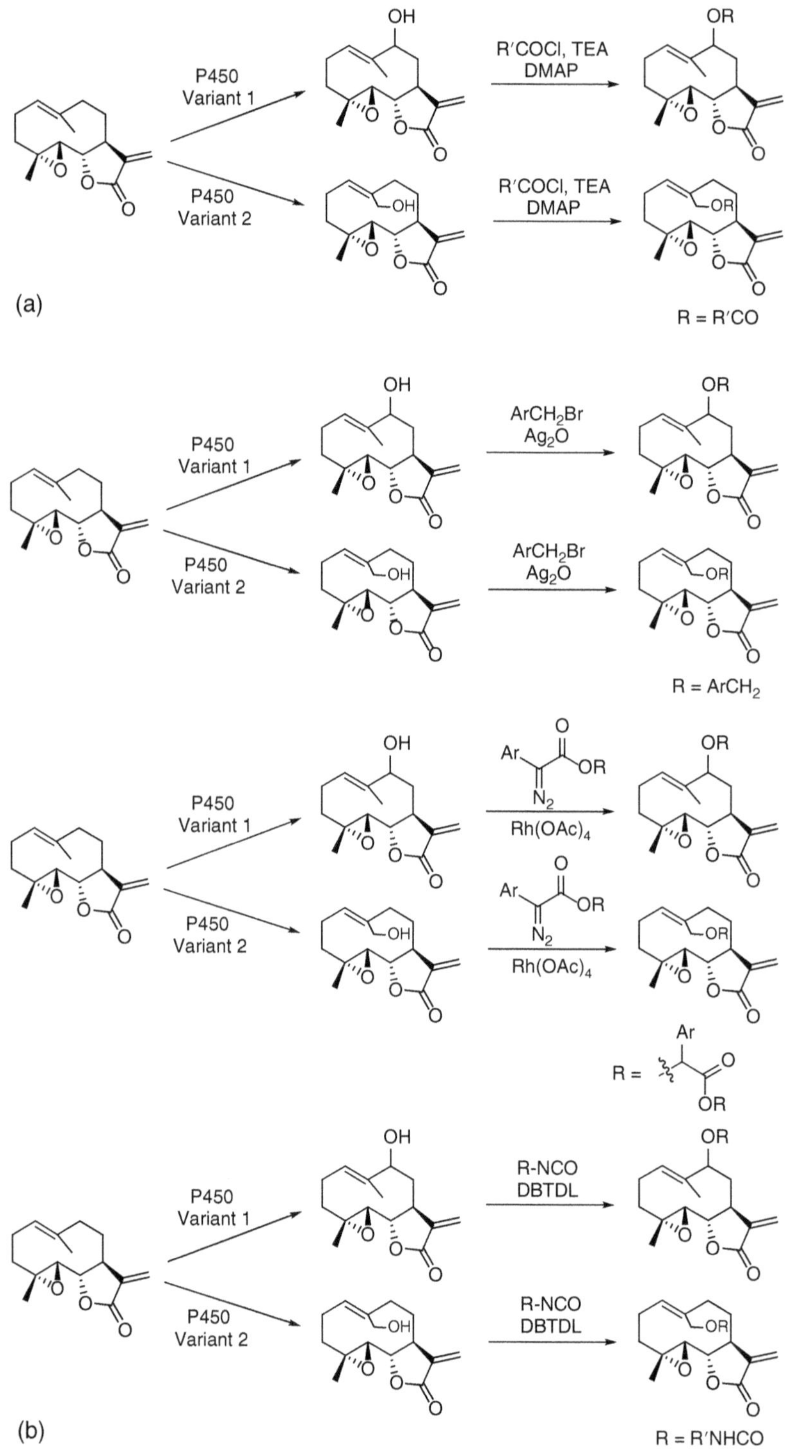

Scheme 2.55 Chemoenzymatic synthesis of 9- and 14-substituted parthenolide derivatives.

antileukemic activity, and one of these compounds showed significantly improved antileukemic potency against primary acute myeloid leukemia cells with low toxicity.

P450-catalyzed hydroxylation has also been successfully applied in the selective late-stage functionalization of cembranoid scaffolds, a large family of 14-membered oxygenated macrocyclic diterpenoids as potential therapeutic agents. A set of 14-membered macrocycles were prepared by metal-catalyzed RCM and evaluated toward the P450 BM3-catalyzed hydroxylation. The results revealed that the regioselectivity increased with increasing ring rigidity as well as size and polarity of the exocyclic substituents. In addition, the regioselectivity was also dependent on the enzyme that could be improved by active site mutagenesis. For example, dimethyl (3*E*,7*E*,11*E*)-9(*R*/*S*)-hydroxy-4,8-dimethylcyclotetradeca-3,7,11-triene-1,1-dicarboxylate was synthesized via RCM using Grubbs second-generation catalyst [59]. The subsequent hydroxylation of the (*R*)- or (*S*)-enantiomer by the V78A/F87A variant proceeded regioselectively at 5-carbon, exclusively delivering dimethyl (3*E*,7*E*,11*E*)-(5*R*,9*R*)-5,9-dihydroxy-4,8-dimethylcyclotetradeca-3,7,11-triene-1,1-dicarboxylate or the (5*S*,9*S*)-cembranoid-diols (Scheme 2.56) [60].

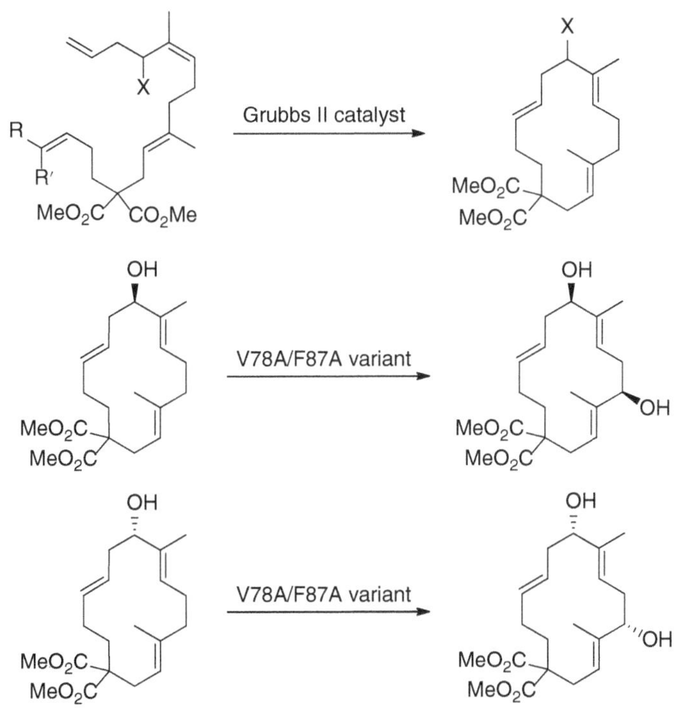

Scheme 2.56 Chemoenzymatic synthesis of *syn*-cembranoid-diols. Source: Adapted from Le-Huu et al. [60].

Alpha-ketoglutarate (αKG)-dependent (also known as 2-oxoglutatarate) hydroxylases are functionally comparable to cytochrome P450 enzymes. These nonheme, iron-containing enzymes also catalyze a wide range of oxygenation reactions, including hydroxylation, demethylations, epoxidation, cyclization, and so on [61]. Leucine 5-hydroxylase is an α-KG-dependent dioxygenase and selectively hydroxylates the 5-carbon of leucine. This enzyme also catalyzes the selective hydroxylation of the δ position of other aliphatic amino acids. For example, GriE, a leucine 5-hydroxylase predicted for the biosynthesis of griselimycin, was found to have excellent catalytic efficiency and substrate promiscuity. This enzymatic selective C–H functionalization at distal positions of amino acids has been integrated into the concise syntheses of a rare alkaloid, manzacidin C, and various L-proline analogs (Scheme 2.57) [62]. GriE catalyzed the hydroxylation of L-leucine yielding 4-hydroxy-L-leucine, which underwent facile intramolecular cyclization following the Boc protection, providing *tert*-butyl ((3*S*,5*R*)-5-methyl-2-oxotetrahydro-2H-pyran-3-yl)carbamate in high yield. The lactone generated could be transformed into a diversity of densely substituted pyrrolidines (Scheme 2.58) [63]. These studies demonstrated the potential applications of enzymatic hydroxylation of L-leucine analogs in the complex molecule synthesis.

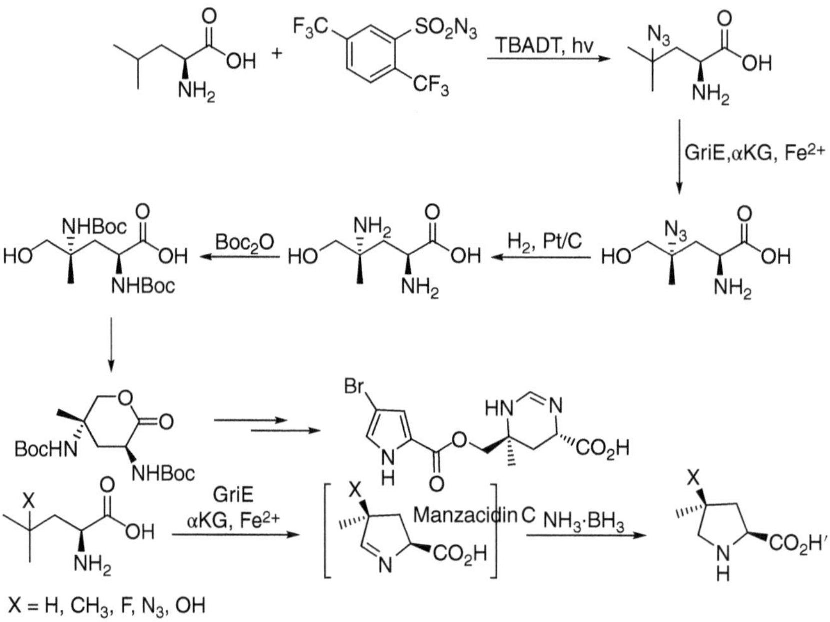

Scheme 2.57 Chemoenzymatic synthesis of manzacidin C and L-proline analogs. Source: Adapted from Zwick et al. [62].

Tambromycin is a nonribosomal peptide natural product isolated from *Streptomyces* strains, with selective antiproliferative activity against some cancerous B- and T-cell lines. Tambroline is an amino acid monomer with two of the four chiral centers in the Tambromycin molecule. The lysine hydroxylase KDO1 catalyzes the regio- and stereoselective C3 hydroxylation of lysine. This hydroxylation was carried out at

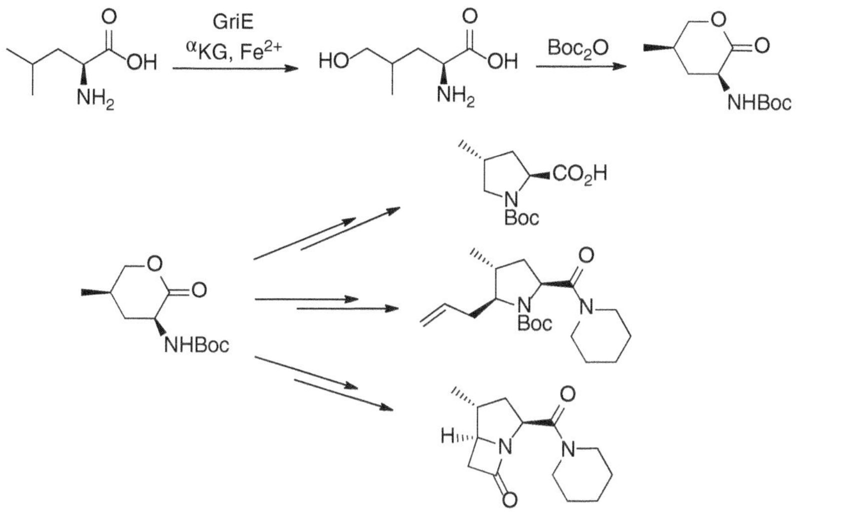

Scheme 2.58 Chemoenzymatic synthesis of substituted pyrrolidines. Source: Adapted from Zwick et al. [63].

gram scale to give C3-hydroxylated lysine, which was transformed into tambroline in a few chemical steps (Scheme 2.59) [64].

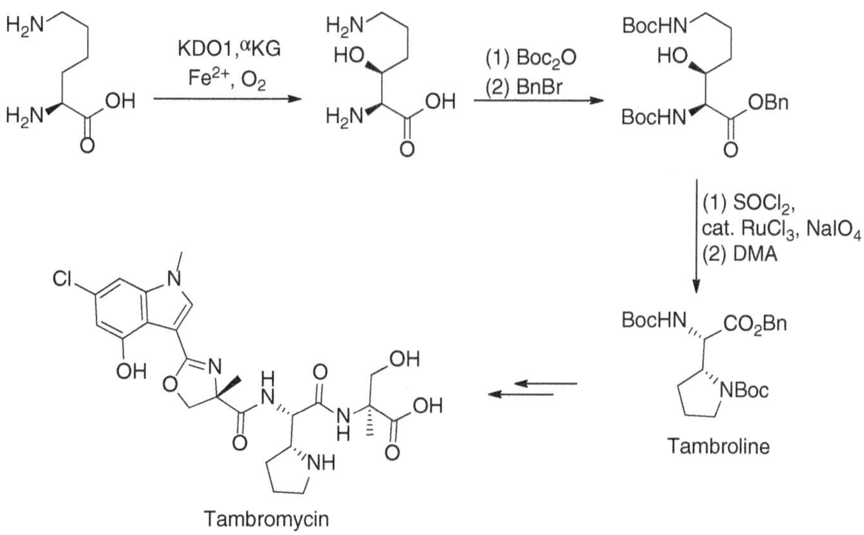

Scheme 2.59 Chemoenzymatic synthesis of Tambroline. Source: Adapted from Zhang et al. [64].

The nonheme 2-oxoglutarate/Fe(II)-dependent dioxygenases (2-ODD) also catalyze the oxidative cyclization for the construction of core structural motifs in natural product biosynthesis [65]. A nonheme dioxygenase 2-ODD-PH catalyzed an intramolecular oxidative C–C coupling of (−)-yatein to furnish

(−)-deoxypodophyllotoxin in 95% yield. The introduction of the alcohol group at C7 was achieved by oxidation of C7 into ketone functionality using CrO_3 as the oxidant followed by reduction with L-Selectride, affording (−)-podophyllotoxin in 58% yield (Scheme 2.60). Other aryltetralin lignans with different substituents on the two benzene rings such as (−)-polygamain, (−)-morelensin, (−)-austrobailignan 1, and (−)-hernandin (24) were also synthesized by employing 2-ODD-PH. The corresponding precursors dibenzylbutyrolactones were obtained by α-alkylation of the enantiopure butyrolactones using substituted benzyl bromide as the alkylating agents (Scheme 2.61). An efficient chemoenzymatic strategy has thus been established for the synthesis of (−)-podophyllotoxin and related aryltetralin lignans [66].

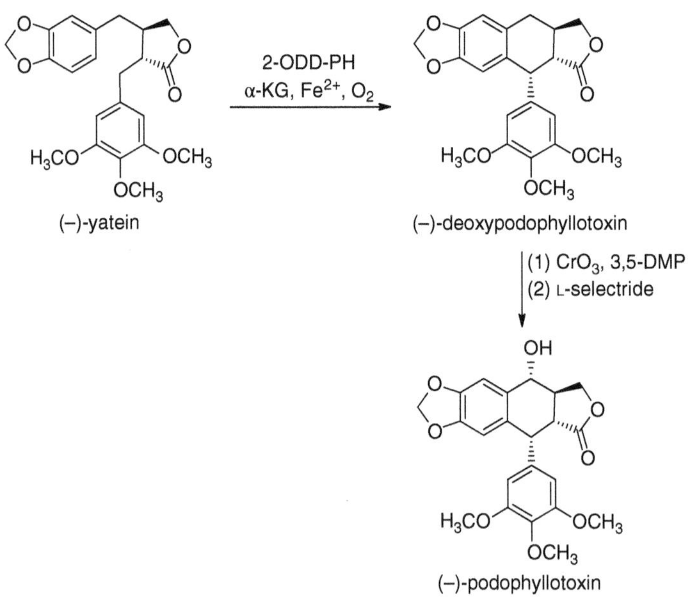

Scheme 2.60 Chemoenzymatic synthesis of (−)-podophyllotoxin.

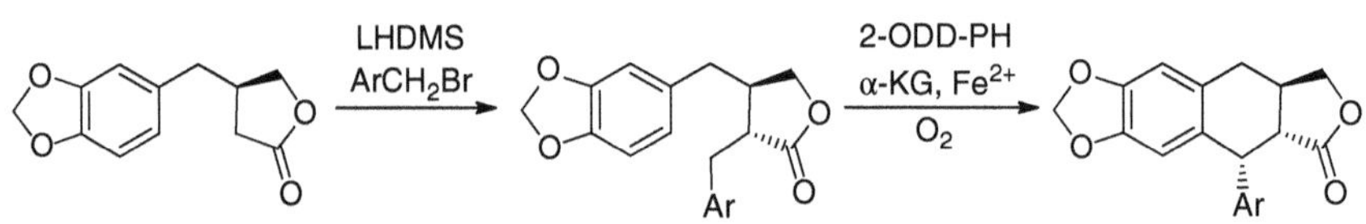

Scheme 2.61 Chemoenzymatic synthesis of aryltetralin lignans.

2.8 Baeyer–Villiger Monooxygenases (BVMOs)

Metal-catalytic and organocatalytic Baeyer–Villiger (BV) reactions, which transform ketones into esters or lactones, usually show limited enantioselectivity except for the activated cyclobutanones with ring strain. Baeyer–Villiger monooxygenases (BVMOs), a class of flavin-dependent enzymes, catalyze the BV transformation using atmospheric oxygen as oxidant and generating water as the only by-product. Enzymatic BV reactions often exhibit high enantioselectivity and chemoselectivity, offering a green alternative to the chemical version of BV reactions. As such, biocatalytic BV reactions have been integrated into the chemical synthesis for the construction of esters or lactone moieties with chiral centers [67, 68].

The metal-catalyzed diastereoselective hydrogenation of α,β-unsaturated ketones has been reported with high chemo and regioselectivity. These chemical reductions can be integrated with biocatalytic reactions for the synthesis of complex molecules. The rhodium catalyst (5% Rh/C) catalyzed the hydrogenation of dihydrojasmone in pentane in the presence of Na_2CO_3, generating the saturated ketone (97:3 *cis*/*trans*) in >95% yield. The subsequent BV oxidation of the resulting ketone would afford the corresponding lactone. Indeed, Aerangis lactone was obtained by utilizing BVMOs. The *cis*-saturated ketone was prepared by using the binary catalyst mixture (Rh/C, Na_2CO_3, 1:5, w/w) in continuous flow reaction setting and transformed into the naturally occurring (5*S*,6*S*)-Aerangis lactone and remaining (2*R*,3*R*)-ketone enantiomer with $E > 200$ and a high reaction rate by employing cyclododecanone monooxygenase (CDMO) from *Rhodococcus ruber* SC1 in a batch reaction. The (5*R*,6*S*)-epimer could be accessed from *trans*-ketone, which was prepared from *cis*-ketone in continuous flow over a glass column packed with Amberlyst 15, with CPMO from *Comamonas* sp. NCIMB 9872 with $E > 200$ (Scheme 2.62) [69].

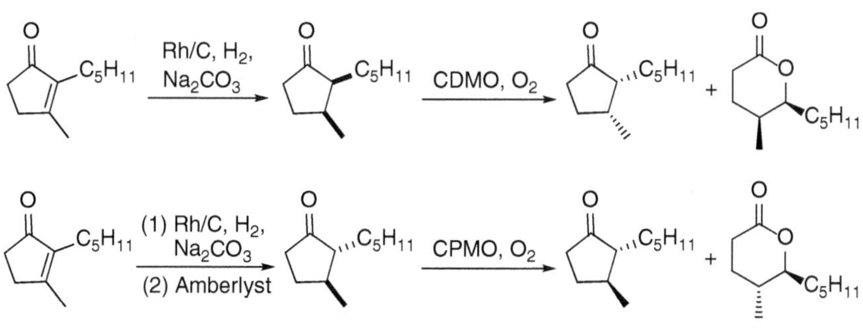

Scheme 2.62 Chemoenzymatic synthesis of Aerangis lactone. Source: Adapted from Fink et al. [69].

Chiral lactones with tetra-substituted stereocenters and cyclic ketones bearing α-quaternary centers are high-value chiral building blocks for the synthesis of complex molecules. However, such stereocenters are hard to build and present a great challenge in organic synthesis. Stereoselective BV transformation of α,α-dialkyl cyclic ketones delivers enantiomerically enriched lactones and α,α-dialkyl cyclic ketones bearing such α-quaternary stereocenters via kinetic resolution. A CHMO from *Acinetobacter calcoaceticus* (NCIMB 9871) was found to effectively catalyze the kinetic resolution of cycloketones bearing an α-quaternary stereocenter to give the corresponding lactones and unreacted ketones with up to >99% ee and complete chemoselectivity. The resulting ketones were transformed into enantiomerically enriched cycloheptan- and cycloctan-1,4-diols via an SmI_2-mediated cyclization process. SmI_2 also mediated the cyclization of enantiomerically enriched ketones, generating cyclobutanol products with up to >99% ee (Scheme 2.63). A chemoenzymatic divergent approach was thus developed by the combination of CHMO-catalyzed BV kinetic resolution with SmI_2-mediated radical cyclization reactions, offering effective access to important carbocyclic scaffolds in enantiomerically enriched form [70].

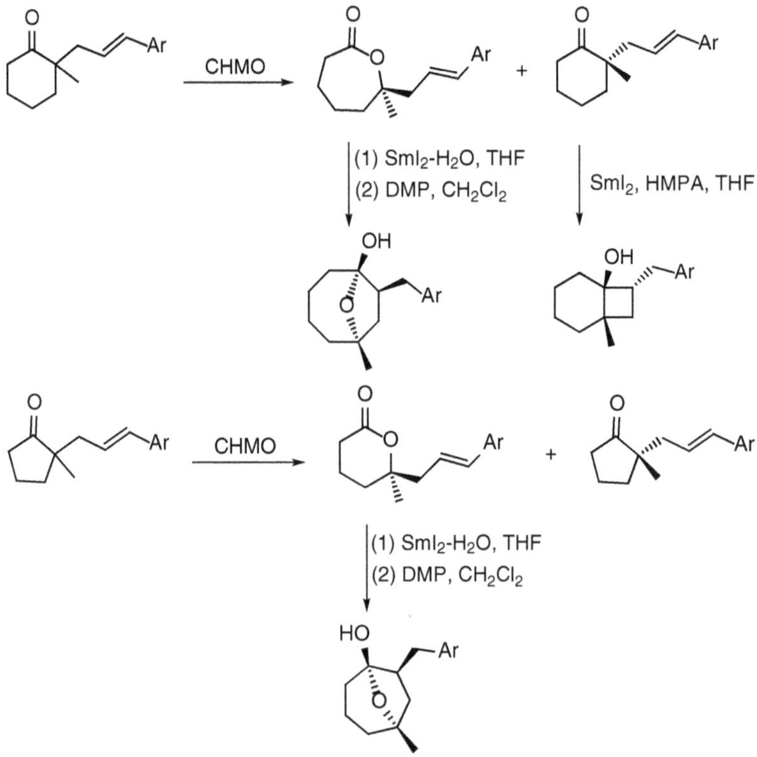

Scheme 2.63 Chemoenzymatic construction of carbocyclic scaffolds.

BVMOs catalyze the oxidation of symmetric ketones, resulting in desymmetrization of the ketones to give chiral lactones. Camphor monooxygenase (CAMO) from the ascomycete *Cylindrocarpon radicicola* and a recombinant 2-oxo-D3-4,5,5-trimethylcyclopentenylacetyl-coenzyme A monooxygenase (OTEMO) from *Pseudomonas putida* catalyzed the oxidation of 3-vinylcyclobutanone, furnishing (*R*)-(−)-Taniguchi lactone in 57 and 60% isolated yield, respectively. 3-Vinylcyclobutanone was prepared via a [2+2] cycloaddition of 1,3-butadiene with dichloroketone, generated *in situ* from trichloroacetyl chloride in the presence of activated zinc, followed by the reduction of the dichlorocyclobutanone (Scheme 2.64) [71]. While (*S*)-(+)-Taniguchi lactone had been applied in the synthesis of various natural products such as salicifoliol and purine analogs, (*R*)-(−)-Taniguchi lactone could be a useful building block for the synthesis of *ent*-quinine and *ent*-salicifoliol.

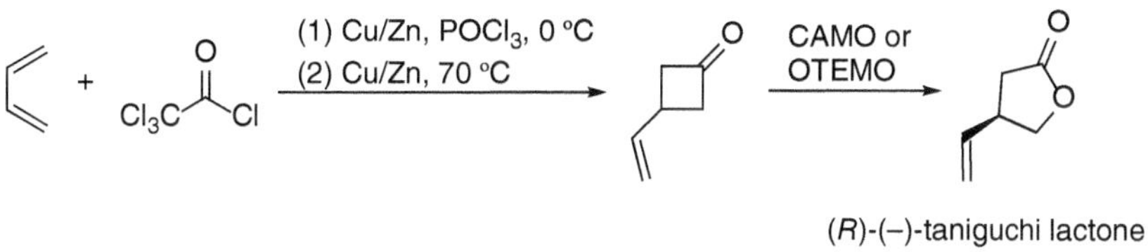

Scheme 2.64 Chemoenzymatic synthesis of (*R*)-(−)-Taniguchi lactone. Source: Adapted from Rudroff et al. [71].

The [4 + 3] cycloaddition of furan and 1,1,3,3-tetrabromoacetone in the presence of freshly prepared Cu/Zn-couple and catalytic amount of dibromoethane under sonication afforded 2,4-dibromo-8-oxabicyclo[3.2.1]oct-6-en-3-one, which was transformed to 8-oxabicyclo[3.2.1]oct-6-en-3-one by a debromination step with Cu/Zn couple in the presence of NH_4Cl. *Comamonas* sp. CPMO ($CPMO_{Coma}$) catalyzed the BV oxidation of the seven-membered bicyclic ketones, generating the corresponding lactone (1*S*,6*S*)-3,9-dioxabicyclo[4.2.1]non-7-en-4-one in 70% yield and 95% ee (Scheme 2.60). Similarly, the cycloaddition of cyclopentadiene with the tetrabromoacetone generated 2,4-dibromobicyclo[3.2.1]oct-6-en-3-one, and further debromination produced bicyclo[3.2.1]oct-6-en-3-one. The BV oxidation of the corresponding carba-ketone by $CPMO_{Coma}$ and $CHMO_{Xantho}$ from a *Xanthobacter* afforded both antipodes of the carba-lactone 3-oxabicyclo[4.2.1]non-7-en-4-one in 63% yield with 89% ee and 61% yield with >99% ee, respectively (Scheme 2.65). These lactones are key building blocks for the synthesis of goniofufurone and their carba analogs. Therefore, these chemoenzymatic reactions involving BVMOs offer an efficient strategy to access optically pure THF-based natural products and their carba analogs from readily available starting materials such as furan, cyclopentadiene, and tetrabromoacetone [72].

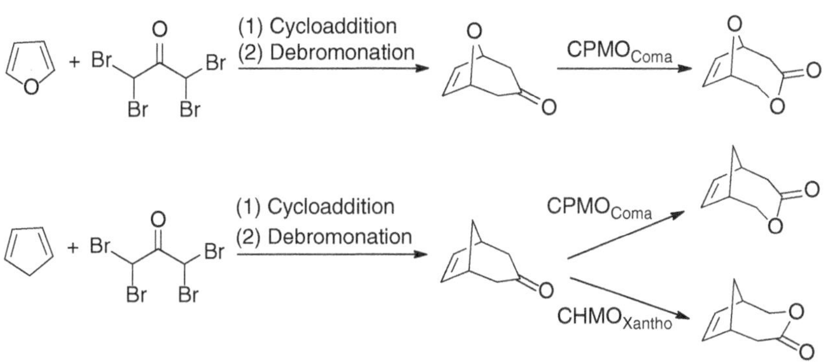

Scheme 2.65 Chemoenzymatic synthesis of bicyclic lactones.

2.9 Aldolases

Aldolases catalyze asymmetric aldol reactions under mild conditions with impressive selectivities. While dihydroxyacetone phosphate (DHAP)-dependent enzymes require the expensive and unstable donor DHAP, dihydroxyacetone (DHA)-dependent aldolases such as D-fructose-6-phosphate aldolase (FSA) and rhamnulose-L-phosphate aldolase (RhuA) accept inexpensive donors such as DHA, hydroxyacetone (HA), or glycolaldehyde, offering tremendous opportunities for the synthesis of carbohydrates and their derivatives with multiple chiral centers from simple, readily available achiral starting materials.

FSA catalyzes the reversible aldol addition reaction of DHA to D-glyceraldehyde-3-phosphate. An FSA from *E. coli* catalyzed the aldol addition of DHA to N-Cbz-3-aminopropanal, furnishing (3*S*,4*R*)-6-[(benzyloxycarbonyl)amino]-5,6-dideoxyhex-2-ulose in 69% isolated yield. The catalytic reductive amination of the aldol adduct by Pd/C and H_2 at 50 psi afforded D-fagomine in 89% isolated yield and 93:7 dr. The N-alkylated derivatives of D-fagomine could be obtained by carrying out the reductive amination reaction in the presence of the corresponding aldehyde (Scheme 2.66) [73].

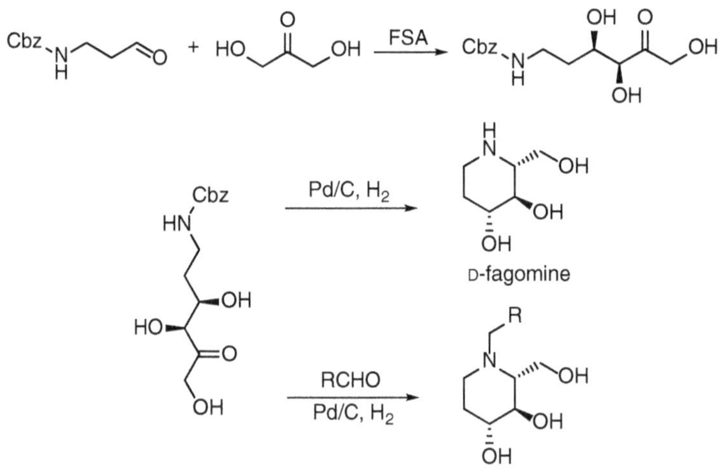

Scheme 2.66 Chemoenzymatic synthesis of D-fagomine and its N-alkylated derivatives. Source: Adapted from Castillo et al. [73].

FSA has been engineered by a structure-guided redesign approach for the addition of DHA, HA, and glycolaldehyde to various linear, branched, and cyclic N-Cbz-aminoaldehydes. Some variant FSA enzymes showed good activity toward the tested substrates, furnishing the corresponding D-*threo*-configured aldol adduct in >95:5 dr. The aldol addition products were further converted into the corresponding functionalized heterocyclic compounds by reductive cycloamination as shown in Scheme 2.67 [74].

Wild-type FSA and variants also catalyzed the aldol addition of DHA, HA, or glycolaldehyde to the N-Cbz-β-hydroxy aminoaldehydes. The resulting aldol adducts were purified and treated with H_2 in the presence of Pd/C, giving the corresponding pipecolic acid derivatives (Scheme 2.68) [75]. Similarly, the aldol addition of DHA, HA, or glycolaldehyde to (2*R*,3*R*)-3-N-benzyloxycarbonylamino-2,4-dihydro-

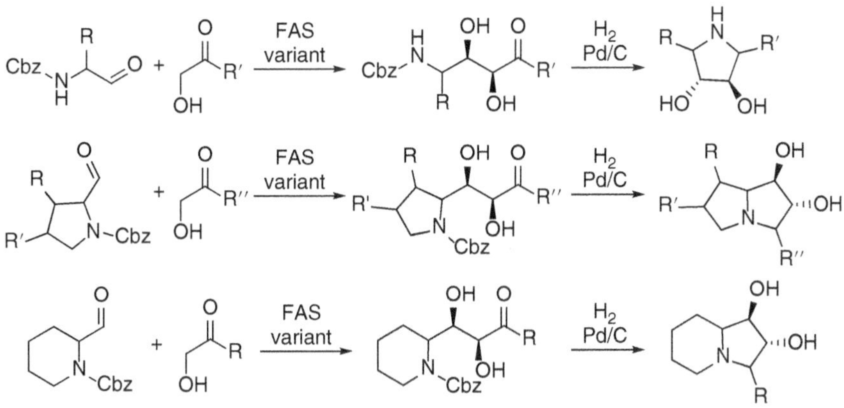

Scheme 2.67 Chemoenzymatic synthesis of functionalized heterocyclic compounds. Source: Adapted from Soler et al. [74].

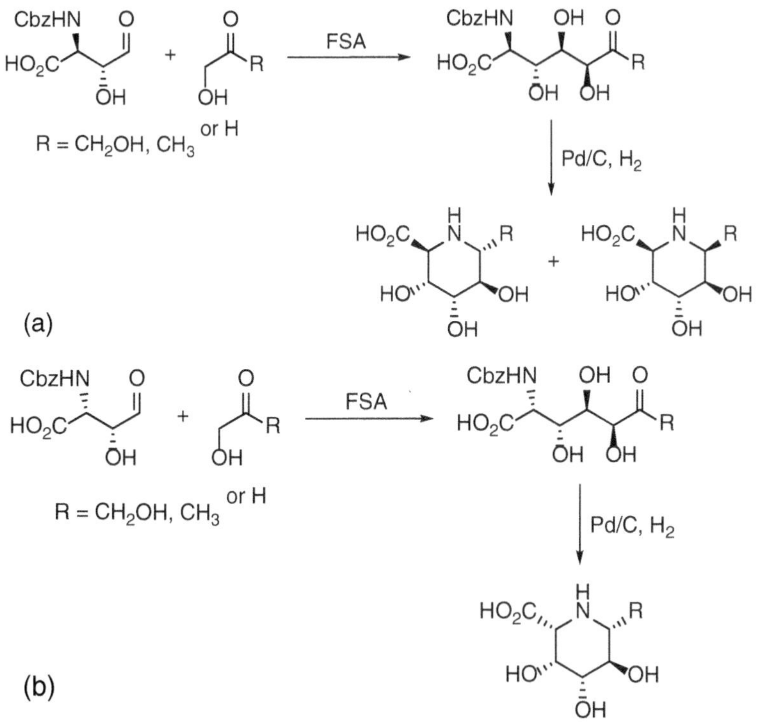

Scheme 2.68 Chemoenzymatic synthesis of polyhydroxylated pipecolic acids. Source: Adapted from Soler et al. [75].

xybutanal using FSA or variants as the biocatalyst gave the corresponding *syn*-configured aldol adducts in 68–90% yields. The subsequent reductive cycloamination under H_2 in the presence of Pd/C afforded the corresponding homoiminocyclitols (Scheme 2.69).

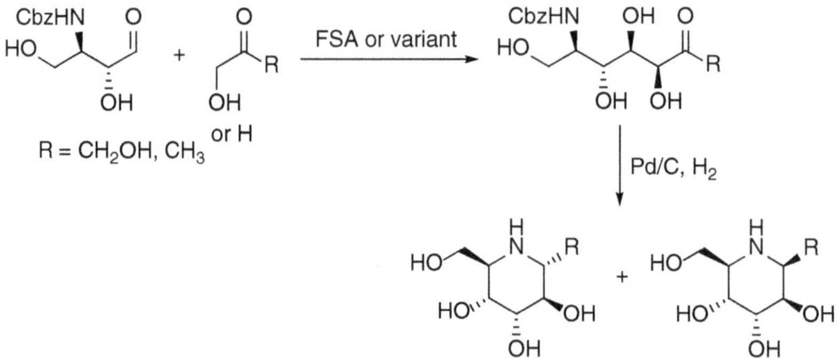

Scheme 2.69 Chemoenzymatic synthesis of homoiminocyclitols.

FSA or its variants do not accept (2R,3S)-2-(N-benzyloxycarbonylamino)-3-hydroxy-4,4-dimethoxybutyraldehyde as the substrate. However, aldolases FucA and RhuA catalyzed the aldol addition of DHAP to (2*R*,3*S*)-2-(*N*-benzyloxycarbonylamino)-3-hydroxy-4,4-dimethoxybutyraldehyde in an excellent stereocontrolled manner, giving the *trans* and *syn* adduct, respectively, which were transformed into the pyrrolidine iminocyclitol derivatives as mixtures of epimers by treating with acid phosphatase from potato type II, and subsequently by H_2 in the presence of Pd/C (Scheme 2.70).

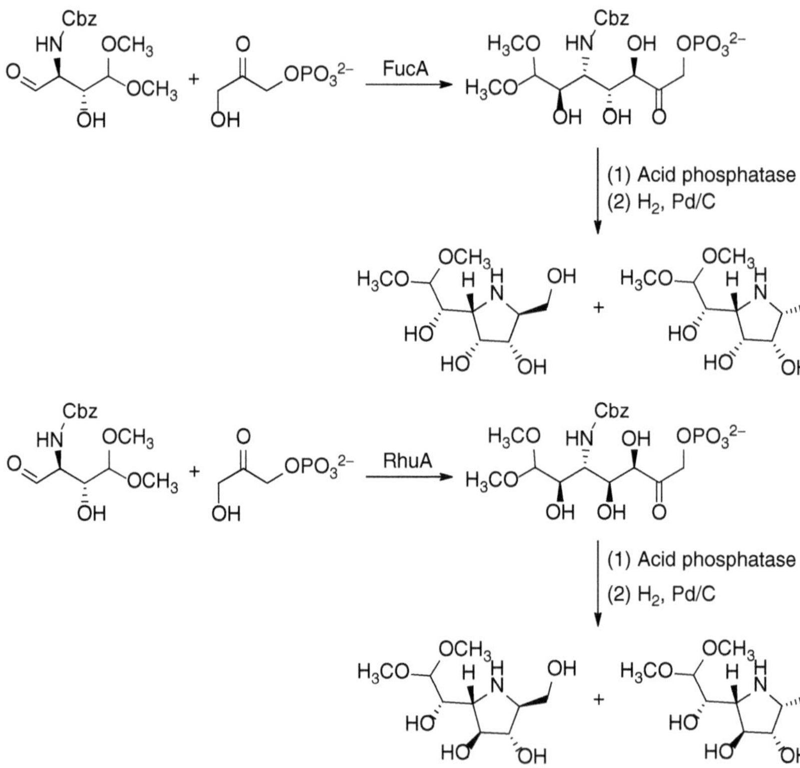

Scheme 2.70 Chemoenzymatic synthesis of pyrrolidine iminocyclitols.

The FSA-catalyzed aldol reaction has been applied in the chemoenzymatic synthesis of biologically relevant thiosugars. FSA catalyzed the aldol condensation of DHA with 2-haloacetaldehyde. The aldol product was treated with NaSH, generating the thiosugar by halogen substitution (Scheme 2.71). This offers a straightforward approach to access thiosugar scaffolds [76].

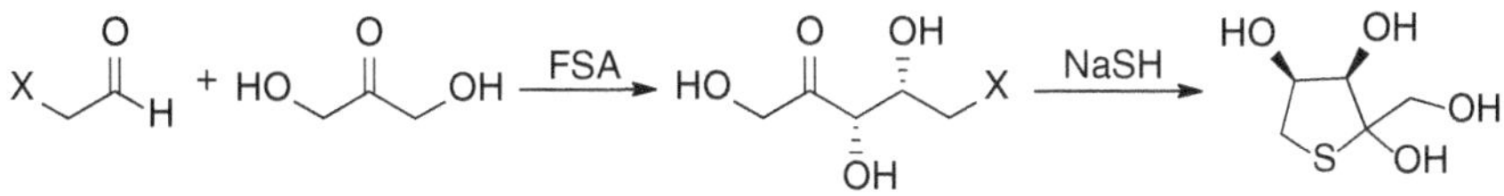

Scheme 2.71 Chemoenzymatic synthesis of thiosugar scaffolds.

The FSA-catalyzed aldol reaction was also combined with a thiamine diphosphate (ThDP)-dependent carboligase in a chemoenzymatic process to synthesize 5- or 6-C-aryl carbohydrates. In a four-step process, aromatic aldehydes reacted with dimethoxyacetaldehyde under the catalysis of benzaldehyde lyase from *P. fluorescens* (BAL) to give the 2*R*-hydroxyketones, which was reduced by $NaBH_4$ in MTBE. The hydrolysis of the resulting (1*S*,2*S*)-1-aryl-3,3-dimethoxypropane-1,2-diol was carried out in an aqueous 2 M H_2SO_4/acetonitrile (5:1) mixture affording (2*S*,3*S*)-3-aryl-2,3-dihydroxypropanal. FSA catalyzed the aldol addition of (2*S*,3*S*)-3-aryl-2,3-dihydroxypropanal with DHA, HA, or glycolaldehyde, generating the corresponding (6*S*)-6-C-aryl-α-L-sorbose derivatives or (5*S*)-5-C-aryl-α/β-L-xylose derivatives with high stereoselectivity (Scheme 2.72). L-Rhamnulose-1-phosphate aldolase from *E. coli* (RhuA) catalyzed the aldol reaction of (2*S*,3*S*)-3-aryl-2,3-dihydroxypropanal with DHA, resulting in the corresponding 6-C-arylated L-fructoses and L-tagatoses with epimerization at C4-carbon [77].

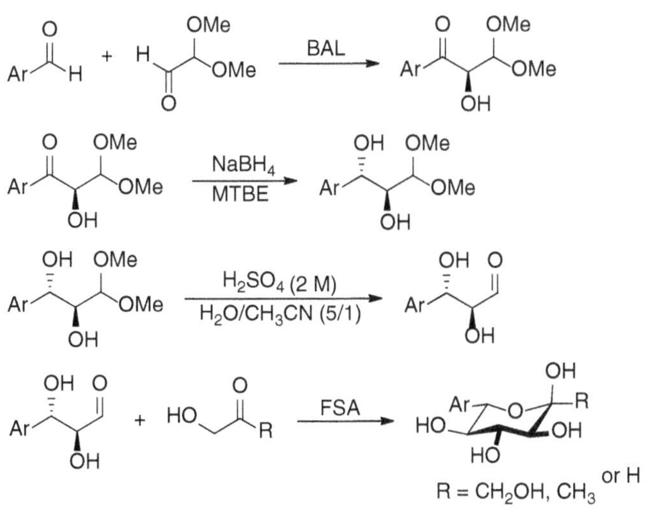

Scheme 2.72 Chemoenzymatic synthesis of 5- or 6-C-Aryl carbohydrates.

2.10 Epoxide Hydrolases

Epoxide hydrolases catalyze the stereoselective hydrolysis of epoxides, yielding enantiopure vicinal diols under mild conditions. The enzymatic hydrolysis of epoxides is usually highly regio- and stereoselective, and has been exploited in the synthesis of important complex molecules since 1990s. For example, whole cells of *A. niger* catalyzed the hydrolysis of *p*-nitrostyrene oxide, giving the (*R*)-diol and (*S*)-epoxide. The latter can be converted to the (*R*)-diol by acid-catalyzed hydrolysis. By carefully controlling the enzymatic and acidic hydrolysis, racemic *p*-nitrostyrene oxide was hydrolyzed to give the (*R*)-diol, which was then transformed into (*R*)-Nifenalol (Scheme 2.73) [78].

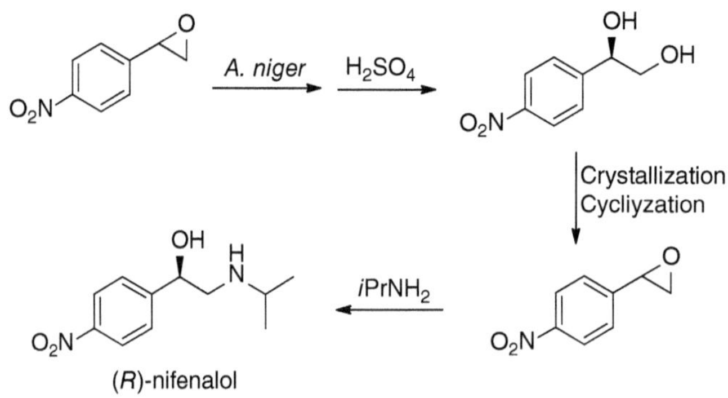

Scheme 2.73 Chemoenzymatic synthesis of (*R*)-Nifenalol. Source: Adapted from Pedragosa-Moreau et al. [78].

Similarly, after assessing the selectivity of a large variety of epoxide hydrolases on geraniol-derived oxiranes, a chemoenzymatic method involving stereoconvergent enzymatic and acidic hydrolysis was developed for the stereoselective synthesis of furanoid linalool oxide. An epoxide hydrolase (E4) catalyzed the hydrolysis of *rac*-(*E*)-5-(3,3-dimethyloxiran-2-yl)-3-methylpent-2-en-1-yl pivalate in a stereoconvergent manner, affording (*R*,*E*)-6,7-dihydroxy-3,7-dimethyloct-2-en-1-yl pivalate in 52% isolated yield and 97% ee after the product mixture was extracted and treated with aqueous acid. Subsequent palladium-catalyzed stereoselective Tsuji–Trost reaction generated *trans*-(2*R*,5*R*)-configured linalool oxide in 75% yield, 97% diastereomeric excess (de), and 97% ee (Scheme 2.74) [79].

Epoxide hydrolases catalyze the kinetic resolution of racemic epoxides, generating one enantiomer of the vicinal diols with maximum 50% yield, posing a major drawback for synthetic applications. In the above cases, this problem was solved by coupling the epoxide hydrolase-catalyzed reaction with the stereoconvergent acid-catalyzed hydrolysis. It can also be addressed by using two enantio- and regiocomplementary enzymes. Epoxide hydrolase from *Solanum tuberosum L.* (StEH) catalyzed the hydrolysis of the (*S*)-enantiomer of *p*-chlorostyrene oxide to furnish the (*R*)-diol, while epoxide hydrolase from *A. niger* (AnEH) catalyzed the hydrolysis of the (*R*)-enantiomer of *p*-chlorostyrene oxide to give the (*R*)-diol.

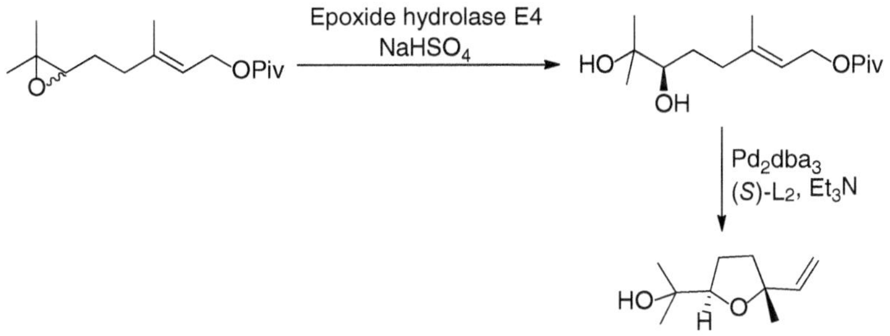

Scheme 2.74 Chemoenzymatic synthesis of optically pure linalool oxide. Source: Adapted from van Lint et al. [79].

When racemic *p*-chlorostyrene oxide was hydrolyzed by using StEH and AnEH sequentially, the corresponding (*R*)-diol was obtained in 93% yield with 96% ee. The resulting (*R*)-diol was used as the building block for the preparation of (*R*)-Eliprodil (Scheme 2.75) [80].

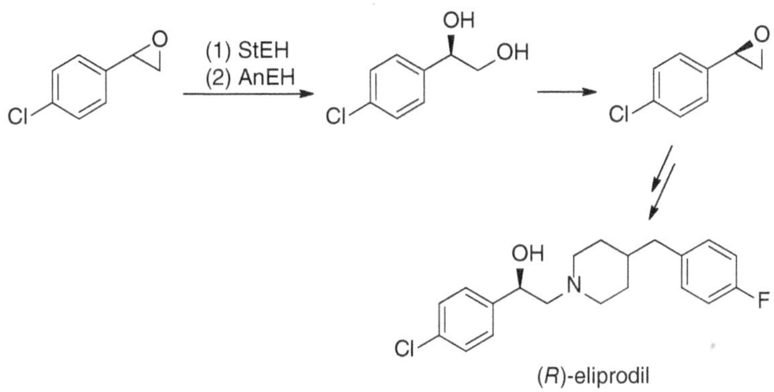

Scheme 2.75 Chemoenzymatic synthesis of (*R*)-Eliprodil. Source: Adapted from Manoj et al. [80].

Chiral chromanemethanols are the key building blocks for the synthesis of biologically important α-tocopherols and tocotrienols. An epoxide hydrolase together with a halohydrin dehalogenase had been applied for the enantiocomplementary chemoenzymatic synthesis of optically pure chromanemethanol from racemic 2,2-disubstituted oxirane. Epoxide hydrolases of whole cells of *R. ruber* CBS 717.73 (RrEH) catalyzed the kinetic resolution of 2-benzyloxymethyl-2-methyloxirane, furnishing the (*R*)-diol and unreacted (*R*)-2-benzyloxymethyl-2-methyloxirane. The halohydrin dehalogenase HheC from *Agrobacterium radiobacter* AD1 hydrolyzed 2-benzyloxymethyl-2-methyloxirane to give the enantiocomplementary (*S*)-diol and unreacted (*S*)-2-benzyloxymethyl-2-methyloxirane. The enantiopure 2-benzyloxymethyl-2-methyloxiranes obtained were chemically transformed to (*S*)- and (*R*)-chromanemethanol, respectively (Scheme 2.76) [81].

Racemic 2-benzyloxymethyl-2-methyloxirane was also hydrolyzed by an engineered epoxide hydrolase from *Bacillus subtilis* (BSEH), followed by the

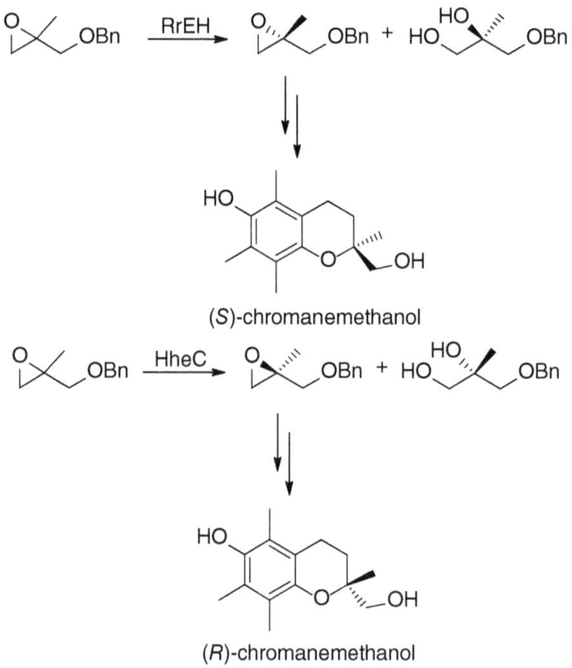

Scheme 2.76 Chemoenzymatic synthesis of (*S*)- and (*R*)-chromanemethanol. Source: Adapted from Fuchs et al. [81].

acid-catalyzed stereoconvergent hydrolysis, affording the (*R*)-diol in optically pure form. The (*R*)-2-methylpropane-1,2,3-triol monobenzyl ether obtained was chemically transformed into (*R*)-bicalutamide, thus establishing a chemoenzymatic route to access (*R*)-bicalutamide, which has antiandrogenic activity and is used as an antineoplastic agent for treating prostate cancer (Scheme 2.77) [82].

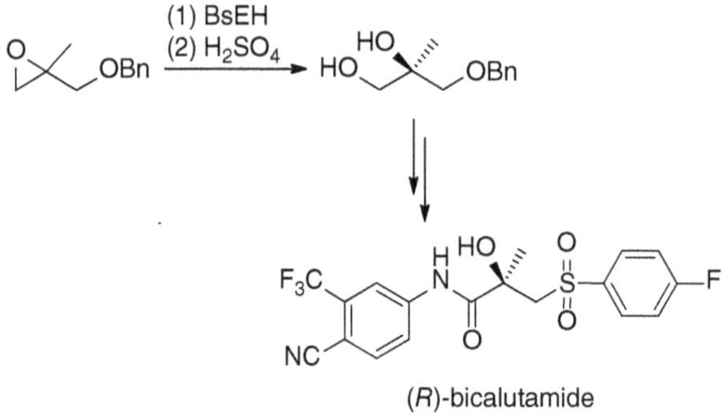

Scheme 2.77 Chemoenzymatic synthesis of (*R*)-bicalutamide. Source: Adapted from Fujino et al. [82].

(*R*)-Marmin is a natural product that has an isoprenylated coumarin structure and is isolated from the bark of *Aegle marmelos* and grapefruit peel oil. Selective epoxidation of geranyl chloride by an epoxide hydrolase utilizing *m*-chloroperbenzoic acid as oxidant afforded (*E*)-3-(5-chloro-3-methylpent-3-enyl)-2,2-dimethyloxirane, which reacted with 7-hydroxycoumarin in the presence of NaH in dimethylformamide (DMF) furnished 7-[5-(3,3-dimethyloxiranyl)-3-methylpent-2*E*-enyloxy]-chromen-2-one (epoxyaurapten). The resulting racemic epoxide was treated with *R. ruber* DSM 44539 under argon atmosphere, and (*R*)-Marmin was obtained in 95% ee (Scheme 2.78) [83].

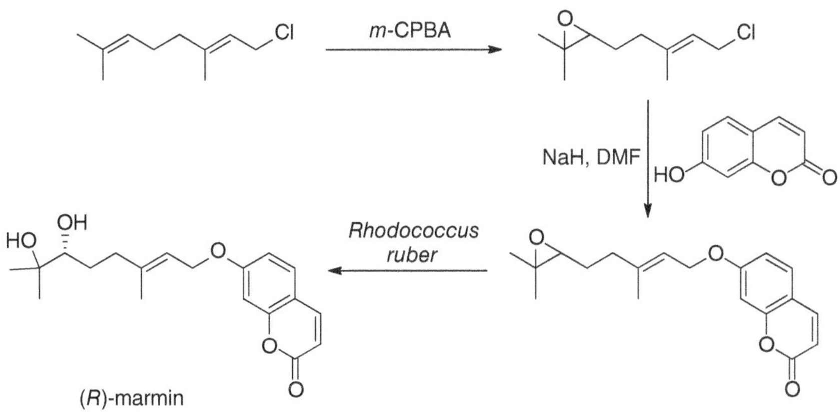

Scheme 2.78 Chemoenzymatic synthesis of (*R*)-Marmin. Source: Modified from Edegger et al. [83].

2.11 Other Enzymes

Laccases (benzenediol: oxygen oxidoreductase) catalyze the oxidation of phenols, polyphenols, or aromatic amines using aerial oxygen as oxidizing agent under mild reaction conditions. The oxidized product can instantly undergo a Michael addition with the nucleophilic C-3 carbon of the 3,4-dihydrocoumarins, resulting in the formation of a new class of 3-arylated 3,4-dihydrocoumarins. The dihydrocoumarins are prepared by the hydrogenation of coumarins carried out in a flow system. The coumarin derivatives with substituent at the six position can in turn be synthesized in good yields by Knoevenagel condensation of salicylaldehydes with diethyl malonate. The chemical synthesis of coumarins through a Knoevenagel condensation followed by a quantitative hydrogenation in a flow system when combined with the laccase-catalyzed arylation led to successful preparation of a diversity of arylated dihydrocoumarins bearing an all-carbon quaternary center in the three position in moderate to excellent yields, as shown in Scheme 2.79 [84].

Equisetin is a fungal tetramate natural product with potent anti-infectious activity. Its biosynthesis involves a Diels–Alderase that stereoselectively constructs the *trans*-decalin system of the molecule. The precursor of the Diels–Alderase-catalyzed reaction, a polyenoyltretamic acid, was synthesized in seven steps from the easily available (+)-citronellal. A concise chemoenzymatic route was then developed by employing a Diels–Alderase fsa2 from *Fusarium* sp. FN080326 to prepare equisetin

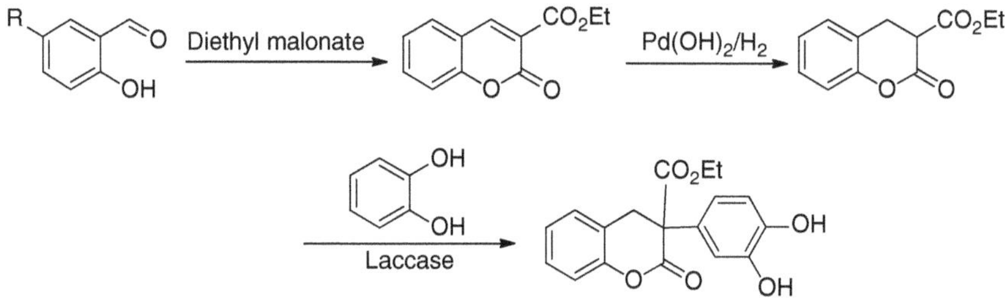

Scheme 2.79 Chemoenzymatic synthesis of arylated dihydrocoumarins. Source: Adapted from Suljić et al. [84].

(Scheme 2.80). This exemplifies that the combination of chemical and biological methods is an efficient approach to access structurally complex cyclic natural products and their derivatives [85].

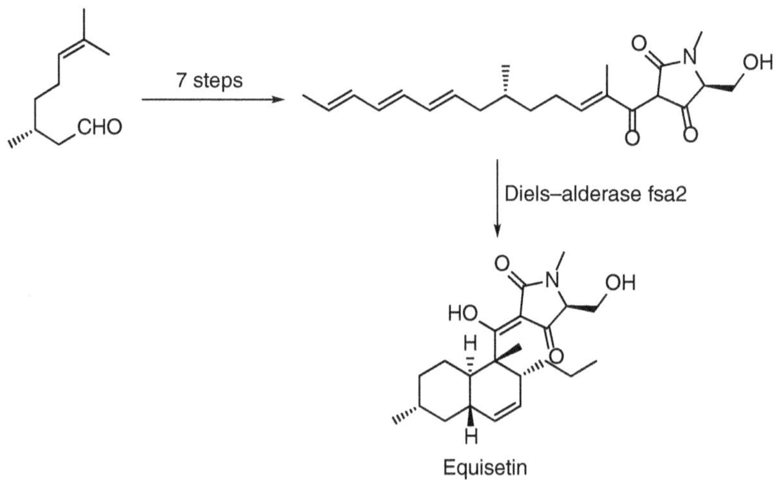

Scheme 2.80 Chemoenzymatic synthesis of equisetin.

Phenylalanine ammonia lyases (PALs) catalyze the asymmetric hydroamination of arylpropenoic acids to furnish the corresponding optically pure L-arylalanines. A chemoenzymatic approach was developed for the synthesis of a range of N-protected nonnatural L-biarylalanine derivatives from 4-bromocinnamic acid by combining the PAL-catalyzed hydroamination and the Pd-catalyzed coupling with arylboronic acids after Boc protection (Scheme 2.81). Similarly, an evolved D-amino acid dehydrogenase (DAADH) was also applied to prepare both enantiomers of 4-bromophenylalanine from 4-bromophenylpyruvic acid, which were subsequently protected with Boc group and coupled with arylboronic acids to afford the desired D-biarylalanines (Scheme 2.82) [86].

Haloalkane dehalogenases catalyze the hydrolytic cleavage of carbon–halide bonds, giving the product with inversion of configuration. A haloalkane dehalogenase from *B. japonicum* USDA110 catalyzed the kinetic resolution

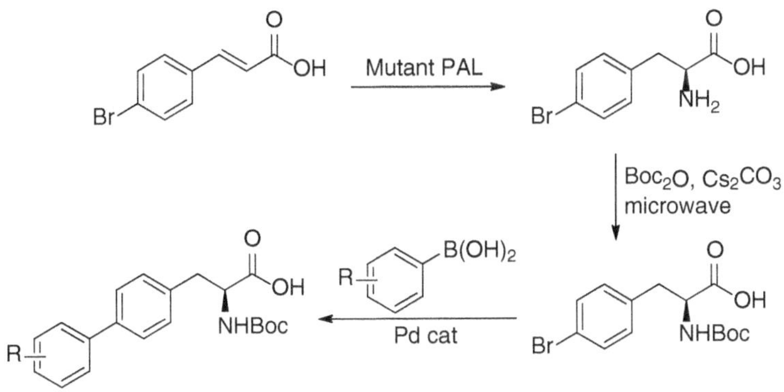

Scheme 2.81 Chemoenzymatic synthesis of L-biarylalanines.

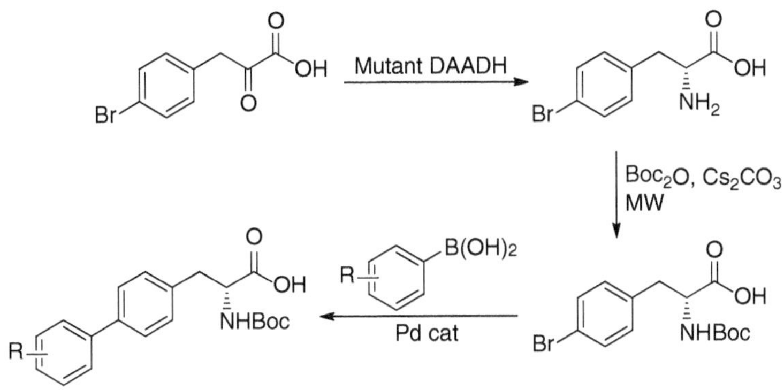

Scheme 2.82 Chemoenzymatic synthesis of D-biarylalanines. Source: Adapted from Ahmed et al. [86].

of racemic α-bromoamide with excellent enantioselectivity, giving a mixture of (*S*)-α-hydroxyamide and (*S*)-α-bromoamide. Without separating the two compounds, the hydroxy group of (*S*)-α-hydroxyamide was converted into methanesulfonyl esters, (*S*)-α-mesyloxyamide. Both (*S*)-α-bromoamide and (*S*)-α-mesyloxyamide reacted with nucleophiles, such as azide and benzylamine, in an S_N2-type process, furnishing the corresponding (*R*)-enantiomer of the α-substituted amides (Scheme 2.83). By employing haloalkane dehalogenases-catalyzed enantioinverting reaction in combination with simple chemical processes, readily available racemic α-bromoamides were transformed into versatile chiral α-substituted amides in high overall yields and excellent ee values [87].

The chemoenzymatic flow synthetic strategy had also been applied to the preparation of enantiomerically pure Captopril, a widely used antihypertensive drug. In the first step, the cheap prochiral 2-methyl-1,3-propandiol was regio- and stereoselectively oxidized under continuous flow conditions using the immobilized *Acetobacter aceti* MIM 2000/28 whole cells packed in a glass column. The isolated (*R*)-3-hydroxy-2-methylpropanoic acid was chlorinated to give (*R*)-3-chloro-2-methylpropanoyl chloride, which was converted into Captopril

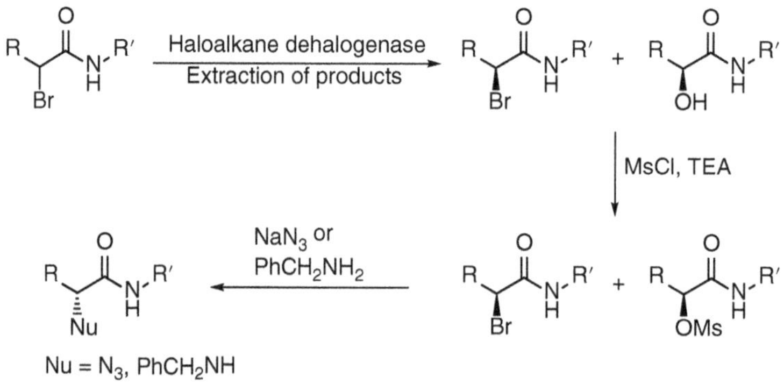

Scheme 2.83 Chemoenzymatic synthesis of chiral α-substituted amides.

[(*S*)-1-[(*S*)-3-chloro-2-methylpropanoyl]pyrroli-dine-2-carboxylic acid] by coupling with L-proline followed by nucleophilic substitution of chloride with NaSH (Scheme 2.84). The three sequential chemical transformations were performed in a multistep continuous flow process without isolation of the intermediates [88].

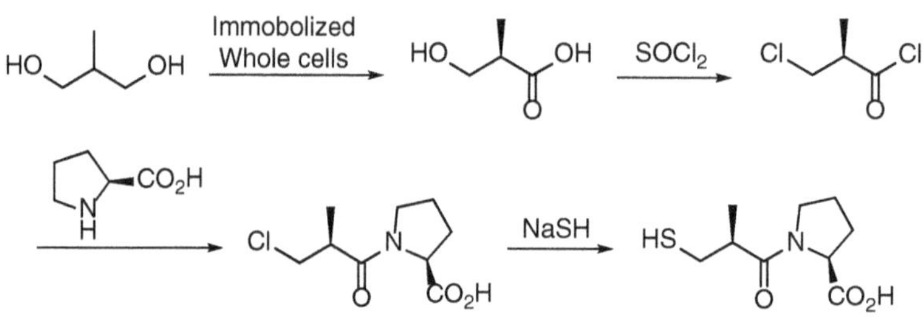

Scheme 2.84 Chemoenzymatic synthesis of Captopril under continuous flow conditions.

Hydroxynitrile lyase (HNL) catalyzes the enantioselective hydrocyanation of aldehydes and ketones, generating the corresponding chiral hydroxynitriles. A (*R*)-HNL of *Badamu* (*Prunus communis* L. var. *dulcis* Borkh, almond from Xinjiang, China) was found to catalyze the hydrocyanation of 4-methylsulfanyl-benzaldehyde, affording (*R*)-4-methylsulfanyl-mandelonitrile in 86% yield with 99% ee. The enantiomerically pure (*R*)-4-methylsulfanyl-mandelonitrile obtained was then applied for the stereoselective synthesis of thiamphenicol and florfenicol (Scheme 2.85), which are broad-spectrum antibiotics against many gram-negative and gram-positive microorganisms [89].

Recently, a select group of flavin-dependent monooxygenases (FDMO) had been found to catalyze the asymmetric dearomatization of phenols, offering distinct advantages over small-molecule catalysts in terms of site selectivity and over-oxidation. These biocatalysts showed complementary site- and stereoselectivity, and were employed in the stereodivergent, chemoenzymatic synthesis of azaphilone natural products. A FDMO AzaH catalyzed the hydroxylation of (*E*)-2,4-dihydroxy-3-methyl-6-(2-oxopent-3-en-1-yl)benzaldehyde, giving (*R*)-(*E*)-7-

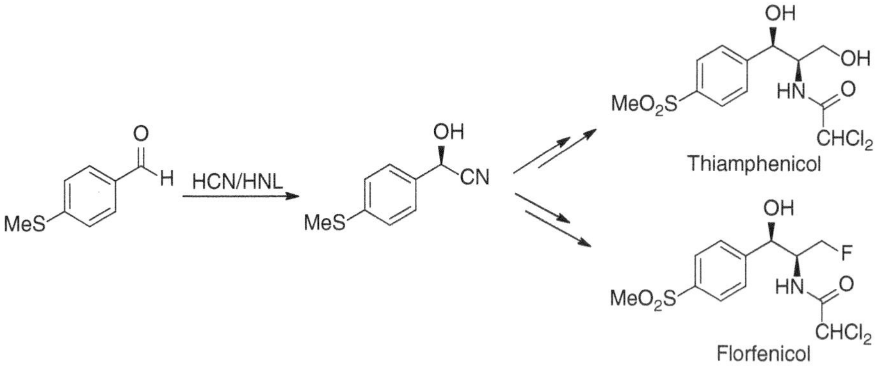

Scheme 2.85 Chemoenzymatic synthesis of Thiamphenicol and Florfenicol.

hydroxy-7-methyl-3-(prop-1-en-1-yl)-6*H*-isochromene-6,8(7*H*)-dione in 96% yield and >99% ee. In contrast, the (*S*)-counterpart was obtained in 83% yield and 98% ee by utilizing an FDMO AfoD as the biocatalyst for the hydroxylation. The (*R*)- and (*S*)-(*E*)-7-hydroxy-7-methyl-3-(prop-1-en-1-yl)-6*H*-isochromene-6,8(7*H*)-diones prepared were then transformed into (*S*)-trichoflectin and (*R*)-trichoflectin, respectively, by acylation with the acylketene derived from 2,2,6-trimethyl-4*H*-1,3-dioxin-4-one followed by Knoevenagel condensation (Scheme 2.86). Similarly, AzaH catalyzed the hydroxylation of 2,4-dihydroxy-3-methyl-6-(2-oxopropyl)benzaldehyde to furnish (*R*)-(*E*)-7-hydroxy-7-methyl-3-methyl-6*H*-isochromene-6,8(7*H*)-dione in 95% yield and >99% ee. The acylation with the acylketene generated *in situ* from 6-heptyl-2,2-dimethyl-4*H*-1,3-dioxin-4-one and subsequent Knoevenagel condensation produced (*S*)-deflectin-1a in 87% yield (Scheme 2.87). Lunatoic acid A was also synthesized by the chemical transformations of (*R*)-(*E*)-7-hydroxy-7-methyl-3-(prop-1-en-1-yl)-6*H*-isochromene-6,8(7*H*)-dione as shown in Scheme 2.88 [90].

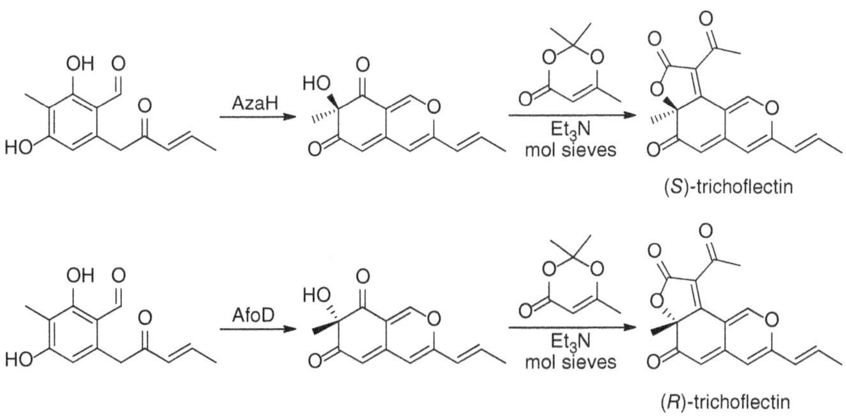

Scheme 2.86 Chemoenzymatic synthesis of (*S*)-trichoflectin and (*R*)-trichoflectin.

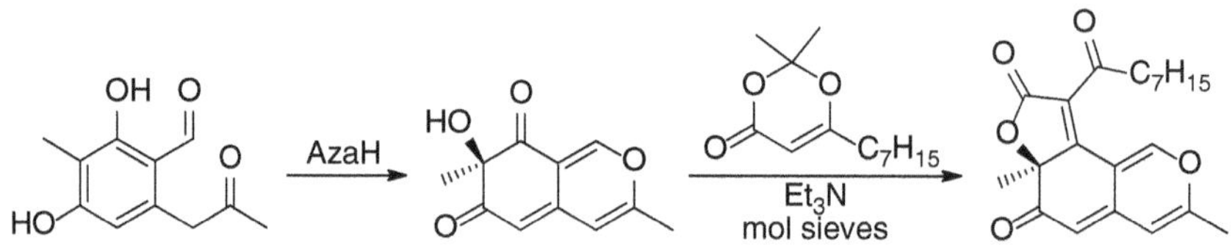

(*S*)-deflectin-1a

Scheme 2.87 Chemoenzymatic synthesis of (*S*)-deflectin-1a.

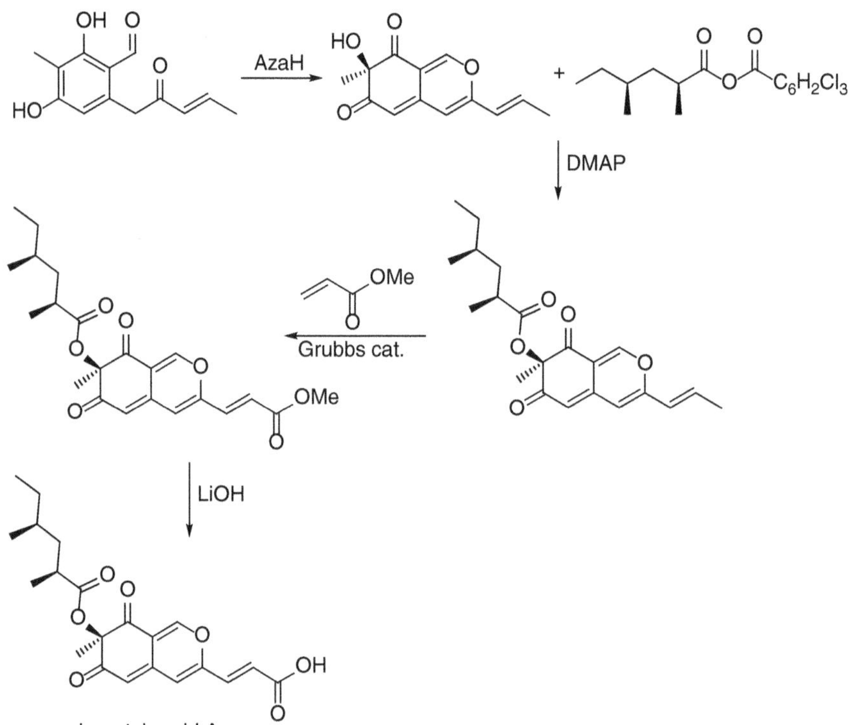

Lunatoic acid A

Scheme 2.88 Chemoenzymatic synthesis of Lunatoic acid A. Source: Adapted from Pyser et al. [90].

Pictet–Spenglerase or strictosidine synthase (STR1) catalyzes the Pictet–Spengler reaction of tryptamine with secologanin to give 3α-(*S*)-strictosidine, a key intermediate for the biosynthesis of indole alkaloids. The acid-catalyzed Pictet–Spengler reaction of tryptamine with secologanin generated a diastereomeric mixture of strictosidine. Replacing tryptamine with 1*H*-indole-1-ethanamine resulted in the formation of 3α-(*S*)-piperazino[1,2-α]indolyl strictosidine. This enzyme also used 1*H*-indole-4-ethanamine as a substrate, furnishing 4*α*-(*S*)-1*H*-azepino[3,4,5-*cd*]indolyl-strictosidine (>98% de) as the product that could be converted to 4α-(*S*)-1*H*-Azepino-[3,4,5-cd]indolyl-strictosidine lactam in 10% Na_2CO_3 solution (Scheme 2.89). When the reaction was conducted in

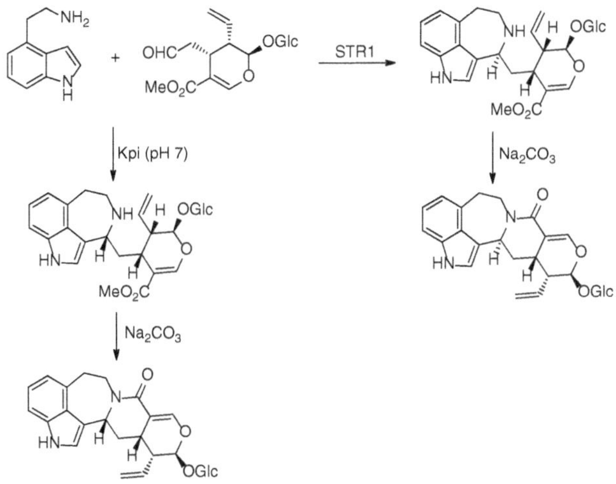

Scheme 2.89 Chemoenzymatic synthesis of 1*H*-Azepino-[3,4,5-cd]indolyl-strictosidine lactam.

potassium phosphate buffer (pH 7) at 70 °C, 4β-(*R*)-1*H*-azepino[3,4,5-cd]indolyl-vincoside was furnished in 85% de, and similar lactamization afforded 4β-(*R*)-1*H*-Azepino-[3,4,5-cd]indolyl-vincoside lactam in 70% yield. The lactams were hydrogenated using Pd/C as the catalyst, followed by enzymatic deglycosylation with β-glucosidase, affording the corresponding dihydro-aglycones, which were further transformed into a series of 1*H*-azepino-indole alkaloids (Scheme 2.90).

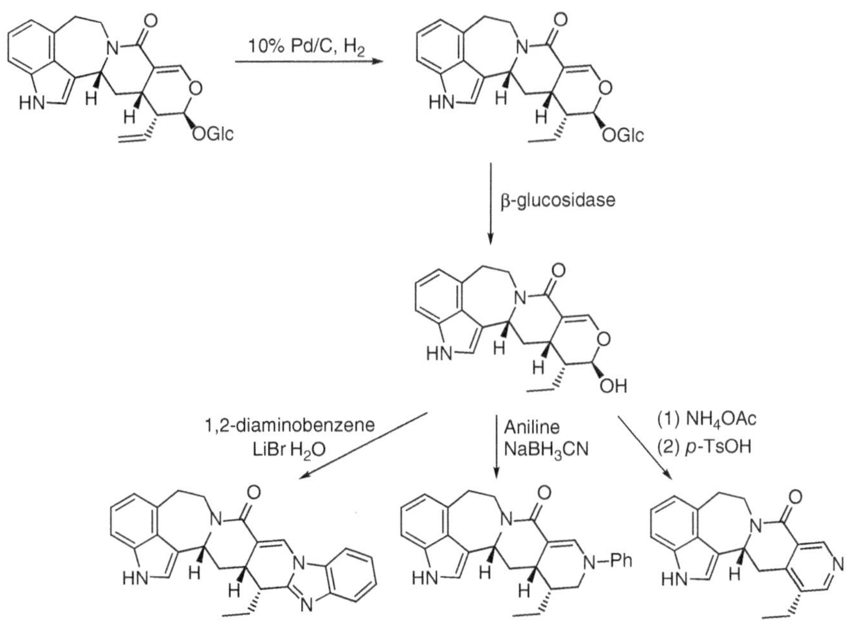

Scheme 2.90 Chemoenzymatic synthesis of 1*H*-azepino-indole alkaloids.

For example, hydrogenated 4α-(S)-1*H*-azepino-[3,4,5-cd]indolyl-vincoside dihydroaglycone was transformed into (15*R*, 15a*R*, 16a*S*)-15-ethyl-4,5,15,15a,16,16a-hexahydrobenzo[4,5]imidazo[1,2-b]indolo[3′,4′:3,4,5]azepino[2,1-g][2,7]naphthyridin-7(18H)-one, a azepino-indole alkaloid with antimalarial activity ($IC_{50} \approx 3.4\,\mu M$) [91].

2.12 Integration of Multienzyme Cascade with Chemical Transformation

In recent years, one-pot multienzyme (OPME) cascades with two or more enzymes have been rapidly developed with increasing diversity and complexity of cascades. This is attributed to the fact that enzymes are generally active under similar reaction conditions and advances in biotechnologies that make enzymes more easily accessible. In this context, multienzyme cascades have increasingly combined with classic chemical reactions and integrated into organic synthesis of complex molecules [92].

Poly-ε-caprolactone is a well-known biodegradable polyester most commonly used for the production of specialty polyurethanes. Wedde et al. designed a chemoenzymatic synthetic route to access poly-ε-caprolactone starting from phenol (Scheme 2.91). After the chemical hydrogenation, the rhodium catalyst was removed and the crude product was used for subsequent biocatalytic double oxidation by an alcohol dehydrogenase and a BVMO; the latter are well-known reactions in the degradation pathway of cyclohexanol in some bacteria. Incorporating an *in situ* product recovery strategy to relieve product inhibition, the ε-caprolactone formed was extracted from the aqueous reaction medium with an organic solvent through a permeable polydimethylsiloxane membrane. This was followed by lipase-catalyzed polymerization in the organic phase to give a fraction that contained poly-ε-caprolactone [93].

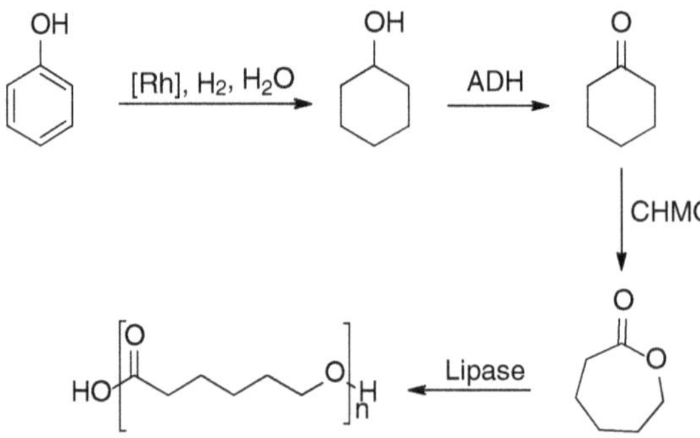

Scheme 2.91 Chemoenzymatic synthesis of poly-ε-caprolactone.

The OPME cascade reactions have also been applied to the synthesis of carbohydrates. 2,4-Diazido-2,4,6-trideoxymannose (6deoxyMan2,4-diN3) was prepared from D-fucose in eight steps in 60% overall yield and tested as starting material for OPME sialylation systems with three different acceptors: *para*-nitrophenyl β-galactoside (Galβ*p*NP), thiotolyl β-galactoside (GalβSTol), and lactosyl β-propyl chloride (LacβProCl). A PmAldolase catalyzed the aldol reaction of 6deoxyMan2,4-diN3 with pyruvate generating the diazido derivative of Leg (Leg5,7diN3), which was converted into CMP-Leg5,7diN3 by an NmCSS-catalyzed reaction with cytidine 5′-triphosphate. The resulting CMP-Leg5,7diN3 acted as substrate for a sialyltransferase (PmST1_M144D or Psp2,6ST). The α2-3/6-linked Leg5,7diN3-containing glycosides produced were transformed into α2-3/6-linked Leg5,7Ac2-glycosides, the bacterial nonulosonic acid (NulO) analogs of sialic acid, by azide-to-amine reduction and the selective acetylation of the amine group (Scheme 2.92) [94].

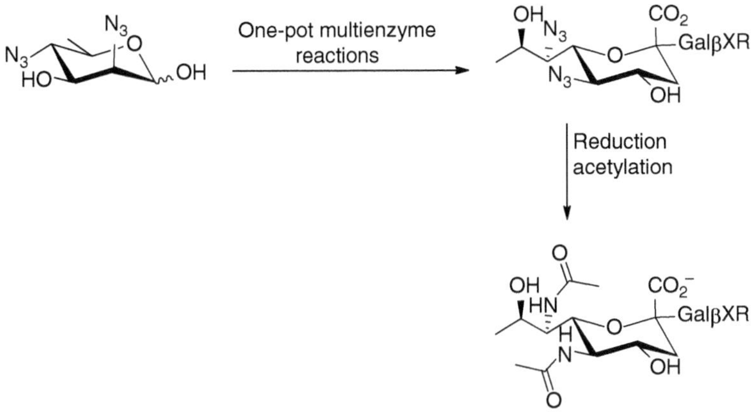

Scheme 2.92 Chemoenzymatic approach to access α2-3/6-linked Leg5,7Ac2-glycosides. Source: Adapted from Santra et al. [94].

Artemisinin derivatives have potent *anti*malarial activity, and artemisinin-based combination treatments (ACTs) are designated as first-line antimalarial treatment by the World Health Organization in 2002. The supply of artemisinin used to rely predominantly on the extraction from the plant *Artemisia annua* and the market price was highly volatile, ranging from US $350 to $1700 per kilogram. Recently, an efficient synthetic route to produce artemisinin has been developed by the combination of a biosynthetic process with chemical steps. By expressing the mevalonate pathway in *Saccharomyces cerevisiae*, 40 g/l of amorphadiene was obtained. Further engineering of the amorphadiene-producing strains enabled the production of artemisinic acid at 25 g/l titer. Artemisinic acid could be converted to artemisinin by a four-step chemical process or an alternative high-efficiency photochemical process (Scheme 2.93) [95].

E. coli strains were also widely engineered for the production of chemicals, which were further biocatalytically or chemically transformed to value-added products. Fermentational production of important industrial chemicals from glucose offers a sustainable manufacturing route from renewable resources instead of the fossil feedstocks that are associated with environmental, health, and safety concerns.

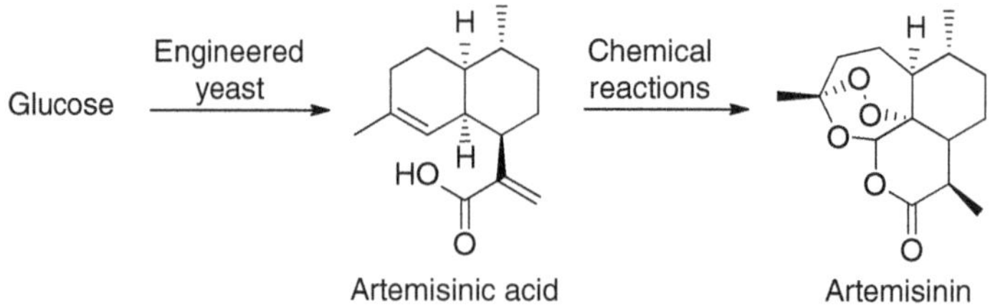

Scheme 2.93 Semisynthesis of artemisinin involving microbial production and chemical transformation. Source: Adapted from Paddon and Keasling [95].

In this context, an *E. coli* strain was engineered to produce *cis,cis*-muconic acid by expressing three genes: the 3-dehydroshikimate (DHS) dehydratase gene (aroZ) from *Klebsiella pneumoniae*, the PCA decarboxylase gene (aroY) from *K. pneumoniae,* and the catechol 1,2-dioxygenase gene (catA) from *A. calcoaceticus*. The recombinant *E. coli* strain could synthesize 36.8 g/l *cis*, *cis*-muconic acid from glucose at titer of 36.8 g/l in 22% (mol/mol) yield within 48 h in a fed-batch fermentation process. After removal of the cells and the protein, the resulting solution was hydrogenated with Pt/C (10%) catalyst at ambient temperature, resulting in the conversion of *cis,cis*-muconic acid into adipic acid in 97% yield (Scheme 2.94) [96]. Adipic acid is a precursor to nylon-6,6 synthesis, a type of polyamide that is widely used in textile and plastic industries. Otherwise prepared from benzene, the biocatalytic approach eliminated the use of a carcinogenic agent as feedstock, and avoided the generation of nitrous oxide, a long-lived greenhouse gas, as a by-product of nitric acid oxidation, intrinsic to the chemical production of adipic acid. The chemoenzymatic process provides a green alternative to its current commercial production from benzene by a chemical route.

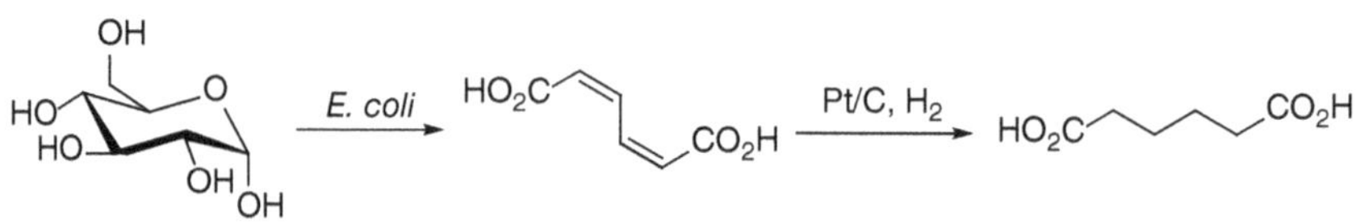

Scheme 2.94 Synthesis of adipic acid involving microbial production and chemical transformation. Source: Adapted from Niu et al. [96].

Recently, a bioprocess was developed to produce mucic acid from D-galacturonic acid by using an engineered *E. coli* strain. By deleting the native catabolism pathway of D-galacturonic acid and mucic acid in *E. coli*, mucic acid was produced with 95.4% of D-galacturonic acid at a titer of 10.3 g/l, using a mixture of glucose, L-arabinose, and D-galacturonic acid (10 g/l of each) as the substrate, which mimicked the composition of sugar beet residue. The sugar beet residue was enzymatically hydrolyzed to give a mixture of 22% glucose, 17.4% L-arabinose, and 16.5% D-galacturonic acid. By utilizing this hydrolysate for the whole-cell transformation, 86% of D-galacturonic acid was converted into mucic acid with 14.2% overall yield based on sugar beet residue. The resulting mucic acid was easily isolated from the reaction mixture. NH_4ReO_4 was found to catalyze the deoxydehydration reaction of mucic acid, generating muconic acid in 72% yield, which was hydrogenated with a Pt/C catalyst to afford adipic acid in 92% yield (Scheme 2.95). This demonstrates a new chemoenzymatic process for the production of adipic acid directly from sugar beet residue, an inexpensive renewable feedstock [97].

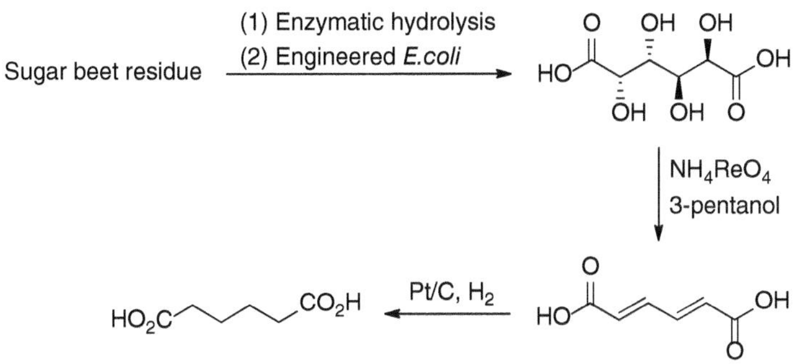

Scheme 2.95 Chemoenzymatic production of adipic acid from sugar beet residue.

2.13 Summary and Outlook

As discussed above, enzyme catalysts have been increasingly integrated into the synthesis of high-value fine chemicals due to their high chemo, regio, and stereoselectivity, which often enables a shorter and more effective synthetic route to the target compound, thus reducing the production cost and environmental impact. This has been made possible by the recent advances in the methodologies for enzyme discovery, protein engineering, and design, which help discover or create numerous new enzymes with new activity and/or enhanced properties such as catalytic activity, selectivity, and stability. The constant advances in modern biotechnology will continuously accelerate the development pace of novel enzymes tailor-made for the desired organic transformations in the years to come. In addition, retrosynthetic design involving both traditional chemical reaction and biotransformation will result in much more efficient synthetic pathways for the target compounds, which would not have been expected before. The rapid development in the integration of enzymes into organic synthesis will greatly benefit both chemical industry and the environment, thus helping us address the challenges that we will face in resources and environment in the coming generations.

References

1 Donova, M.V. and Egorova, O.V. (2012). Microbial steroid transformations: current state and prospects. *Applied Microbiology and Biotechnology* 94 (6): 1423–1447.

2 Hansen, K.B., Hsiao, Y., Xu, F. et al. (2009). Highly efficient asymmetric synthesis of sitagliptin. *Journal of the American Chemical Society* 131 (25): 8798–8804.

3 Savile, C.K., Janey, J.M., Mundorff, E.C. et al. (2010). Biocatalytic asymmetric synthesis of chiral amines from ketones applied to sitagliptin manufacture. *Science* 329 (5989): 305–309.

4 Fishman, A. and Zviely, M. (1998). Chemo-enzymatic synthesis of (S)-α-cyano-3-phenoxybenzyl alcohol. *Tetrahedron: Asymmetry* 9 (1): 107–118.

5 Kimura, M., Kuboki, A., and Sugai, T. (2002). Chemo-enzymatic synthesis of enantiomerically pure (R)-2-naphthylmethoxyacetic acid. *Tetrahedron: Asymmetry* 13 (10): 1059–1068.

6 Borowiecki, P., Paprocki, D., Dudzik, A., and Plenkiewicz, J. (2016). Chemoenzymatic synthesis of proxyphylline enantiomers. *The Journal of Organic Chemistry* 81 (2): 380–395.

7 Reddipalli, G., Venkataiah, M., and Fadnavis, N.W. (2010). Chemo-enzymatic synthesis of both enantiomers of rugulactone. *Tetrahedron: Asymmetry* 21 (3): 320–324.

8 Tyrikos-Ergas, T., Giannopoulos, V., and Smonou, I. (2018). An efficient chemoenzymatic approach towards the synthesis of rugulactone. *Molecules* 23 (3): 640.

9 Koszelewski, D., Borys, F., Brodzka, A., and Ostaszewski, R. (2019). Synthesis of enantiomerically pure 5,6-dihydropyran-2-ones via chemoenzymatic sequential DKR-RCM reaction. *European Journal of Organic Chemistry* 2019 (7): 1653–1658.

10 Paul, C.E., Gotor-Fernández, V., Lavandera, I. et al. (2012). Chemoenzymatic preparation of optically active 3-(1H-imidazol-1-yl)cyclohexanol-based ionic liquids: application in organocatalysis and toxicity studies. *RSC Advances* 2 (16): 6455–6463.

11 Porcar, R., Sans, V., Ríos-Lombardía, N. et al. (2012). Stereoselective chemoenzymatic synthesis of enantiopure 2-(1H-imidazol-yl)cycloalkanols under continuous flow conditions. *ACS Catalysis* 2 (9): 1976–1983.

12 Poterała, M., Dranka, M., and Borowiecki, P. (2017). Chemoenzymatic preparation of enantiomerically enriched (R)-(−)-mandelic acid derivatives: application in the synthesis of the active agent pemoline. *European Journal of Organic Chemistry* 2017 (16): 2290–2304.

13 Sreelakshmi, C., Bhaskar Rao, A., Lakshmi Narasu, M. et al. (2014). Chemoenzymatic total synthesis of paecilocin A and 3-butyl-7-hydroxyphthalide. *Tetrahedron Letters* 55 (7): 1303–1305.

14 Busto, E., Gotor-Fernández, V., and Gotor, V. (2012). Asymmetric chemoenzymatic synthesis of ramatroban using lipases and oxidoreductases. *The Journal of Organic Chemistry* 77 (10): 4842–4848.

15 Ríos-Lombardía, N., Busto, E., Gotor-Fernández, V., and Gotor, V. (2011). Chemoenzymatic asymmetric synthesis of optically active pentane-1,5-diamine fragments by means of lipase-catalyzed desymmetrization transformations. *The Journal of Organic Chemistry* 76 (14): 5709–5718.

16 Iacazio, G. and Réglier, M. (2005). Chemo-enzymatic synthesis of all four diastereoisomers of 1-fluoro-2-amino-indane. *Tetrahedron: Asymmetry* 16 (22): 3633–3639.

17 Mukherjee, H. and Martinez, C.A. (2011). Biocatalytic route to chiral precursors of β-substituted-γ-amino acids. *ACS Catalysis* 1 (9): 1010–1013.

18 Feng, Y., Wang, Z., Luo, Z. et al. (2019). Development of an efficient and scalable biocatalytic route to (1S,4R)-8-hydroxy-1,2,3,4-tetrahydro-1,4-methanonaphthalen-5-yl propionate via enantioselective enzymatic desymmetrization of a prochiral diester. *Organic Process Research & Development* 23 (6): 1243–1251.

19 Moni, L., Banfi, L., Basso, A. et al. (2015). Ugi and Passerini reactions of biocatalytically derived chiral aldehydes: application to the synthesis of bicyclic pyrrolidines and of antiviral agent telaprevir. *The Journal of Organic Chemistry* 80 (7): 3411–3428.

20 Schröder, H., Strohmeier, G.A., Leypold, M. et al. (2013). Racemization-free chemoenzymatic peptide synthesis enabled by the ruthenium-catalyzed synthesis of peptide enol esters via alkyne-addition and subsequent conversion using alcalase-cross-linked enzyme aggregates. *Advanced Synthesis & Catalysis* 355 (9): 1799–1807.

21 Bergeron, S., Chaplin, D.A., Edwards, J.H. et al. (2006). Nitrilase-catalysed desymmetrisation of 3-hydroxyglutaronitrile: preparation of a statin side-chain intermediate. *Organic Process Research & Development* 10 (3): 661–665.

22 Benz, P., Muntwyler, R., and Wohlgemuth, R. (2007). Chemoenzymatic synthesis of chiral carboxylic acids via nitriles. *Journal of Chemical Technology & Biotechnology* 82 (12): 1087–1098.

23 Xie, Z., Feng, J., Garcia, E. et al. (2006). Cloning and optimization of a nitrilase for the synthesis of (3S)-3-cyano-5-methyl hexanoic acid. *Journal of Molecular Catalysis B: Enzymatic* 41 (3): 75–80.

24 Duan, Y., Yao, P., Ren, J. et al. (2014). Biocatalytic desymmetrization of 3-substituted glutaronitriles by nitrilases. A convenient chemoenzymatic access to optically active (S)-Pregabalin and (R)-Baclofen. *Science China Chemistry* 57 (8): 1164–1171.

25 Yu, S., Yao, P., Li, J. et al. (2019). Improving the catalytic efficiency and stereoselectivity of a nitrilase from *Synechocystis* sp. PCC6803 by semi-rational engineering en route to chiral γ-amino acids. *Catalysis Science & Technology* 9 (6): 1504–1510.

26 Hadi, T., Dı′ az-Rodrı′ guez, A., Khan, D. et al. (2018). Identification and implementation of biocatalytic transformations in route discovery: synthesis of chiral 1,3-substituted cyclohexanone building blocks. *Organic Process Research & Development* 22 (7): 871–879.

27 Ankati, H., Yang, Y., Zhu, D. et al. (2008). Synthesis of optically pure 2-azido-1-arylethanols with Isolated enzymes and conversion to triazole-containing β-blocker analogues employing click chemistry. *The Journal of Organic Chemistry* 73 (16): 6433–6436.

28 Goswami, J., Bezbaruah, R.L., Goswami, A., and Borthakur, N. (2002). A convenient stereoselective synthesis of (R)-(−)-denopamine and (R)-(−)-salmeterol. *Tetrahedron: Asymmetry* 12 (24): 3343–3348.

29 Simon, R.C., Busto, E., Richter, N. et al. (2014). Chemoenzymatic synthesis of enantiomerically pure syn-configured 1-aryl-3-methylisochroman derivatives. *European Journal of Organic Chemistry* 2014 (1): 111–121.

30 Nelson, T.D., LeBlond, C.R., Frantz, D.E. et al. (2004). Stereoselective synthesis of a potent thrombin inhibitor by a novel P2–P3 lactone ring opening. *The Journal of Organic Chemistry* 69 (11): 3620–3627.

31 D'Arrigo, P., Pedrocchi-Fantoni, G., and Servi, S. (2010). Chemo-enzymatic synthesis of ethyl (R)-2-hydroxy-4-phenylbutyrate. *Tetrahedron: Asymmetry* 21 (8): 914–918.

32 Díaz-Rodríguez, A., Borzęcka, W., Lavandera, I., and Gotor, V. (2014). Stereodivergent preparation of valuable γ- or δ-hydroxy esters and lactones through one-pot cascade or tandem chemoenzymatic protocols. *ACS Catalysis* 4 (2): 386–393.

33 Mittmann, E., Hu, Y., Peschke, T. et al. (2019). Chemoenzymatic synthesis of O-containing heterocycles from α-diazo esters. *ChemCatChem* 11 (22): 5519–5523.

34 Cui, B., Yang, M., Shan, J. et al. (2017). Chemoenzymatic synthesis of β-hydroxyl-sulfoxides by a two-step reaction of enzymatic reduction using *Pseudomonas monteilii* species and sulfoxidation with chiral titanium complexe. *Tetrahedron* 73 (34): 5200–5206.

35 Chen, X., Liu, Z.-Q., Lin, C.-P., and Zheng, Y.-G. (2016). Chemoenzymatic synthesis of (S)-duloxetine using carbonyl reductase from *Rhodosporidium toruloides*. *Bioorganic Chemistry* 65: 82–89.

36 Winkler, C.K., Faber, K., and Hall, M. (2018). Biocatalytic reduction of activated CC-bonds and beyond: emerging trends. *Current Opinion in Chemical Biology* 43: 97–105.

37 Pietruszka, J. and Schölzel, M. (2012). Ene reductase-catalysed synthesis of (R)-profen derivatives. *Advanced Synthesis & Catalysis* 354 (4): 751–756.

38 Winkler, C.K., Clay, D., Davies, S. et al. (2013). Chemoenzymatic asymmetric synthesis of pregabalin precursors via asymmetric bioreduction of β-cyanoacrylate esters using ene-reductases. *The Journal of Organic Chemistry* 78 (4): 1525–1533.

39 Winkler, C.K., Clay, D., Turrini, N.G. et al. (2014). Nitrile as activating group in the asymmetric bioreduction of β-cyanoacrylic acids catalyzed by ene-reductases. *Advanced Synthesis & Catalysis* 356 (8): 1878–1882.

40 Biermann, M., Gruß, H., Hummel, W., and Gröger, H. (2016). Guerbet alcohols: from processes under harsh conditions to synthesis at room temperature under ambient pressure. *ChemCatChem* 8 (5): 895–899.

41 Slabu, I., Galman, J.L., Lloyd, R.C., and Turner, N.J. (2017). Discovery, engineering, and synthetic application of transaminase biocatalysts. *ACS Catalysis* 7 (12): 8263–8284.

42 Fuchs, M., Koszelewski, D., Tauber, K. et al. (2010). Chemoenzymatic asymmetric total synthesis of (S)-Rivastigmine using ω-transaminases. *Chemical Communications* 46 (30): 5500–5502.

43 Meadows, R.E., Mulholland, K.R., Schürmann, M. et al. (2013). Efficient synthesis of (S)-1-(5-fluoropyrimidin-2-yl)ethylamine using an ω-transaminase biocatalyst in a two-phase system. *Organic Process Research & Development* 17 (9): 1117–1122.

44 Midelfort, K.S., Kumar, R., Han, S. et al. (2012). Redesigning and characterizing the substrate specificity and activity of *Vibrio fluvialis* aminotransferase for the synthesis of imagabalin. *Protein Engineering, Design and Selection* 26 (1): 25–33.

45 Mangion, I.K., Sherry, B.D., Yin, J., and Fleitz, F.J. (2012). Enantioselective synthesis of a dual orexin receptor antagonist. *Organic Letters* 14 (13): 3458–3461.

46 Girardin, M., Ouellet, S.G., Gauvreau, D. et al. (2013). Convergent kilogram-scale synthesis of dual orexin receptor antagonist. *Organic Process Research & Development* 17 (1): 61–68.

47 Molinaro, C., Bulger, P.G., Lee, E.E. et al. (2012). CRTH2 antagonist MK-7246: a synthetic evolution from discovery through development. *The Journal of Organic Chemistry* 77 (5): 2299–2309.

48 Marx, L., Ríos-Lombardía, N., Farnberger, J.F. et al. (2018). Chemoenzymatic approaches to the synthesis of the calcimimetic agent cinacalcet employing transaminases and ketoreductases. *Advanced Synthesis & Catalysis* 360 (11): 2157–2165.

49 Zhang, Y.-H., Chen, F.-F., Li, B.-B. et al. (2020). Stereocomplementary synthesis of pharmaceutically relevant chiral 2-aryl-substituted pyrrolidines using imine reductases. *Organic Letters* 22 (9): 3367–3372.

50 Yao, P., Xu, Z., Yu, S. et al. (2019). Imine reductase-catalyzed enantioselective reduction of bulky α,β-unsaturated imines en route to a pharmaceutically important morphinan skeleton. *Advanced Synthesis & Catalysis* 361 (3): 556–561.

51 France, S.P., Aleku, G.A., Sharma, M. et al. (2017). Biocatalytic routes to enantiomerically enriched dibenz[c,e]azepines. *Angewandte Chemie International Edition* 56 (49): 15589–15593.

52 Zawodny, W., Montgomery, S.L., Marshall, J.R. et al. (2018). Chemoenzymatic synthesis of substituted azepanes by sequential biocatalytic reduction and organolithium-mediated rearrangement. *Journal of the American Chemical Society* 140 (51): 17872–17877.

53 Rentmeister, A., Arnold, F.H., and Fasan, R. (2008). Chemo-enzymatic fluorination of unactivated organic compounds. *Nature Chemical Biology* 5: 26.

54 Ilie, A., Harms, K., and Reetz, M.T. (2018). P450-catalyzed regio- and stereoselective oxidative hydroxylation of 6-iodotetralone: preparative-scale synthesis of a key intermediate for Pd-catalyzed transformations. *The Journal of Organic Chemistry* 83 (14): 7504–7508.

55 Lewis, J.C., Bastian, S., Bennett, C.S. et al. (2009). Chemoenzymatic elaboration of monosaccharides using engineered cytochrome $P450_{BM3}$ demethylases. *Proceedings of the National Academy of Sciences* 106 (39): 16550–16555.

56 Kolev, J.N., O'Dwyer, K.M., Jordan, C.T., and Fasan, R. (2014). Discovery of potent parthenolide-based antileukemic agents enabled by late-stage P450-mediated C—H functionalization. *ACS Chemical Biology* 9 (1): 164–173.

57 Tyagi, V., Alwaseem, H., O'Dwyer, K.M. et al. (2016). Chemoenzymatic synthesis and antileukemic activity of novel C9- and C14-functionalized parthenolide analogs. *Bioorganic & Medicinal Chemistry* 24 (17): 3876–3886.

58 Alwaseem, H., Frisch, B.J., and Fasan, R. (2018). Anticancer activity profiling of parthenolide analogs generated via P450-mediated chemoenzymatic synthesis. *Bioorganic & Medicinal Chemistry* 26 (7): 1365–1373.

59 Heidt, T., Baro, A., Köhn, A., and Laschat, S. (2015). Synthesis of cembranoid analogues through ring-closing metathesis of terpenoid precursors: a challenge regarding ring-size selectivity. *Chemistry – A European Journal* 21 (35): 12396–12404.

60 Le-Huu, P., Rekow, D., Krüger, C. et al. (2018). Chemoenzymatic route to oxyfunctionalized cembranoids facilitated by substrate and protein engineering. *Chemistry – A European Journal* 24 (46): 12010–12021.

61 Peters, C. and Buller, M.R. (2019). Industrial application of 2-oxoglutarate-dependent oxygenases. *Catalysts* 9 (3): 221.

62 Zwick, C.R. and Renata, H. (2018). Remote C–H hydroxylation by an α-ketoglutarate-dependent dioxygenase enables efficient chemoenzymatic synthesis of manzacidin C and proline analogs. *Journal of the American Chemical Society* 140 (3): 1165–1169.

63 Zwick, C.R. and Renata, H. (2018). Evolution of biocatalytic and chemocatalytic C–H functionalization strategy in the synthesis of manzacidin C. *The Journal of Organic Chemistry* 83 (14): 7407–7415.

64 Zhang, X., King-Smith, E., and Renata, H. (2018). Total synthesis of tambromycin by combining chemocatalytic and biocatalytic C–H functionalization. *Angewandte Chemie International Edition* 57 (18): 5037–5041.

65 Chang, W.-C., Yang, Z.-J., Tu, Y.-H., and Chien, T.-C. (2019). Reaction mechanism of a nonheme iron enzyme catalyzed oxidative cyclization via C–C bond formation. *Organic Letters* 21 (1): 228–232.

66 Li, J., Zhang, X., and Renata, H. (2019). Asymmetric chemoenzymatic synthesis of (−)-podophyllotoxin and related aryltetralin lignans. *Angewandte Chemie International Edition* 58 (34): 11657–11660.

67 Fürst, M.J.L.J., Gran-Scheuch, A., Aalbers, F.S., and Fraaije, M.W. (2019). Baeyer–Villiger monooxygenases: tunable oxidative biocatalysts. *ACS Catalysis* 9: 11207–11241.

68 Leisch, H., Morley, K., and Lau, P.C.K. (2011). Baeyer–Villiger monooxygenases: more than just green chemistry. *Chemical Reviews* 111 (7): 4165–4222.

69 Fink, M.J., Schön, M., Rudroff, F. et al. (2013). Single operation stereoselective synthesis of aerangis lactones: combining continuous flow hydrogenation and biocatalysts in a chemoenzymatic sequence. *ChemCatChem* 5 (3): 724–727.

70 Morrill, C., Jensen, C., Just-Baringo, X. et al. (2018). Biocatalytic conversion of cyclic ketones bearing α-quaternary stereocenters into lactones in an enantioselective radical approach to medium-sized carbocycles. *Angewandte Chemie International Edition* 57 (14): 3692–3696.

71 Rudroff, F., Fink, M.J., Pydi, R. et al. (2017). First chemo-enzymatic synthesis of the (R)-Taniguchi lactone and substrate profiles of CAMO and OTEMO, two new

Baeyer–Villiger monooxygenases. *Monatshefte für Chemie - Chemical Monthly* 148 (1): 157–165.

72 Rudroff, F., Bianchi, D.A., Moran-Ramallal, R. et al. (2016). Synthesis of tetrahydrofuran-based natural products and their carba analogs via stereoselective enzyme mediated Baeyer–Villiger oxidation. *Tetrahedron* 72 (46): 7212–7221.

73 Castillo, J.A., Calveras, J., Casas, J. et al. (2006). Fructose-6-phosphate aldolase in organic synthesis: preparation of d-fagomine, N-alkylated derivatives, and preliminary biological assays. *Organic Letters* 8 (26): 6067–6070.

74 Soler, A., Gutiérrez, M.L., Bujons, J. et al. (2015). Structure-guided engineering of D-fructose-6-phosphate aldolase for improved acceptor tolerance in biocatalytic aldol additions. *Advanced Synthesis & Catalysis* 357 (8): 1787–1807.

75 Soler, A., Garrabou, X., Hernández, K. et al. (2014). Sequential biocatalytic aldol reactions in multistep asymmetric synthesis: pipecolic acid, piperidine and pyrrolidine (homo)iminocyclitol derivatives from achiral building blocks. *Advanced Synthesis & Catalysis* 356 (14–15): 3007–3024.

76 Fanton, J., Camps, F., Castillo, J.A. et al. (2012). Enzymatic and organocatalyzed asymmetric aldolization reactions for the synthesis of thiosugar scaffolds. *European Journal of Organic Chemistry* 2012 (1): 203–210.

77 Hernández, K., Parella, T., Joglar, J. et al. (2015). Expedient synthesis of C-aryl carbohydrates by consecutive biocatalytic benzoin and aldol reactions. *Chemistry – A European Journal* 21 (8): 3335–3346.

78 Pedragosa-Moreau, S., Morisseau, C., Baratti, J. et al. (1997). Microbiological transformations 37. An enantioconvergent synthesis of the β-blocker (R)-Nifénalol® using a combined chemoenzymatic approach. *Tetrahedron* 53 (28): 9707–9714.

79 van Lint, M.J., Gümüs, A., Ruijter, E. et al. (2019). Enantioselective bio-hydrolysis of geranyl-derived rac-epoxides: a chemoenzymatic route to trans-furanoid linalool oxide. *Advanced Synthesis & Catalysis* 361 (4): 813–825.

80 Manoj, K.M., Archelas, A., Baratti, J., and Furstoss, R. (2001). Microbiological transformations. Part 45: a green chemistry preparative scale synthesis of enantiopure building blocks of Eliprodil: elaboration of a high substrate concentration epoxide hydrolase-catalyzed hydrolytic kinetic resolution process. *Tetrahedron* 57 (4): 695–701.

81 Fuchs, M., Simeo, Y., Ueberbacher, B.T. et al. (2009). Enantiocomplementary chemoenzymatic asymmetric synthesis of (R)- and (S)-chromanemethanol. *European Journal of Organic Chemistry* 2009 (6): 833–840.

82 Fujino, A., Asano, M., Yamaguchi, H. et al. (2007). *Bacillus subtilis* epoxide hydrolase-catalyzed preparation of enantiopure 2-methylpropane-1,2,3-triol monobenzyl ether and its application to expeditious synthesis of (R)-bicalutamide. *Tetrahedron Letters* 48 (6): 979–983.

83 Edegger, K., Mayer, S.F., Steinreiber, A., and Faber, K. (2004). Chemo-enzymatic enantio-convergent asymmetric synthesis of (R)-(+)-Marmin. *Tetrahedron* 60 (3): 583–588.

84 Suljić, S. and Pietruszka, J. (2014). Synthesis of 3-arylated 3,4-dihydrocoumarins: combining continuous flow hydrogenation with laccase-catalysed oxidation. *Advanced Synthesis & Catalysis* 356 (5): 1007–1020.

85 Li, X., Zheng, Q., Yin, J. et al. (2017). Chemo-enzymatic synthesis of equisetin. *Chemical Communications* 53 (34): 4695–4697.

86 Ahmed, S.T., Parmeggiani, F., Weise, N.J. et al. (2015). Chemoenzymatic synthesis of optically pure l- and d-biarylalanines through biocatalytic asymmetric amination and palladium-catalyzed arylation. *ACS Catalysis* 5 (9): 5410–5413.

87 Szymański, W., Westerbeek, A., Janssen, D.B., and Feringa, B.L. (2011). A simple enantioconvergent and chemoenzymatic synthesis of optically active α-substituted amides. *Angewandte Chemie International Edition* 50 (45): 10712–10715.

88 De Vitis, V., Dall'Oglio, F., Pinto, A. et al. (2017). Chemoenzymatic synthesis in flow reactors: a rapid and convenient preparation of captopril. *ChemistryOpen* 6 (5): 668–673.

89 Lu, W., Chen, P., and Lin, G. (2008). New stereoselective synthesis of thiamphenicol and florfenicol from enantiomerically pure cyanohydrin: a chemo-enzymatic approach. *Tetrahedron* 64 (33): 7822–7827.

90 Pyser, J.B., Baker Dockrey, S.A., Benítez, A.R. et al. (2019). Stereodivergent, chemoenzymatic synthesis of azaphilone natural products. *Journal of the American Chemical Society* 141 (46): 18551–18559.

91 Cai, Y., Shao, N., Xie, H. et al. (2019). Stereocomplementary chemoenzymatic Pictet–Spengler reactions for formation of rare azepino-indole frameworks: discovery of antimalarial compounds. *ACS Catalysis* 9 (8): 7443–7448.

92 Sperl, J.M. and Sieber, V. (2018). Multienzyme cascade reactions—status and recent advances. *ACS Catalysis* 8 (3): 2385–2396.

93 Wedde, S., Rommelmann, P., Scherkus, C. et al. (2017). An alternative approach towards poly-ε-caprolactone through a chemoenzymatic synthesis: combined hydrogenation, bio-oxidations and polymerization without the isolation of intermediates. *Green Chemistry* 19 (5): 1286–1290.

94 Santra, A., Xiao, A., Yu, H. et al. (2018). A diazido mannose analogue as a chemoenzymatic synthon for synthesizing di-N-acetyllegionaminic acid-containing glycosides. *Angewandte Chemie International Edition* 57 (11): 2929–2933.

95 Paddon, C.J. and Keasling, J.D. (2014). Semi-synthetic artemisinin: a model for the use of synthetic biology in pharmaceutical development. *Nature Reviews Microbiology* 12 (5): 355–367.

96 Niu, W., Draths, K.M., and Frost, J.W. (2002). Benzene-free synthesis of adipic acid. *Biotechnology Progress* 18 (2): 201–211.

97 Zhang, H., Li, X., Su, X. et al. (2016). Production of adipic acid from sugar beet residue by combined biological and chemical catalysis. *ChemCatChem* 8 (8): 1500–1506.

3

One-pot Sequential Chemoenzymatic Reactions

"None-one-pot" chemoenzymatic transformation can be easily implemented and integrated into the synthetic route of a target compound because the reaction condition compatibility of a chemical reaction with biotransformation is not required. While integration of biotransformations into organic synthesis can often shorten the synthetic routes and improve the synthetic efficiency because of the high selectivities of biotransformations, "none-one-pot" chemoenzymatic processes still require the isolation of the intermediates. However, one-pot sequential or concurrent reactions avoid purification of intermediates, and thus further improve overall efficiency and reduce production cost. Therefore, there is an increasing interest in combining chemical and biocatalytic reactions in a one-pot process. Although the development of such a process faces many challenges such as incompatible reaction conditions, catalyst inhibition, undesired side reactions, and poor solubility of substrates, more and more one-pot chemoenzymatic reactions have emerged in recent years [1]. These transformations can occur sequentially or concurrently. In the former case, chemical and biocatalytic reactions are carried out in sequence, and will be referred to as "one-pot sequential chemoenzymatic reactions" in this book. The advances in this area will be discussed in this chapter. The "one-pot concurrent chemoenzymatic reactions," in which the chemical and biocatalytic reactions occur simultaneously, will be presented in the following chapters. For the sequential one-pot chemoenzymatic process, chemical and biological transformations can be operated under different conditions by changing pH and/or temperature and adding additional solvents and/or reagents between the reaction steps. The presentation will be organized by enzyme types.

3.1 Lipases and Esterases

In addition to their applications in numerous industrial processes including oils and fats, detergents, food, and leather and paper processing, lipases and esterases are widely used in organic synthesis by taking advantage of their ability to catalyze the chemo, regio, and/or stereoselective hydrolysis of carboxylic acid esters or the reverse reaction in organic solvents. The lipase- or esterase-catalyzed reactions

Chemo-Enzymatic Cascade Reactions, First Edition. Dunming Zhu.

have been combined with various metal-catalyzed or organocatalytic reactions in a "one-pot" sequential manner.

The recombinant pig liver esterases (PLEs) catalyzed the selective hydrolysis of one of two methyl ester groups of dimethyl cyclohex-4-ene-cis-1,2-dicarboxylate. The desymmetrization generated (1*S*, 2*R*)-1-(methoxycarbonyl)cyclohex-4-ene-2-carboxylic acid. The *meso*-diester was synthesized from the corresponding diacid or *meso*-anhydride. Esterification of the economically favored *meso*-anhydride was combined with the highly selective PLE isozyme 6 (ECS-PLE06)-catalyzed desymmetrization reaction, and a chemoenzymatic process was developed for the "one-pot" production of (1*S*,2*R*)-1-(methoxycarbonyl)cyclohex-4-ene-2-carboxylic acid with 84% isolated yield and >99% ee (Scheme 3.1) [2].

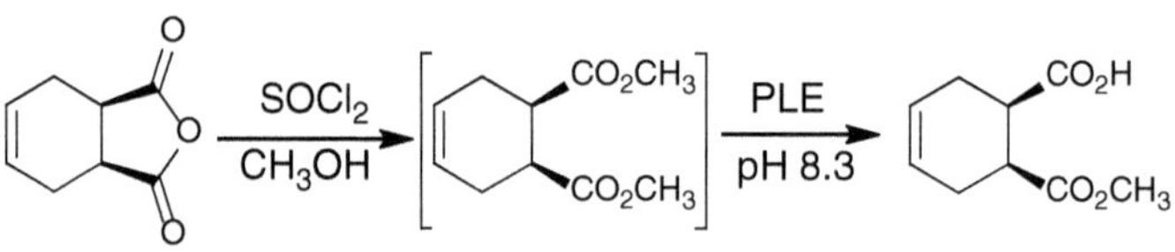

Scheme 3.1 Chemoenzymatic synthesis of (1*S*,2*R*)-1-(methoxycarbonyl)cyclohex-4-ene-2-carboxylic acid. Source: Adapted from Süss et al. [2].

The metal-catalyzed metathesis reaction of double olefin compounds is one of the most efficient and atom-economical synthetic methods that has found important applications on the industrial scale. As a result of intensive research efforts made to develop metathesis reactions in water, tailor-made water-soluble Grubbs catalysts became available. As such, development of new one-pot synthetic processes by combining metal-catalyzed olefin metathesis with a subsequent biotransformation in aqueous media became feasible. This was first demonstrated by the preparation of cyclic malonic acid monoesters in aqueous media based on the combination of a ruthenium-catalyzed metathesis reaction and subsequent biocatalytic hydrolysis with a PLE (Scheme 3.2). The reaction conditions were optimized to achieve low loading of the metal catalyst for the metathesis reaction and high compatibility of the esterase enzyme with the metal catalyst [3].

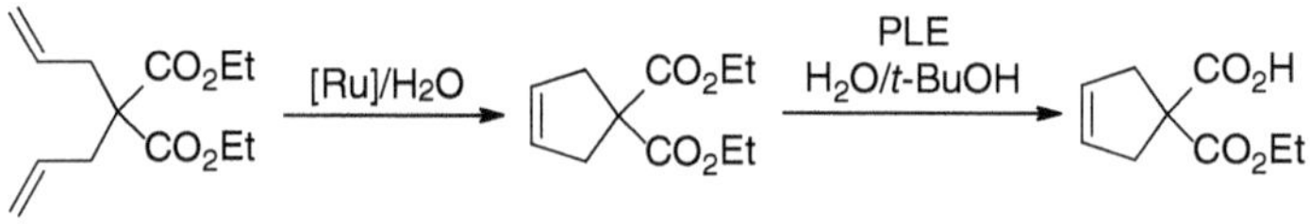

Scheme 3.2 One-pot sequential ruthenium-catalyzed metathesis and PLE-catalyzed hydrolysis.

Recently, the same research group encapsulated the Grubbs catalyst and PLE separately in octyl-grafted alginate amide and chitosan-coated calcium alginate beads, respectively. The encapsulated Grubbs catalyst and enzyme were evaluated individually in different reaction media. After selecting suitable solvents, the ring-closing metathesis reaction was conducted in water, and the subsequent enzymatic hydrolysis was carried out in HEPES buffer with *t*BuOH. The sequential one-pot chemoenzymatic reactions resulted in isolation of the product in 84% yield (Scheme 3.3). The encapsulated Grubbs catalyst and PLE enzyme in biopolymer-based hydrogels could be recycled, and maintained >80% conversion for seven runs [4].

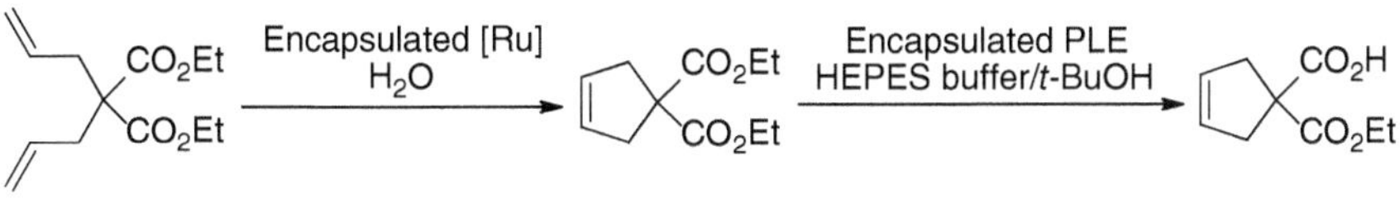

Scheme 3.3 One-pot sequential ring-closing metathesis and enzymatic hydrolysis by encapsulated Grubbs catalyst and PLE enzyme.

Candida antarctica lipase-B (CALB) has been found to catalyze the hydrolysis of (*S*)-*O*-acetylcyanohydrins, leaving the (*R*)-enantiomer unreacted. A bimetallic chiral (salen)titanium complex was used to convert aldehydes into *O*-acetylcyanohydrins with (*R*)-enantiomer as the major product (61–93% ee). The lipase-catalyzed resolution was employed to remove the unwanted (*S*)-enantiomer. The two reactions were carried out in a one-pot sequential mode, affording (*R*)-*O*-acetylcyanohydrins in 75–96% overall yields and 80 to >99% ee (Scheme 3.4). A chemoenzymatic one-pot sequential process has thus been developed for the preparation of (*R*)-*O*-acetylcyanohydrins from the corresponding aldehydes [5].

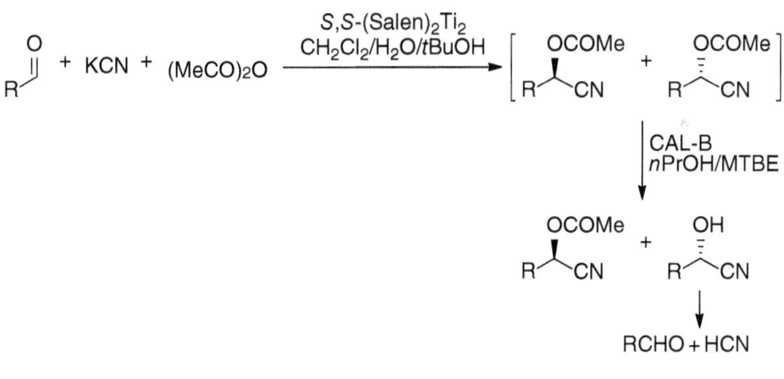

Scheme 3.4 One-pot chemoenzymatic synthesis of (*R*)-*O*-acetylcyanohydrins from aldehydes.

The chemical transformation of aldehydes to the *O*-acetylcyanohydrins and the subsequent lipase-catalyzed kinetic resolution were also carried out in the ionic liquid reaction media. Optically active (*R*)-*O*-acetylcyanohydrins were synthesized in 73–98% ee by the kinetic resolution of the *in situ* generated racemic *O*-acetyl cyanohydrins using CALB as biocatalyst. The *in situ* generated racemic cyanohydrins were converted into the corresponding (*S*)-*O*-acetylcyanohydrins in 60–92% ee by utilizing Amano Lipase PS from *Pseudomonas cepacia* and vinyl acetate as the acyl donor (Scheme 3.5) [6].

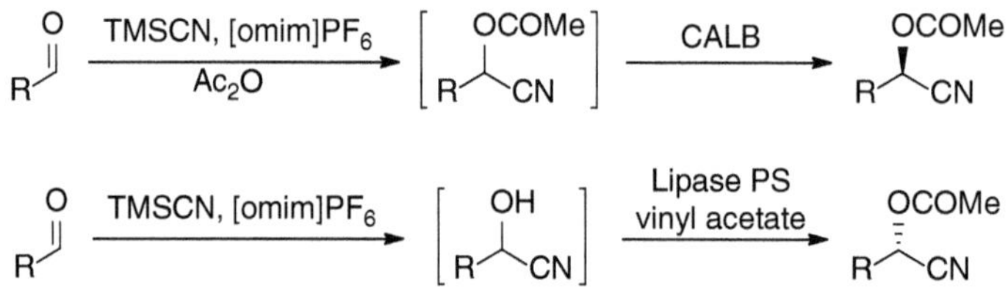

Scheme 3.5 One-pot chemoenzymatic synthesis of *O*-acetylcyanohydrins from aldehydes in ionic liquid medium. Source: Adapted from Shen et al. [6].

In the above cases, the esters are prepared by chemical methods, followed by esterase- or lipase-catalyzed regio- or stereoselective hydrolysis, affording the desired products. In addition, the enzymatic hydrolysis of esters has also been combined with the chemical transformations of the resulting alcohols or acids. The Cu(I)/bipyridine-mediated oxidation of alcohols into aldehydes is usually carried out in organic solvents, and a small amount of water can result in a drastic reduction of product yield. Recently, this reaction was performed in an aqueous medium by site-isolation of the Cu(I)/bipyridine catalyst in micelles prepared from polymeric amphiphiles. This offered a possibility of combining enzymatic ester hydrolysis with the micellar catalytic oxidation in a one-pot process to convert esters into the aldehydes, thus enabling a milder and greener process to replace the traditional methods using harsh reaction conditions. However, this simple micellar site-isolation of the Cu(I)/bipyridine catalyst was not effective enough, since only 9% benzaldehyde was generated when benzyl acetate was tested as a model substrate. Benzyl acetate was quantitatively hydrolyzed to give benzyl alcohol, suggesting inhibition of the Cu(I)/bipyridine catalyst in the presence of the enzyme, which was most likely due to the dynamic nature of micellar aggregates. Therefore, the compartmentalization of Cu/bipyridine catalyst in a core–shell like nanoparticle was used to solve this problem. The Cu/bipyridine functionalized nanoparticles were successfully combined with CALB in a one-pot two-step process to convert various acetate ester substrates to the corresponding aldehydes under aerobic conditions in aqueous media (Scheme 3.6) [7].

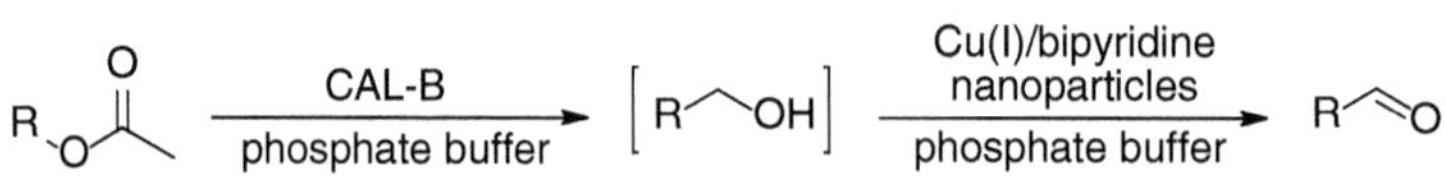

Scheme 3.6 One-pot chemoenzymatic transformation of acetate esters to aldehydes. Source: Based on Sand and Weberskirch et al. [7].

The chemoenzymatic dynamic kinetic resolution of secondary alcohols utilizing lipase in combination with homogeneous ruthenium-based racemization catalysts has emerged as a powerful tool for the synthesis of enantiomerically pure alcohols. The racemic alcohols can be prepared by metal-catalyzed hydrogenation of the corresponding ketones. For example, the regioselective hydrogenation of 1,2-indanedione catalyzed by Pd/Al_2O_3 catalyst gave *rac*-2-hydroxy-1-indanone, from which (*R*)-2-acetoxy-1-indanol was produced by chemoenzymatic dynamic kinetic resolution by using $Ru(OH)_3/Al_2O_3$ as the racemization catalyst in combination with lipase AK. Furthermore, a "one-pot" process was developed by combining the hydrogenation and dynamic kinetic resolution, and the readily available 1,2-indanedione was converted to a useful chiral building block with high regioselectivity and moderate stereoselectivity (Scheme 3.7) [8].

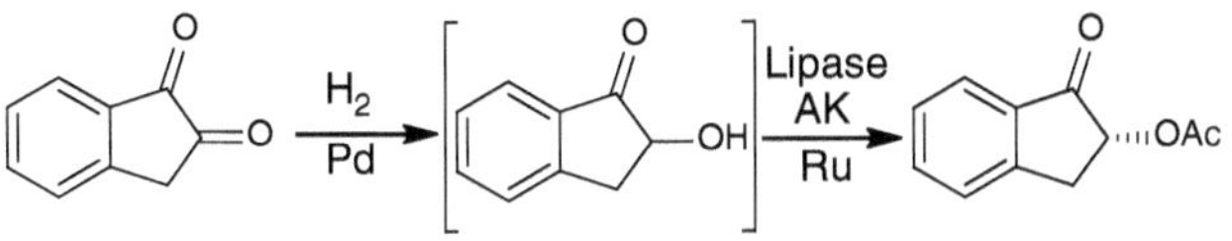

Scheme 3.7 One-pot chemoenzymatic hydrogenation and dynamic kinetic resolution. Source: Based on Langvik et al. [8].

Under carefully controlled (Mitsunobu) reaction conditions, (*R*)-1,5-undecadien-3-ol was transformed into (*S*)-dictyoprolene. This chemical stereoinversion combined with lipase-catalyzed kinetic resolution of racemic 1,5-undecadien-3-ol resulted in (*S*)-dictyoprolene, subsequent hydrolysis of which produced (*S*)-1,5-undecadien-3-ol in 96% yield and 90% ee (Scheme 3.8). A single enantiomer of this allylic-homoallylic alcohol was prepared in less than 50% yield by a lipase-catalyzed dynamic kinetic resolution due to its incompatibility with Ru-complexes usually used for chiral alcohol racemization [9].

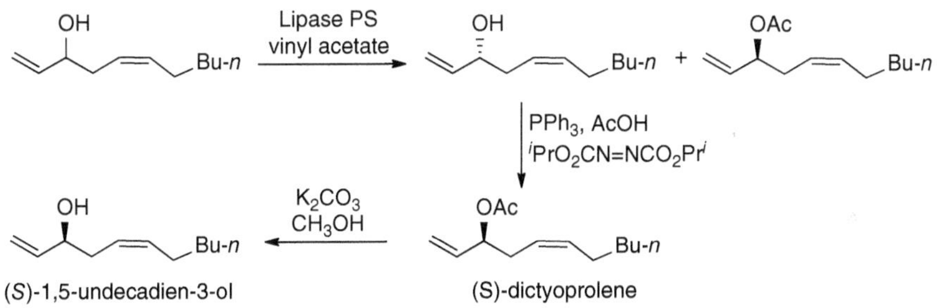

Scheme 3.8 One-pot chemoenzymatic preparation of (*S*)-dictyoprolene and (*S*)-1,5-undecadien-3-ol.

Lipases also catalyze amidation reactions such as the CALB-assisted amidation of carboxylic methyl esters with propargyl amine. The resulting propargyl amides reacted consecutively with azide via a Cu(I)-catalyzed click reaction, furnishing 1,2,3-triazoles in 51–85% yields. The enzymatic amidation and Cu(I)-catalyzed click reaction were carried out in a one-pot sequential manner (Scheme 3.9) [10].

In addition, the researchers carried out lipase-catalyzed amidation of methyl carboxylates and Sonogashira coupling in a consecutive one-pot manner. CALB catalyzed the amidation of carboxylic methyl esters with propargyl amine, and the resulting propargyl amides were reacted with (hetero)aryliodides in the presence of catalytic amounts of $Pd(PPh_3)_4$ and CuI, affording the corresponding arylated

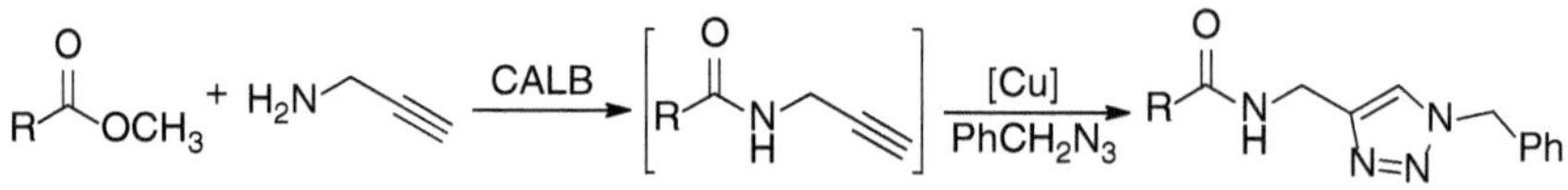

Scheme 3.9 One-pot chemoenzymatic synthesis of 1,2,3-triazole derivatives via sequential enzymatic amidation and click reaction. Source: Based on Hassan et al. [10].

propargyl amides (Scheme 3.10). This one-pot chemoenzymatic cascade was further concatenated with Cu-catalyzed click reaction to synthesize 1,4-disubstituted 1,2,3-triazole with arylated propargyl amides, as shown in Scheme 3.11 [11].

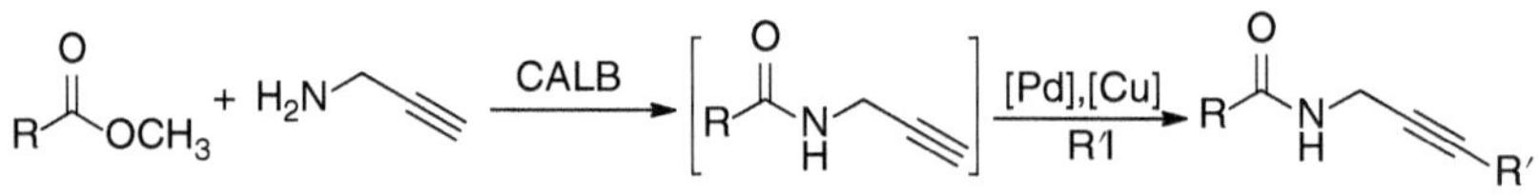

Scheme 3.10 One-pot sequential lipase-catalyzed amidation and Sonogashira coupling reaction.

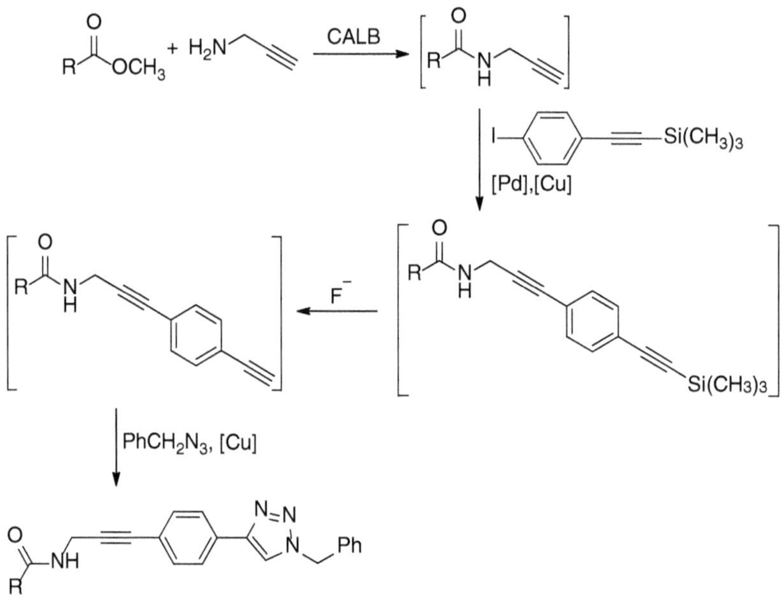

Scheme 3.11 One-pot chemoenzymatic synthesis of 1,4-disubstituted 1,2,3-triazole with arylated propargyl amides. Source: Based on Hassan et al. [11].

The Leuckart reaction converts aldehydes or ketones to amines by reductive amination using ammonium formate as amino donor and hydrogen source. Recently, metal-catalyzed Leuckart reactions have attracted increasing attention and some progress has been made to address the chemoselectivity issues of this reaction. In this context, after screening various palladium catalysts using vanillin as the model substrate and solid ammonium formate as the amino donor under different reaction conditions, Palo-Nieto et al. discovered that the reductive amination of vanillin in the presence of palladium(0)-aminopropyl-mesocellular foam (Pd^0-AmP-MCF, 5 mol%) in toluene at 80 °C proceeded smoothly with >99% conversion to afford vanillylamine and 4-hydroxy-3-methoxytoluene in the ratio of 94 : 6, without formation of vanillylalcohol. Vanillylamine was isolated in 87% yield. A set of aldehydes with diverse structures were reductively aminated under these conditions to furnish the desired amines in good to excellent isolated yields. The highly chemoselective heterogeneous Pd-catalyzed reductive amination of aldehydes was then coupled with lipase-catalyzed amide formation to synthesize a variety of amides such as nonivamide and capsaicin. After the reductive amination of vanillin was completed, addition of nonanoic acid and the commercially available CALB immobilized on a macroporous resin into the reaction mixture resulted in the isolation of nonivamide in 74 yield (Scheme 3.12). Similarly, capsaicin and phenylcapsaicin were synthesized in 73% and 78% overall yield, respectively, by using the corresponding acids in the enzymatic amidation step. This one-pot sequential process accepted aromatic, heterocyclic, and aliphatic aldehydes as well as various acids to produce a variety of amides in good overall yields; the reaction scope seemed to be determined by the substrate specificity of the lipase CALB [12].

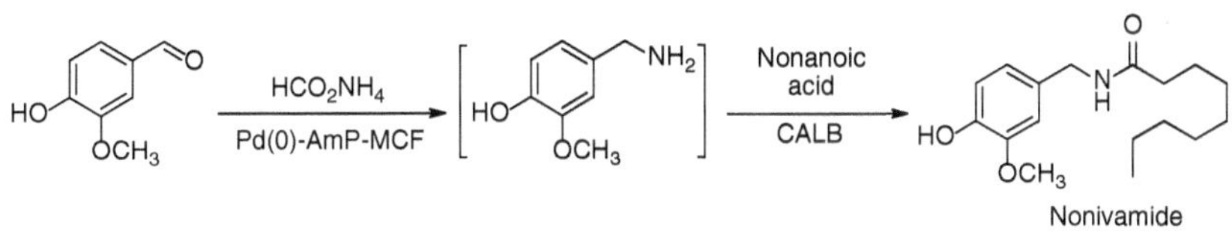

Scheme 3.12 One-pot chemoenzymatic coupling of Leuckart reaction and lipase-catalyzed amidation.

However, the heterogeneously Pd(0)-Amp-MCF-catalyzed reductive amination of acetophenone with ammonium formate in toluene generated 1-phenylethanol as the major product. Further studies showed that the reductive amination of acetophenone occurred in methanol with a decreased catalyst loading, and 1-phenylethylamine and 1-phenylethanol were obtained in the ratio of 88 : 12. Several aromatic ketones and cyclohexanone were examined under the same reaction conditions and the corresponding amination products were produced in even higher ratio related to the reduction product alcohols. The subsequent lipase-catalyzed amidation with ethyl methoxyacetate as acyl donor gave the (*R*)-enantiomer of the corresponding amide product in 59% yield and 97% ee. It has been known that Pd-nanoparticles could be used in combination with lipase for the dynamic kinetic resolution of secondary amines. As such, after the reductive amination of ketones was finished, the Pd catalyst loading was increased and H_2 was added in the enzymatic amide formation step. The one-pot reaction sequences produced the (*R*)-configurated amides in good overall yields with high ees from the corresponding ketones (Scheme 3.13) [12].

The palladium(0)-aminopropyl-controlled pore glass (Pd(0)-AmP-CPG) catalyzed the oxidation of alcohols to aldehydes under similar reaction conditions. It was combined with the above chemoenzymatic sequential reaction to convert alcohols to

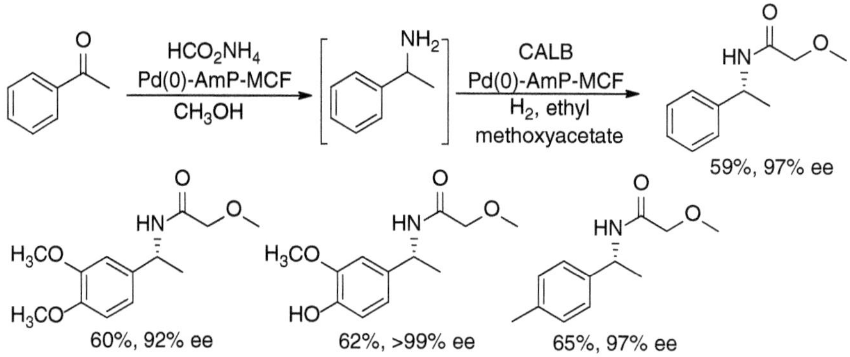

Scheme 3.13 One-pot chemoenzymatic synthesis of (*R*)-configurated amides. Source: Based on Palo-Nieto et al. [12].

the corresponding amides. By using this multicatalyst system benzyl alcohol was transformed into the corresponding nonivamide in one-pot with 49% overall yield (Scheme 3.14).

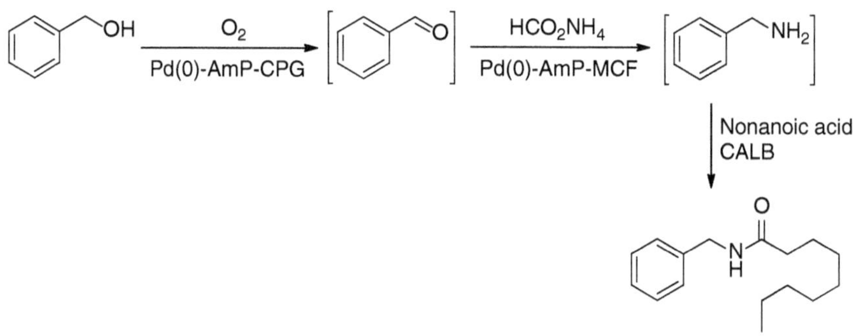

Nonivamide

Scheme 3.14 One-pot chemoenzymatic conversion of benzyl alcohol to nonivamide.

Lipase-catalyzed aminolysis of esters had also been coupled with chemical transformation in one-pot processes. The aza-Michael addition of benzylamine to enoates generated the corresponding β-amino esters, which could be enantioselectively aminolyzed by using lipase as the catalyst, resulting in the resolution of the racemic esters to furnish optically active β-amino esters and β-amino amides. The aza-Michael addition and enantioselective aminolysis were carried out in a one-pot sequential process under solvent-free conditions, affording the β-amino esters with excellent enantioselectivities of up to 99% ee (Scheme 3.15) [13, 14]. This offered a practical, highly enantioselective method for the synthesis of short-chain aliphatic β-amino acid esters from easily accessible substrates and demonstrated the continuously operating production of ethyl (*S*)-3-(benzylamino)butanoate in a packed-bed reactor [15].

The Ugi four-component reaction (U-4CR) between an aldehyde, an amine, a carboxylic acid, and an isocyanide offered a highly efficient synthetic strategy

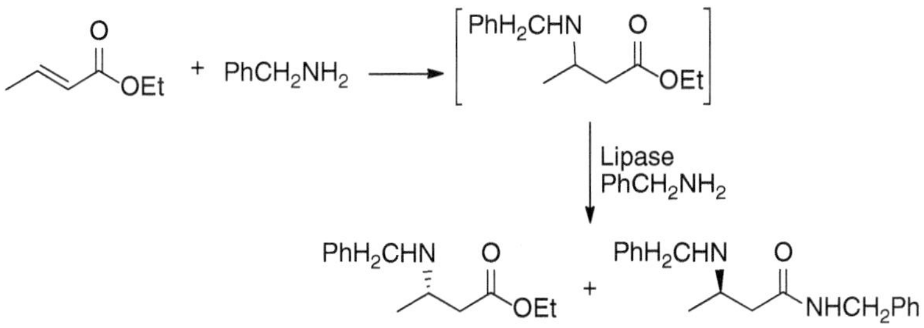

Scheme 3.15 One-pot chemoenzymatic transformation involving aza-Michael addition and enzymatic aminolysis. Source: Weiß et al. [13].

for the preparation of α-aminoacyl amide derivatives in a minimum of reaction steps. The U-4CR products constitute peptidomimetics with structural diversity and potential pharmaceutical applications. Furthermore, they also serve as substrates for subsequent reactions, allowing for further diversity oriented structural modification that generates compound libraries for screening lead structures in pharmaceutical and medicinal chemistry. In this context, the U-4CR reaction was concatenated with lipase CALB-catalyzed aminolysis in a one-pot manner. Benzaldehyde, benzylamine, acetic acid, and methyl 2-isocyanoacetate reacted in methanol, furnishing the U-4CR product in 94% yield. Lipase CALB catalyzed the aminolysis of the U-4CR product with benzylamine and propargylamine in methyl *tert*-butyl ether, affording the corresponding benzyl amide and propargyl amide (Scheme 3.16). The CALB-catalyzed aminolysis reached complete conversion in methanol after 24 hours. The U-4CR reaction and lipase CALB-catalyzed aminolysis were then carried out sequentially in a one-pot manner (Scheme 3.17), and the corresponding amides were prepared in good yield [16].

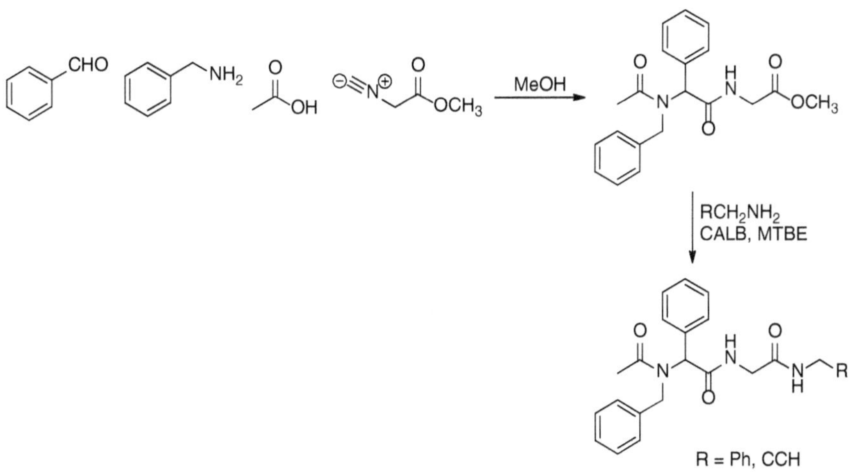

Scheme 3.16 U-4CR reaction and lipase CALB-catalyzed aminolysis.

In the case of propargylamine, the copper-catalyzed alkyne–azide cycloaddition could be integrated into this one-pot cascade. Indeed, addition of substituted

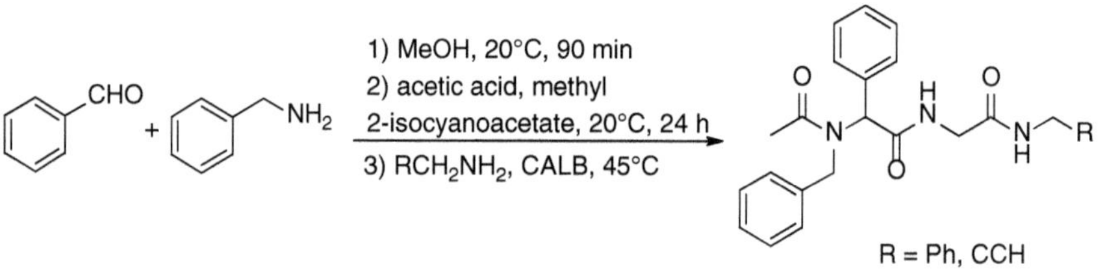

Scheme 3.17 One-pot sequential U-4CR reaction and lipase CALB-catalyzed aminolysis.

benzylazide, Cu_2O, and benzoic acid into the U-4CR-CALB-catalyzed aminolysis reaction mixture generated the corresponding triazoles in high yields (Scheme 3.18). When iodobenzyl amine was used in the CALB-catalyzed aminolysis reaction, the iodo functional group allowed for further concatenation of Suzuki cross-coupling into the U-4CR-CALB-catalyzed aminolysis sequence. A variety of triamide functionalized biaryl scaffolds were synthesized in 41–60% yields by this consecutive one-pot process (Scheme 3.19).

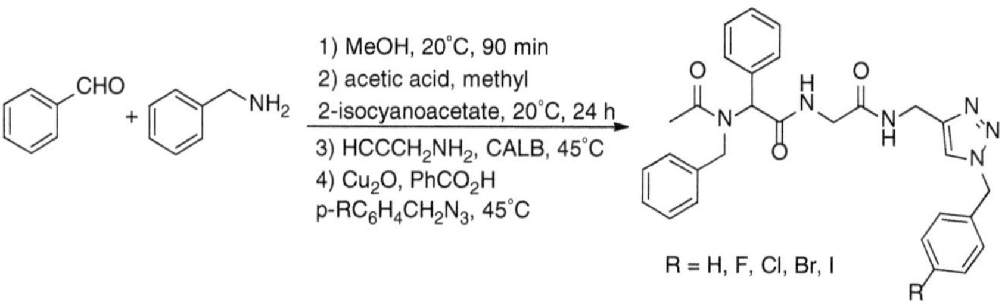

Scheme 3.18 One-pot sequential U-4CR reaction, lipase CALB-catalyzed aminolysis, and copper-catalyzed alkyne–azide cycloaddition.

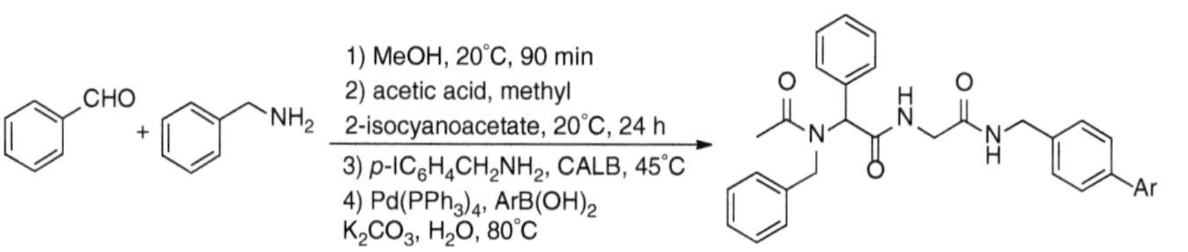

Scheme 3.19 One-pot sequential U-4CR reaction, lipase CALB-catalyzed aminolysis, and Pd-catalyzed Suzuki cross-coupling reaction.

3.2 Carbonyl Reductases

Carbonyl reductases (alcohol dehydrogenases) catalyze the asymmetric reduction of ketones to furnish the chiral alcohol. This enzymatic reduction has been widely integrated into the synthesis of high-value chemicals. The biocatalytic reduction is combined with various kinds of metal- or organocatalytic reactions and the chemoenzymatic transformations can be carried out in one-pot manner.

Asymmetric aldol reaction is one of the most powerful synthetic tools for carbon–carbon bond-forming reactions, providing a beneficial route to access chiral β-hydroxy carbonyl compounds, which are versatile synthetic motifs found in biologically active natural products and pharmaceutically attractive intermediates. β-Hydroxy ketones can be asymmetrically reduced to afford chiral 1,3-diols. As such, the combination of asymmetric aldol reaction and biocatalytic reduction leads to the sequential construction of both stereogenic centers of 1,3-diols, which are also important building blocks in the synthesis of pharmaceutically active compounds and chiral ligands for asymmetric synthesis in addition to a structural moiety in natural products. This chemoenzymatic synthetic approach made it possible to efficiently access all four stereoisomers in enantiomerically pure forms of 1,3-diols from simple starting materials. Since the end of last century organocatalytic aldol reactions have been extensively explored, and the reaction can be carried out under reaction conditions that are compatible with a direct subsequent enzymatic reduction without the need for a workup of the aldol reaction. A two-step one-pot process was successfully established for the synthesis of enantiomerically pure 1,3-diols by combining the asymmetric organocatalytic aldol reaction with biocatalytic reduction. The organocatalysts (*S*,*S*)-**1** and (*R*,*R*)-**1** catalyzed the aldol addition of *para*-chlorobenzaldehyde with acetone to give the corresponding *R*- or *S*-configurated β-hydroxyketone with 82–83% ee, respectively. The *R*-configurated β-hydroxyketone was subsequently reduced with an *S*-selective or *R*-selective alcohol dehydrogenase to generate the (1*R*,3*S*) and (1*R*,3*R*) diastereomers of 1,3-diol product, respectively. The (1*S*,3*S*) and (1*S*,3*R*) diastereomers were prepared by the reduction of *S*-configurated β-hydroxyketone catalyzed by an *S*-selective or *R*-selective alcohol dehydrogenase. The (*S*,*S*)-**1**-catalyzed aldol addition of *para*-chlorobenzaldehyde with acetone and the subsequent reduction of the aldol adduct with an alcohol dehydrogenase from *Lactobacillus kefir* were carried out in one-pot, and the desired 1,3-diol was obtained in (1*R*,3*S*) configuration with a conversion of 80% over two steps, a syn/anti ratio of 1 : 10, and >99% ee (Scheme 3.20) [17].

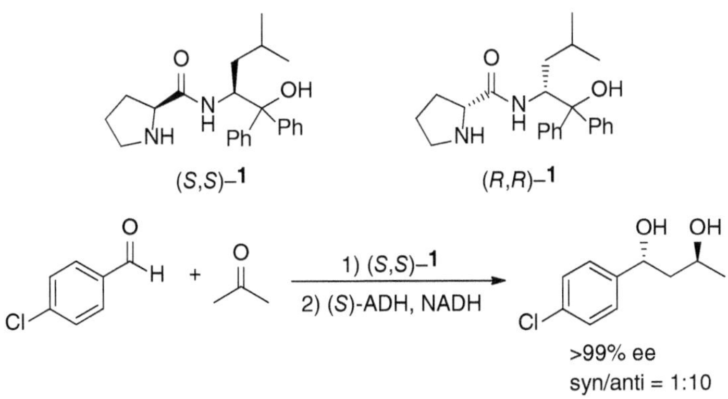

Scheme 3.20 One-pot chemoenzymatic synthesis of chiral 1,3-diols. Source: Based on Baer et al. [17].

When *meta*-chlorobenzaldehyde was used as the substrate, the one-pot synthesis of this type of 1,3-diols with *meta*-chloro-substituent was realized with diastereomeric ratios of >25 : 1 and excellent enantioselectivities of 99% ee [18]. Furthermore, the diastereomeric ratios of this 1,3-diol was improved to >35 : 1 when the one-pot transformation was performed in cyclohexane and the bioreduction was catalyzed by the superabsorber-based co-immobilized alcohol dehydrogenase (RsADH) from *Rhodococcus* sp. with cofactor NAD^+. This chemoenzymatic synthesis of chiral 1,3-diols was also achieved in a "one-pot like" process in organic media based on immobilized proline-derivative and RsADH/NAD^+. The proline-derivative-catalyzed aldol reaction and subsequent reduction of the aldol adduct were carried out in two separate compartments without isolating the intermediate, affording the chiral 1,3-diol in 89% overall yield with >99% ee and >35 : 1 dr (Scheme 3.21) [19]. A suitable "process window" was identified for the combination of an asymmetric organocatalytic aldol reaction and subsequent biocatalytic reduction in aqueous medium, which thus enabled the enantio- and diastereoselective synthesis of 1,3-diols in a tandem-type, one-pot process. A key feature of this one-pot synthesis was the high 500 mM loading of the aldehyde substrate as the starting material [20].

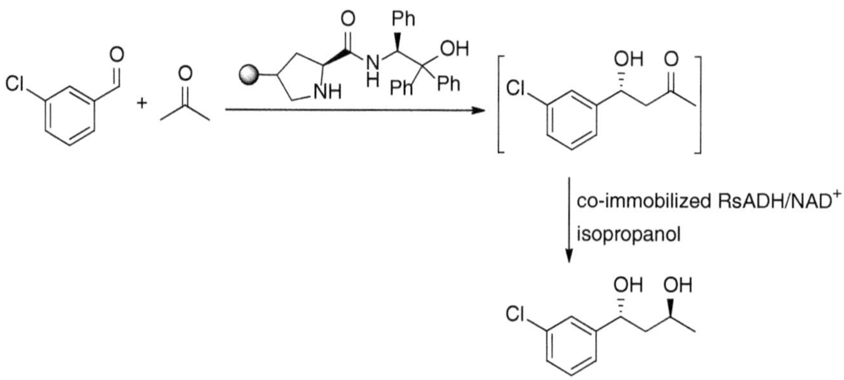

Scheme 3.21 "One-pot like" chemoenzymatic synthesis of chiral 1,3-diols. Source: Adapted from Heidlindemann et al. [19].

It had been reported that chiral Zn^{2+} complexes of L-prolylpendant [12]aneN_4 (L-ZnL^1) and L-valyl-pendant [12]aneN_4 (L-ZnL^2) catalyzed the enantioselective aldol reactions of acetone and some aldehydes in aqueous solution. The combination of this enantioselective aldol reactions and the enzymatic reduction of the aldol products was investigated. All four stereoisomers of some chiral 1,3-diols were produced via the one-pot chemoenzymatic synthesis in an aqueous solvent system at room temperature by selection of an appropriate chiral Zn^{2+}-complex and a commercially available oxidoreductase with the regeneration of the NADH cofactor (Scheme 3.22) [21].

Aldehydes reacted with amines to form aldimines; L/D-proline and derivatives catalyzed the asymmetric Mannich reaction of aldimines with acetone furnishing

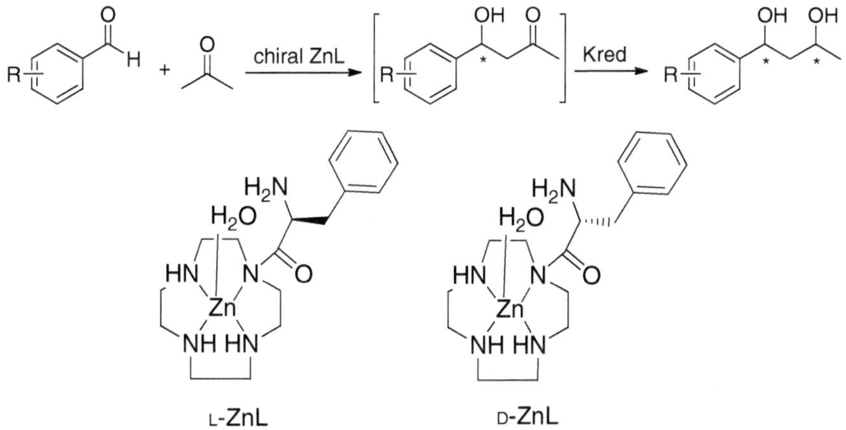

R	ZnL	Enzyme	Yield (%)	Product ratio 1*R*,3*R*	1*S*,3*R*	1*R*,3*S*	1*S*,3*S*
2-Cl	L-ZnL	E001	88	<1	<1	96	4
2-Cl	L-ZnL	E039	88	95	5	<1	<1
2-Cl	D-ZnL	E001	84	<1	<1	4	96
2-Cl	D-ZnL	E039	92	5	95	<1	<1
4-Cl	L-ZnL	E001	68	<1	<1	96	4
4-Cl	L-ZnL	E039	60	96	4	<1	<1
4-Cl	D-ZnL	E001	60	<1	<1	5	95
4-Cl	D-ZnL	E039	48	4	96	<1	<1

Scheme 3.22 One-pot chemoenzymatic synthesis of four diastereomers of 1,3-diols. Source: Based on Sonoike et al. [21].

enantiomerically enriched β-amino ketones. Glyoxylic acid ethyl ester formed an aldimine with *para*-methoxyaniline. The resulting PMP (*para*-methoxyphenyl)-protected aldimine reacted with acetone to give (*R*)- or (*S*)-enantiomer of ethyl γ-oxo-α-(PMP-protected-amino)pentanoate, which was reduced by (*S*)-selective or (*R*)-selective alcohol dehydrogenases to generate all four diastereomers of ethyl γ-hydroxy-α-(PMP-protected-amino)pentanoate. The three reaction steps were integrated in a one-pot process, using 2-PrOH as the solvent and reducing agent. The resulting product was then converted to PMP-protected α-amino-γ-butyrolactone via a transesterification–lactonization cascade. In one case, the whole cascade transformation afforded the (3*S*,5*R*)-configured lactone in 47% yield with a syn/anti ratio of 86/14 and >99% ee (Scheme 3.23) [22].

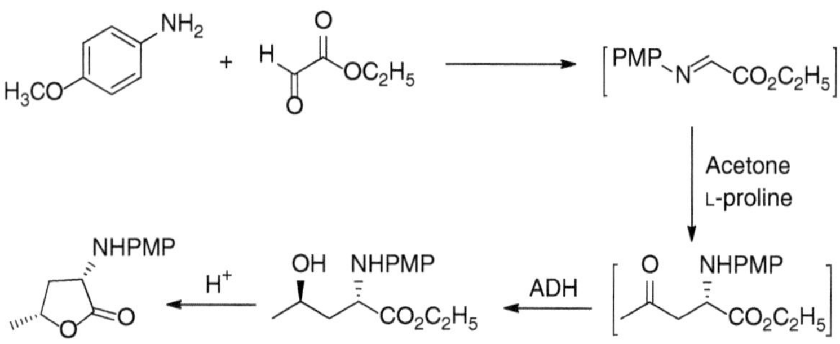

Scheme 3.23 Chemoenzymatic synthesis of PMP-protected α-amino-γ-butyrolactone. Source: Based on Simon et al. [22].

Palladium-catalyzed cross-coupling reactions are one of the important C–C bond formation tools in synthetic organic chemistry. It is known that palladium complex $Pd(PPh_3)_2Cl_2$ catalyzes the Suzuki cross-coupling of phenylboronic acid with bromoacetophenone to give the biaryl ketone in aqueous media. Gröger et al. reported the first studies on the combination of this Suzuki cross-coupling reaction with enzymatic reduction. Several chiral biaryl-substituted ethanols were prepared with high conversion and excellent ee by reducing the *in situ* formed biaryl ketones in a one-pot manner using an alcohol dehydrogenase from *Rhodococcus* sp. as the biocatalyst (Scheme 3.24) [23]. Furthermore, the Pd-catalyzed Suzuki cross-coupling reaction of acetylphenylboronic acid with *para*-bromoacetophenone afforded biaryl-containing diketones. The asymmetric reduction of both ketone moieties proceeded smoothly by using the alcohol dehydrogenase from *Rhodococcus* sp., resulting in the (*S*,*S*)-diols with 99% yield, high diastereoselectivity (d.r. > 25 : 1), and 99% ee [24].

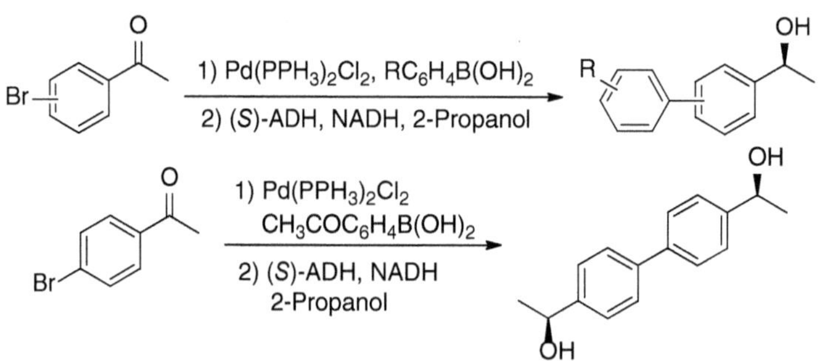

Scheme 3.24 One-pot sequential Suzuki cross-coupling and enzymatic reduction. Source: Based on Burda et al. [23].

This research team also developed a water-soluble palladium catalyst using ligand TPPTS (tris(3-sulfonatophenyl)phosphine hydrate, sodium salt containing 10–15% oxide), enabling the Suzuki cross-coupling reaction of 4′-iodoacetophenone with phenylboronic acid at room temperature in a reaction medium compatible with the biocatalyst. After reaction condition optimization, the cross-coupling reaction was carried out in an aqueous phase containing 50% of isopropanol (v/v) in the presence of an excess of sodium carbonate. As the coupling was completed, the pH was adjusted to 7, and the alcohol dehydrogenase from *Rhodococcus* sp. was added, resulting in the formation of the desired (*S*)-configurated alcohol with >95% yield and >99% ee [25].

The one-pot sequential palladium-catalyzed Suzuki cross-coupling and enzymatic reduction was also carried out in the medium containing deep eutectic solvents (DESs), and a series of chiral biaryl alcohols were prepared in good yields and excellent ee (>99 %) [26].

Palladium nanoparticles were prepared within the Dps protein (DNA binding protein from starved cells) from the thermophilic bacterium *Thermosynechoccus elongatus* (Te-Dps), and showed good catalytic activity in Suzuki–Miyaura cross-coupling reactions under aerobic conditions in water. This prompted the authors to investigate the combination of this palladium-catalyzed process with an asymmetric enzymatic reduction catalyzed by alcohol dehydrogenase (LBADH) from *Lactobacillus brevis*. As such, the chiral biaryl alcohols were prepared in high to excellent yields and an enantiomeric excess higher than 99% in a two-step one-pot process (Scheme 3.25) [27].

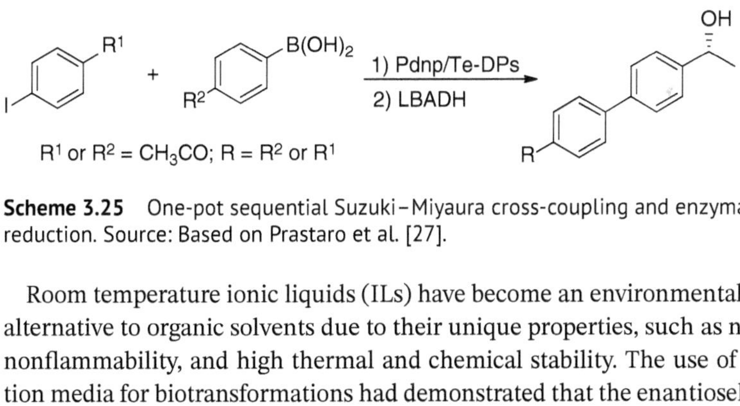

Scheme 3.25 One-pot sequential Suzuki–Miyaura cross-coupling and enzymatic reduction. Source: Based on Prastaro et al. [27].

Room temperature ionic liquids (ILs) have become an environmentally attractive alternative to organic solvents due to their unique properties, such as nonvolatility, nonflammability, and high thermal and chemical stability. The use of ILs as reaction media for biotransformations had demonstrated that the enantioselectivity and enzyme activity are generally equivalent to those observed in organic solvents. Using different ionic liquid media the palladium-catalyzed Suzuki coupling reaction and asymmetric enzyme reduction of ketones were studied and the two reactions were combined to achieve one-pot chemoenzymatic cascade in a biphasic solvent system containing an imidazolium-based IL and an aqueous phase. When the one-pot process was applied to different substrates in an aqueous medium containing 30% [bmim][NTf_2], a broad range of biaryl alcohols were prepared with high conversion and excellent enantioselectivity. *Escherichia coli* whole cells expressing an alcohol dehydrogenase from *Rhodococcus ruber* DSM 44541 was used as biocatalyst [28].

However, when this one-pot two-step process was applied to synthesize the chiral biaryl alcohol intermediate of Odanacatib, a potent orally active selective cathepsin K inhibitor for the treatment of postmenopausal osteoporosis, the reaction sequence had to be adjusted because the alcohol dehydrogenase was not active toward 4′-(4-methylsulfonylphenyl)-2,2,2-trifluoroacetophenone. Subsequently, 4′-bromo-2,2,2-trifluoroacetophenone was reduced using the *E. coli* whole cells expressing the alcohol dehydrogenase from *R. ruber* to produce (*R*)-1-(4-bromophenyl)-2,2,2-trifluoroethanol, which was then coupled with 4-(methanesulfonyl)phenylboronic acid via palladium-catalyzed Suzuki–Miyaura reaction to afford the desired biaryl alcohol intermediate with a total yield of 73% and enantiomeric excess of 99% (Scheme 3.26) [29].

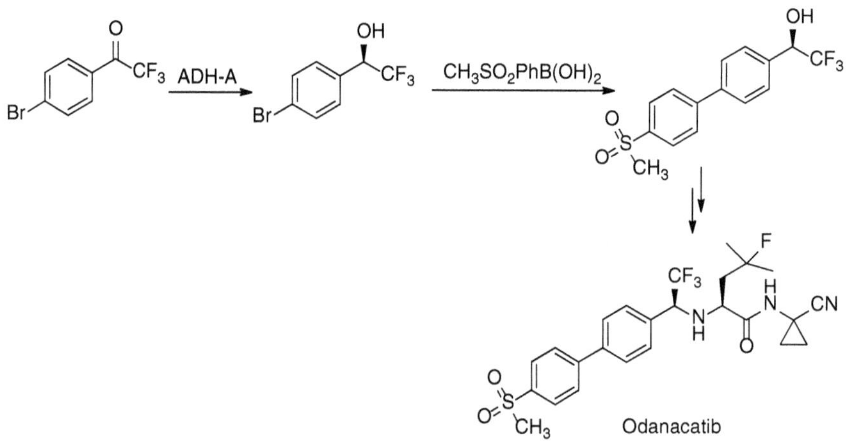

Scheme 3.26 One-pot sequential chemoenzymatic synthesis of the chiral biaryl alcohol intermediate of Odanacatib. Source: Based on de Oliveira Lopes et al. [29].

Recently, this one-pot reaction sequence was reversed to synthesize (*R*)-2,2,2-trifluoro-1-(4′-(methylsulfonyl)-[1,1′-biphenyl]-4-yl)ethanol by using an alcohol dehydrogenase (RasADH) from *Ralstonia* species. The palladium-catalyzed Suzuki cross-coupling of 1-(4-bromophenyl)-2,2,2-trifluoroethanone with 4-(methylsulfonyl)phenylboronic acid was conducted in aqueous medium. After the reaction was completed, the RasADH enzyme and cofactor regeneration system were added for the reduction of the resulting 2,2,2-trifluoro-1-(4′-(methylsulfonyl)-[1,1′-biphenyl]-4-yl)ethanone without isolation of the ketone. The enantiomerically pure (*R*)-2,2,2-trifluoro-1-(4′-(methylsulfonyl)-[1,1′-biphenyl]-4-yl)ethanol was obtained in 85 % yield with 128 g/ l/day productivity (Scheme 3.27) [30].

Another recent advance is the production of diaryl ketones by Suzuki–Miyaura coupling of aryl amides with aryl boronates in aqueous medium catalyzed by $Ni(cod)_2$. This C–C coupling reaction was then combined with an asymmetric biocatalytic reduction to access diarylmethanol derivatives, a class of common building blocks in stereoselective synthesis of bioactive molecules. The Suzuki–Miyaura coupling of aryl amides with aryl boronates was carried out in

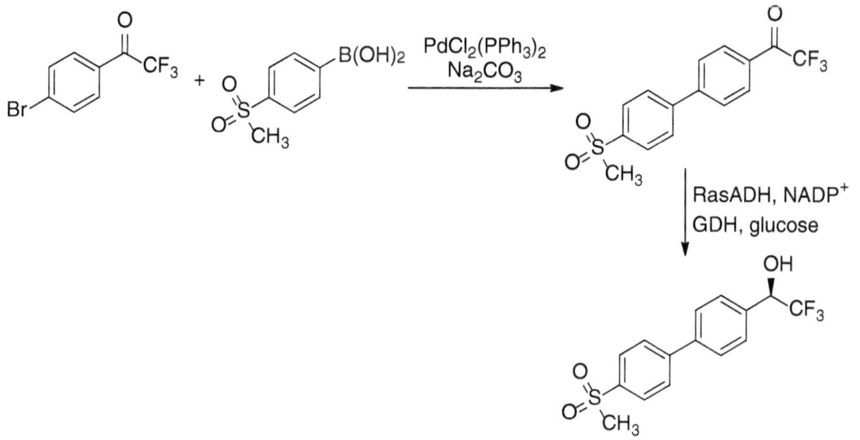

Scheme 3.27 One-pot sequential Pd-catalyzed Suzuki coupling and bioreduction. Source: Based on González-Martínez et al. [30].

aqueous medium at 60 °C for 24 hours, and after quenching with 1 M HCl, the enzymatic reduction was performed at 35 °C for 24 hours. A series of diarylmethanol derivatives were obtained in high yields and enantiomeric excesses (Scheme 3.28). Both enantiomers of the pharmaceutical orphenadrine were prepared by employing enantiocomplementary ketoreductases (Scheme 3.29) [31].

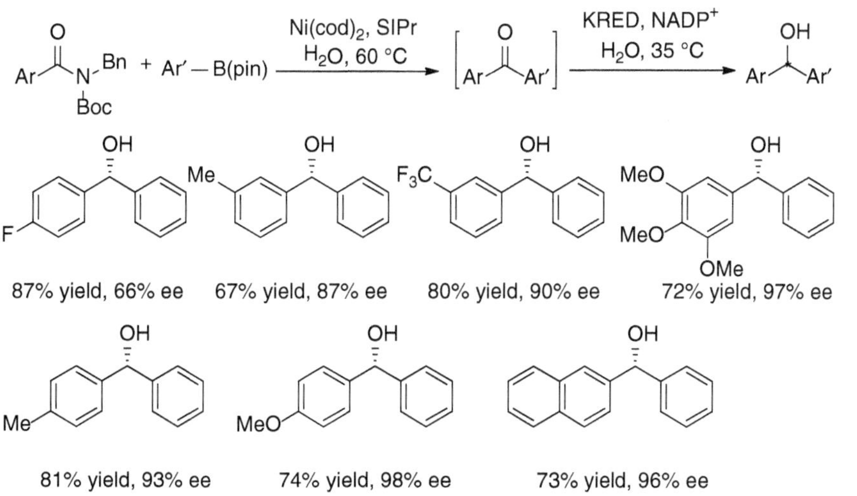

Scheme 3.28 One-pot sequential Ni-catalyzed Suzuki–Miyaura coupling and enzymatic ketone reduction.

The palladium-catalyzed Heck reaction of aryl halides with allylic alcohols is one of the most convenient synthetic protocols for the formation of 3-arylaldehydes and ketones, which can be reduced and serve as useful synthetic intermediates. The combination of palladium-catalyzed Heck reaction with enantioselective enzymatic

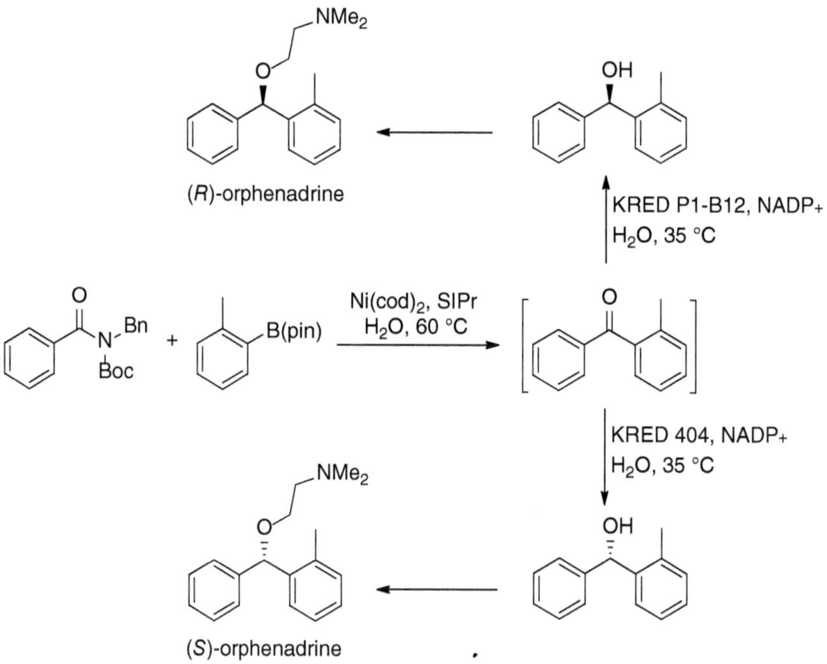

Scheme 3.29 One-pot chemoenzymatic synthesis of (*R*)- and (*S*)-orphenadrine. Source: Based on Dander et al. [31].

reduction was first demonstrated in the synthesis of (*R*)-(−)-rhododendrol. In the course of finding the compatible reaction conditions for the Heck reaction and the enzymatic reduction, phosphine-free perfluoro-tagged palladium nanoparticles, immobilized on fluorous silica gel both through fluorous–fluorous interactions and covalent bonding, have been successfully used in the Heck reaction of aryl iodides with allylic alcohols under aerobic phosphine-free conditions in water. Under these reaction conditions, *para*-iodophenol reacted with 3-buten-2-ol to give 4-(4-hydroxyphenyl)-butan-2-one, which was reduced with an alcohol dehydrogenase from *L. brevis* [(*R*)-LBADH], by adding the enzyme, NADPH, and 2-propanol into the reaction mixture of the first step. (*R*)-(−)-Rhododendrol was isolated in 90% yield and >99% ee (Scheme 3.30). The one-pot two-step process was successfully extended to the preparation of other chiral alcohols [32].

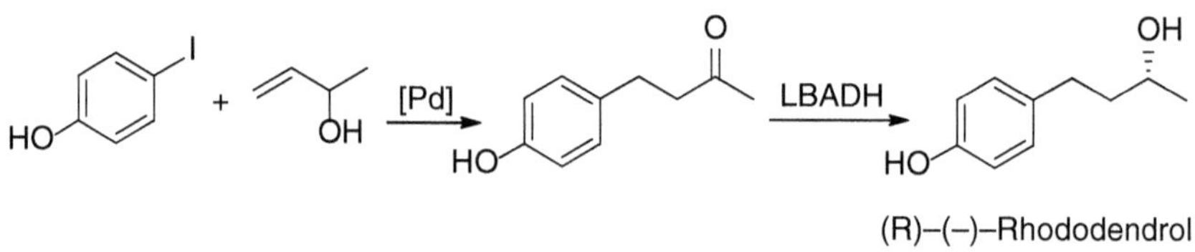

Scheme 3.30 One-pot sequential chemoenzymatic synthesis of (*R*)-(−)-rhododendro.

In the presence of $EtN(i\text{-}Pr)_2$, dimethylacetamide (DMA), and LiCl, $Pd(OAc)_2$ catalyzed the coupling of aryl iodides with acetic anhydride, affording the corresponding acetophenones. Combination of this Pd-catalyzed C–C formation and enzymatic reduction of the resulting ketones with the alcohol dehydrogenase (LBADH) from *L. brevis* was applied to prepare (*R*)-1-arylethanols in high to excellent yields (Scheme 3.31) [33].

Scheme 3.31 One-pot sequential chemoenzymatic synthesis of (*R*)-1-arylethanols.

Wittig reaction is a well-established methodology for C=C double bond formation in the construction of complex molecules. After the successful preparation of α,β-unsaturated esters and ketones by Wittig reaction in pure water was reported, various benzylidene acetone derivatives were also prepared with >95% conversion and a high E/Z ratio of >95 : 5 in a phosphate buffer/2-propanol [25% (v/v)] mixture. The *in situ*-formed α,β-unsaturated ketones were reduced by using alcohol dehydrogenases (ADHs) from *L. kefir* or *Rhodococcus* sp. as biocatalysts, producing the corresponding (*R*)- or (*S*)-enantiomeric alcohols with 31–90% conversions and >99% ee, respectively (Scheme 3.32) [34]. The enantioselective synthesis of hydrophobic allylic alcohols was achieved in a one-pot two-step process by combining a Wittig reaction with the bioreduction of the α,β-unsaturated ketone intermediates.

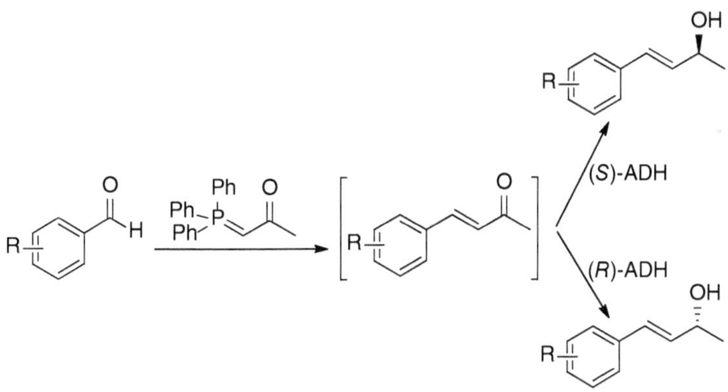

Scheme 3.32 One-pot sequential Wittig reaction and enzymatic reduction. Source: Adapted from Kraußer et al. [34].

Arylketones were brominated to generate α-bromoarylketones, which could be transformed into α-acetoxymethyl aryl ketones. Subsequent hydrolysis produced α-hydroxyarylketones that could be enantioselectively reduced to give (*R*)- or (*S*)-1-aryl-1,2-ethanediols. It was found that α-bromination of arylketones proceeded smoothly using polymer-bound pyridinium tribromide (PBPTB) as the brominating agent in acetonitrile, giving α-bromoarylketones in quantitative yields. Subsequent transformation of α-bromoarylketones was carried out with sodium acetate in a phase-transfer system with 18-crown-6 (18C6), furnishing α-acetoxymethyl aryl ketones. CALB catalyzed the ester transfer of α-acetoxymethyl aryl ketones with methanol to quantitatively yield the α-hydroxyarylketones, which was reduced with baker's yeast cells, affording the corresponding enantiomerically enriched (*R*)- or (*S*)-1-aryl-1,2-ethanediols (Scheme 3.33). These chemical and enzymatic transformations were undertaken in a one-pot sequential procedure, and a range of (*R*)- and (*S*)-1-aryl-1,2-ethanediols were prepared in 68–82% yields and 77–98% ee from the corresponding arylethanones [35].

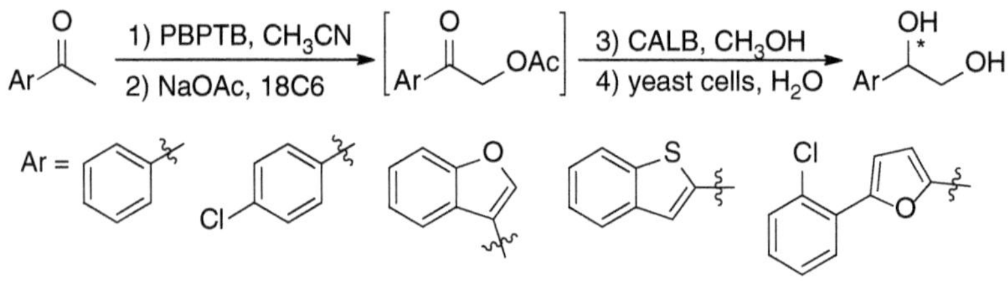

Scheme 3.33 One-pot chemoenzymatic synthesis of (*R*)- and (*S*)-1-aryl-1,2-ethanediols.

The direct asymmetric hydration of non-activated alkenes, such as styrene, into the corresponding secondary alcohols represented a great challenge in organic chemistry. Recently, Gröger et al. reported an indirect asymmetric hydration of styrene by incorporating the Wacker–Tsuji oxidation of olefins with the enzymatic reduction of the resulting ketones. After screening various reaction conditions, they found that the desired acetophenone could be formed with >99% conversion and 90% selectivity at a high substrate concentration of 1.3 M of styrene when a mixture of methanol and water (7 : 1, v/v) was used as solvent for the oxidation. Acetophenone was reduced using an alcohol dehydrogenase from *L. kefir* in the presence of 2-propanol (25% (v/v)) to give (*R*)-1-phenylethanol with 94% conversion and >99% ee. These two efficient separate steps were thus combined into a one-pot process, which corresponded to the desired "formal" hydration process. Detailed studies revealed that a palladium species formed in the Wacker–Tsuji oxidation in the aqueous phase exerted a negative influence on the biotransformation. Addition of 2 mol% 2,2′-bipyridine, thiourea, or ethylenediaminetetraacetic acid (EDTA) into the reaction mixture prevented the inhibiting effect of the palladium species on the enzymatic reduction; the desired (*R*)-1-phenylethanol was formed in the one-pot synthesis with excellent conversion of 92–93% (Scheme 3.34) [36].

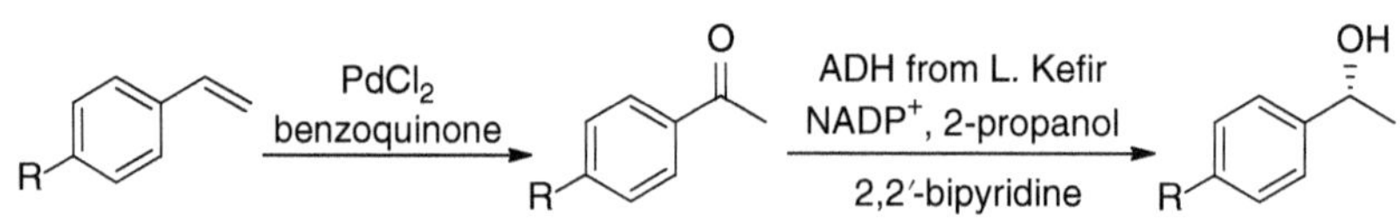

Scheme 3.34 Asymmetric hydration of styrenes via one-pot sequential Wacker–Tsuji oxidation and enzymatic reduction. Source: Adapted from Schnapperelle et al. [36].

In order to avoid the use of a stoichiometric amount of benzoquinone, Wacker oxidation could be performed with molecular oxygen as the oxidation agent, and palladium chloride and copper chloride as the suitable catalyst system. However, the Cu salts strongly inhibited the subsequent enzymatic reduction, thus making the combination of these two reactions into a one-pot tandem process inefficient. To circumvent this, a site-isolation method was adopted to separate $PdCl_2/CuCl$ from the ADH by charging the chemical catalyst in a polydimethylsiloxane (PDMS) thimble. Owing to the hydrophobic properties of the PDMS membrane, acetophenone formed *in situ* in the interior could flux through this membrane into the exterior to react with the enzyme, thus preventing the impact of copper ions on enzymatic reduction. A number of styrene analogs were smoothly transformed into (*R*)-1-phenylethanols with high overall conversion and 98–99% ee via this one-pot process using a (*R*)-selective alcohol dehydrogenase (LKADH) from *L. kefir* (Scheme 3.35). This represented a general approach to combining incompatible chemo and biocatalytic reactions toward one-pot processes [37].

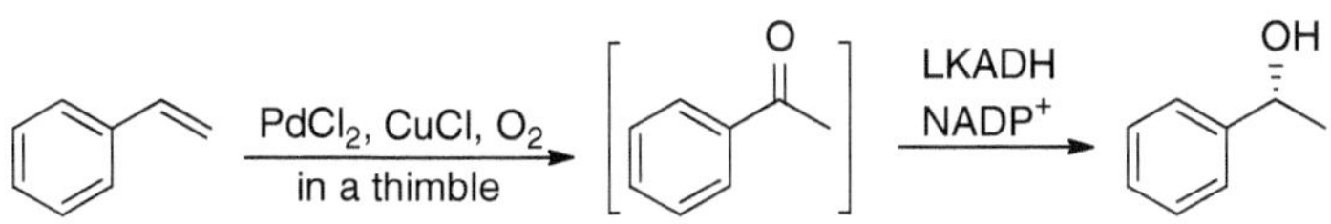

Scheme 3.35 Asymmetric hydration of styrenes via one-pot sequential oxidation and enzymatic reduction using a catalyst separation technique.

It is well known that a family of bis(allyl)-ruthenium(IV) complexes catalyze the redox isomerization of allylic alcohols into the corresponding saturated ketones in quantitative yields under mild reaction conditions in water. This offered a good opportunity for combining the metal-catalyzed isomerization with further biotransformations of the carbonyl functional group in one-pot synthetic process. Indeed, the bis(allyl)(acetate)Ru(IV) complex retained its high catalytic activity and selectivity in the isomerization of a set of allylic alcohols under the reaction conditions required by the ketoreductases. In addition, various racemic allylic alcohols were successfully transformed into the corresponding chiral alcohols with high conversion and ee values via the one-pot sequential process of the Ru(IV)-catalyzed redox isomerization of allylic alcohols and the bioreduction of the *in situ*-formed prochiral ketones with commercially available ketoreductases (Scheme 3.36) [38]. Similarly, allyl alcohol was converted to propanol via the Ru(II)-catalyzed olefin isomerization of allyl alcohol followed by reduction of propanal by horse-liver alcohol dehydrogenase [39].

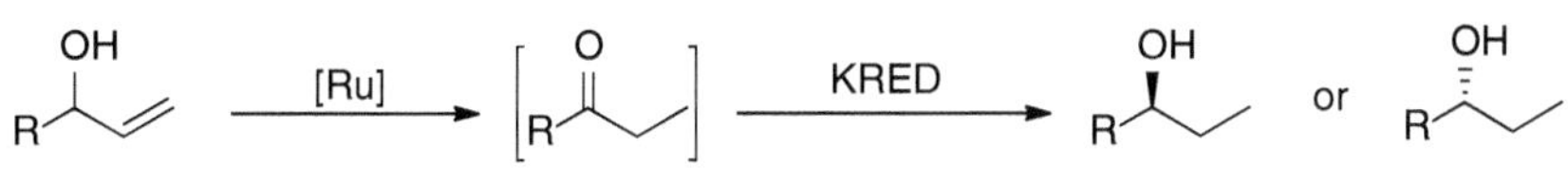

Scheme 3.36 One-pot sequential Ru-catalyzed redox isomerization of allylic alcohols and enzymatic reduction. Source: Adapted from Ríos-Lombardía et al. [38].

Palladium(II) complexes were reported to catalyze the cycloisomerization of alkynols such as 4-pentyn-1-ol and 3-pentyn-1-ol in water under aerobic conditions, followed by the concomitant nucleophilic attack of water to afford 5-hydroxy-2-pentanone. The hydroxyl ketone was reduced by commercially available ketoreductases (KREDs) to furnish the corresponding 1,4-pentanediol. The Pd-catalyzed isomerization and enzymatic reduction were performed in one-pot two-step mode, and both enantiomers of 1,4-pentanediol were obtained in good yields (up to 90%) and excellent ee values (up to >99%) (Scheme 3.37). This process was further coupled with the laccase-catalyzed oxidation of the resulting enantiopure 1,4-pentanediol to give γ-valerolactone in a one-pot three-step sequence [40].

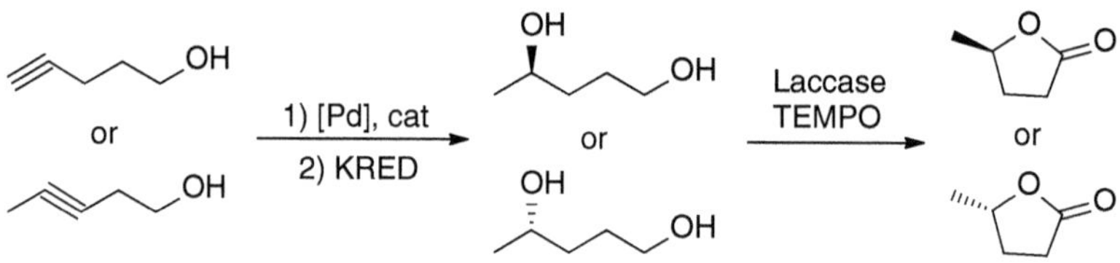

Scheme 3.37 One-pot chemoenzymatic conversion of alkynols to optically active γ-valerolactones.

After having examined various palladium and gold complexes, a Spanish research group has found that $KAuCl_4$ catalyzed the cycloisomerization of 4-pentynoic acid giving rise to the corresponding five-membered enol-lactones, which were hydrolyzed to levulinic acid in air and aqueous media. The resulting levulinic acid was subsequently reduced by KREDs without isolation or purification steps. Both enantiomers of γ-hydroxyvaleric acid were obtained with excellent ee values (up to >99%) in a one-pot procedure in aqueous media with 5% v/v of dimethyl sulfoxide (DMSO) by choosing the adequate KRED [40]. Interestingly, when the $KAuCl_4$-catalyzed cycloisomerization of 4-pentynoic acid was carried out in the presence of isopropanol, the subsequent KRED-mediated bioreduction of the resulting isopropyl levulinate gave rise to enantiopure γ-valerolactone instead of γ-hydroxyvaleric acid (Scheme 3.38).

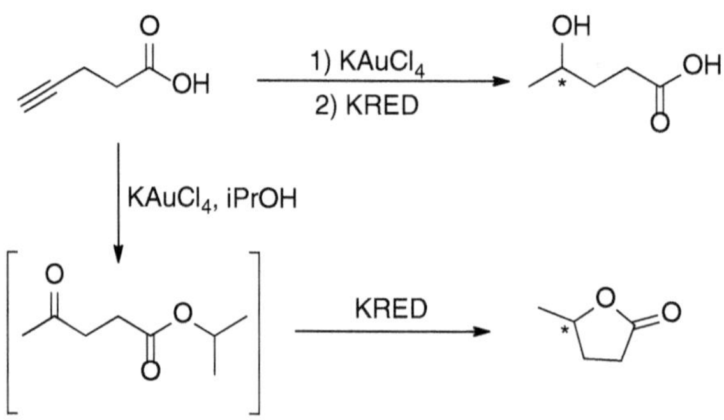

Scheme 3.38 One-pot sequential Au-catalyzed isomerization and enzymatic ketone reduction.

When the one-pot combination was extended to a metal-catalyzed cycloisomerization of alkynyl amides, $KAuCl_4$ was found to catalyze the cycloisomerization/hydrolysis of *N*-tosylpent-4-ynamide under the same conditions to afford the corresponding 4-oxo-*N*-tosylpentanamide, although longer time was required to achieve full conversion. However, for the bioreduction of the produced 4-oxo-*N*-tosylpentanamide, most of the tested KREDs led to low conversions and ee values up to 40%. This might be due to the bulky tosyl group, which exceeded the steric requirements of these ketoreductases. The $KAuCl_4$-catalyzed cycloisomerization of 4-pentynoic acid and concomitant aminolysis of the enol-lactone with aniline was then explored. As expected, the desired 4-oxo-*N*-phenylpentanamide was obtained in almost quantitative yield, and transformed into the corresponding γ-hydroxy amide by majority of the tested KREDs in the one-pot process (Scheme 3.39). By employing suitable KREDs, both enantiomers of γ-hydroxy amide were produced with ee values in the 99 to >99% range [40].

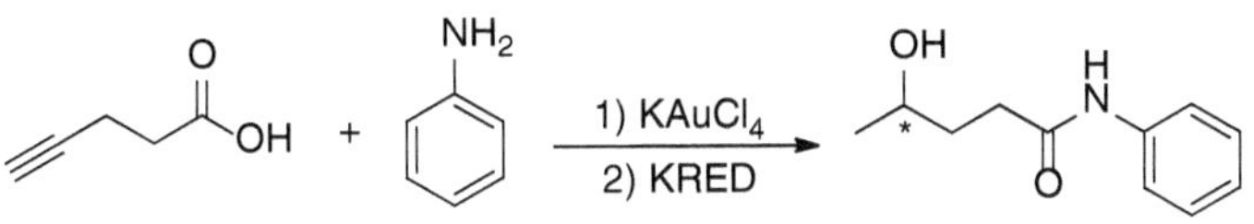

Scheme 3.39 One-pot chemoenzymatic synthesis of γ-hydroxy amides.

The hydration of aromatic terminal alkynes affords the acetophenone analogs, which could be reduced to the corresponding chiral alcohols. The combination of metal-catalyzed hydration and enzymatic reduction in a one-pot process was adopted to convert aromatic terminal alkynes to the corresponding (*S*)-enantiomer of phenyl-substituted 1-(phenyl)ethanols by using $AuCl_3$ and lyophilized cells of *E. coli* expressing an alcohol dehydrogenase gene from *L. kefir* (ADH-A) (Scheme 3.40) [41].

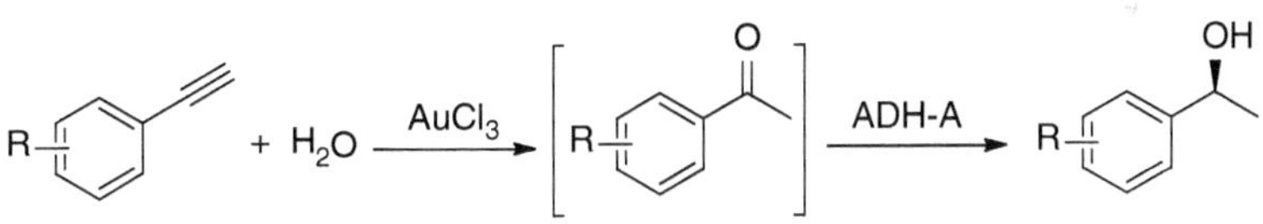

Scheme 3.40 One-pot chemoenzymatic transformation of aromatic terminal alkynes into substituted phenylethanols. Source: Adapted from Schaaf et al. [41].

The chemical oxidation of secondary alcohols using 2,2,6,6-tetramethylpiperidin-1-oxyl (TEMPO)/iodine system as oxidizing agent and the bioreduction of the resulting ketones could be carried out in a sequential manner, offering a new strategy for the deracemization of secondary alcohols. Sonication greatly facilitated the chemical process to give the corresponding ketone intermediates in quantitative yield. After simple destruction of iodine, bioreduction of the ketones was performed in the same pot by using either (*R*)- or (*S*)-selective alcohol dehydrogenases, allowing for the preparation of the enantiopure 1-arylethanols and lineal aliphatic alcohols in excellent yields (Scheme 3.41) [42].

The one-pot two-step chemoenzymatic approach had also been successfully applied to deracemize 2,2-dichloro-1-phenylethanol, but the laccase from *Trametes*

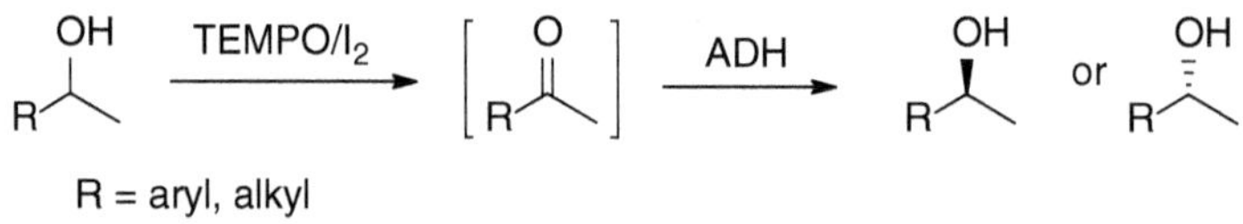

Scheme 3.41 One-pot chemoenzymatic deracemization of secondary alcohols. Source: Adapted from Méndez-Sánchez et al. [42].

versicolor/TEMPO pair was used as the oxidizing agent instead of the TEMPO/iodine system. Under optimized conditions, the racemic 2,2-dichloro-1-phenylethanol was oxidized by *T. versicolor* laccase and TEMPO to give α,α-dichloroacetophenone. After the oxidation was complete and the buffer pH was adjusted to 7.5, *E. coli* whole cells expressing an alcohol dehydrogenase gene from *R. ruber* (*E. coli*/ADH-A) and 2-propanol as hydrogen donor were added into the reaction mixture. (*R*)-2,2-dichloro-1-phenylethanol was obtained in total conversion with 97% ee (Scheme 3.42) [43].

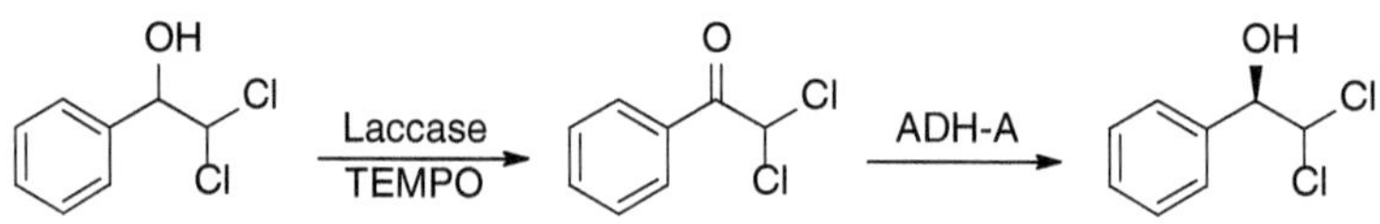

Scheme 3.42 One-pot chemoenzymatic deracemization of 2,2-dichloro-1-phenylethanol. Source: Adapted from Kedziora et al. [43].

Similarly, the nonselective oxidation of alcohols by *T. versicolor* laccase and TEMPO was coupled with the dynamic reductive kinetic resolution of the resulting aldehyde intermediate, leading to a cascade protocol useful for the deracemization of 2-phenyl-1-propanol derivatives. In this context, aerobic oxidation of 2-phenyl-1-propanol mediated by laccase from *T. versicolor* and TEMPO afforded the racemic aldehyde intermediate, which was reduced by using two stereocomplementary ADHs, i.e. commercially available HLADH and Evo-1.1.200, under dynamic conditions (Scheme 3.43). Both enantiomers of 2-phenyl-1-propanol were accessible with high ee values by this mild one-pot methodology [44].

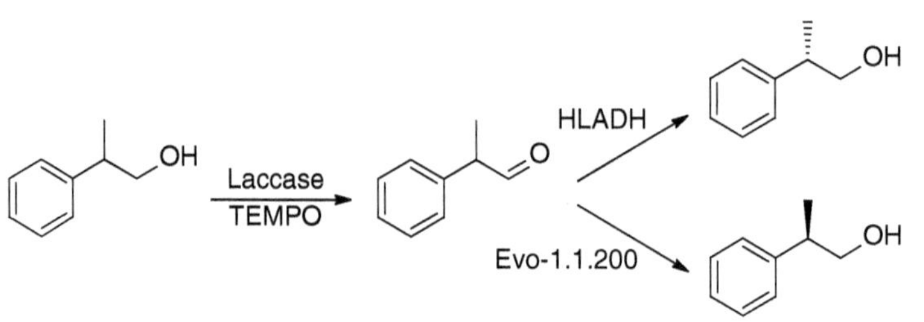

Scheme 3.43 One-pot chemoenzymatic deracemization of 2-phenyl-1-propanol.

Secondary alcohols could also be oxidized by using sodium hypochlorite (NaOCl) and 2-azaadamantane *N*-oxyl (AZADO), generating the corresponding ketones.

Coupling of this organocatalytic oxidation with a stereoselective enzymatic reduction of the intermediate ketone resulted in the deracemization of racemic secondary alcohols. A series of secondary alcohols with diverse structural features were deracemized by using this one-pot sequential process. The corresponding enantiopure alcohols were isolated in 76–98% yields by employing suitable KREDs with the organocatalytic oxidation (Scheme 3.39). It is worth noting that the stereoisomeric rac/meso mixture of diols could also be converted into essentially single stereoisomer by this one-pot chemoenzymatic oxidation–reduction process. For example, (1′*R*,1″*R*)-2,6-bis(1-hydroxyethyl)pyridine was obtained with 95% yield, >99% ee, and >99 : <1 dr (Scheme 3.44) [45].

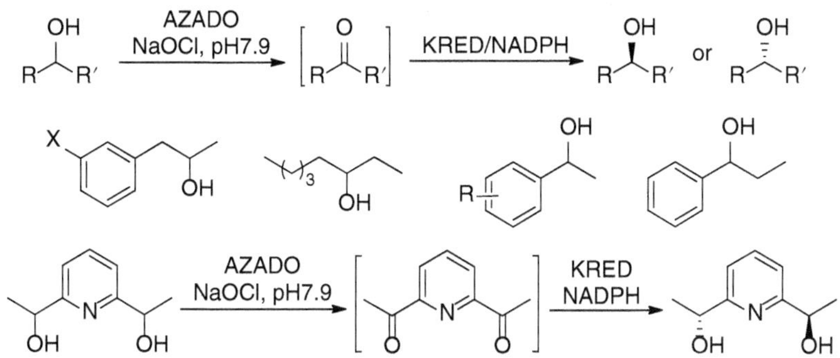

Scheme 3.44 One-pot sequential organocatalytic oxidation–enzymatic reduction. Source: Adapted from Liardo et al. [45].

L-α-Amino acids could be reduced to (*S*)-amino alcohols via amino acid carboxyl reduction using $NaBH_4$–H_2SO_4 system. The resulting amino alcohols underwent transaminase-catalyzed deamination, giving α-hydroxy ketones. Enzymatic reduction of α-hydroxy ketones afforded the corresponding optically pure vicinal 1,2-diols (Scheme 3.45), which are valuable chiral intermediates for the production of fine chemicals and pharmaceuticals. These three reactions were carried out in a one-pot three-step procedure, and chiral vicinal 1,2-diols were obtained in 69–90% yields and 91 to >99% ee from L-α-amino acids, a class of inexpensive and renewable feedstocks for the production of high value-added chemicals [46].

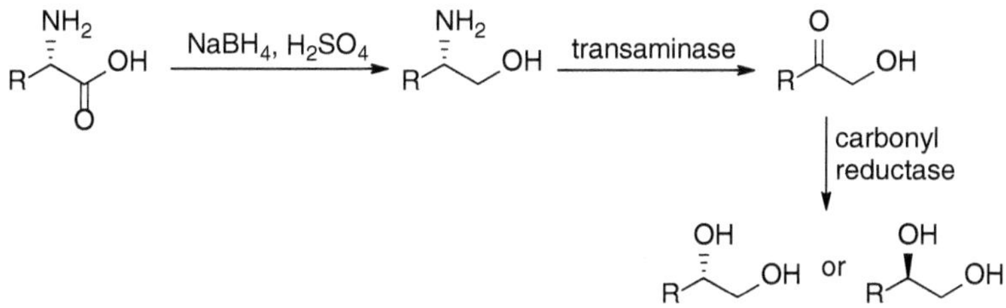

Scheme 3.45 One-pot chemoenzymatic conversion of L-α-amino acids into chiral vicinal 1,2-diols.

In the above text, we have discussed the combination of the chemical preparation of ketones with the enzymatic ketone reduction in a one-pot sequential process to prepare chiral alcohols. Biocatalytic ketone reduction has been also incorporated with the chemical transformations of the functional groups in the *in situ*-formed chiral alcohol molecules. It has been well documented that ADHs effectively catalyze asymmetric reduction of aromatic 2-azido ketones to furnish 2-azido-1-arylethanol. Schrittwieser et al. reported the combination of this reaction and Pd nanoparticle-catalyzed hydrogenation of the resulting azido alcohols. By this one-pot process both enantiomers of aromatic 2-azido alcohols were accessed in high yields and excellent optical purity (ee >99%) using highly (*R*)-selective ADH from *R. ruber* DSM 44541 or commercially available (*S*)-selective enzyme, and converted to the corresponding aromatic 1,2-amino alcohols. Encouraged by these positive results, the authors further extended the one-pot process to include the formation of aromatic 2-azido ketones from the corresponding halo ketones and the benzoylation of aromatic 1,2-amino alcohols. As such, (*S*)-tembamide, a naturally occurring benzamide derivative with antiviral (HIV) activity was prepared from 1-(4′-methoxylphenyl)-2-bromo-ethanone with 73% yield in a four-step one-pot operation (azidolysis, ADH reduction, hydrogenation, and benzoylation, Scheme 3.46) [47].

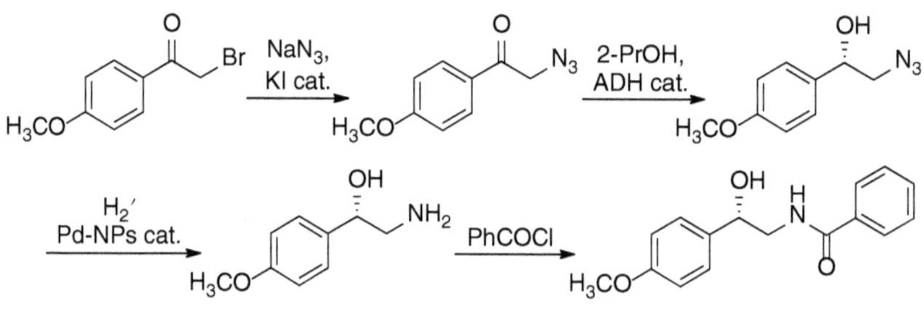

Scheme 3.46 One-pot chemoenzymatic synthesis of (*S*)-tembamide. Source: Adapted from Schrittwieser et al. [47].

1,2,3-Triazoles are very important pharmacophores existing in various biologically active compounds, such as potent β-adrenergic receptor blockers. The copper(I)-catalyzed Huisgen cycloaddition between an alkyne and an azido compound is an efficient method to construct the triazole moiety with excellent atomic economy, and is perhaps the most outstanding example of the so-called "click" chemistry. The azido alcohols obtained from the enzymatic reduction of the corresponding azido ketones have been reported to react with alkynes under Cu(I) catalysis furnishing chiral 1,2,3-triazole alcohols. The biocatalytic ketone reduction had been combined with the Cu-catalyzed "click" reaction in a one-pot two-step manner. In this approach, different chiral 1,2,3-triazole-derived diols with two chiral centers were synthesized in high yields and excellent enantio- and diastereoselectivities. The stereoselective bioreduction of an α-azido-ketone and an alkynyl ketone were concurrently carried out in one-pot, followed by the Cu(I)-catalyzed cycloaddition of the *in situ*-formed enantioenriched azido- and alkynyl alcohols (Scheme 3.47) [48].

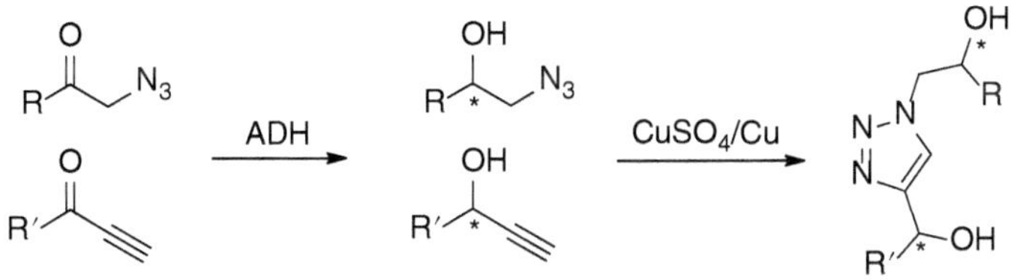

Scheme 3.47 One-pot chemoenzymatic synthesis of chiral 1,2,3-triazole-derived diols. Source: Adapted from Cuetos et al. [48].

γ- and δ-Lactones are valuable building blocks for fuels, biodegradable polymers, flavor, aroma constituents, and natural products such as polyketides, prostaglandins, pheromones, and other important compounds. ADHs catalyzed the asymmetric reduction of γ and δ ketoesters to the corresponding hydroxyl esters. For example, the ADHs from *R. ruber* (ADH-A) and *L. brevis* (LBADH) reduced several γ and δ ketoesters with excellent stereoselectivities. The resulting hydroxyl esters could be transformed to the corresponding γ- and δ-lactones, respectively, after adding 1 M HCl solution into the reaction mixtures (Scheme 3.48) [49].

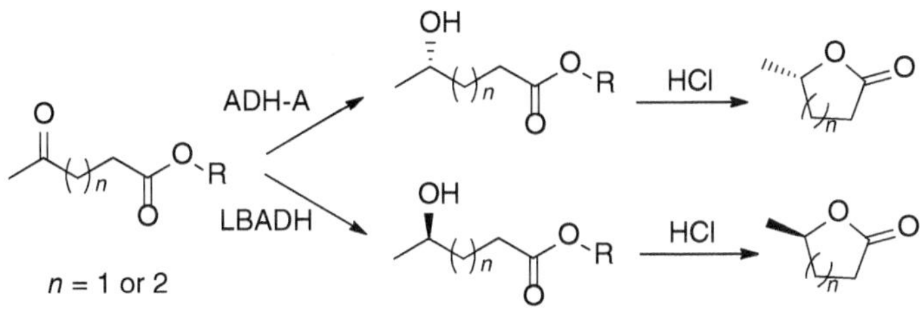

Scheme 3.48 One-pot chemoenzymatic synthesis of γ- and δ-lactones. Source: Adapted from Díaz-Rodríguez et al. [49].

For the bulkier ketoesters such as methyl 4-oxo-4-phenylbutanoate, the ADH from *Ralstonia* sp. (RasADH) with the glucose/glucose dehydrogenase (GDH) system for cofactor regeneration served as an effective biocatalyst, affording methyl (*R*)-4-hydroxy-4-phenylbutanoate with excellent conversion (>97%) and ee values (>99%). After treatment with 1 M HCl, the corresponding 5-membered ring lactone with identical ee value was obtained. Similar results were obtained for methyl 4-oxo-4-(*p*-bromophenyl)butanoate. The resulting brominated lactone was further functionalized to construct more complex and potentially bioactive lactones by different palladium-catalyzed C–C coupling reactions, such as Heck coupling, Suzuki coupling, and Sonogashira coupling reactions (Scheme 3.49). These enzymatic ketone reduction, lactonation, and C–C coupling reactions were carried out in one-pot multistep manner without isolating the brominated lactone intermediate [49].

Alcohol dehydrogenases usually showed excellent chemo and regioselectivity, generating alcohol products with other functional groups intact, thus allowing their further transformation. An ADH from *L. kefir* (LKADH) catalyzed the selective

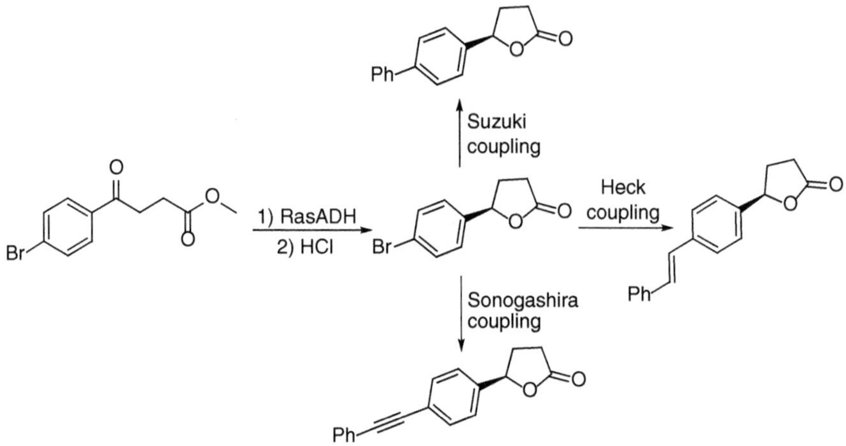

Scheme 3.49 One-pot multistep chemoenzymatic synthesis of γ- and δ-lactones.

reduction of methyl keto group of (3*E*)-non-3-ene-2,8-dione in eight minutes leaving the methyl enone group untouched. The resulting (3*E*)-8(*R*)-hydroxy-non-3-en-2-one partially cyclized to give 1-((2*R*, 6*R*)-6-methyltetrahydro-2*H*-pyran-2yl)propan-2-one via intramolecular oxa-Michael reaction (IMOMR), which could be facilitated to completion with ethereal hydrogen chloride solution to give the substituted tetrahydropyran (THP) in 75% yield and 12 : 1 *cis/trans* ratio. The enzymatic reduction and subsequent IMOMR were applied to the synthesis of some THPs and tetrahydrofurans in good yields and high ee and dr (Scheme 3.50), especially the preparation of an analog of the fungal antioxidant

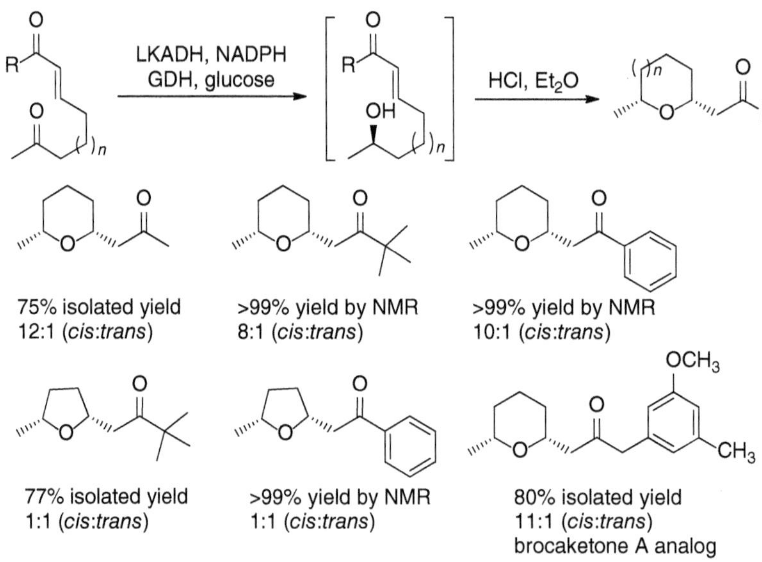

Scheme 3.50 One-pot sequential enzymatic reduction and intramolecular oxa-Michael reaction.

brocaketone A, and the (−)-(*R*,*R*)-enantiomer of the natural product, (+)-(*S*,*S*)-(*cis*-6-methyltetrahydropyran-2-yl)acetic acid (Scheme 3.51) [50].

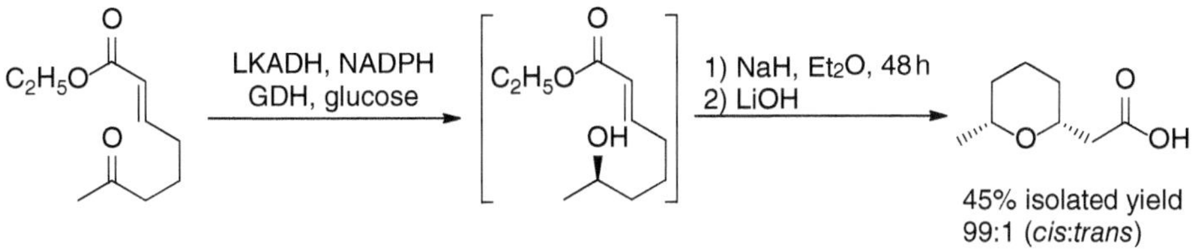

Scheme 3.51 One-pot synthesis of (−)-(*R*,*R*)-(*cis*-6-methyltetrahydropyran-2-yl)acetic acid. Source: Adapted from Eastman et al. [50].

3.3 Ene Reductases

Ene-reductases (ERs) are a family of NAD(P)H-dependent enzymes, which catalyze the asymmetric reduction of activated C=C bonds to generate up to two stereogenic centers. This biocatalytic alkene reduction has shown potential applications in generating valuable chemicals such as chiral aldehydes, ketones, nitriles, nitro compounds, mono- and dicarboxylic acids, and esters [51]. The activated alkenes are readily accessible by the Wittig reaction of aldehydes with ylides. Therefore, the combination of Wittig reaction in aqueous media and subsequent *in situ* reduction of the resultant C=C bonds would offer a new one-pot process methodology for increasing the molecular complexity. Indeed, the Wittig reaction of aryl aldehydes and ylides with carbonyl group proceeded smoothly in aqueous media to give the corresponding α,β-unsaturated ketones. The subsequent *in situ* reduction by the enoate reductase (GoER) from *G. oxydans* gave the desired saturated ketone in >95% overall conversion (Scheme 3.52) [52].

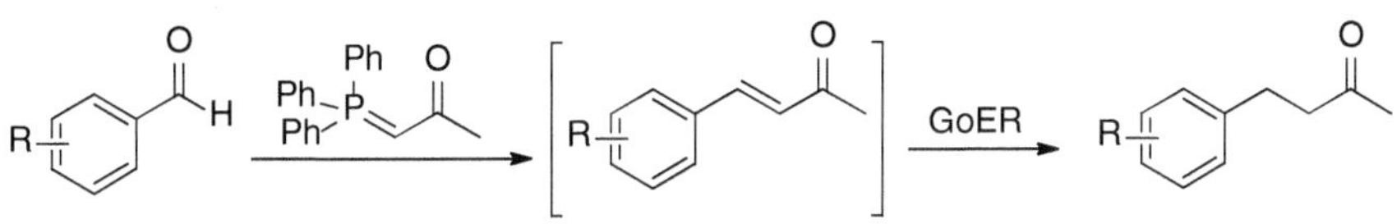

Scheme 3.52 One-pot sequential Wittig reaction and alkene bioreduction. Source: Adapted from Kraußer et al. [52].

Rhodium(II) complexes [$Rh_2(OPiv)_4$] and [$Rh_2(Oct)_4$] catalyzed the cross-coupling of aryldiazo compounds with diazoacetate derivatives in dichloromethane, affording predominantly the dicarbonyl compounds with high selectivity for the formation of *E*-alkenes. After the coupling reaction was finished, the solvent was removed and all reagents and the ER for the reduction reaction were added into the crude mixture. The bioreduction of the coupling products proceeded smoothly to give the enantioenriched 2-aryl succinic acid derivatives in >70% yield in one-pot process (Scheme 3.53) [53].

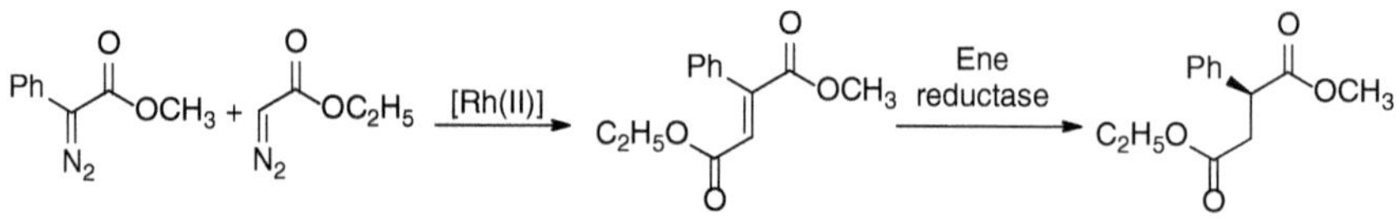

Scheme 3.53 One-pot chemoenzymatic synthesis of chiral 2-aryl succinic acid derivatives. Source: Adapted from Wang et al. [53].

For the reduction of (*Z*)-β-aryl-β-cyanoacrylates, mutants of Old Yellow Enzyme 1 (OYE1) were active but not the wild-type enzyme. While the mutant W116A catalyzed the reduction to give the (*S*)-enantiomer, W116L flipped the substrate binding mode to afford the corresponding (*R*)-configurated product. The resulting optically active cyanopropanoates were converted into the corresponding (*S*)- and (*R*)-β-aryl-γ-lactams by reduction of the nitrile with $NiCl_2 \cdot 6H_2O$ and $NaBH_4$ in a sequential one-pot procedure. Subsequent hydrolysis produced β-aryl-γ-amino acids (Scheme 3.54) [54].

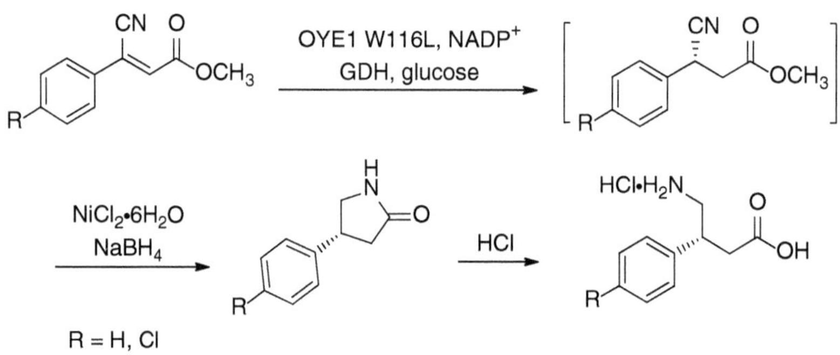

Scheme 3.54 One-pot chemoenzymatic synthesis of (*R*)-β-aryl-γ-lactams. Source: Adapted from Brenna et al. [54].

3.4 Transaminases

ω-Transaminases (ω-TAs) (E.C. 2.6.1.X) catalyze the transfer of an amino group between an amine donor (amino acid or amine) and an acceptor (ketone or aldehyde) with cofactor pyridoxal-5′-phosphate (PLP) serving as molecular shuttle for ammonia and electrons. The reaction includes two steps, namely, the enantioselective oxidative deamination of amino donor and the stereoselective reductive amination of ketone. As such, ω-TAs have been utilized in the kinetic resolution of racemic chiral amines and the asymmetric synthesis of chiral amines from prochiral ketones. TAs have been proved as one of the most promising biocatalysts for the preparation of chiral amines [55].

The oxidation of secondary alcohols followed by the ω-TA-catalyzed reductive amination of the resultant ketones offers an attractive methodology for the asymmetric conversion of racemic alcohols into enantiomerically pure amines. The alcohol oxidation can be realized by chemical and biocatalytic methods. While the chemical oxidation usually requires toxic reagents, the biocatalytic approaches often show high catalytic specificity and selectivity, with ADHs and alcohol oxidases (AOx) being the most commonly used biocatalysts [56]. Many cascade biotransformations have been developed for the production of optically pure chiral amines and alcohols by combining enzymatic oxidation of racemic alcohols and the biotransformation of the resultant ketone intermediates [57]. In this context, the *T. versicolor* laccase/TEMPO system oxidized racemic (hetero)aromatic alcohols to furnish the corresponding ketones, of which reductive aminations were catalyzed by using different commercially available TAs. Both enantiomers of the aromatic chiral amines were prepared by employing this one-pot two-step process (Scheme 3.55) [58].

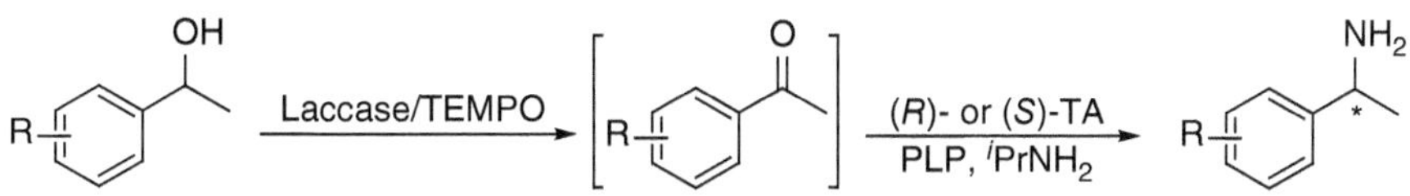

Scheme 3.55 One-pot transformation of racemic alcohols into optically active amines. Source: Adapted from Martinez-Montero et al. [58].

By using an organocatalyst, AZADO and an oxidant, NaOCl, various secondary alcohols including 2-(benzyloxy)cyclohexanol and 2-(benzyloxy)cyclopentanol were efficiently oxidized, furnishing the corresponding ketones. When the oxidation was completed, $iPrNH_2$, PLP, and the ω-TA were added into the reaction mixture, and the reductive amination of the ketone intermediates proceeded smoothly. A diversity of synthetically and pharmacologically interesting amines were thus obtained with high isolated yields and excellent stereomeric purity by carefully choosing the oxidation conditions and biocatalyst. Particularly, for 2-(benzyloxy)cyclohexanol and 2-(benzyloxy)cyclopentanol, ω-TA-catalyzed reductive amination of the resulting ketone intermediate would generate two chiral centers with four possible diastereomers. By selecting suitable reaction pH and TA, dynamic asymmetric transaminations of 2-(benzyloxy)cyclohexanone and 2-(benzyloxy)cyclopentanone were achieved to give three of the four possible stereoisomers with high enantio- and diastereomeric ratio (Scheme 3.56). Furthermore, nonselective organocatalytic oxidation and biocatalytic dynamic

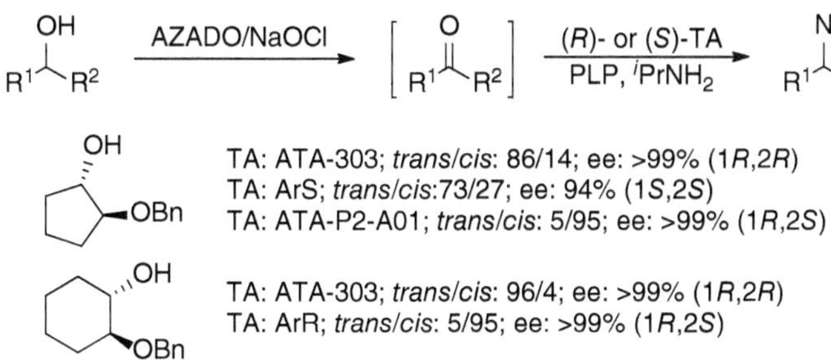

Scheme 3.56 One-pot chemoenzymatic synthesis of different diastereomers of chiral amino alcohols.

asymmetric transamination could be implemented in a one-pot sequential process in aqueous medium, enabling the practical synthesis of optically active amino alcohol derivatives by means of quantitative and simple processes [59].

Metal-catalyzed isomerization of allylic alcohols into ketones or aldehydes is a widely studied catalytic reaction, which has recently been carried out in nonconventional solvents such as water and ionic liquids. This provides the opportunity to implement isomerization of allylic alcohols and asymmetric bioamination of the resulting ketones or aldehydes into a one-pot process in aqueous media. Indeed, isomerization of a set of allylic alcohols catalyzed by the Ru(IV) complex was carried out in a phosphate buffer (100 mM, pH 8.5 and 1 M $^{i}PrNH_2$) to afford prochiral ketone intermediates. After having tested 28 commercially available ω-TAs from the Codexis Transaminase Screening Kit using isopropylamine as an amino donor, suitable ω-TAs were identified to exhibit excellent enantioselectivity toward the corresponding ketones. Ruthenium-catalyzed isomerization and ω-TA-catalyzed bioamination were then integrated into a one-pot process, and the allylic alcohols were converted into the corresponding chiral amines with high overall yields (>70%) and excellent enantiomeric excesses (>99% ee, Scheme 3.57) [60].

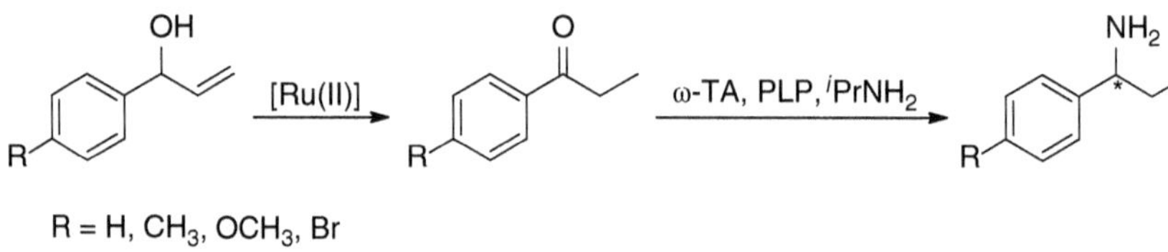

Scheme 3.57 One-pot sequential Ru-catalyzed isomerization of allylic alcohols and TA-catalyzed amino transfer reaction. Source: Adapted from Rios-Lombardia et al. [60].

The chemical oxidation of primary alcohols was also coupled with TA-catalyzed amination of the aldehyde intermediates and the resulting one-pot processes were integrated into the synthesis of complex molecules. Pd(0) nanoparticles on controlled pore glass (CPG) and Pd on *meso*-cellular foam (MCF) effectively catalyzed the aerobic oxidation of vanillyl alcohol to vanillin and >99% conversion was achieved within two hours. Vanillin was then converted to vanillylamine using a three-enzyme system including a TA, an L-alanine dehydrogenase, and a GDH, which catalyzed the amination of vanillin using L-alanine as amino donor, the regeneration of L-alanine, and the recycling of cofactor NADH, respectively. When the chemoenzymatic cascade was carried out in one-pot manner, vanillyl alcohol was converted to vanillylamine with 87% yield (Scheme 3.58) [61]. This process can also be used for the synthesis of other amines from the corresponding primary alcohols.

The vanillylamine generated was converted into capsaicin and nonivamide in 91% and 94% isolated yields, respectively, by reacting with the corresponding acyl chloride under Schotten–Baumann conditions. Furthermore, the total synthesis of vanillyl nonivamide from vanillin was achieved by converting vanillin to vanillylamine using the three-enzyme system, followed by addition of sodium bicarbonate, chloroform, and the acyl chloride into the reaction mixture (Scheme 3.59). Pure

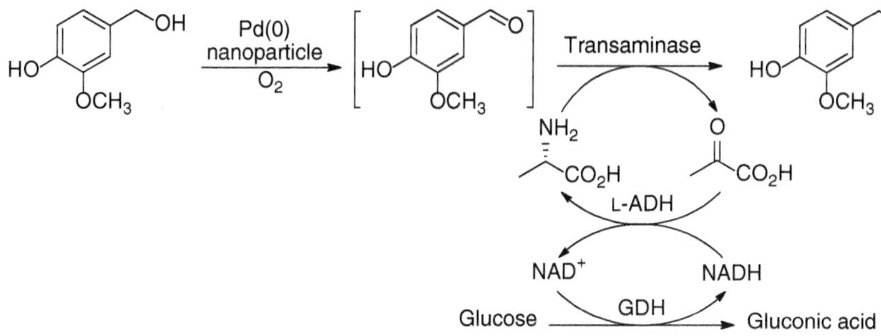

Scheme 3.58 One-pot sequential Pd-catalyzed alcohol oxidation and TA-catalyzed amino transfer reactions. Source: Adapted from Mattias et al. [61].

nonivamide was obtained in an overall yield of 92% in this one-pot process after silica gel column chromatography [61].

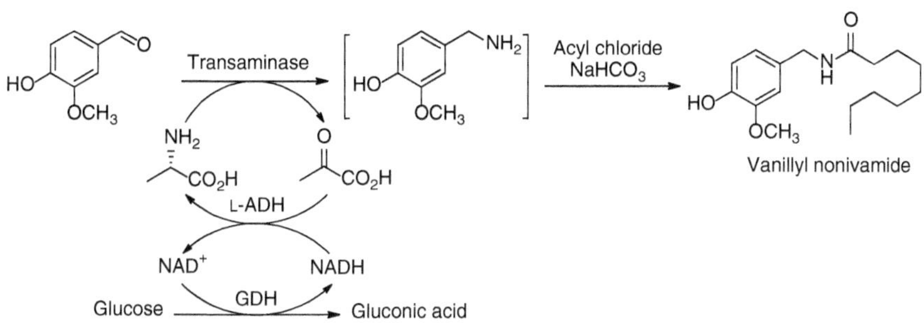

Scheme 3.59 One-pot chemoenzymatic synthesis of vanillyl nonivamide.

Wacker oxidation reaction involves the Pd(II)-catalyzed oxidative nucleophilic addition of water to olefins in the presence of Cu(II) salts and molecular oxygen, which reoxidize Pd(0) to Pd(II). The Wacker oxidation of styrenes generates the corresponding acetophenones and has been combined with enzymatic reduction for the one-pot sequential chemoenzymatic synthesis of optically pure alcohols. This metal-catalyzed oxidation of styrenes was also coupled with TA-catalyzed amination of the resulting acetophenones, leading to the direct transformation of styrenes into 1-phenylethylamines. However, compartmentation of the catalytic systems was needed to avoid deactivation of the TA by the Cu-component of the Pd/Cu-catalyzed Wacker oxidation. Therefore, PDMS thimbles were used to separate the two reactions. These thimbles are permeable for hydrophobic components but non-permeable for water-soluble ones. As such, the metal catalysts and enzymes can remain separate, but the intermediate acetophenones can go through the PDMS membrane for the next-step reaction. Wacker oxidation of styrenes and the subsequent enzymatic transamination were thus performed in one-pot sequential manner by using this compartmentation technique, and a broad range of 1-phenylethylamines were synthesized with 72–93% conversions and 97–99 % ee (Scheme 3.60) [62].

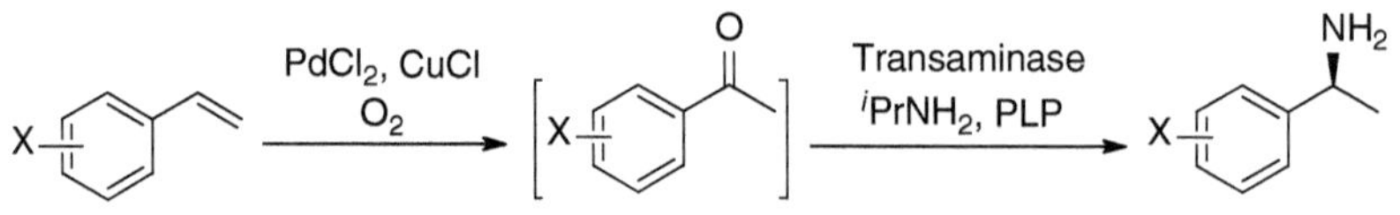

Scheme 3.60 One-pot sequential Wacker oxidation of styrenes and TA-catalyzed amino transfer reactions. Source: Adapted from Uthoff et al. [62].

Recently, it was found that using palladium (II) complex as catalyst and iron (III) salt as terminal oxidant for the Wacker–Tsuji oxidation of allylbenzenes, when the reaction was carried out in aqueous/acetonitrile media the metal-catalyzed Wacker oxidation of allylbenzenes could be combined with the amine-transaminase-catalyzed transamination of the resulting 1-arylpropan-2-ones in a one-pot sequential manner without compartmentation. Both enantiomers of several 1-arylpropan-2-amines were obtained with 74–92% overall yields and >99% ee by employing enantiocomplementary amine TAs (Scheme 3.61) [63].

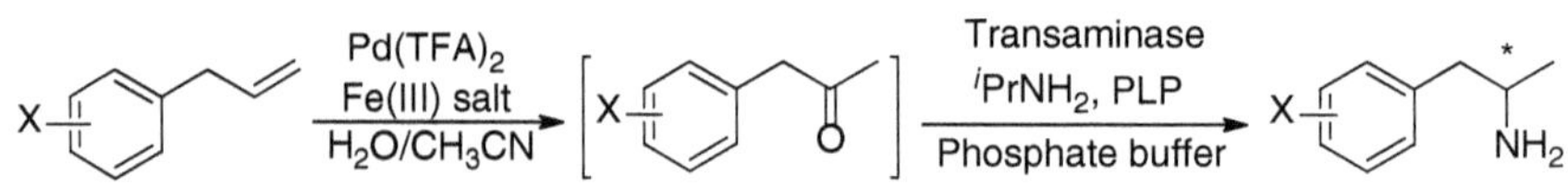

Scheme 3.61 One-pot sequential Wacker oxidation of allylbenzenes and TA-catalyzed amino transfer reactions. Source: Adapted from González-Martínez et al. [63].

Palladium catalyzes the Suzuki–Miyaura cross-coupling reaction between arylhalides and arylboronic acids, leading to biaryl compounds. This C–C bond formation was originally performed in organic solvents. Recently, it was reported that Suzuki–Miyaura reaction could proceed in phosphine-free and water-solvent systems such as *N*,*N*-dimethylformamide (DMF)/water mixture. For example, Suzuki–Miyaura cross-coupling of 5-bromo-3-acetyl-pyridine or 4′-bromoacetophenone with arylboronic acids proceeded smoothly in the DMF/water (v/v, 1/1) mixture with $PdCl_2$ and sodium carbonate. This reaction was combined with TA-catalyzed amino transfer between the resulting biarylacetones and amino donor. A one-pot two-step process was thus established (Scheme 3.62), and the corresponding chiral biaryl amines were obtained in >99 % ee and up to 84 % total conversion by utilizing a mutant TA from *Asperguillus fumigatus* (4CHI-TA) [64].

DESs consist of two or three compounds with an extensive H-bond network throughout the solvent, and the melting point is far below those of the individual components. Since the compounds are usually from renewable sources, DESs are cheaper, tunable, highly biodegradable, and are thus known as environmentally benign solvents. Recently, a TA from *Exophiala xenobiotica* (EX-ωTA) was found to be stable in the KPi buffer–DES mixtures consisting of choline chloride (ChCl) and glycerol (Gly). A chemoenzymatic cascade was established in the DES-buffer system by coupling Pd-catalyzed Suzuki reaction with an enantioselective transamination. The Pd-catalyzed reaction was carried out at 200 mM loading of substrate and the subsequent TA-catalyzed biotransformation was performed at 25 mM substrate

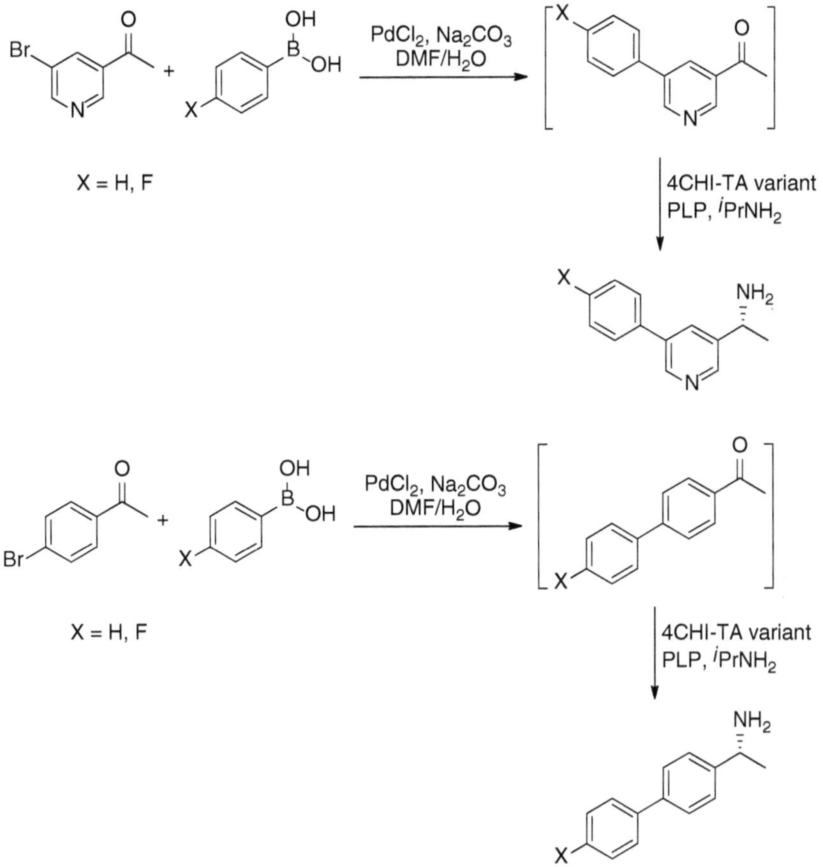

Scheme 3.62 One-pot sequential Suzuki–Miyaura cross-coupling and TA-catalyzed amino transfer reactions.

concentration (Scheme 3.63). A few biaryl amines were prepared in up to 90% isolated yields and >99% ee by using a mutant ATA from *E. xenobiotica* (EX-STA) [65].

The TA CV2025 from *Chromobacterium violaceum* showed activity toward 3,4-dihydroxyphenethylamine, which was converted to 3,4-dihydoxyphenylacetaldehyde with pyruvate as the amino acceptor. During the reaction, the unreacted 3,4-dihydroxyphenethylamine coupled with 3,4-dihydoxyphenylacetaldehyde to form Norlaudanosoline (tetrahydropapaveroline) via a the nonenzymatic Pictet–Spengler condensation. Addition of a plant enzyme norcoclaurine synthase (NCS) into the reaction system facilitated the Pictet–Spengler condensation, and dopamine was converted to norlaudanosoline with 87% yield and 99% ee under the optimized reaction conditions. At a preparative scale, 86% conversion was achieved in two hours and norlaudanosoline was isolated in 62% yield with >95% ee.

Furthermore, the one-pot concurrent transformation was extended to the synthesis of tetrahydroprotoberberine alkaloids (THPBs, berbines). After the transformation of dopamine to norlaudanosoline had proceeded for three hours, formaldehyde

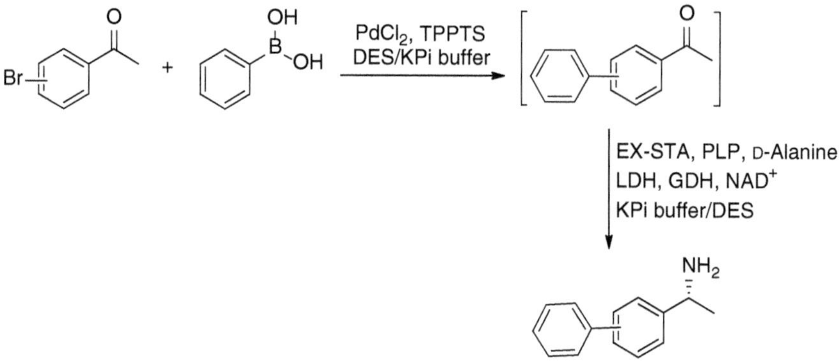

Scheme 3.63 One-pot sequential Suzuki C–C coupling and TA-catalyzed amino transfer reactions in the DES-buffer system.

was added to the reaction mixture, triggering the second Pictet–Spengler condensation to form 10,11-dihydroxy-(*S*)-THPBs and 9,10-dihydroxy-(*S*)-THPBs in a ratio of approximately 7 : 1 with an overall 64% conversion (56% and 8%, Scheme 3.64) [66]. The regiochemistry of this transformation with 10,11-dihydroxy major regioisomer was found to be different from that of the plant berberine-bridge enzyme (BBE), which typically formed 9,10-dihydroxy-THPBs from an *N*-methylated BIA precursor, thus providing complementary methods to access pharmaceutically important 10,11-dihydroxy-(*S*)-THPBs and 9,10-dihydroxy-(*S*)-THPBs, respectively.

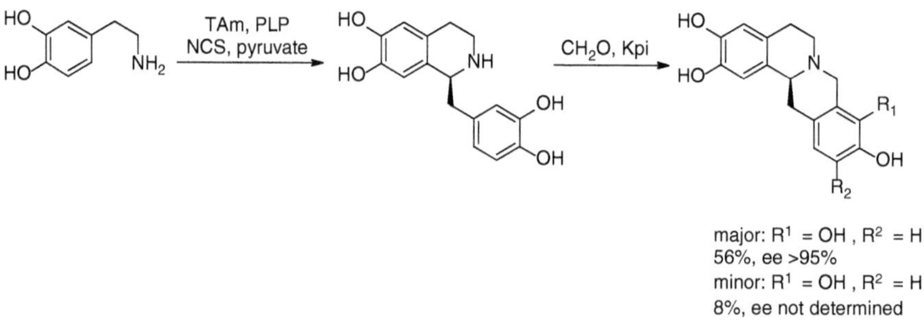

Scheme 3.64 One-pot sequential chemoenzymatic synthesis of 10,11-dihydroxy-(*S*)-THPBs and 9,10-dihydroxy-(*S*)-THPBs. Source: Adapted from Lichman et al. [66].

The tetrahydroisoquinoline (THIQ) moiety exists as a "privileged scaffold" in many natural products and synthetic pharmaceutical drugs with wide ranging bioactivities, such as antitumor, anticholinergic, and antiparasitic properties. Both enzymatic and chemoenzymatic three-step one-pot processes that enable the synthesis of stereochemically complementary trisubstituted THIQs of biological importance from inexpensive starting materials such as 3-hydroxybenzaldehyde and pyruvate were developed. In these cascade transformations, the first step was the carboligation of 3-hydroxybenzaldehyde and pyruvate catalyzed by the ThDP-dependent acetohydroxy acid synthase I from *E. coli* (EcAHAS-I). EcAHAS-I catalyzed the decarboxylation of pyruvate to form acetaldehyde, and the subsequent condensation with 3-hydroxybenzaldehyde to generate (*R*)-1-hydroxy-1-(3-hydroxyphenyl)propan-2-one in the same active site. The following transamination of the hydroxyketone intermediate was carried out without purification of the intermediate by using a TA from *C. violaceum* (Cv2025) as biocatalyst and isopropylamine as amino donor. (1*R*, 2*S*)-2-Amino-1-(3-hydroxyphenyl) propan-1-ol (metaraminol) was obtained with excellent yield (91%) and stereoselectivity over four possible isomers (97%). The subsequent Pictet–Spengler reaction of (1*R*, 2*S*)-2-amino-1-(3-hydroxyphenyl)propan-1-ol with the carbonyl co-substrate phenylacetaldehyde could be performed with either enzymatic or phosphate catalysis. In the enzymatic reaction with a variant norcoclaurine synthase (NCS, TfNCS-A79I) from *Thalictrum flavum* as biocatalyst, (1*S*, 3*S*, 4*R*)-1-benzyl-3-methyl-1,2,3,4-tetrahydroisoquinoline-4,6-diol was isolated as HCl salt in excellent yield (92%) and purity (de 99%). For the overall one-pot three-step enzymatic process, the desired trisubstituted THIQ product was obtained with 88% overall conversion and >97% isomer content. In the transformation catalyzed by inorganic phosphate, three isomers of the THIQ product were produced, with the (1*S*, 3*S*, 4*R*)-isomer being one of the minor products. In contrast, when 2-bromobenzaldehyde was used as the co-substrate, the phosphate-catalyzed Pictet–Spengler reaction of (1*R*, 2*S*)-2-amino-1-(3-hydroxyphenyl)propan-1-ol gave rise to the product (1*S*,3*S*,4*R*)-1-(2-bromophenyl)-1,2,3,4-tetrahydroisoquinoline-4,6-diol, which was isolated as the HCl salt in 54% yield (97% de, Scheme 3.65). The configuration of the C1 stereocenter was opposite to that of THIQs produced by NCS-catalyzed reaction, although priority naming rules dictated it to be (1*S*). When the two enzymatic and one chemical reactions were carried out sequentially in one-pot manner, the desired trisubstituted THIQ product was obtained with 77% overall conversion and excellent isomer content (>97%) [67].

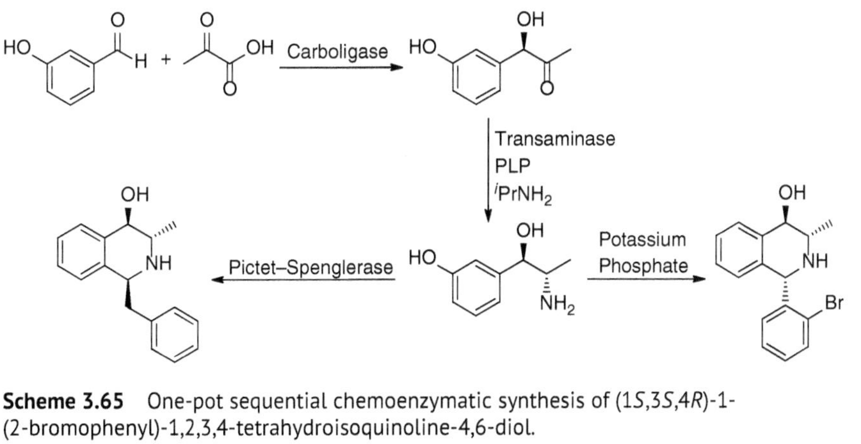

Scheme 3.65 One-pot sequential chemoenzymatic synthesis of (1*S*,3*S*,4*R*)-1-(2-bromophenyl)-1,2,3,4-tetrahydroisoquinoline-4,6-diol.

In plant biosynthetic pathways, NCSs catalyze the Pictet–Spengler reaction between dopamine and 4-hydroxyphenylacetaldehyde to generate (*S*)-norcoclaurine. In addition to aromatic aldehydes, aliphatic substrates could also be accepted by NCSs [68]. Therefore, the Pictet–Spengler reaction of dopamine with methyl

4-oxobutanoate in the presence of NCS gave methyl (1*S*)-3-(6,7-dihydroxy-1,2,3,4-tetrahydroisoquinolin-1-yl)propanoate, which was converted into alkaloid trolline in 74% isolated yield and 95% ee by treatment with sodium carbonate at pH 7.5 without isolating the tetrahydroisoquinoline alkaloids (THIA) intermediate. Similarly, when methyl 5-oxopantanoate was used as the aldehyde substrate, the Pictet–Spengler condensation was followed by cyclization to form 9,10-dihydroxy-1,2,3,6,7,11b-hexahydro-4*H*-pyrido[2,1-a]isoquinolin-4-one in 87% isolated yield and 96% ee. NCS also catalyzed the Pictet–Spengler condensation of dopamine with methyl 6-oxohexanoateto giving the linear THIA, methyl 5-(6,7-dihydroxy-1,2,3,4-tetrahydroisoquinolin-1-yl)pentanoate, in 92% yield and in very high optical purity (>99% ee). However, subsequent treatment with sodium carbonate at pH 7.5 did not result in the expected cyclization, and methyl 5-(6,7-dihydroxy-1,2,3,4-tetrahydroisoquinolin-1-yl)pentanoate was isolated in 63% yield (>99% ee, Scheme 3.66) [69].

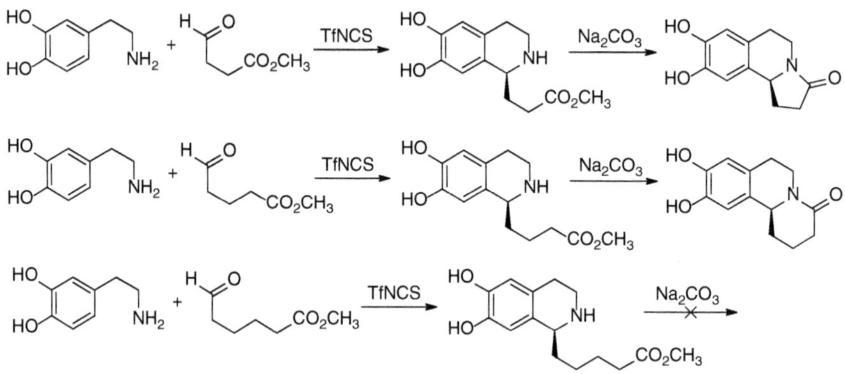

Scheme 3.66 One-pot sequential NCS-catalyzed Pictet–Spengler condensation and chemical cyclization. Source: Adapted from Zhao et al. [69].

This one-pot two-step cascade was further extended to the Pictet–Spengler reaction of dopamine analogs, 2-(3-hydroxyphenyl)ethylamine, 2-(4-fluoro-3-hydroxyphenyl)ethylamine, and metaraminol with methyl 4-oxobutanoate. For 2-(3-hydroxyphenyl)ethylamine and 2-(4-fluoro-3-hydroxyphenyl)ethylamine, the NCS-catalyzed reaction followed by cyclization afforded 8-hydroxy-1,5,6,10b-tetrahydropyrrolo [2,1-a] isoquinolin-3(2*H*)-one and 9-fluoro-8-hydroxy-1,5,6,10b-tetrahydropyrrolo [2,1-a] isoquinolin-3(2*H*)-one in 51% yield (93% ee) and 57% yield (87% ee), respectively. The reaction of metaraminol with methyl 4-oxobutanoate gave trolline derivatives in 76% yield, with (5*S*, 6*R*, 10b*S*)-6,8-dihydroxy-5-methyl-1,5,6, 10b–tetrahydropyrrolo[2,1-a]isoquinolin-3(2*H*)-one being the major isomer (ratio 10 : 1). The minor isomers came from the (*R*)-enantiomer at C-1 and the *ortho*-product (ratio ~1 : 1) of the NCS-catalyzed Pictet–Spengler reaction (Scheme 3.67). This suggested that the NCS has higher stereoselectivity toward the reaction of metaraminol and phenylacetaldehyde than with nonaromatic aldehyde [67].

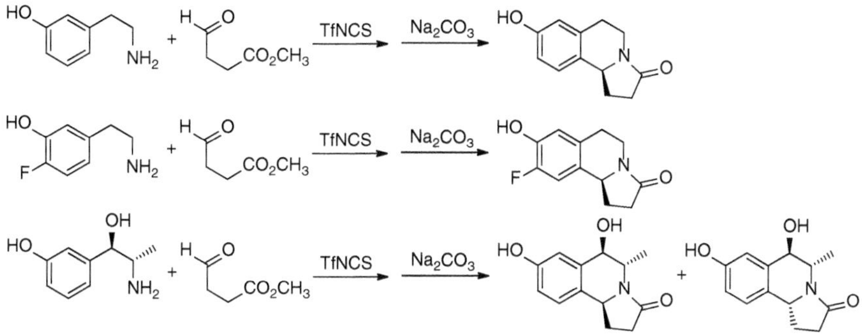

Scheme 3.67 One-pot sequential NCS-catalyzed Pictet–Spengler condensation and chemical cyclization.

Cinacalcet, *N*-[(1*R*)-1-(naphthyl)ethyl]-3-[3-(trifluoromethyl)-phenyl]propane-1-amine, a calcimimetic for the treatment of secondary hyperparathyroidism and hypercalcemia, was mainly synthesized from the chiral precursor, (*R*)-(+)-1-(1-naphthyl)ethylamine. The chiral amine could be prepared by the asymmetric amination of 1-acetonaphthone using TAs. Recently, it was reported that the commercially available TA, ArRmut11-M117F/G279A, catalyzed the amination of 1-acetonaphthone with isopropyl amine as amino donor at 50 mM substrate concentration to render enantiopure (*R*)-(+)-1-(1-naphthyl)ethylamine in 77% yield, which was alkylated with 3-[3-(trifluoromethyl)phenyl]propan-1-ol mediated by the iridium catalyst $[Cp^*IrCl_2]_2$ in toluene to give cinacalcet. After the enzymatic step, the reaction mixture was centrifuged. The iridium catalyst $[Cp^*IrCl_2]_2$ and 3-[3-(trifluoromethyl)phenyl]propan-1-ol were added into the resultant supernatant, and enantiopure cinacalcet was produced with an overall yield of 50% (Scheme 3.68) [70].

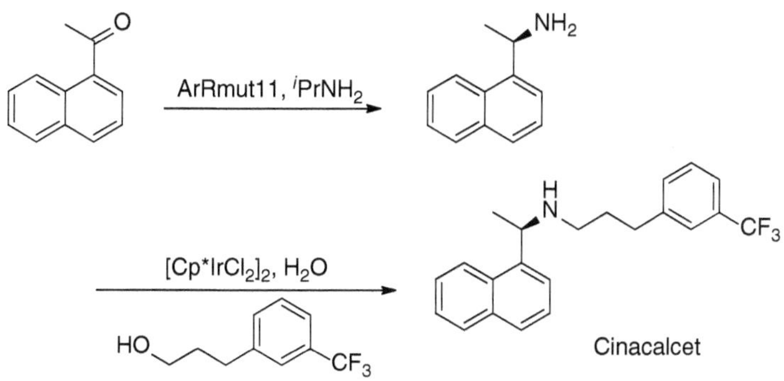

Scheme 3.68 One-pot sequential chemoenzymatic synthesis of Cinacalcet. Source: Adapted from Marx et al. [70].

The TA-catalyzed reaction was also applied to the kinetic resolution of racemic amines, which could be synthesized by the reductive amination of ketones. For example, a heterogeneous metal catalyst, Pd(0)-Amp-MCF, catalyzed the reductive amination of acetophenone and its derivatives to afford the racemic amines, which were resolved by the amino transfer reaction with commercially available TAs. By choosing (*R*)- or (*S*)-selective TA as biocatalyst, the chemoenzymatic reductive amination/kinetic resolution one-pot relay sequence produced the (*S*)- or (*R*)-enantiomer of the chiral amines from the corresponding ketones and ammonium formate with >90% yields and >85% ee values via dynamic kinetic resolution (Scheme 3.69) [12].

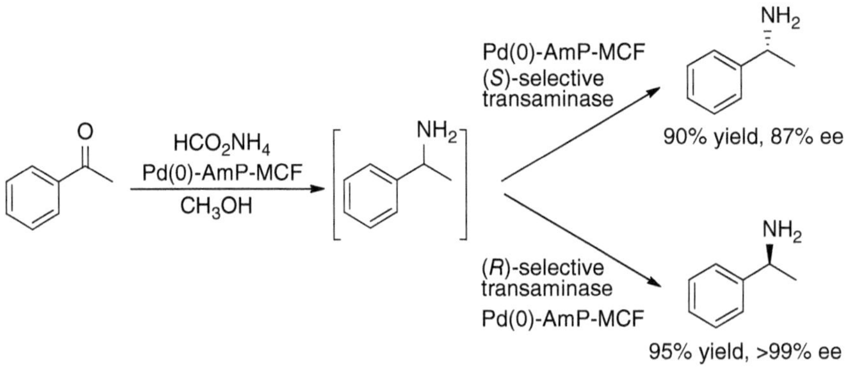

Scheme 3.69 One-pot sequential chemoenzymatic dynamic kinetic amination of ketones. Source: Based on Palo-Nieto et al. [12].

3.5 Epoxide Hydrolases (EHs)

Enantiomerically pure vicinal diols are important intermediates for the synthesis of valuable fine chemicals and active pharmaceutical ingredients, and are also widely used as chiral ligands for asymmetric catalysis. A number of synthetic methods for these compounds have been developed. Among them, the asymmetric dihydroxylation of olefins provides one of the most efficient methods for the preparation of chiral diols. The most famous reaction is Sharpless asymmetric dihydroxylation with osmium tetroxide in the presence of a chiral quinine ligand, which converts olefins into the corresponding vicinal diols via *cis*-dihydroxylation [71]. Dioxygenases also catalyze the *cis*-dihydroxylation of alkenes [72]. The complementary *trans*-dihydroxylation of olefins can lead to the vicinal diols with different configurations, and has been achieved by different approaches [73]. For example, the sequential epoxidation and hydrolysis of olefins have been employed for the preparation of optically active vicinal diols. In this context, a bacterial strain *Sphingomonas* sp. HXN-200 was found to be a good biocatalyst for the *trans*-dihydroxylation of some olefins with high enantioselectivity, in which the monooxygenase and epoxide hydrolase were responsible for the epoxidation of the C=C bond and the hydroxylation of the epoxide, respectively [74]. However, its efficiency and generality were limited by the intrinsic properties of monooxygenase and epoxide hydrolase in the wild-type strain [75]. The epoxidation of olefins could be realized by using other enzymes or chemical catalysts. Although chemical epoxidation of olefins with 3-chloroperoxybenzoic acid (*m*-CPBA) as the oxidant was usually carried out in dichloromethane, this reaction was found to proceed smoothly in aqueous buffer. *N*-benzyloxycarbonyl-3-pyrroline (0.2 M) was completely epoxidized with *m*-CPBA (0.3 M) in aqueous potassium phosphate buffer (KPi, 50 mM, pH = 8.0) in only four hours. Following the completion of the epoxidation reaction, Na_2SO_3 was added into the reaction mixture to remove the unreacted *m*-CPBA. The cells of *Sphingomonas* sp. HXN-200 were then added, and the *trans*-diol (3*R*,4*R*)-*N*-benzyloxycarbonyl-3,4-dihydroxy-pyrrolidine was obtained in 96% conversion and 92% ee (Scheme 3.70). A one-pot chemoenzymatic process for the asymmetric *trans*-dihydroxylation of olefins was thus developed by combining aqueous epoxidation of olefins with the stereoselective biocatalytic hydroxylation of epoxides [76].

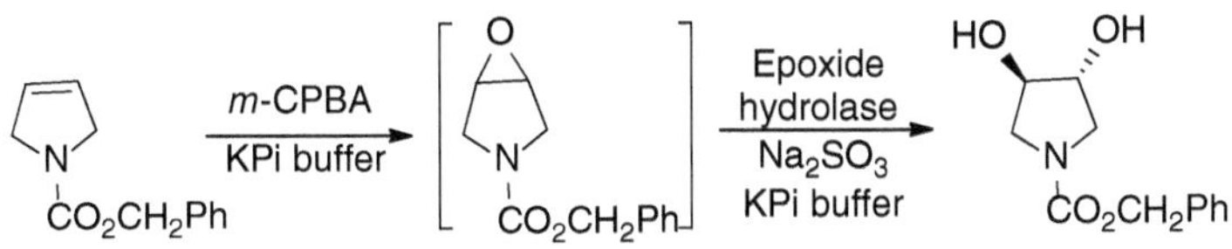

Scheme 3.70 One-pot chemoenzymatic *trans*-dihydroxylation of olefins.

Similar sequential epoxidation–hydrolysis reaction was applied to the selective dihydroxylation of (−)-limonene via the epoxidation using Ti deposited on ordered or amorphous mesoporous silica (Ti/SBA-15 or Ti/SiO_2) catalysts and tBuO_2H as oxidant in PhCF_3 solvent, followed by the selective hydrolysis of the resulting epoxide with a recombinant limonene 1,2-epoxide hydrolase from *Rhodococcus erythropolis* (Re-LEH) in potassium phosphate buffer. Without isolating the reaction intermediate, a single (1*R*,2*R*,4*S*)-menthene-1,2-diol isomer was prepared with good yield (Scheme 3.71). Furthermore, (+)-limonene was also transformed into the opposite diol enantiomer by this one-pot process of heterogeneous chemical and enzymatic catalysis [77].

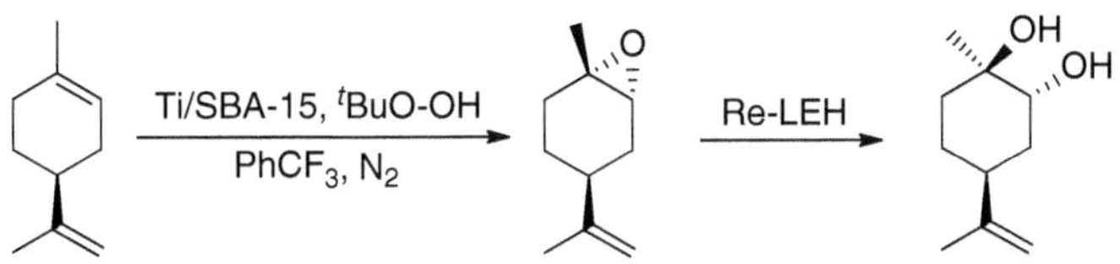

Scheme 3.71 One-pot chemoenzymatic *trans*-dihydroxylation of olefins.

The epoxide hydrolase-catalyzed hydrolysis of epoxides usually operated via a kinetic resolution pattern, affording the enantioenriched vicinal diol and epoxide, both in 50% theoretical yield. A one-pot two-step chemoenzymatic method was developed to solve this intrinsic problem of kinetic resolutions for (±)-2,2-disubstituted epoxides. The enantioselective enzymatic hydrolysis of racemic (±)-2,2-disubstituted epoxides with *Nocardia* spp. provided the (*S*)-diol and (*R*)-oxirane. By carefully controlling the reaction conditions, the remaining

(*R*)-oxirane was hydrolyzed with complete inversion of configuration, generating the (*S*)-diol. As such, racemic (±)-2,2-disubstituted epoxides were transformed into vicinal (*S*)-diols in up to 98% yields and 98% ee via this enantioconvergent two-step bio- and acidic hydrolysis procedure (Scheme 3.72) [78].

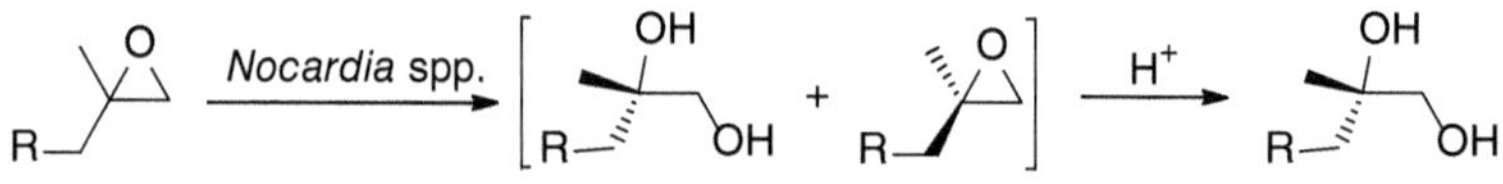

Scheme 3.72 One-pot chemoenzymatic deracemization of (±)-2,2-disubstituted epoxides. Source: Adapted from Orru et al. [78].

Similarly, by employing *Rhodococcus* sp. R312 with (*R*)-selective epoxide hydrolase activity, (*R*)-1-benzyloxy-2-methylpropane-2,3-diol was prepared with 78% yield and >97% ee from racemic (±)-2-methylglycidyl benzyl ether via enantioselective biohydrolysis (with configuration retention) followed by acid-catalyzed hydrolysis with configuration inversion (Scheme 3.73) [79].

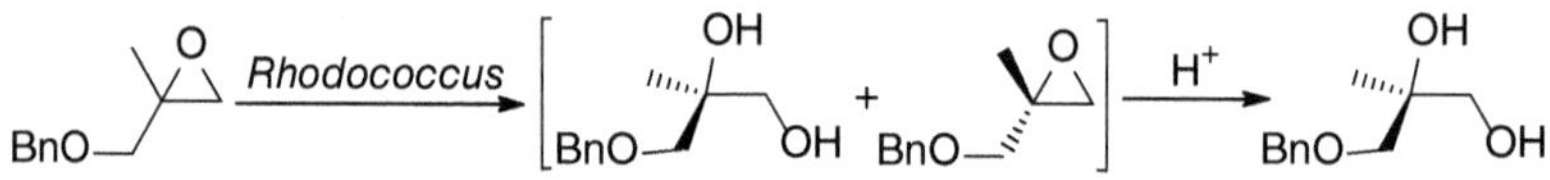

Scheme 3.73 One-pot chemoenzymatic deracemization of (±)-2-methylglycidyl benzyl ether. Source: Adapted from Simeó et al. [79].

An epoxide hydrolase from *Vigna radiate* (VrEH2) was found to catalyze the enantioconvergent hydrolysis of racemic *para*-nitrostyrene oxide, affording (*R*)-*p*-nitrophenyl glycol in 85% ee. By employing the aforementioned chemoenzymatic enantioconvergent hydrolysis procedure, (*R*)-*p*-nitrophenyl glycol was obtained in 90% ee, and with improved ee of 99.0% and 71.5% overall yield after single recrystallization of the crude product (Scheme 3.74) [80].

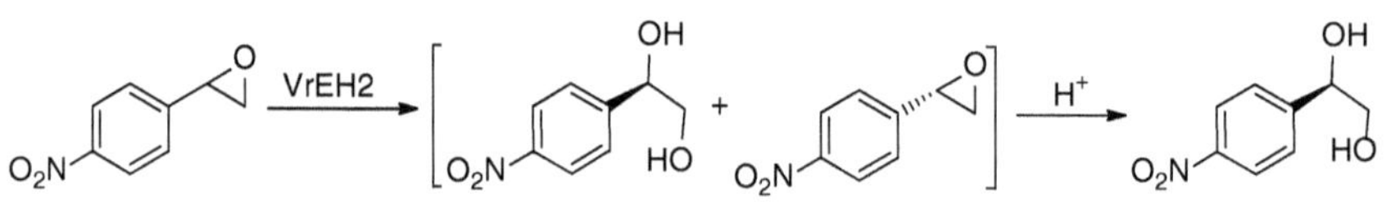

Scheme 3.74 One-pot chemoenzymatic deracemization of racemic *para*-nitrostyrene oxide. Source: Adapted from Wu et al. [80].

3.6 Other Enzymes

3.6.1 Aldolases

Aldolases catalyze the C–C bond formation reactions, and thus are a family of important enzymes for the construction of molecular skeletons. For example,

N-acetylneuraminic acid lyase (NAL) catalyzes the reversible aldol condensation of *N*-acetylmannosamine with pyruvate to give *N*-acetylneuraminic acid. The wild-type and mutant NAL can accept a range of aldehydes as substrate to furnish the corresponding aldol products. A diamine organocatalyst promoted the condensation of a glyoxylamide and acetaldehyde, yielding an aldol product that could serve as the aldehyde substrate of an NAL variant. Both organocatalytic and enzymatic aldol reactions operated in aqueous buffers. Accordingly, the two aldol reactions were performed in one-pot sequence mode, enabling the synthesis of a range of heterocyclic products via the three-component condensation of glyoxylamides, acetaldehyde, and pyruvate, involving two C–C bond-forming steps (Scheme 3.75) [81].

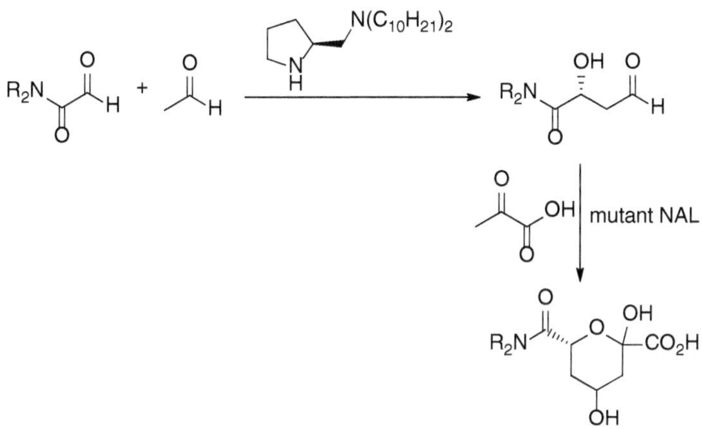

Scheme 3.75 One-pot sequential organocatalytic and enzymatic aldol reactions. Source: Adapted from Kinnell et al. [81].

2-Oxoacid aldolases catalyze the aldol addition of 2-oxoacids with methanol to give 3-substituted 4-hydroxy-2-oxocarboxylic acids, which can be oxidatively decarboxylated to furnish chiral 2-substituted 3-hydroxycarboxylic acid derivatives, a family of valuable intermediates for the synthesis of biologically active compounds. By utilizing enantiocomplementary 2-keto-3-deoxy-L-rhamnonate aldolase double mutant (MBP-YfaU, EC 4.1.2.53) from *E. coli* and 3-methyl-2-oxobutanoate hydroxymethyltransferase (KPHMT, also acted as a Class II aldolase), both (*S*)- and (*R*)-3-substituted 4-hydroxy-2-oxocarboxylic acids were obtained from the aldol reaction of 2-oxoacids with formaldehyde. They were transformed into the corresponding (*S*)- and (*R*)-2-substituted 3-hydroxycarboxylic acids, respectively, by oxidative decarboxylation with hydrogen peroxide in a one-pot sequential process. The subsequent esterification of the resulting acids afforded *S*-configurated esters in 69–80% isolated yields with 94–99% ee, and the *R*-enantiomers in 57–88% isolated yields with 88–98% ee for the three-step chemoenzymatic procedure (Scheme 3.76) [82].

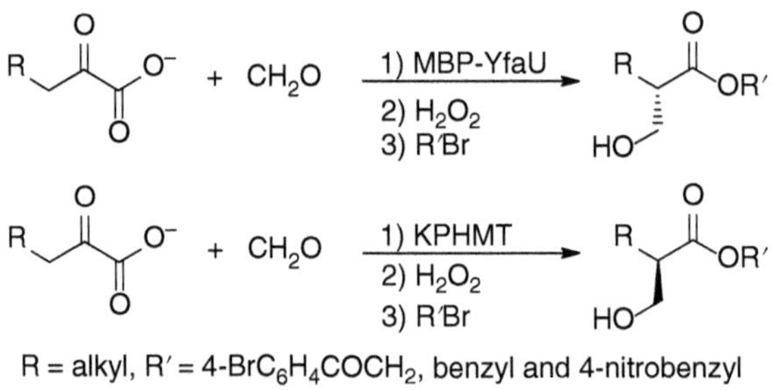

Scheme 3.76 One-pot sequential chemoenzymatic synthesis of chiral 2-substituted 3-hydroxycarboxylic esters. Source: Adapted from Marín-Valls et al. [82].

3.6.2 Halohydrin Dehalogenases

In the carbonyl reductase section, we discussed the one-pot cascade reaction involving enzymatic reduction of α-azidoacetophenone derivatives and the copper-catalyzed azide–alkyne cycloaddition (CuAAC) for the preparation of optically pure triazoles from α-azidoacetophenone derivatives. The intermediate chiral azido alcohols could also be produced via biocatalytic azidolysis of epoxides by using a halohydrin dehalogenase as biocatalyst. In this context, the halohydrin dehalogenase from *Agrobacterium radiobacter* (HheC) was applied to the azidolysis of the substituted styrene oxides in the presence of sodium azide to generate the corresponding chiral 2-azido-1-substituted phenylethanol with high optical purity. The following [3 + 2] cycloaddition of the azido functional group with alkynes under copper catalysis was conducted in one-pot manner to afford the corresponding optically pure 1,2,3-triazoles (Scheme 3.77) [83]. These compounds are promising pharmacophores, β-adrenergic receptor blocker analogs, and potential imaging agents.

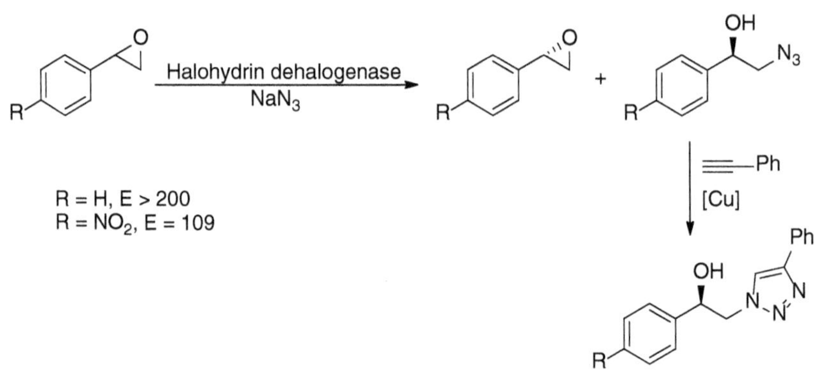

Scheme 3.77 One-pot sequential biocatalytic azidolysis of epoxides and click reaction. Source: Adapted from Campbell-Verduyn et al. [83].

The one-pot ring opening and click reaction can also proceed under copper-free conditions. Therefore, cyclooctyne was added into the reaction mixture of the halohydrin dehalogenase-catalyzed azidolysis of *para*-nitrostyrene oxides without the copper catalyst, and the corresponding triazole was obtained with 96% ee after 48 hours, and the remaining epoxide had 47% ee (Scheme 3.78) [83].

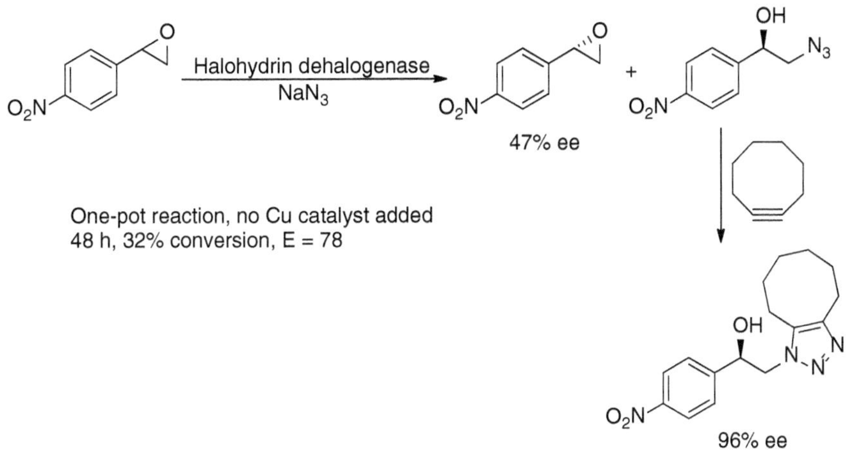

Scheme 3.78 One-pot sequential biocatalytic azidolysis of epoxides and click reaction. Source: Adapted from Campbell-Verduyn et al. [83].

Actually, the enantiopure β-hydroxy azides could be produced from chiral β-halo alcohols via the halohydrin dehalogenase-catalyzed cascade of ring closure toward the epoxide and subsequent ring opening by the azide anion [84]. Furthermore, the ADH catalyzed the reduction of prochiral α-halo ketones to afford the enantiopure β-halo alcohol. As such, recombinant *E. coli* strains were constructed to co-express both enzymes with complementary stereoselectivity. One recombinant *E. coli* strain (CT cells) expressed AdhT, an *R*-selective ADH from *Thermoanaerobacter* sp., and HheC, the *R*-selective halohydrin dehalogenase from *A. radiobacter* AD1. The other *E. coli* cells (BL cells) expressed AdhL, the *S*-selective ADH from *L. brevis*, and HheBGP1, the halohydrin dehalogenase from *Mycobacterium* sp. GP1. These whole cell biocatalysts effectively catalyzed the transformation of various prochiral α-halo ketones into the corresponding β-azido alcohols with 96–99% ee and 35–70% isolated yields. The whole-cell cascade of three biocatalytic reactions was then combined with the click reaction in a one-pot process. Following the biocatalytic cascade, phenylacetylene, $CuSO_4$, sodium ascorbate, and a MonoPhos ligand were added into the reaction mixture; the expected β-hydroxytriazoles were produced in 18–65% isolated yields and 97–99% ee (Scheme 3.79). These results demonstrated the synthetic applicability of this one-pot chemoenzymatic process consisting of four enzymatic reactions and one chemical transformation [85].

3.6.3 Phenylalanine Ammonia Lyases

In nature, phenylalanine ammonia lyase (PAL) catalyzes the deamination of L-phenylalanine to cinnamic acid and the reverse reaction. Various L-phenylalanine derivatives could be prepared by the enantioselective addition of ammonia to

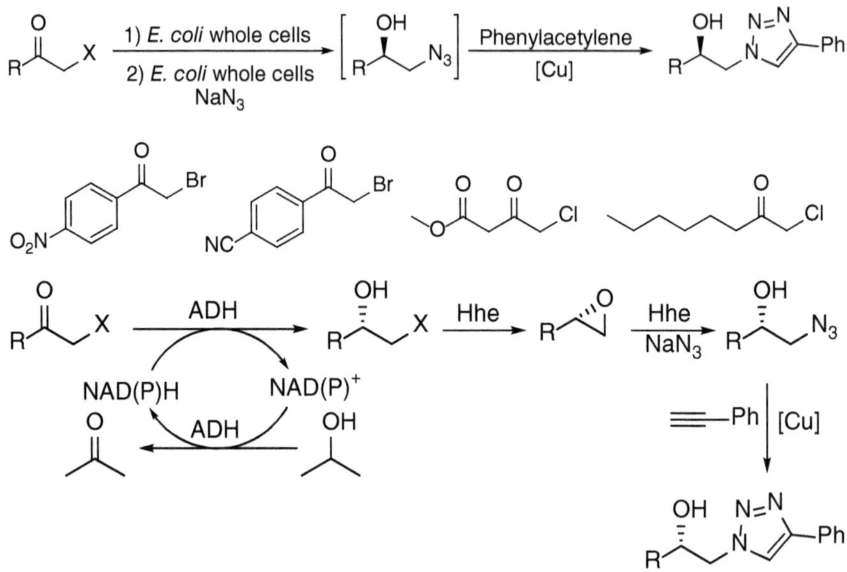

Scheme 3.79 One-pot chemoenzymatic synthesis of both enantiomers of β-hydroxytriazoles.

cinnamic acid derivatives catalyzed by this enzyme [86]. The α,β-unsaturated acids were produced from commercially available and inexpensive substituted benzaldehydes by reacting with malonic acid under organocatalysis to form the product with loss of a molecule of CO_2. The so-called Knoevenagel–Doebner condensation was incorporated with PAL-catalyzed amination to synthesize L-arylalanines with high yield and optical purity in a one-pot chemoenzymatic process (Scheme 3.80) [87]. The PAL-catalyzed synthesis of (*S*)-arylalanines was also integrated with the copper-catalyzed ring closure of (*S*)-*ortho*-chlorophenylalanine and (*S*)-*ortho*-bromophenylalanine to prepare (*S*)-2-indolinecarboxylic acid (Scheme 3.81), a key intermediate for angiotensin 1 converting enzyme (ACE) inhibitors that are used for the treatment of hypertension [88].

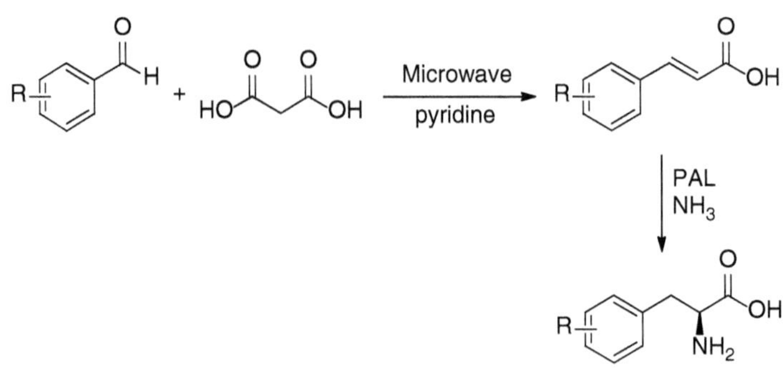

Scheme 3.80 One-pot chemoenzymatic synthesis of L-arylalanines. Source: Adapted from Parmeggiani et al. [87].

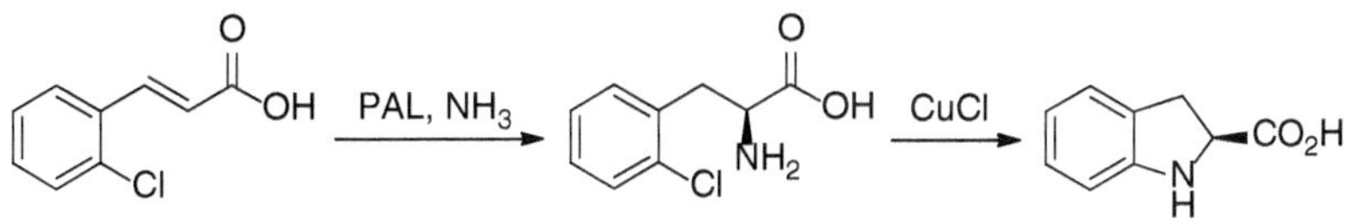

Scheme 3.81 One-pot chemoenzymatic synthesis of (*S*)-2-indolinecarboxylic acid.

When 4-bromocinnamic acid was used as the substrate of PAL, the product (*S*)-4-bromophenylalanine could undergo Suzuki coupling with arylboronic acids to give the biaryl alanines. Indeed, a mutant F107A of AvPAL from the cyanobacterium *Anabaena variabilis* catalyzed the hydroamination of 4-bromocinnamic acid to afford the L-bromophenylalanine. Following the removal of ammonium salts and unreacted 4-bromocinnamic acid from the reaction mixture by adsorption on an ion exchange resin, Boc protection and Pd-catalyzed cross-coupling with a range of arylboronic acids were performed in a one-pot sequential procedure, and the corresponding Boc-protected biaryl alanines were prepared in 33–65% overall isolated yields (Scheme 3.82) [89].

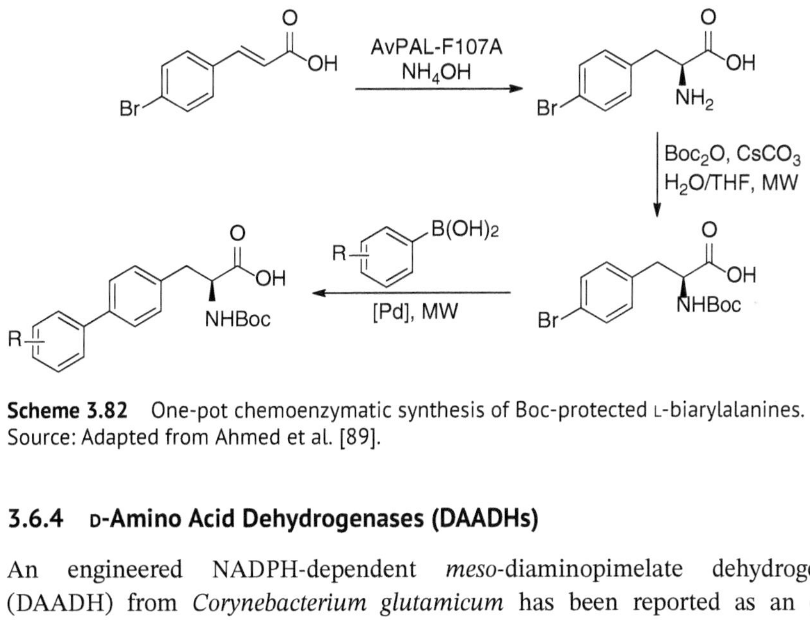

Scheme 3.82 One-pot chemoenzymatic synthesis of Boc-protected L-biarylalanines. Source: Adapted from Ahmed et al. [89].

3.6.4 D-Amino Acid Dehydrogenases (DAADHs)

An engineered NADPH-dependent *meso*-diaminopimelate dehydrogenase (DAADH) from *Corynebacterium glutamicum* has been reported as an effective catalyst for the reductive amination of a wide range of aliphatic and aromatic α-ketoacids. The evolved DAADH also catalyzed the reductive amination of 4-bromophenylpyruvic acid using glucose/GDH system for the regeneration of NADPH cofactor, giving D-4-bromophenylalanine with >99% ee. The 4-bromophenylalanine was subsequently Boc-protected and coupled with a panel of arylboronic acids with Pd catalysis. The whole chemoenzymatic sequence was successfully run in a one-pot system to furnish the Boc-protected biaryl alanine derivatives in D-configuration with high overall yield and optical purity (Scheme 3.83) [89].

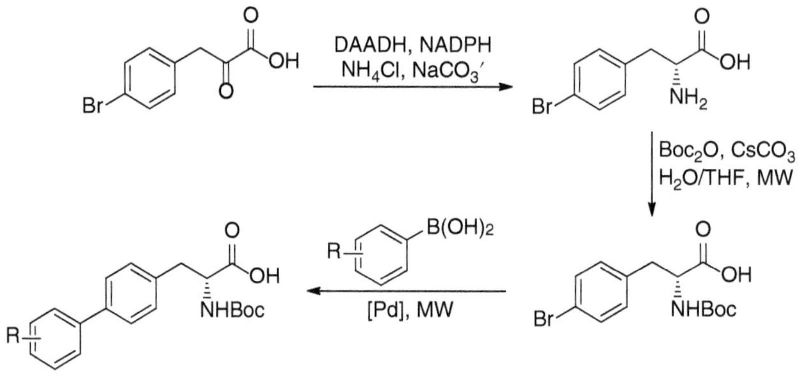

Scheme 3.83 One-pot chemoenzymatic synthesis of Boc-protected D-biarylalanines. Source: Adapted from Ahmed et al. [89].

Earlier in this chapter, we discussed combination of Wacker oxidation and TA-catalyzed reductive amination via catalyst compartmentalization to realize the direct transformation of styrenes into 1-phenylethylamines in a one-pot two-step process. Amine dehydrogenase from *Exigobacterium sibiricum* was used to replace the TA for the reductive amination of the *in situ*-generated acetophenone intermediate. As such, styrene was converted to (*R*)-1-phenylethylamine in 76% isolated yield and 99% ee via this overall one-pot asymmetric hydroamination of styrene with ammonia (Scheme 3.84) [90].

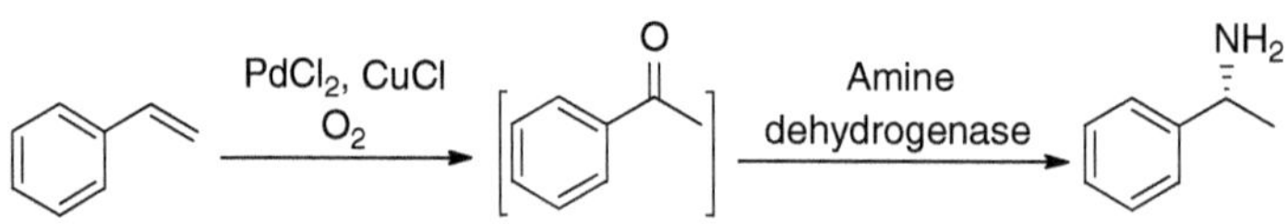

Scheme 3.84 One-pot sequential Wacker oxidation of styrene and amine dehydrogenase-catalyzed reductive amination. Source: Adapted from Uthoff and Gröger [90].

3.6.5 Halogenases

Flavin adenine dinucleotide (FAD)-dependent halogenases catalyze the regioselective halogenation of aryl C–H bonds in the presence of benign halide salts such as NaCl or NaBr and oxygen, providing a promising sustainable approach to access aryl halides. The biocatalytic halogenation installs a synthetic useful C–X bond that allows for further modification of the molecule via a wide range of different bond (C–C, C–N, C–O) formation. The enzyme preparations of tryptophan 5-halogenase PyrH, 6-halogenase Thal, and 7-halogenase RebH using the CLEA methodology catalyzed the regioselective bromination of tryptophan at room temperature in the presence of NaBr and O_2. After the separation of the biocatalyst by filtration, direct Suzuki–Miyaura cross-coupling reaction was performed by adding the Pd catalyst, base, and arylboronic acid into the reaction mixture. After the subsequent *in situ tert*-butoxycarbonyl (Boc) protection, a series of Boc-protected C5, C6, or C7 aryl-substituted derivatives were obtained in up to 94% yields (Scheme 3.85) [91].

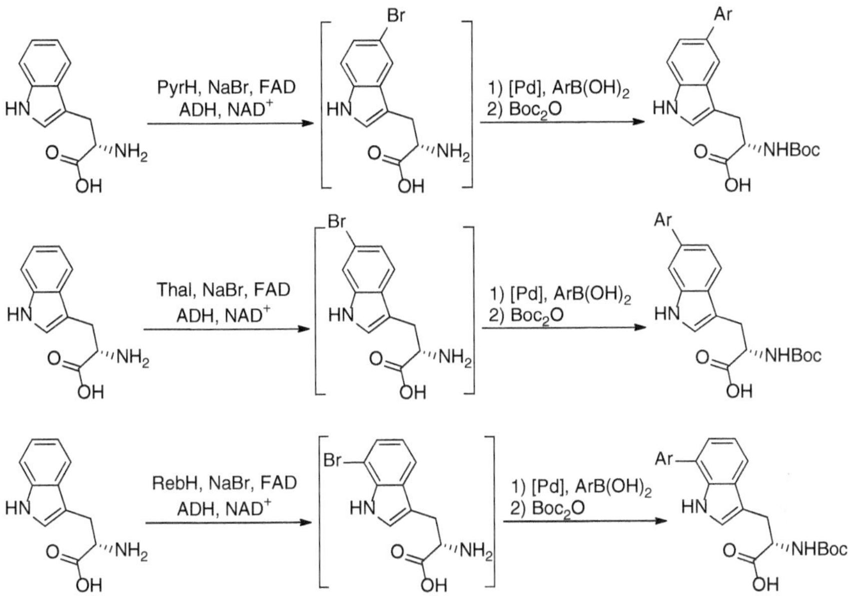

Scheme 3.85 One-pot chemoenzymatic synthesis of Boc-protected C5, C6, or C7 aryl-substituted derivatives. Source: Based on Frese et al. [91].

The variant 3-SS of RebH showed high activity toward tryptoline and converted it to 5-bromotryptoline. By using a combination of $Pd(OAc)_2$ and water-soluble SPhos, the subsequent Suzuki reaction of arylboronic acids with the crude bioconversion extract proceeded smoothly, producing the corresponding arylated tryptoline derivatives in good to high isolated yields (Scheme 3.86). In addition to tryptoline, carvedilol and pindolol were also accepted by variant 3-SS as the substrate for biohalogenation and the following Suzuki coupling reaction. The crude extracts of 5-bromotryptoline from enzymatic halogenation were also tested for Buchwald–Hartwig amination and alkoxylation. The *in situ*-formed 5-bromotryptoline underwent smooth coupling with aminopyridine derivatives,

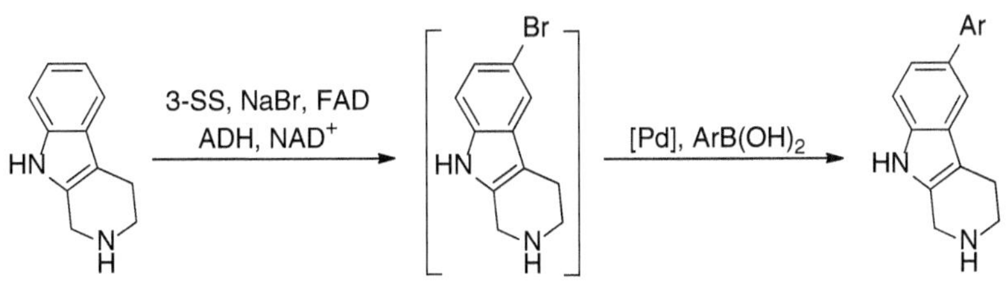

Scheme 3.86 One-pot sequential enzymatic halogenation and Suzuki–Miyaura cross-coupling reaction.

generating the corresponding secondary amines in 51–87% yields via a two-step C–H amination (Scheme 3.87). Although the two-step C–H alkoxylation on tryptoline was unsuccessful, thenalidine underwent two-step chemoenzymatic C–H alkoxylation using a mutant 4-V as the biocatalyst, providing trifluoroethoxythenalidine in 33% yield (Scheme 3.88) [92]. The combination of biohalogenation and chemical reactions has also been applied in the synthetic modification of the natural products to enhance their structural diversity [93, 94].

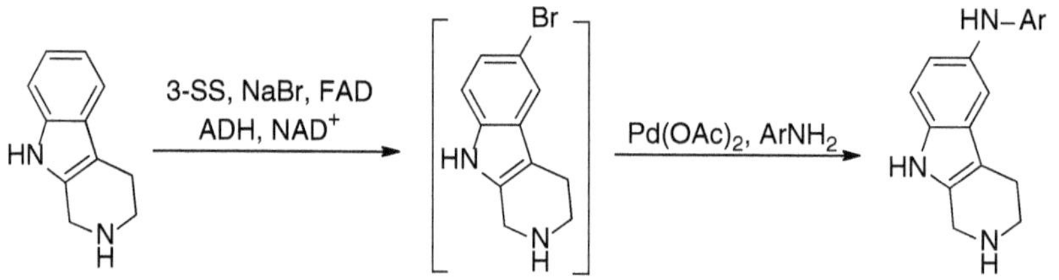

Scheme 3.87 One-pot sequential enzymatic halogenation and Buchwald–Hartwig amination.

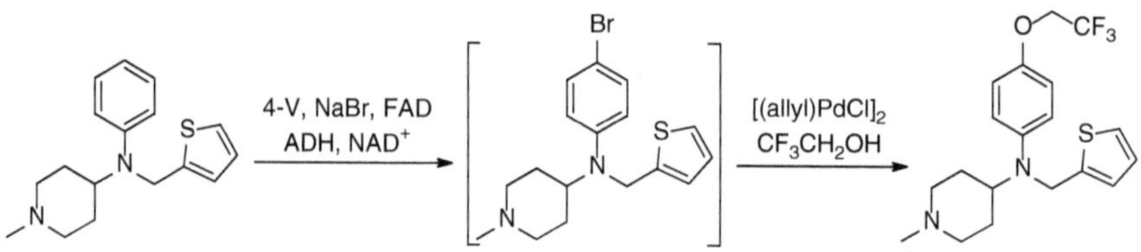

Scheme 3.88 One-pot sequential enzymatic halogenation and Buchwald–Hartwig alkoxylation. Source: Based on Durak et al. [92].

3.6.6 Imine Reductases

The Asinger-type multicomponent reaction, in which an α-halogenated carbonyl compound reacts with sodium hydrosulfide (NaSH) and the *in situ*-formed thiol reacts with another carbonyl compound and ammonia to give a 3-thiazoline, is an effective method for the synthesis of 3-thiazolines and other related heterocycles. This cyclic imine could be reduced to afford 3-thiazolidine. Recently, imine reductases were increasingly applied for the reduction of various imines. In this context, an imine reductase from *Mycobacterium smegmatis* was tested toward the combination of Asinger reaction in a one-pot process. While methylene chloride was usually used to dilute the α-chlorinated ketone during Asinger-type multicomponent reaction, the reaction of sodium hydrosulfide monohydrate, cyclopentanone, aqueous ammonia solution, and 3-chloro-3-methylbutan-2-one in DMSO was found to proceed smoothly, furnishing 2,2,3-trimethyl-1-thia-4-azaspiro[4.4]non-3-ene. Methylene chloride is hardly compatible with biocatalysts, but DMSO has been widely used in biocatalytic reactions to increase the substrate solubility. When Asinger-type multicomponent condensation reaction and the subsequent enzymatic reduction of 3-thiazoline were conducted in a sequential mode,

(*S*)-2,2,3-trimethyl-1-thia-4-azaspiro[4.4]nonane was obtained with 26% yield and 99% ee (Scheme 3.89) [95].

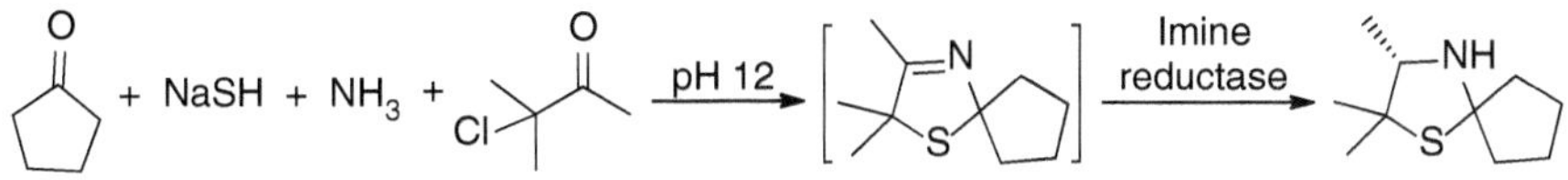

Scheme 3.89 One-pot sequential Asinger-type multicomponent reaction and enzymatic reduction. Source: Adapted from Zumbrägel and Gröger [95].

3.6.7 Decarboxylases

A phenolic acid decarboxylase from *Bacillus subtilis* (bsPAD) had been reported to catalyze the enzymatic decarboxylation of hydroxycinnamic acids, yielding vinylphenols. Metal-catalyzed metathesis of two vinylpenol molecules would produce symmetric stilbene derivatives such as 4,4′-dihydroxy-*trans*-stilbene. Because hydroxycinnamic acid derivatives are abundant in lignocellulosic biomass, this synthetic methodology makes them very attractive resources for bio-based polymers and other materials. When the enzymatic decarboxylation and the metathesis reaction were conducted under their individually optimized conditions in a two-pot process, the vinylphenols were obtained in high yields of 80–95%, but their tendency to undergo spontaneous polymerization required immediate conducting of metathesis reaction and reduced the overall yields of stilbene derivatives. In order to alleviate this limitation, the enzymatic decarboxylation and metal-catalyzed metathesis reaction were carried out in a one-pot cascade using a biphasic system comprising buffer and isooctane. Although enzymatic decarboxylation proceeded smoothly to full conversion, the metathesis reaction did not exceed 20% conversion in the biphasic buffer/isooctane system due to the lability of the metathesis catalysts toward water and air. The decarboxylase bsPAD was then encapsulated in poly(vinyl alcohol)/poly(ethylene glycol) (PVA/PEG) cryogels, and the encapsulated bsPAD enzyme showed good performance in nonpolar solvents such as methyl *tert*-butyl ether (MTBE) or isooctane. The enzymatic decarboxylation could achieve >99% conversion in MTBE; following the enzyme recovery after the reaction and an intermediate drying step, subsequent Ru-catalyzed metathesis was performed in the same solvent, producing 4,4′-dihydroxystilbene in 90% overall yield (Scheme 3.90) [96].

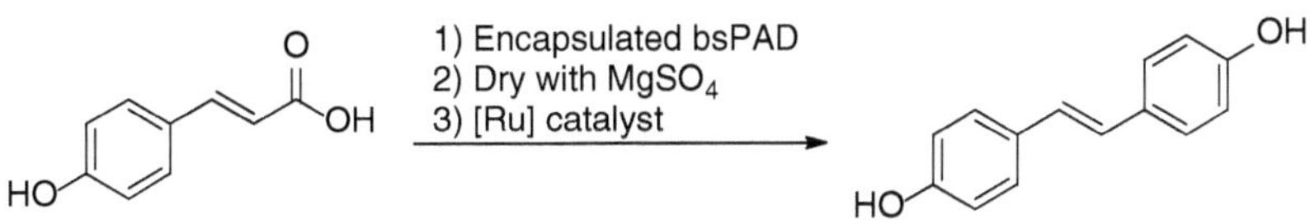

Scheme 3.90 One-pot sequential enzymatic decarboxylation and Ru-catalyzed metathesis. Source: Based on Gómez Baraibar et al. [96].

Transition metal catalyst could be embedded in a non-active protein scaffold, forming a biohydrid catalyst (BHC). This artificial metalloprotein enabled organometallic catalysts to be active in an aqueous medium. Grubbs–Hoveyda catalyst was covalently anchored on a variant of Ferric hydroxamate uptake protein component A (FhuA) of *E. coli*. The resulting BHC was combined with ferulic acid decarboxylase (FDC1) from *Saccharomyces cerevisiae* for the preparation of stilbene derivatives from cinnamic acids. In this two-step one-pot cascade process, substituted cinnamic acids were enzymatically decarboxylated to give styrenes in >98% conversion, which were converted into the corresponding stilbenes with 74% and 64% conversions by olefin cross-metathesis (Scheme 3.91) [97].

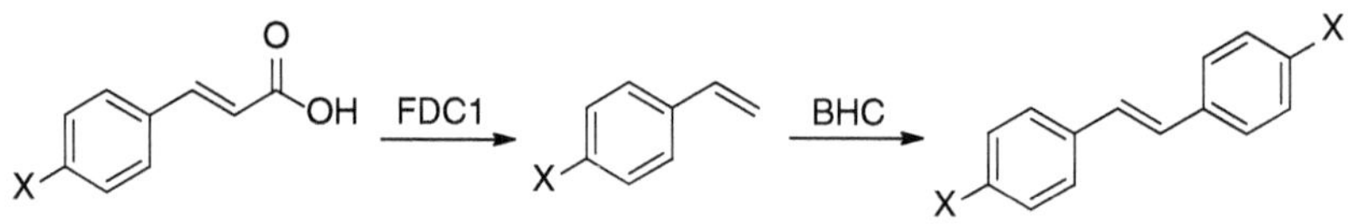

Scheme 3.91 One-pot sequential enzymatic decarboxylation and BHC-catalyzed metathesis. Source: Based on Mertens et al. [97].

Arylmalonate decarboxylase (AMDase) from *Bordetella bronchiseptica* catalyzed the decarboxylation of prochiral arylmalonic acids to give the corresponding (*R*)-monoacids. This enzyme and its mutants also catalyzed the decarboxylation of methylvinylmalonic acid derivatives to generate (*R*)- or (*S*)-2-methylalk-3-enoic acid intermediates, which were reduced with hydrazine monohydrate in the presence of catalytic amount of $CuCl_2$. Both (*R*)- and (*S*)-2-methylalkanoic acids were prepared with up to 83% yields and >99% ee values via this chemoenzymatic one-pot sequential cascade using *R*- or *S*-selective mutant AMDase (Scheme 3.92) [98].

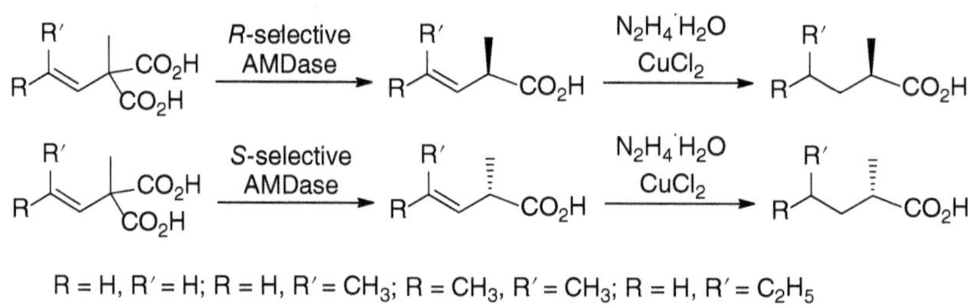

Scheme 3.92 One-pot sequential enzymatic decarboxylation and chemical reduction. Source: Based on Enoki et al. [98].

3.6.8 Cytochrome P450s

The regio- and stereoselective hydroxylation of saturated fatty acids is a useful transformation because it produces high-value chiral hydroxyacids and lactones. However, it still represents a synthetic challenge due to the difficulty in the activation of unreactive C–H bonds and the need for regio- and stereoselectivity. A cytochrome P450 monooxygenase (CYP116B46) from *Tepidiphilus thermophilus* was recently found to catalyze the regio- and enantioselective hydroxylation at C5 of decanoic acid, affording (*S*)-5-hydroxydecanoic acid, which underwent acid-catalyzed lactonization to generate (*S*)-δ-decalactone, a high-value fragrance compound, with greater than 90% ee (Scheme 3.93) [99].

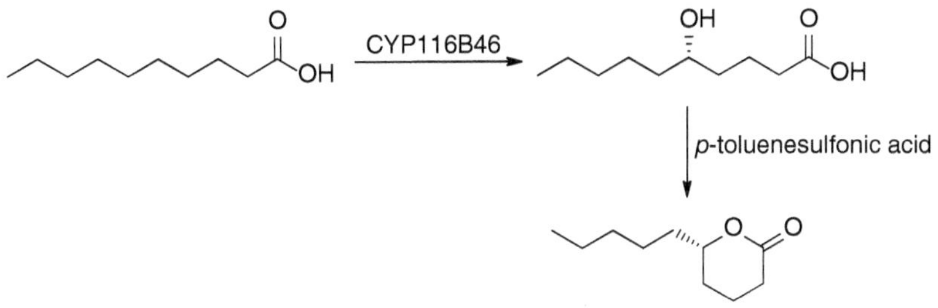

Scheme 3.93 One-pot chemoenzymatic synthesis of (*S*)-δ-decalactone. Source: Based on Manning et al. [99].

Recently, a chemoenzymatic strategy involving Rh-catalyzed asymmetric transfer hydrogenation and whole-cell-mediated asymmetric hydroxylation was reported for the construction of a series of chiral 2-substituted-tetrahydroquinoline-4-ols. The two reactions were conducted in aqueous buffer in a one-pot two-step process; the corresponding 2-substituted-tetrahydroquinoline-4-ols bearing two stereocenters were prepared from 2-substituted-quinolines in up to 47% yield, 99 : 1 dr, and >99% ee (Scheme 3.94) [100].

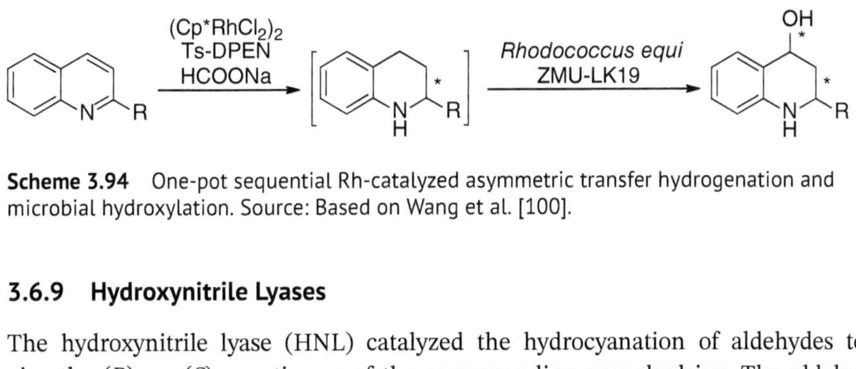

Scheme 3.94 One-pot sequential Rh-catalyzed asymmetric transfer hydrogenation and microbial hydroxylation. Source: Based on Wang et al. [100].

3.6.9 Hydroxynitrile Lyases

The hydroxynitrile lyase (HNL) catalyzed the hydrocyanation of aldehydes to give the (*R*)- or (*S*)-enantiomer of the corresponding cyanohydrins. The aldehydes could be prepared from alcohols using TEMPO/PhI(OAc)$_2$ as the oxidant. A chemoenzymatic process was thus developed by combining the oxidation of γ,δ-unsaturated alcohols using TEMPO/PhI(OAc)$_2$ with the hydrocyanation catalyzed by a hydroxynitrile lyase from *Hevea brasiliensis* (HbHNL), and the corresponding (*S*)-cyanohydrin derivatives were prepared in good overall yields and high ee values after *in situ* acetylation (Scheme 3.95). This protocol avoided the isomerization of the double bond during the isolation of aldehyde intermediate. The immobilization of TEMPO on silica improved the optical purity of the products compared to the soluble TEMPO [101].

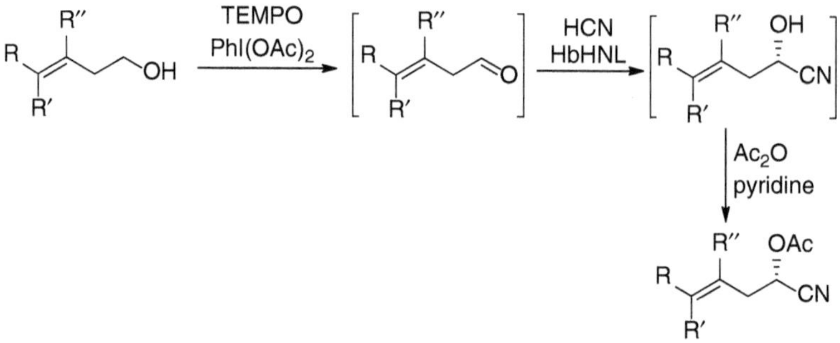

Scheme 3.95 One-pot chemoenzymatic synthesis of acetylated (*S*)-cyanohydrins.

3.6.10 Nitrilases

Various nitrilases have been reported to catalyze the selective hydrolysis of dinitriles to give mono-cyano carboxylic acids. An efficient chemoenzymatic one-pot process involving nitrilase was developed for the synthesis of gabapentin. 1-Cyanocycloalkaneacetonitrile was hydrolyzed by an engineered nitrilase from *Acidovorax facilis* ZJB09122, furnishing 1-cyanocyclohexaneacetic acid in 94% yield with a space-time productivity of 461 g/l/day. Without further purification, the resulting aqueous 1-cyanocycloalkaneacetic acid was directly hydrogenated on Raney-nickel catalyst to produce 2-azaspiro[4.5]decan-3-one, which was subsequently converted into gabapentin by simple chemical hydrolysis (Scheme 3.96). Gabapentin was prepared in high purity and 77.3% overall yield from 1-cyanocyclohexylacetonitrile [102].

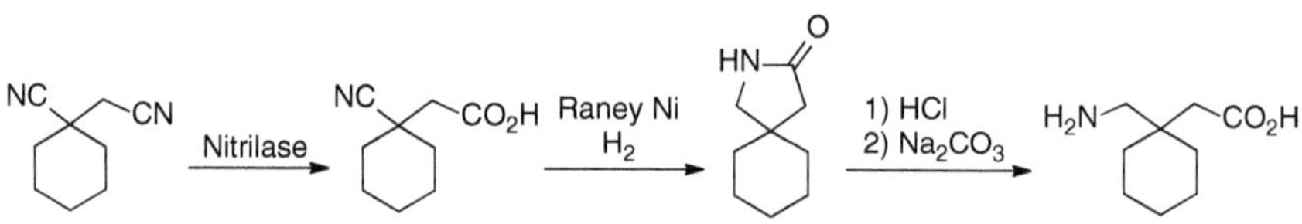

Scheme 3.96 One-pot chemoenzymatic synthesis of gabapentin.

3.6.11 Laccases

Lewis acids catalyze domino intermolecular 1,2-addition and intramolecular 1,4-addition between (*Z*)-3-hexene-2,5-dione and 1,3-dicarbonyls followed by elimination of water, generating 3-methyl-6,7-dihydrobenzofuran-4(5*H*)-ones exclusively. For example, $Gd(OTf)_3$ catalyzed the reaction of (*Z*)-3-hexene-2,5-dione with a diversity of 1,3-dicarbonyls, affording the corresponding 3-methyl-6,7-dihydrobenzofuran-4(5H)-ones in up to 82% isolated yields. The Lewis acid-catalyzed transformation had been combined with the laccase-catalyzed formation of (*Z*)-3-hexene-2,5-dione by oxidative cleavage of 2,5-dimethylfuran. This reaction cascade could not be achieved in the one-pot concurrent mode, but proceeded smoothly when Lewis acid and 1,3-dicarbonyls were added after the completion of the laccase-catalyzed oxidative cleavage of 2,5-dimethylfuran (Scheme 3.97). A series of 6,7-dihydrobenzofuran-4-(5*H*)-ones were prepared directly from 2,5-dimethylfuran and 1,3-dicarbonyls in 45–60% isolated yields by employing this chemoenzymatic one-pot sequential process [104].

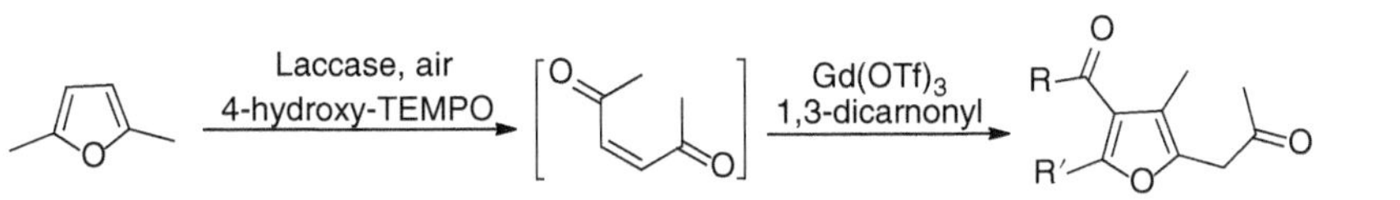

Scheme 3.97 One-pot chemoenzymatic synthesis of 6,7-dihydrobenzofuran-4-(5*H*)-ones. Source: Adapted from Akagawa et al. [103].

Laccase catalyzed the α-oxygenation of aldehydes in the presence of TEMPO. This was combined with a peptide-catalyzed Friedel–Crafts-type alkylation of α,β-unsaturated aldehydes with indoles in a one-pot two-step process. A resin-supported peptide catalyst **A** carried out Friedel–Crafts-type alkylation of α,β-unsaturated aldehydes with several substituted indoles in a mixture of H_2O and THF (v/v, 2 : 1), followed by laccase-catalyzed oxygenation at the α-position of aldehydes, affording highly functionalized indole derivatives in a highly enantioselective manner (Scheme 3.98) [103].

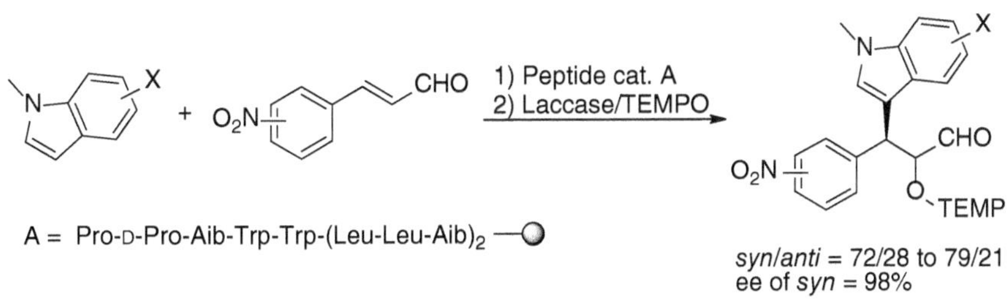

Scheme 3.98 One-pot chemoenzymatic synthesis of highly functionalized indole derivatives.

3.6.12 Transglutaminases

Microbial transglutaminase (MTG) catalyzes the acyl transfer between the γ-carboxamide group of a glutamine-containing substrate and the ε-amino group of a lysine-containing substrate. MTG displays flexibility toward chemically diverse amines as the substitute for lysine. For example, the model peptide *N*-benzyloxycarbonyl-L-glutaminylglycine (ZQG) reacted with the azido- or alkyne-bearing amines such as azidopropylamine or propargylamine under the action of MTG. The resulting peptides could undergo CuAAC with an alkyne or azide functional group. Indeed, when the MTG-catalyzed acyl transfer and CuAAC reaction were conducted in a one-pot stepwise manner, the corresponding amidotriazole peptides were obtained in 53% and 76% yields (Schemes 3.99 and 3.100), offering a versatile peptide and protein conjugation methodology that could be useful for targeted protein labeling [105, 106].

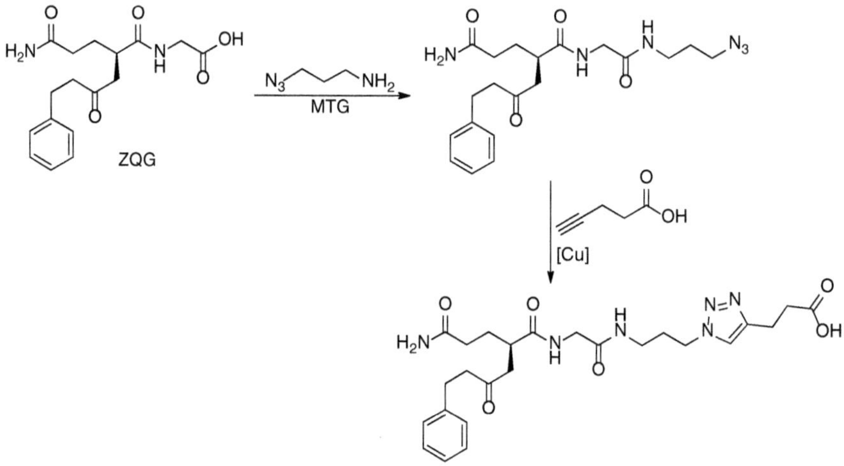

Scheme 3.99 One-pot chemoenzymatic synthesis of amidotriazole peptides.

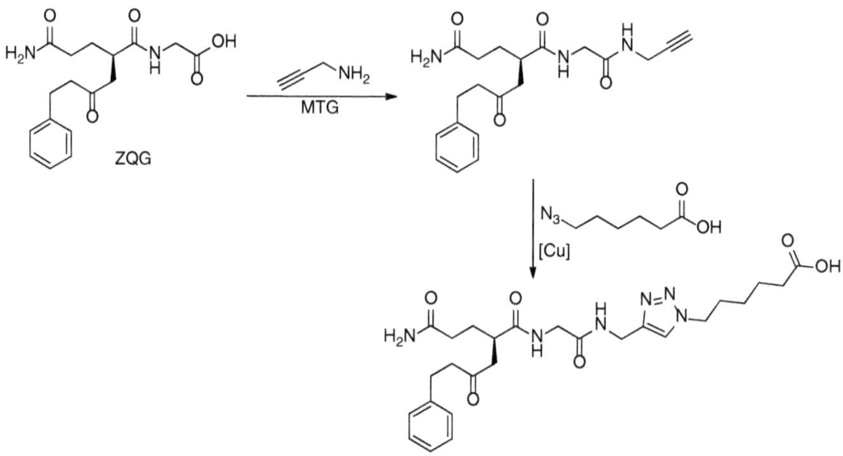

Scheme 3.100 One-pot chemoenzymatic synthesis of amidotriazole peptides.

3.6.13 α-Ketoglutarate (α-KG)-dependent Non-heme Iron Oxygenases

There are several biocatalytic approaches to accomplish benzylic hydroxylation by utilizing microbes or monooxygenases. α-Ketoglutarate (α-KG)-dependent non-heme iron oxygenase catalyzed the selective oxidation of benzylic C–H bonds affording benzyl alcohol. Two fungal non-heme iron oxygenases CitB (from *Monascus ruber*) and ClaD (*Penicillium crustosum*) were found to catalyze benzylic hydroxylation of *o*-cresol substrates. Heating of the resulting mixture resulted in the formation of *o*-quinone methides, which could be intercepted by a nucleophile or a dienophile to enable the synthesis of an array of new compounds. Indeed, after biocatalytic C–H hydroxylation of 2,4-dihydroxy-3,6-dimethylbenzaldehyde using CitB as biocatalyst, heating the reaction mixture to 40 °C in the presence of thiophenol resulted in the quantitative transformation of 2,4-dihydroxy-3-(hydroxymethyl)-6-methylbenzaldehyde into 2,4-dihydroxy-3-((phenylthiol)methyl)-6-methylbenzaldehyde through the *o*-quinone methide intermediate. This chemoenzymatic transformation of C–H bond to C–S bond was applied to a panel of *o*-cresol substrates using CitB and ClaD, and the corresponding thioethers were obtained in 26–40% yields. Thiophenol was also replaced by other nucleophiles such as alcohols and amines, resulting in the direct transformation of C–H bond into C–O and C–N bonds (Scheme 3.101).

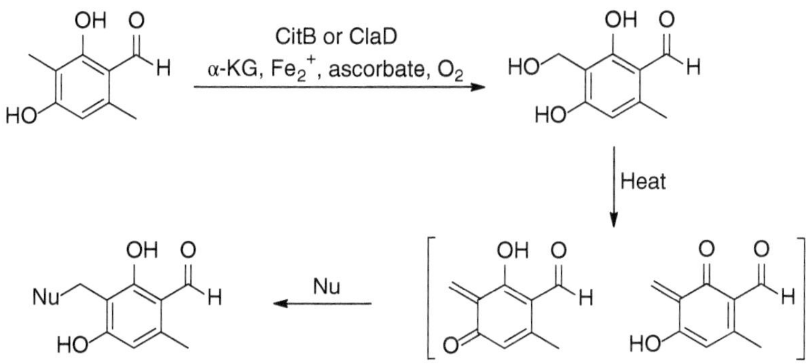

Nu = Methanol, 59% yield; *iso*-Propanol, 41% yield; *n*-Butanol, 45% yield; Thiophenol, 71% yield; Benzyl mercaptan, 41% yield; Morpholine, 51% yield

Scheme 3.101 One-pot chemoenzymatic transformation of benzylic C–H bond into C–O, C–S, and C–N bonds.

The *o*-quinone methide intermediate could also be intercepted with various dienophiles via inverse electron-demand Diels–Alder (IEDDA) reactions. Following the biocatalytic benzylic hydroxylation, the reaction mixture was heated to 45–65 °C in the presence of a dienophile such as ethyl vinyl ether; two isomeric products were obtained in varied ratio depending on the dienophile (Scheme 3.102).

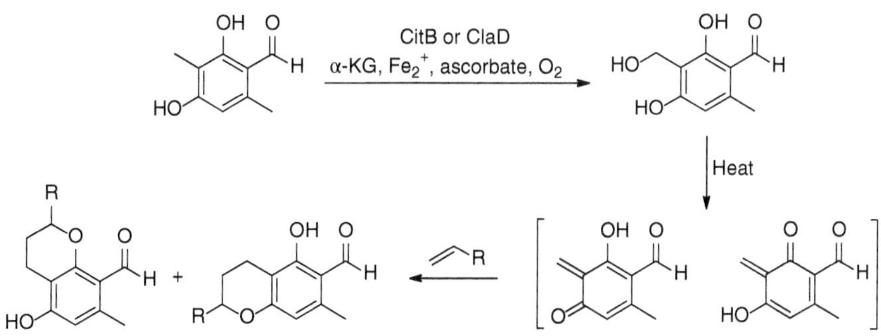

R = OEt, 55% yield (1:2); SEt, 44% yield (1:4); Ph, 33% yield (3:1); *o*-MeOPh, 52% yield (3:1); *p*-MeOPh, 56% yield (4:1); dienophile = 2,3-dihydrofuran, 39% yield (1:4).

Scheme 3.102 One-pot sequential enzymatic benzylic C–H hydroxylation and inverse electron-demand Diels–Alder reaction.

The two isomers were derived from the two possible *o*- quinone methide intermediates. This provided a simple, one-pot sequential process for the direct transformation of benzylic C–H bond into C–O, C–N, C–S, and C–C bonds, enabling ready access to a diversity of important chemical scaffolds. This was highlighted in the chemoenzymatic synthesis of (−)-xyloketal D, a chroman natural product found in fungi *Xylaria* sp. (Scheme 3.103) and the selective modification of peptides [107].

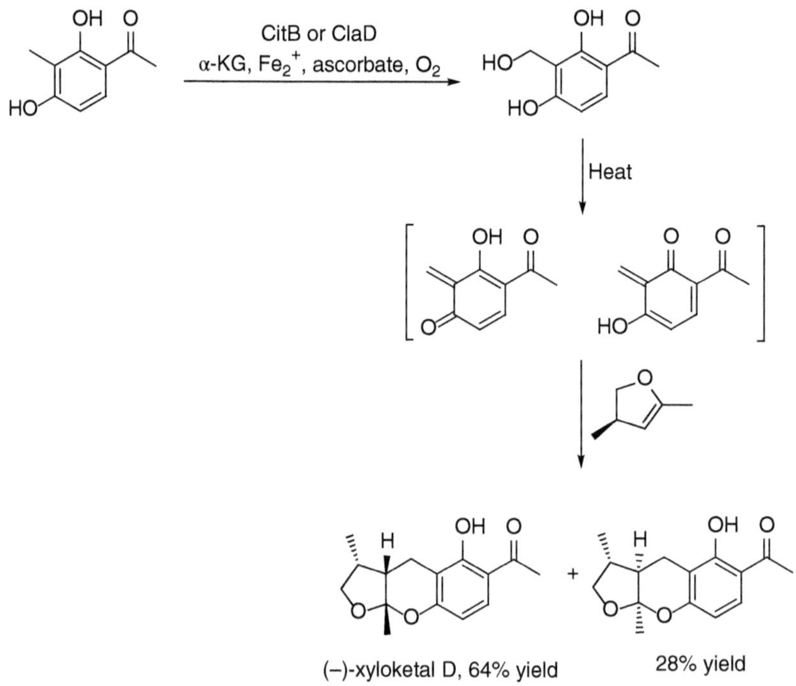

Scheme 3.103 One-pot chemoenzymatic synthesis of (−)-xyloketal D.

3.6.14 Galactose Oxidases

Galactose oxidase from *Fusarium* NRRL 2903 is known as an efficient biocatalyst for the oxidation of alcohols. Lyophilized *E. coli* whole cells containing this enzyme effectively catalyzed the oxidation of benzyl alcohols in the presence of horseradish peroxidase and 2,2′-azino-bis(3-ethylbenzothiazoline-6-sulfonic acid) (ABTS) in phosphate buffer (pH 7.0). The resulting benzaldehydes could be allylated via indium(0)-mediated Barbier-type coupling with allyl bromide or by coupling with allylboronic acid pinacol ester without any workup step in between (Scheme 3.104). This one-pot, two-step chemoenzymatic process was applied to the synthesis of a variety of homoallylic sec-alcohols in water from the corresponding (hetero)-benzylic and cinnamic alcohols in up to 96% isolated yield [108].

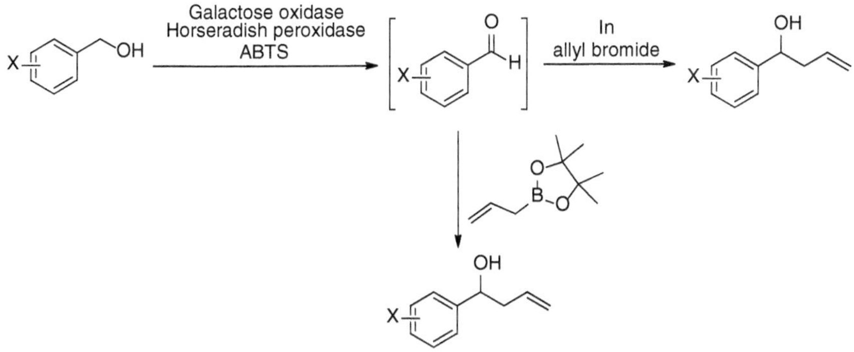

Scheme 3.104 One-pot chemoenzymatic oxidation and allylation cascade.

3.6.15 FAD-dependent Monooxygenases

FAD-dependent monooxygenases are known to catalyze the site- and stereoselective oxidative dearomatization of phenol and resorcinol substrates. For example, the FAD-dependent monooxygenase (TropB) in the gene cluster encoding the biosynthetic pathway of fungal tropolone natural product stipitatonic acid catalyzed the oxidative dearomatization of 2,4-dihydroxy-3,6-dimethylbenzaldehyde to give (*R*,*Z*)-2-hydroxy-6-(hydroxymethylene)-2,5-dimethylcyclohex-4-ene-1,3-dione. A similar FAD-dependent monooxygenase AzaH from *Aspergillus niger* ATCC 1015 catalyzed the oxidative dearomatization of 2,4-dihydroxy-6-(4-hydroxy-2-oxopentyl)-3-methylbenzaldehyde to afford 2-hydroxy-6-(hydroxymethylene)-5-(4-hydroxy-2-oxopentyl)-2-methylcyclohex-4-ene-1,3-dione. Another FAD-dependent monooxygenase (SorbC) involved in the sorbicillactone A synthetic pathway oxidized (2*E*,4*E*)-1-(2,4-dihydroxy-3,5-dimethylphenyl)hexa-2,4-dien-1-one (sorbicillin) at 5-position of the benzene ring with different site- and facial selectivity to furnish the C5-hydroxylation product, the precursor to sorbicillactone A. These three enzymes also catalyzed the site- and stereoselective oxidative dearomatization of various phenol substrates to generate *ortho*-quinol products that could be chemically or biocatalytically transformed into complex molecules. For example, AzaH catalyzed the dearomatization of 2,4-dihydroxy-3-methyl-6-(2-oxopropyl)benzaldehyde; the resulting enol intermediate spontaneously cyclized to the natural product (*R*)-7-hydroxy-3,7-dimethyl-6*H*-isochromene-6,8(7*H*)-dione in 54% isolated yield and >99% ee (Scheme 3.105) [109].

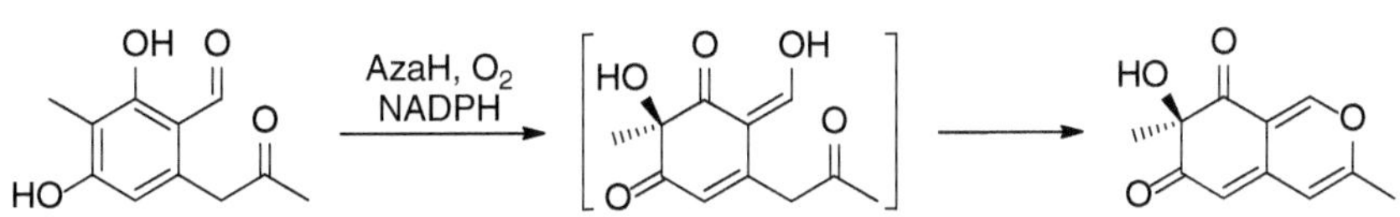

Scheme 3.105 One-pot chemoenzymatic synthesis of natural product (*R*)-7-hydroxy-3,7-dimethyl-6H-isochromene-6,8(7H)-dione. Source: Adapted from Baker Dockrey et al. [109].

The SorbC-catalyzed dearomatization of sorbicillin generated sorbicillinol, which was quite stable in the aqueous buffer system. When the enzymatic reaction mixture was treated with dichloromethane, sorbicillinol was dimerized to furnish bisorbicillinol in 27% yield [110]. Sorbicillinol had also been chemically transformed *in situ* into other natural products. The natural product urea sorbicillinoid was also prepared by coupling SorbC-catalyzed oxidation of sorbicillin with a [4+2] cycloaddition. The *in situ*-generated sorbicillinol underwent facile addition with 1,3-diacetyl-2,3-dihydro-1*H*-imidazol-2-one, delivering urea sorbicillinoid in 21% yield after addition of LiOH into the reaction mixture (Scheme 3.106) [109]. The enzymatically produced sorbicillinol could undergo Diels–Alder reaction with a set of dienophiles to give the corresponding unnatural sorbicillinoids. After simple isolation, the Diels–Alder reaction of sorbicillinol with 2-methoxy-4-vinylphenol led to the formation of sorbicatechol A in 30% yield (Scheme 3.107). Similarly, rezishanone C and rezishanone B were prepared by the Diels–Alder reaction of sorbicillinol with vinyl ethyl ether and vinyl *n*-butyl ether, respectively. When sorbicillinol was treated with KOH and *t*-butylperoxide in DMF at 0 °C, (+)-epoxysorbicillinol was obtained in 25% yield (Scheme 3.108). Sorbicillinol also underwent Michael addition with an alanine-derived C-nucleophile to generate sorbicillactone A [111].

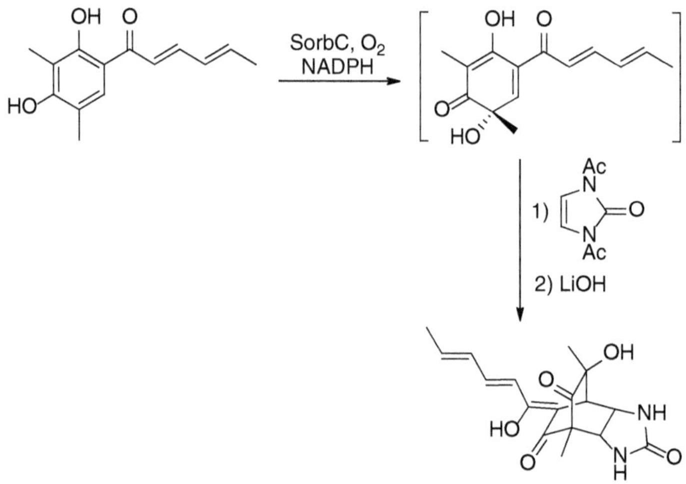

Scheme 3.106 One-pot chemoenzymatic synthesis of natural product urea sorbicillinoid. Source: Adapted from Baker Dockrey et al. [109].

3.7 Summary and Outlook

Compared to the non-one-pot process, one-pot sequential cascade shows improved overall efficiency and operational simplicity because purification of intermediate is not needed. In addition, the chemical and biological transformations can be operated under different conditions by changing pH and/or temperature and adding additional solvents and/or reagents between the reaction steps, offering some flexibility

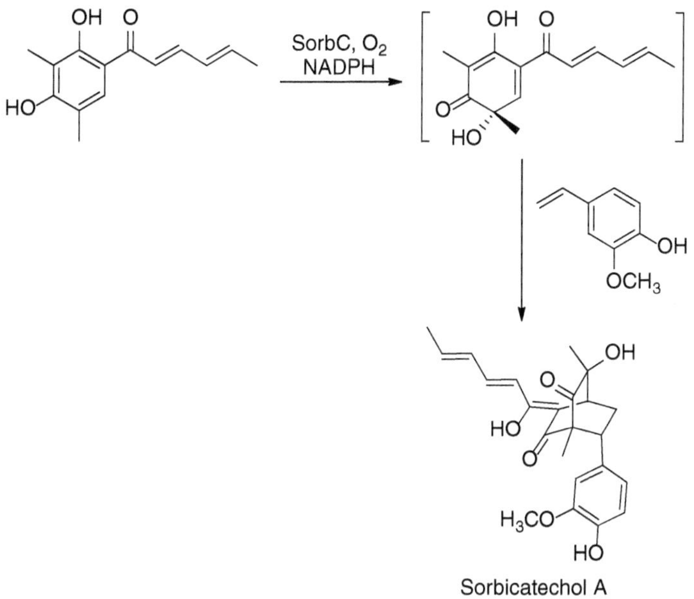

Scheme 3.107 One-pot chemoenzymatic synthesis of sorbicatechol A.

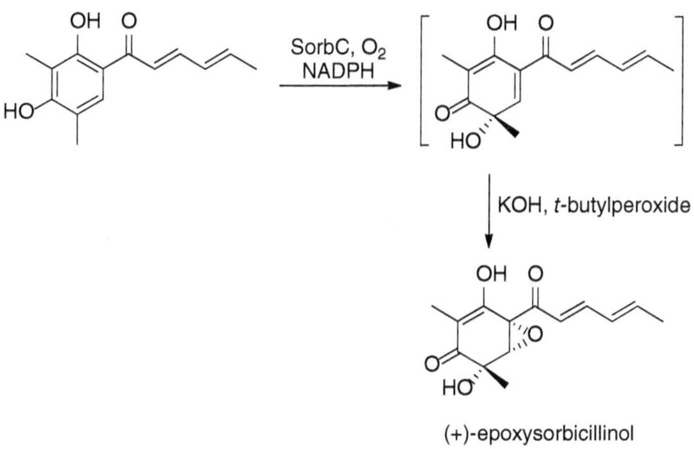

Scheme 3.108 One-pot chemoenzymatic synthesis of (+)-epoxysorbicillinol.

in implementation that one-pot concurrent cascade lacks. From the above discussion, we can see that to date chemoenzymatic one-pot sequential cascades mainly involve lipase, carbonyl reductase, and transaminase. This is largely due to the facts that these enzymes are readily available and that their catalytic properties have been widely studied. In recent years, other enzymes such as ER, epoxide hydrolase, and P450 have also been applied to one-pot sequential cascades, realizing a wide range of unique transformations. With continuing advances in the technologies for enzyme discovery, protein engineering, and design, more and more novel enzymes will be tailor-made for the desired biotransformations. This will in turn bring about rapid development of chemoenzymatic one-pot sequential cascades in the years to come, which will provide extraordinary benefits of operational simplicity and flexibility for implementing the synthesis of a target molecule.

References

1 Dumeignil, F., Guehl, M., Gimbernat, A. et al. (2018). From sequential chemoenzymatic synthesis to integrated hybrid catalysis: taking the best of both worlds to open up the scope of possibilities for a sustainable future. *Catalysis Science & Technology* 8 (22): 5708–5734.

2 Süss, P., Borchert, S., Hinze, J. et al. (2015). Chemoenzymatic sequential multistep one-pot reaction for the synthesis of (1S,2R)-1-(methoxycarbonyl) cyclohex-4-ene-2-carboxylic acid with recombinant pig liver esterase. *Organic Process Research & Development* 19 (12): 2034–2038.

3 Tenbrink, K., Seßler, M., Schatz, J., and Gröger, H. (2011). Combination of olefin metathesis and enzymatic ester hydrolysis in aqueous media in a one-pot synthesis. *Advanced Synthesis & Catalysis* 353 (13): 2363–2367.

4 Pauly, J., Gröger, H., and Patel, A.V. (2019). Catalysts encapsulated in biopolymer hydrogels for chemoenzymatic one-pot processes in aqueous media. *ChemCatChem* 11 (5): 1504–1510.

5 Belokon, Y.N., Blacker, A.J., Clutterbuck, L.A. et al. (2006). An asymmetric, chemo-enzymatic synthesis of *O*-acetylcyanohydrins. *European Journal of Organic Chemistry* 2006 (20): 4609–4617.

6 Shen, Z.-L., Zhou, W.-J., Liu, Y.-T. et al. (2008). One-pot chemoenzymatic syntheses of enantiomerically-enriched O-acetyl cyanohydrins from aldehydes in ionic liquid. *Green Chemistry* 10 (3): 283–286.

7 Sand, H. and Weberskirch, R. (2017). Chemoenzymatic one-pot reaction of noncompatible catalysts: combining enzymatic ester hydrolysis with Cu(i)/bipyridine catalyzed oxidation in aqueous medium. *RSC Advances* 7 (53): 33614–33626.

8 Langvik, O., Sandberg, T., Warna, J. et al. (2015). One-pot synthesis of (R)-2-acetoxy-1-indanone from 1,2-indanedione combining metal catalyzed hydrogenation and chemoenzymatic dynamic kinetic resolution. *Catalysis Science & Technology* 5 (1): 150–160.

9 Wallner, A., Mang, H., Glueck, S.M. et al. (2003). Chemo-enzymatic enantio-convergent asymmetric total synthesis of (S)-(+)-dictyoprolene using a kinetic resolution—stereoinversion protocol. *Tetrahedron: Asymmetry* 14 (16): 2427–2432.

10 Hassan, S., Tschersich, R., and Müller, T.J.J. (2013). Three-component chemoenzymatic synthesis of amide ligated 1,2,3-triazoles. *Tetrahedron Letters* 54 (35): 4641–4644.

11 Hassan, S., Ullrich, A., and Muller, T.J.J. (2015). Consecutive three-component synthesis of (hetero)arylated propargyl amides by chemoenzymatic

aminolysis-Sonogashira coupling sequence. *Organic & Biomolecular Chemistry* 13 (5): 1571–1576.

12 Palo-Nieto, C., Afewerki, S., Anderson, M. et al. (2016). Integrated heterogeneous metal/enzymatic multiple relay catalysis for eco-friendly and asymmetric synthesis. *ACS Catalysis* 6 (6): 3932–3940.

13 Strompen, S., Weiß, M., Ingram, T. et al. (2012). Kinetic investigation of a solvent-free, chemoenzymatic reaction sequence towards enantioselective synthesis of a β-amino acid ester. *Biotechnology and Bioengineering* 109 (6): 1479–1489.

14 Weiß, M. and Gröger, H. (2009). Practical, highly enantioselective chemoenzymatic one-pot synthesis of short-chain aliphatic β-amino acid esters. *Synlett* 2009 (08): 1251–1254.

15 Strompen, S., Weiß, M., Gröger, H. et al. (2013). Development of a continuously operating process for the enantioselective synthesis of a β-amino acid ester via a solvent-free chemoenzymatic reaction sequence. *Advanced Synthesis & Catalysis* 355 (11-12): 2391–2399.

16 Gesse, P. and Müller, T.J.J. (2019). Consecutive five-component Ugi-4CR-CAL B-catalyzed aminolysis sequence and concatenation with transition metal catalysis in a one-pot fashion to substituted triamides. *European Journal of Organic Chemistry* 2019 (11): 2150–2157.

17 Baer, K., Kraußer, M., Burda, E. et al. (2009). Sequential and modular synthesis of chiral 1,3-diols with two stereogenic centers: access to all four stereoisomers by combination of organo- and biocatalysis. *Angewandte Chemie International Edition* 48 (49): 9355–9358.

18 Rulli, G., Duangdee, N., Baer, K. et al. (2011). Direction of kinetically versus thermodynamically controlled organocatalysis and its application in chemoenzymatic synthesis. *Angewandte Chemie International Edition* 50 (34): 7944–7947.

19 Heidlindemann, M., Rulli, G., Berkessel, A. et al. (2014). Combination of asymmetric organo- and biocatalytic reactions in organic media using immobilized catalysts in different compartments. *ACS Catalysis* 4 (4): 1099–1103.

20 Rulli, G., Duangdee, N., Hummel, W. et al. (2017). First tandem-type one-pot process combining asymmetric organo- and biocatalytic reactions in aqueous media exemplified for the enantioselective and diastereoselective synthesis of 1,3-diols. *European Journal of Organic Chemistry* 2017 (4): 812–817.

21 Sonoike, S., Itakura, T., Kitamura, M., and Aoki, S. (2012). One-pot chemoenzymatic synthesis of chiral 1,3-diols using an enantioselective aldol reaction with chiral Zn^{2+} complex catalysts and enzymatic reduction using oxidoreductases with cofactor regeneration. *Chemistry – An Asian Journal* 7 (1): 64–74.

22 Simon, R.C., Busto, E., Schrittwieser, J.H. et al. (2014). Stereoselective synthesis of [gamma]-hydroxynorvaline through combination of organo- and biocatalysis. *Chemical Communications* 50 (99): 15669–15672.

23 Burda, E., Hummel, W., and Gröger, H. (2008). Modular chemoenzymatic one-pot syntheses in aqueous media: combination of a palladium-catalyzed cross-coupling with an asymmetric biotransformation. *Angewandte Chemie International Edition* 47 (49): 9551–9554.

24 Burda, E., Bauer, W., Hummel, W., and Gröger, H. (2010). Enantio- and diastereoselective chemoenzymatic synthesis of C2-symmetric biaryl-containing diols. *ChemCatChem* 2 (1): 67–72.
25 Borchert, S., Burda, E., Schatz, J. et al. (2012). Combination of a Suzuki cross-coupling reaction using a water-soluble palladium catalyst with an asymmetric enzymatic reduction towards a one-pot process in aqueous medium at room temperature. *Journal of Molecular Catalysis B: Enzymatic* 84: 89–93.
26 Paris, J., Ríos-Lombardía, N., Morís, F. et al. (2018). Novel insights into the combination of metal- and biocatalysis: cascade one-pot synthesis of enantiomerically pure biaryl alcohols in deep eutectic solvents. *ChemCatChem* 10 (19): 4417–4423.
27 Prastaro, A., Ceci, P., Chiancone, E. et al. (2009). Suzuki-Miyaura cross-coupling catalyzed by protein-stabilized palladium nanoparticles under aerobic conditions in water: application to a one-pot chemoenzymatic enantioselective synthesis of chiral biaryl alcohols. *Green Chemistry* 11 (12): 1929–1932.
28 Gauchot, V., Kroutil, W., and Schmitzer, A.R. (2010). Highly recyclable chemo-/biocatalyzed cascade reactions with ionic liquids: one-pot synthesis of chiral biaryl alcohols. *Chemistry – A European Journal* 16 (23): 6748–6751.
29 de Oliveira Lopes, R., de Miranda, A.S., Reichart, B. et al. (2014). Combined batch and continuous flow procedure to the chemo-enzymatic synthesis of biaryl moiety of odanacatib. *Journal of Molecular Catalysis B: Enzymatic* 104: 101–107.
30 González-Martínez, D., Gotor, V., and Gotor-Fernández, V. (2019). Chemoenzymatic synthesis of an odanacatib precursor through a Suzuki-Miyaura cross-coupling and bioreduction sequence. *ChemCatChem* 11 (23): 5800–5807.
31 Dander, J.E., Giroud, M., Racine, S. et al. (2019). Chemoenzymatic conversion of amides to enantioenriched alcohols in aqueous medium. *Communications Chemistry* 2 (1): 82.
32 Boffi, A., Cacchi, S., Ceci, P. et al. (2011). The Heck reaction of allylic alcohols catalyzed by palladium nanoparticles in water: chemoenzymatic synthesis of (R)-(−)-rhododendrol. *ChemCatChem* 3 (2): 347–353.
33 Cacchi, S., Cirilli, R., Fabrizi, G. et al. (2009). (R)-1-arylethanols from aryl iodides through a two-step one-pot enantioselective chemoenzymatic process. *Journal of Molecular Catalysis B: Enzymatic* 61 (3): 184–187.
34 Kraußer, M., Hummel, W., and Gröger, H. (2007). Enantioselective one-pot two-step synthesis of hydrophobic allylic alcohols in aqueous medium through the combination of a Wittig reaction and an enzymatic ketone reduction. *European Journal of Organic Chemistry* 2007 (31): 5175–5179.
35 Bencze, L.C., Paizs, C., Toşa, M.I. et al. (2011). Chemoenzymatic one-pot synthesis of both (R)- and (S)-aryl-1,2-ethanediols. *ChemCatChem* 3 (2): 343–346.
36 Schnapperelle, I., Hummel, W., and Gröger, H. (2012). Formal asymmetric hydration of non-activated alkenes in aqueous medium through a "chemoenzymatic catalytic system". *Chemistry – A European Journal* 18 (4): 1073–1076.
37 Sato, H., Hummel, W., and Gröger, H. (2015). Cooperative catalysis of non-compatible catalysts through compartmentalization: Wacker oxidation and

enzymatic reduction in a one-pot process in aqueous media. *Angewandte Chemie International Edition* 54 (15): 4488–4492.

38 Ríos-Lombardía, N., Vidal, C., Liardo, E. et al. (2016). From a sequential to a concurrent reaction in aqueous medium: ruthenium-catalyzed allylic alcohol isomerization and asymmetric bioreduction. *Angewandte Chemie International Edition* 55 (30): 8691–8695.

39 Wang, Z.J., Clary, K.N., Bergman, R.G. et al. (2013). A supramolecular approach to combining enzymatic and transition metal catalysis. *Nature Chemistry* 5: 100.

40 Rodríguez-Álvarez, M.J., Ríos-Lombardía, N., Schumacher, S. et al. (2017). Combination of metal-catalyzed cycloisomerizations and biocatalysis in aqueous media: asymmetric construction of chiral alcohols, lactones, and γ-hydroxy-carbonyl compounds. *ACS Catalysis* 7 (11): 7753–7759.

41 Schaaf, P., Gojic, V., Bayer, T. et al. (2018). Easy access to enantiopure (S)- and (R)-aryl alkyl alcohols by a combination of gold(III)-catalyzed alkyne hydration and enzymatic reduction. *ChemCatChem* 10 (5): 920–924.

42 Méndez-Sánchez, D., Mangas-Sánchez, J., Lavandera, I. et al. (2015). Chemoenzymatic deracemization of secondary alcohols by using a TEMPO–iodine–alcohol dehydrogenase system. *ChemCatChem* 7 (24): 4016–4020.

43 Kedziora, K., Diaz-Rodriguez, A., Lavandera, I. et al. (2014). Laccase/TEMPO-mediated system for the thermodynamically disfavored oxidation of 2,2-dihalo-1-phenylethanol derivatives. *Green Chemistry* 16 (5): 2448–2453.

44 Diaz-Rodriguez, A., Rios-Lombardia, N., Sattler, J.H. et al. (2015). Deracemisation of profenol core by combining laccase/TEMPO-mediated oxidation and alcohol dehydrogenase-catalysed dynamic kinetic resolution. *Catalysis Science & Technology* 5 (3): 1443–1446.

45 Liardo, E., Ríos-Lombardía, N., Morís, F. et al. (2018). A straightforward deracemization of sec-alcohols combining organocatalytic oxidation and biocatalytic reduction. *European Journal of Organic Chemistry* 2018 (23): 3031–3035.

46 Zhang, J.-D., Zhao, J.-W., Gao, L.-L. et al. (2019). One-pot three-step consecutive transformation of L-α-amino acids to (R)- and (S)-vicinal 1,2-diols via combined chemical and biocatalytic process. *ChemCatChem* 11 (20): 5032–5037.

47 Schrittwieser, J.H., Coccia, F., Kara, S. et al. (2013). One-pot combination of enzyme and Pd nanoparticle catalysis for the synthesis of enantiomerically pure 1,2-amino alcohols. *Green Chemistry* 15 (12): 3318–3331.

48 Cuetos, A., Bisogno, F.R., Lavandera, I., and Gotor, V. (2013). Coupling biocatalysis and click chemistry: one-pot two-step convergent synthesis of enantioenriched 1,2,3-triazole-derived diols. *Chemical Communications* 49 (26): 2625–2627.

49 Díaz-Rodríguez, A., Borzęcka, W., Lavandera, I., and Gotor, V. (2014). Stereodivergent preparation of valuable γ- or δ-hydroxy esters and lactones through one-pot cascade or tandem chemoenzymatic protocols. *ACS Catalysis* 4 (2): 386–393.

50 Eastman, H., Ryan, J., Maciá, B. et al. (2019). Alcohol dehydrogenase-triggered oxa-Michael reaction for the asymmetric synthesis of disubstituted tetrahydropyrans and tetrahydrofurans. *ChemCatChem* 11 (16): 3760–3762.

51 Toogood, H.S. and Scrutton, N.S. (2018). Discovery, characterization, engineering, and applications of ene-reductases for industrial biocatalysis. *ACS Catalysis* 8 (4): 3532–3549.

52 Kraußer, M., Winkler, T., Richter, N. et al. (2011). Combination of C-C bond formation by Wittig reaction and enzymatic C-C bond reduction in a one-pot process in water. *ChemCatChem* 3 (2): 293–296.

53 Wang, Y., Bartlett, M.J., Denard, C.A. et al. (2017). Combining Rh-catalyzed diazocoupling and enzymatic reduction to efficiently synthesize enantioenriched 2-substituted succinate derivatives. *ACS Catalysis* 7 (4): 2548–2552.

54 Brenna, E., Crotti, M., Gatti, F.G. et al. (2015). Opposite enantioselectivity in the bioreduction of (Z)-β-aryl-β-cyanoacrylates mediated by the tryptophan 116 mutants of old yellow enzyme 1: synthetic approach to (R)- and (S)-β-aryl-γ-lactams. *Advanced Synthesis & Catalysis* 357 (8): 1849–1860.

55 Guo, F. and Berglund, P. (2017). Transaminase biocatalysis: optimization and application. *Green Chemistry* 19 (2): 333–360.

56 Liu, J., Wu, S., and Li, Z. (2018). Recent advances in enzymatic oxidation of alcohols. *Current Opinion in Chemical Biology* 43: 77–86.

57 Simon, R.C., Richter, N., Busto, E., and Kroutil, W. (2014). Recent developments of cascade reactions involving ω-transaminases. *ACS Catalysis* 4 (1): 129–143.

58 Martinez-Montero, L., Gotor, V., Gotor-Fernandez, V., and Lavandera, I. (2017). Stereoselective amination of racemic sec-alcohols through sequential application of laccases and transaminases. *Green Chemistry* 19 (2): 474–480.

59 Liardo, E., Ríos-Lombardía, N., Morís, F. et al. (2017). Hybrid organo- and biocatalytic process for the asymmetric transformation of alcohols into amines in aqueous medium. *ACS Catalysis* 7 (7): 4768–4774.

60 Rios-Lombardia, N., Vidal, C., Cocina, M. et al. (2015). Chemoenzymatic one-pot synthesis in an aqueous medium: combination of metal-catalysed allylic alcohol isomerisation-asymmetric bioamination. *Chemical Communications* 51 (54): 10937–10940.

61 Mattias, A., Samson, A., Per, B., and Armando, C. (2014). Total synthesis of capsaicin analogues from lignin-derived compounds by combined heterogeneous metal, organocatalytic and enzymatic cascades in one pot. *Advanced Synthesis & Catalysis* 356 (9): 2113–2118.

62 Uthoff, F., Sato, H., and Gröger, H. (2017). Formal enantioselective hydroamination of non-activated alkenes: transformation of styrenes into enantiomerically pure 1-phenylethylamines in chemoenzymatic one-pot synthesis. *ChemCatChem* 9 (4): 555–558.

63 González-Martínez, D., Gotor, V., and Gotor-Fernández, V. (2019). Stereoselective synthesis of 1-arylpropan-2-amines from allylbenzenes through a Wacker-Tsuji oxidation-biotransamination sequential process. *Advanced Synthesis & Catalysis* 361 (11): 2582–2593.

64 Dawood, A.W.H., Bassut, J., de Souza, R.O.M.A., and Bornscheuer, U.T. (2018). Combination of the Suzuki–Miyaura cross-coupling reaction with engineered transaminases. *Chemistry – A European Journal* 24 (60): 16009–16013.

65 Paris, J., Telzerow, A., Ríos-Lombardía, N. et al. (2019). Enantioselective one-pot synthesis of biaryl-substituted amines by combining palladium and enzyme catalysis in deep eutectic solvents. *ACS Sustainable Chemistry & Engineering* 7 (5): 5486–5493.

66 Lichman, B.R., Lamming, E.D., Pesnot, T. et al. (2015). One-pot triangular chemoenzymatic cascades for the syntheses of chiral alkaloids from dopamine. *Green Chemistry* 17 (2): 852–855.

67 Erdmann, V., Lichman, B.R., Zhao, J. et al. (2017). Enzymatic and chemoenzymatic three-step cascades for the synthesis of stereochemically complementary trisubstituted tetrahydroisoquinolines. *Angewandte Chemie International Edition* 56 (41): 12503–12507.

68 Bonamore, A., Calisti, L., Calcaterra, A. et al. (2016). A novel enzymatic strategy for the synthesis of substituted tetrahydroisoquinolines. *ChemistrySelect* 1 (8): 1525–1528.

69 Zhao, J., Lichman, B.R., Ward, J.M., and Hailes, H.C. (2018). One-pot chemoenzymatic synthesis of trolline and tetrahydroisoquinoline analogues. *Chemical Communications* 54 (11): 1323–1326.

70 Marx, L., Ríos-Lombardía, N., Farnberger, J.F. et al. (2018). Chemoenzymatic approaches to the synthesis of the calcimimetic agent cinacalcet employing transaminases and ketoreductases. *Advanced Synthesis & Catalysis* 360: 2157–2165.

71 Kolb, H.C., VanNieuwenhze, M.S., and Sharpless, K.B. (1994). Catalytic asymmetric dihydroxylation. *Chemical Reviews* 94 (8): 2483–2547.

72 Boyd, D.R. and Bugg, T.D.H. (2006). Arene cis-dihydrodiol formation: from biology to application. *Organic & Biomolecular Chemistry* 4 (2): 181–192.

73 Chuan, W. (2018). Vicinal anti-dioxygenation of alkenes. *Asian Journal of Organic Chemistry* 7 (3): 509–521.

74 Chang, D., Heringa, M.F., Witholt, B., and Li, Z. (2003). Enantioselective trans dihydroxylation of nonactivated C–C double bonds of aliphatic heterocycles with *Sphingomonas* sp. HXN-200. *The Journal of Organic Chemistry* 68 (22): 8599–8606.

75 Wu, S., Chen, Y., Xu, Y. et al. (2014). Enantioselective trans-dihydroxylation of aryl olefins by cascade biocatalysis with recombinant *Escherichia coli* coexpressing monooxygenase and epoxide hydrolase. *ACS Catalysis* 4 (2): 409–420.

76 Xu, Y., Li, A., Jia, X., and Li, Z. (2011). Asymmetric trans-dihydroxylation of cyclic olefins by enzymatic or chemo-enzymatic sequential epoxidation and hydrolysis in one-pot. *Green Chemistry* 13 (9): 2452–2458.

77 Chiara, P., Ferrandi, E.E., Carlotta, M. et al. (2016). One-pot selective dihydroxylation of limonene combining metal and enzyme catalysis. *ChemistrySelect* 1 (8): 1795–1798.

78 Orru, R.V.A., Mayer, S.F., Kroutil, W., and Faber, K. (1998). Chemoenzymatic deracemization of (±)-2,2-disubstituted oxiranes. *Tetrahedron* 54 (5): 859–874.

79 Simeó, Y. and Faber, K. (2006). Selectivity enhancement of enantio- and stereo-complementary epoxide hydrolases and chemo-enzymatic deracemization of (±)-2-methylglycidyl benzyl ether. *Tetrahedron: Asymmetry* 17 (3): 402–409.

80 Wu, Y.-W., Kong, X.-D., Zhu, Q.-Q. et al. (2015). Chemoenzymatic enantioconvergent hydrolysis of p-nitrostyrene oxide into (R)-p-nitrophenyl glycol by a newly cloned epoxide hydrolase VrEH2 from *Vigna radiata*. *Catalysis Communications* 58: 16–20.

81 Kinnell, A., Harman, T., Bingham, M. et al. (2012). Development of an organo- and enzyme-catalysed one-pot, sequential three-component reaction. *Tetrahedron* 68 (37): 7719–7722.

82 Marín-Valls, R., Hernández, K., Bolte, M. et al. (2019). Chemoenzymatic hydroxymethylation of carboxylic acids by tandem stereodivergent biocatalytic aldol reaction and chemical decarboxylation. *ACS Catalysis* 9 (8): 7568–7577.

83 Campbell-Verduyn, L.S., Szymanski, W., Postema, C.P. et al. (2010). One pot 'click' reactions: tandem enantioselective biocatalytic epoxide ring opening and [3+2] azide alkyne cycloaddition. *Chemical Communications* 46 (6): 898–900.

84 Schrittwieser, J.H., Iván, L., Birgit, S. et al. (2009). Biocatalytic cascade for the synthesis of enantiopure β-azidoalcohols and β-hydroxynitriles. *European Journal of Organic Chemistry* 2009 (14): 2293–2298.

85 Wiktor, S., Postema, C.P., Chiara, T. et al. (2010). Combining designer cells and click chemistry for a one-pot four-step preparation of enantiopure β-hydroxytriazoles. *Advanced Synthesis & Catalysis* 352 (13): 2111–2115.

86 Weise, N.J., Parmeggiani, F., Ahmed, S.T., and Turner, N.J. (2015). The bacterial ammonia lyase EncP: a tunable biocatalyst for the synthesis of unnatural amino acids. *Journal of the American Chemical Society* 137 (40): 12977–12983.

87 Parmeggiani, F., Ahmed, S.T., Weise, N.J., and Turner, N.J. (2016). Telescopic one-pot condensation-hydroamination strategy for the synthesis of optically pure L-phenylalanines from benzaldehydes. *Tetrahedron* 72 (46): 7256–7262.

88 Ben, d.L., Hyett, D.J., Maas, P.J.D. et al. (2011). Asymmetric synthesis of (S)-2-indolinecarboxylic acid by combining biocatalysis and homogeneous catalysis. *ChemCatChem* 3 (2): 289–292.

89 Ahmed, S.T., Parmeggiani, F., Weise, N.J. et al. (2015). Chemoenzymatic synthesis of optically pure l- and d-biarylalanines through biocatalytic asymmetric amination and palladium-catalyzed arylation. *ACS Catalysis* 5 (9): 5410–5413.

90 Uthoff, F. and Gröger, H. (2018). Asymmetric synthesis of 1-phenylethylamine from styrene via combined Wacker oxidation and enzymatic reductive amination. *The Journal of Organic Chemistry* 83 (16): 9517–9521.

91 Frese, M., Schnepel, C., Minges, H. et al. (2016). Modular combination of enzymatic halogenation of tryptophan with Suzuki–Miyaura cross-coupling reactions. *ChemCatChem* 8 (10): 1799–1803.

92 Durak, L.J., Payne, J.T., and Lewis, J.C. (2016). Late-stage diversification of biologically active molecules via chemoenzymatic C–H functionalization. *ACS Catalysis* 6 (3): 1451–1454.

93 Runguphan, W. and O'Connor, S.E. (2013). Diversification of monoterpene indole alkaloid analogs through cross-coupling. *Organic Letters* 15 (11): 2850–2853.

94 Roy, A.D., Grüschow, S., Cairns, N., and Goss, R.J.M. (2010). Gene expression enabling synthetic diversification of natural products: chemogenetic generation of pacidamycin analogs. *Journal of the American Chemical Society* 132 (35): 12243–12245.

95 Zumbrägel, N. and Gröger, H. (2019). One-pot synthesis of a 3-thiazolidine through combination of an Asinger-type multi-component-condensation reaction with an enzymatic imine reduction. *Journal of Biotechnology* 291: 35–40.

96 Gómez Baraibar, Á., Reichert, D., Mügge, C. et al. (2016). A one-pot cascade reaction combining an encapsulated decarboxylase with a metathesis catalyst for the synthesis of bio-based antioxidants. *Angewandte Chemie International Edition* 55 (47): 14823–14827.

97 Mertens, M.A.S., Sauer, D.F., Markel, U. et al. (2019). Chemoenzymatic cascade for stilbene production from cinnamic acid catalyzed by ferulic acid decarboxylase and an artificial metathease. *Catalysis Science & Technology* 9 (20): 5572–5576.

98 Enoki, J., Mügge, C., Tischler, D. et al. (2019). Chemoenzymatic cascade synthesis of optically pure alkanoic acids by using engineered arylmalonate decarboxylase variants. *Chemistry – A European Journal* 25 (19): 5071–5076.

99 Manning, J., Tavanti, M., Porter, J.L. et al. (2019). Regio- and enantio-selective chemo-enzymatic C–H-lactonization of decanoic acid to (S)-δ-decalactone. *Angewandte Chemie International Edition* 58 (17): 5668–5671.

100 Wang, J., Li, K., Zhou, X. et al. (2017). Asymmetric combinational "metal-biocatalytic system": one approach to chiral 2-subsituted-tetrahydroquinoline-4-ols towards two-step one-pot processes in aqueous media. *Tetrahedron Letters* 58 (23): 2252–2254.

101 Vugts, D.J., Veum, L., al-Mafraji, K. et al. (2006). A mild chemo-enzymatic oxidation–hydrocyanation protocol. *European Journal of Organic Chemistry* 2006 (7): 1672–1677.

102 Xue, Y.-P., Wang, Y.-P., Xu, Z. et al. (2015). Chemoenzymatic synthesis of gabapentin by combining nitrilase-mediated hydrolysis with hydrogenation over Raney-nickel. *Catalysis Communications* 66: 121–125.

103 Akagawa, K., Umezawa, R., and Kudo, K. (2012). Asymmetric one-pot sequential Friedel–Crafts-type alkylation and α-oxyamination catalyzed by a peptide and an enzyme. *Beilstein Journal of Organic Chemistry* 8: 1333–1337.

104 Asta, C., Schmidt, D., Conrad, J. et al. (2013). Combination of enzyme- and Lewis acid-catalyzed reactions: a new method for the synthesis of 6,7-dihydrobenzofuran-4(5H)-ones starting from 2,5-dimethylfuran and 1,3-cyclohexanediones. *Organic & Biomolecular Chemistry* 11 (34): 5692–5701.

105 Oteng-Pabi, S.K., Pardin, C., Stoica, M., and Keillor, J.W. (2014). Site-specific protein labelling and immobilization mediated by microbial transglutaminase. *Chemical Communications* 50 (50): 6604–6606.

106 Rachel, N.M. and Pelletier, J.N. (2016). One-pot peptide and protein conjugation: a combination of enzymatic transamidation and click chemistry. *Chemical Communications* 52 (12): 2541–2544.

107 Doyon, T.J., Perkins, J.C., Baker Dockrey, S.A. et al. (2019). Chemoenzymatic o-quinone methide formation. *Journal of the American Chemical Society* 141 (51): 20269–20277.

108 Fuchs, M., Schober, M., Pfeffer, J. et al. (2011). Homoallylic alcohols via a chemo-enzymatic one-pot oxidation–allylation cascade. *Advanced Synthesis & Catalysis* 353 (13): 2354–2358.

109 Baker Dockrey, S.A., Lukowski, A.L., Becker, M.R., and Narayan, A.R.H. (2018). Biocatalytic site- and enantioselective oxidative dearomatization of phenols. *Nature Chemistry* 10 (2): 119–125.

110 Sib, A. and Gulder, T.A.M. (2017). Stereoselective total synthesis of bisorbicillinoid natural products by enzymatic oxidative dearomatization/dimerization. *Angewandte Chemie International Edition* 56 (42): 12888–12891.

111 Sib, A. and Gulder, T.A.M. (2018). Chemo-enzymatic total synthesis of oxosorbicillinol, sorrentanone, rezishanones B and C, sorbicatechol A, bisvertinolone, and (+)-epoxysorbicillinol. *Angewandte Chemie International Edition* 57 (44): 14650–14653.

4

Chemoenzymatic Dynamic Kinetic Resolution

For "one-pot concurrent chemoenzymatic reactions," the chemical and biocatalytic reactions occur simultaneously without high accumulation of the intermediates. This is the most efficient mode of chemoenzymatic cascades, in which multiple reactions can be realized in one single process operation, presenting tremendous operational and economic advantages. The concurrent reactions are particularly beneficial for situations where the reaction intermediate is unstable or has an inhibitory effect on the catalysts due to the fast consumption and low concentration of the intermediate. At the same time, there are many challenges too. The reaction conditions of chemical reaction and biotransformation must be completely compatible. The inhibition or inactivation of (bio)catalysts has to be eliminated, which often occurs due to the large number of components in the complicated reaction system. Usually, metal- and organocatalyzed reactions are carried out in organic solvents, while biocatalytic transformations require aqueous buffer as the reaction media. In order to solve this incompatibility problem, scientists and engineers are designing chemical catalysts for use in aqueous reaction medium and evolving enzyme catalysts with tolerance to organic solvents, so that these catalysts can work under the same reaction conditions. Reaction engineering has also been carried out for developing efficient and economical chemoenzymatic one-pot one-step processes. An excellent example of "one-pot concurrent chemoenzymatic reactions" is the chemoenzymatic dynamic kinetic resolution (DKR) of chiral molecules. This will be discussed in this chapter, and the subsequent chapters will be dedicated to other "one-pot concurrent chemoenzymatic reactions."

4.1 Enzymatic Kinetic Resolution

When a racemic mixture or racemate reacts with a chiral catalyst or reagent, the two enantiomers may proceed at different reaction rates, resulting in the kinetic resolution of the two enantiomers in a racemic mixture (Scheme 4.1). The first reported kinetic resolution was achieved by Louis Pasteur. When he treated aqueous racemic ammonium tartrate with a *Penicillium glaucum* mold, an excess of (*S*,*S*)-tartrate was found in the isolated tartrate sample, due to the selective metabolization of (*R*,*R*)-tartrate by the chiral microorganisms present in the mold. Since then kinetic

Chemo-Enzymatic Cascade Reactions, First Edition. Dunming Zhu.

resolution has been widely used for the preparation of chiral molecules in organic synthesis [1, 2]. Because the chirality of the active site in enzymes can recognize one enantiomer of the substrate, this enantiomer is converted to the product at a higher rate than the other, resulting in a kinetic resolution of the racemic substrate. As such, enzymes have been commonly used as the biocatalyst for the kinetic resolution of racemates. Especially in the past 30 years, enzymatic kinetic resolution has been greatly developed in the application toward organic synthesis [3].

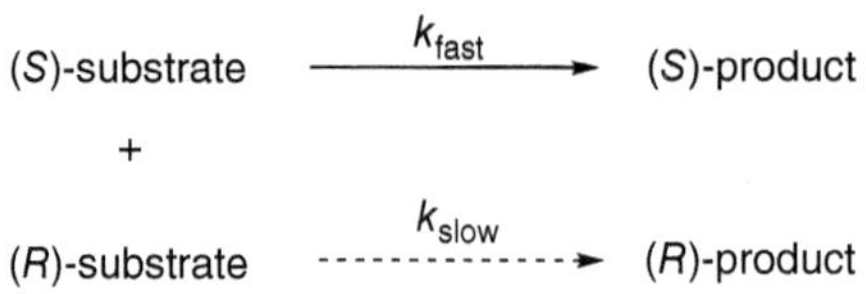

Scheme 4.1 Kinetic resolution of racemate leads to (*S*)-product and (*R*)-substrate.

In an ideal case of enzymatic kinetic resolution, one enantiomer is completely converted to the product, while the other enantiomer remains unreacted, and the reaction stops at 50% conversion. However, this is not the case in most situations. Both enantiomers are converted to the product, just at different reaction rates, and the enantiomeric excess values of both substrate (ee_s) and product (ee_p) change with the degree of conversion. The enantiomeric ratio (E) has thus been introduced to measure the ability of an enzyme to distinguish between two enantiomers. The E value is defined as $(V_A/K_A)/(V_B/K_B)$, where V_A, K_A and V_B, K_B denote maximal velocities and Michaelis constants of fast- and slow-reacting enantiomers, respectively [4]. As an empirical rule, reactions with E values lower than 15 are usually unsuitable for practical purposes [5].

The most extensively used enzyme in kinetic resolution is lipase, which has been widely applied to the kinetic resolution of chiral alcohols, esters, and amines [6]. Chiral alcohols are resolved via lipase-catalyzed esterification, which follows the Kazlauskas' rule [7]. The rule predicts which enantiomer of a racemic secondary alcohol reacts faster, and implies that the most efficiently resolved substrates are those with substituents of significantly different sizes, because enantioselectivity is correlated with the sizes of the substituents at the stereocenter.

However, all kinetic resolution methods including enzymatic kinetic resolution suffer from the limitation of only 50% of the maximum theoretical yield. An effective way to overcome this drawback is to combine the resolution process with *in situ* racemization. This so-called DKR can achieve a theoretical yield of 100%.

4.2 Dynamic Kinetic Resolution

In DKR, the chiral center of the racemic molecule can be easily epimerized so that the (*R*) and (*S*) enantiomers can interconvert throughout the reaction process. As one enantiomer is converted to the product, the unreacted enantiomer is racemized, thus leading to almost 100% yield single enantiomer of the product (Scheme 4.2). One of the classic applications of DKR is Noyori's dynamic asymmetric hydrogenation of ketones with acidic α hydrogen, which allows for easy epimerization at the α chiral center under basic conditions. The (2,2′-bis(diphenylphosphino)-1,1′-binaphthyl)Ruthenium complex catalyzed the hydrogenation of these ketones to give one of the four possible stereoisomers of the corresponding product alcohols, under the control of the sterically bulky phosphorus ligand (Scheme 4.3) [8].

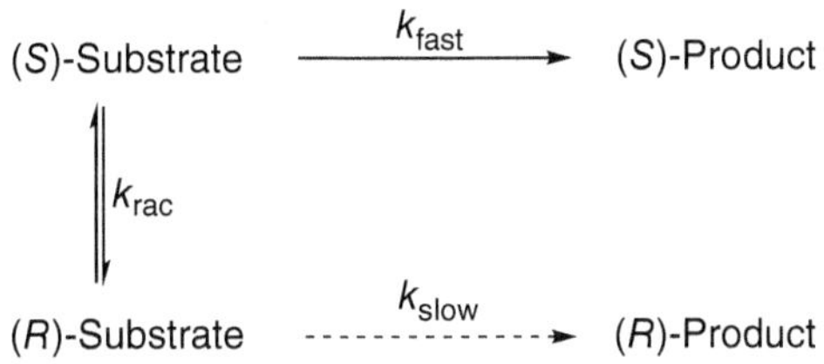

Scheme 4.2 Dynamic kinetic resolution of racemate leads to the (*S*)-product.

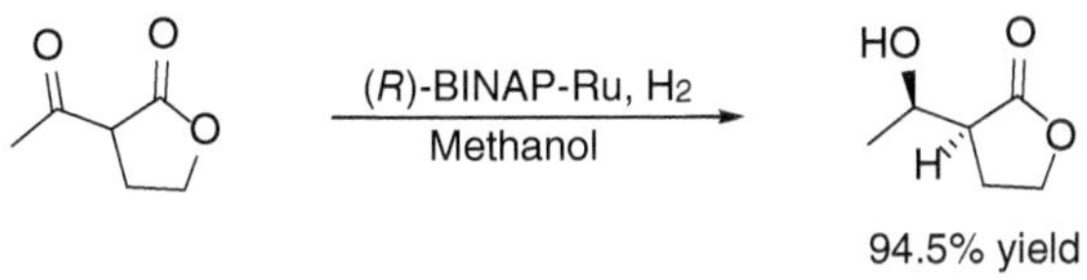

Scheme 4.3 Noyori's dynamic asymmetric hydrogenation of ketones. Source: Adapted from Kitamura et al. [8].

An early example of chemoenzymatic dynamic kinetic resolution was the DKR of an allylic acetate derivative reported by Williams et al. in 1996, in which the allylic acetate was converted to the corresponding allylic alcohol in 81% yield and 96% *ee* in the presence of *Pseudomonas fluorescens* lipase and $PdCl_2(MeCN)_2$ after 19 days (Scheme 4.4) [9]. Since then, chemoenzymatic DKR has been extensively studied and has become an exceedingly useful tool in the synthesis of a wide range of chemical products, such as agrochemicals, food additives, fragrances, and pharmaceuticals. The chemoenzymatic DKRs of various classes of chiral compounds will be discussed in the following sections.

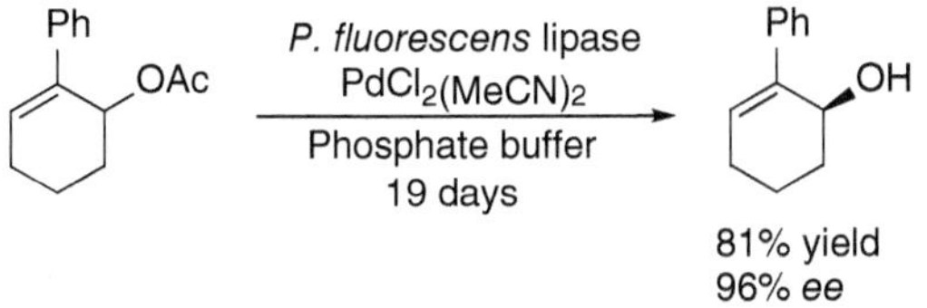

Scheme 4.4 Chemoenzymatic dynamic kinetic resolution of allylic acetate. Source: Adapted from Allen et al. [9].

The principle for a DKR is that two enantiomers, *R* and *S*, are in equilibrium with one another in the reaction system, and one enantiomer, e.g. *R*, reacts faster than the other to give the product, resulting in the equilibrium shift to the *R* enantiomer and the product from the *R* enantiomer as the main one. As such, the basic requirements for an efficient DKR include (i) an efficient kinetic resolution to ensure high enantioselectivity, (ii) a fast racemization than the reaction of the slow enantiomer, and (iii) compatibility of the kinetic resolution and the racemization procedure. The chemical racemization techniques to ensure an effective DKR will be presented in the next section, followed by the chemoenzymatic DKR of chiral alcohols, amines, and other compounds in the subsequent sections.

4.3 Racemization Techniques

Based on the structures of the chiral molecules that are going to be racemized, different racemization techniques have been developed [10]. For the compounds bearing an acidic hydrogen at the chiral center, base-catalyzed racemization has been widely applied. For example, the racemization of ketones with a chiral center at the α-position proceeds via an enolate intermediate under basic condition (Scheme 4.5).

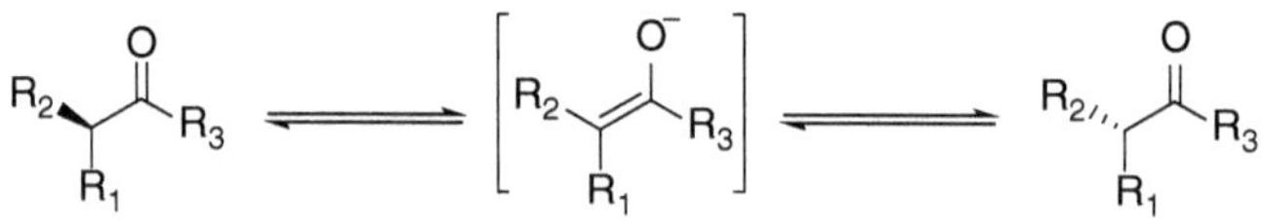

Scheme 4.5 Base-catalyzed racemization of ketones via an enolate intermediate.

Schiff-base-mediated racemization has been often employed in the racemization of compounds bearing a free primary amino group at the chiral center. The primary amino group reacts with an aldehyde or ketone to form an imine, followed by the protonation of the nitrogen atom and proton abstraction at the chiral carbon atom. These reaction steps are reversible, leading to the aldehyde- or ketone-mediated racemization of chiral amines via Schiff base intermediate (Scheme 4.6).

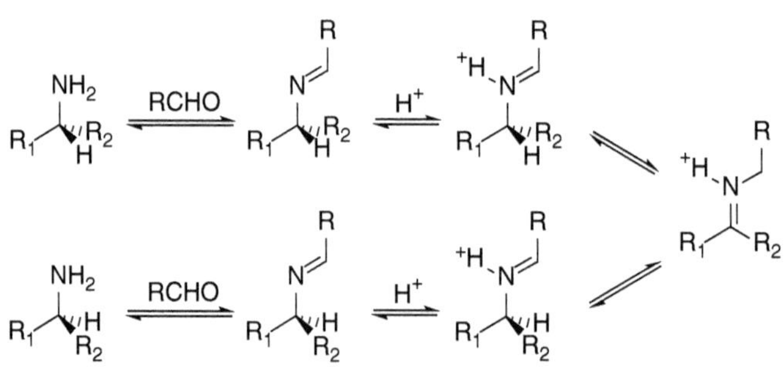

Scheme 4.6 Aldehyde- or ketone-mediated racemization of chiral amines.

Some optically active compounds can be racemized by heating them to an appropriate temperature. This is a simple, cheap, and industrially attractive method.

Some good examples of thermal racemization are biaryls and compounds with a heteroatom chiral center, which are racemized by rotation about a single bond or pyramidal inversion of the heteroatom chiral center, respectively.

For the compounds with a double-bonded heteroatom in the α-position to the chiral center, the heteroatom can be protonated under acidic conditions, followed by abstraction of a hydrogen from the chiral center, resulting in the racemization of the chiral center via keto–enol tautomerism for the heteroatom being oxygen and imine–enamine tautomerism for nitrogen, respectively (Scheme 4.7). The protonated heteroatom enhances the acidity of the α-proton, thus facilitating its abstraction by a weak base. Acid-catalyzed racemization can also be achieved via the reversible formation of a carbocation. In this case, a substituent at the chiral center is protonated under acidic conditions and converted into a good leaving group. The loss of the leaving group forms a carbocation. When the carbocation is stabilized by other functionalities, the leaving group can reversibly attack the carbocation, resulting in racemization (Scheme 4.8). This mechanism is also possible with the compounds lacking a hydrogen attached to the chiral center.

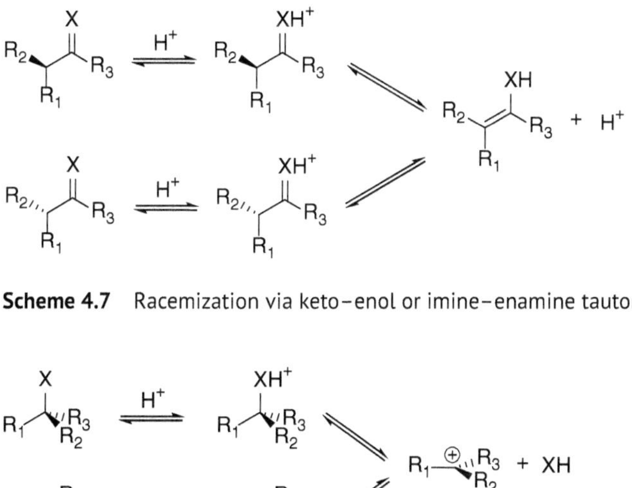

Scheme 4.7 Racemization via keto–enol or imine–enamine tautomerism.

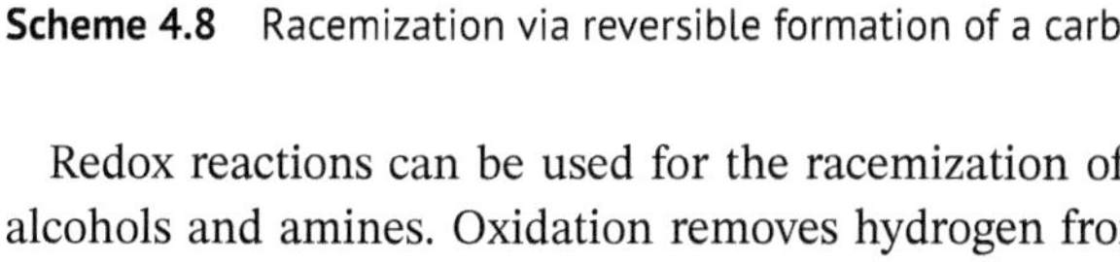

Scheme 4.8 Racemization via reversible formation of a carbocation.

Redox reactions can be used for the racemization of chiral compounds such as alcohols and amines. Oxidation removes hydrogen from the chiral carbon to give a planar intermediate species, which is followed by reduction or hydrogenation to restore the original molecule in a non-stereoselective way, thus generating a racemate (Scheme 4.9). Metal-catalyzed redox racemization has been extensively studied and applied in the DKR of racemic alcohols and amines.

In Nature, there are enzymes called racemases that catalyze racemization of different substrates. For instance, mandelate racemase catalyzes the racemization

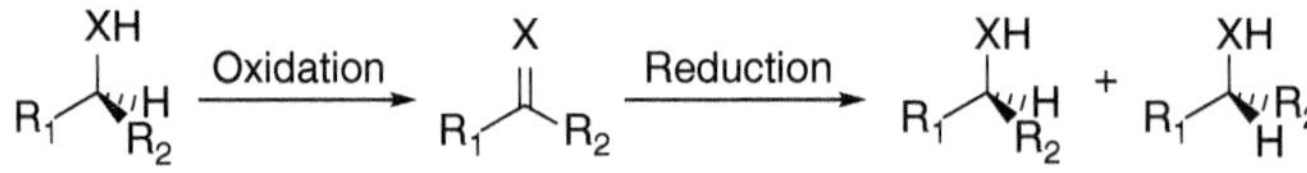

Scheme 4.9 Racemization via redox reactions.

of aromatic α-hydroxycarboxylic acids and has been applied in deracemization of this class of compounds in connection with a kinetic resolution step [11]. Different amino acid racemases have been identified with distinct substrate specificities. They are often named after the first substrate studied, but not necessarily the most suitable substrate. The enzyme-catalyzed racemization and kinetic resolution proceed under compatible reaction conditions. An increasing number of racemases have been reported in combination with resolution in one-pot processes, resulting in DKRs of chiral molecules.

On the other hand, chemical catalysts usually promoted racemization of chiral molecules under quite different environments from enzymatic kinetic resolution. While enzymes usually work under nearly neutral conditions at moderate temperature, chemical racemization often requires relatively harsh reaction conditions. Therefore, the combination of chemical racemization and enzymatic kinetic resolution in a one-pot DKR process is not straightforward, and many parameters such as catalysts, solvent, and temperature must be taken into account to ensure their compatibility. Although the choice of racemization method suitable for DKR is quite challenging, advances in designing chemical catalysts for use in aqueous reaction medium and evolving enzymes with tolerance of organic solvents have made them suitable to work under the same reaction conditions, and efficient chemoenzymatic DKR processes have been achieved for the preparations of optically pure alcohols, amines, carboxylic acids, and their derivatives [5, 12–14].

4.4 DKR of Chiral Alcohols

The enzymatic kinetic resolution of racemic alcohols in general is a straightforward and satisfactory method for the preparation of chiral alcohols and esters in terms of optical purity. This can be achieved by using hydrolases such as lipases and esterases. For example, lipases catalyze the resolution of racemic secondary alcohols via esterification in the presence of an acyl donor in hydrophobic organic solvents such as toluene and *tert*-butyl methyl ether. Two commercially available lipases, *Candida antarctica* lipase B (CALB, Novozym-435) and *Pseudomonas cepacia* lipase (PCL, Lipase PS-C), have been widely used for this purpose due to their broad substrate specificity, high enantioselectivity, and good stability in organic media. However, the application of kinetic resolution is limited because the theoretical maximum yield of the desired enantiomer cannot exceed 50%. This can be overcome by DKR of the racemic alcohols, in which the enzymatic kinetic resolution is combined with an efficient racemization of the alcohol substrate. In such a DKR process where a highly efficient racemization catalyst is combined with a highly enantioselective enzyme, a single enantiomeric alcohol can be obtained theoretically in 100% yield (Scheme 4.10). The use of DKR in the preparation of enantiomerically pure alcohols was introduced by Williams and Backvall in the 1990s. Since then, many metal complexes have been developed as the catalysts for the racemization of chiral alcohols and successfully incorporated with lipase-catalyzed esterification of alcohols, leading to the chemoenzymatic DKR processes for the preparation of a wide range of enantiopure chiral alcohols or their esters.

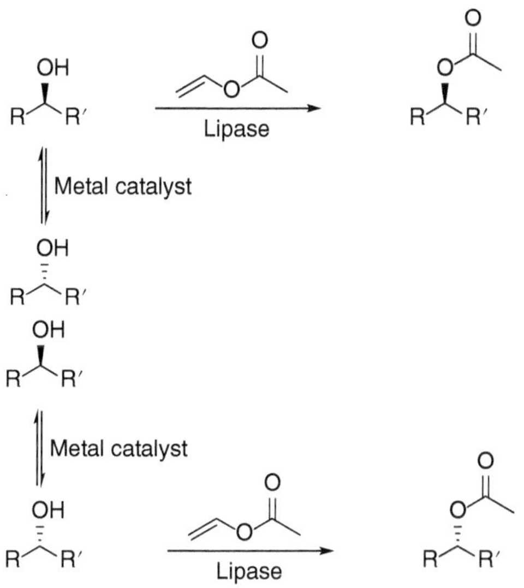

Scheme 4.10 Chemoenzymatic dynamic kinetic resolution (DKR) of racemic alcohols.

For the lipase-catalyzed esterification of alcohols, various acyl donors have been used. The widely used acyl donors in the kinetic resolution of chiral alcohols is alkenyl acetates such as vinyl acetate and isopropenyl acetate, for which the generated enol can isomerize to aldehyde, thus driving the acetyl transfer to completion. Rhodium/phenantroline and iridium/phenantroline catalyzed the racemization of phenethyl alcohol in the presence of stoichiometric quantities of acetophenone and base. When this alcohol racemization was combined with the *P. fluorescens* lipase-catalyzed transesterification using vinyl acetate as the acyl donor (Scheme 4.11), the (*R*)-phenethyl acetate was obtained in 76% conversion with 80% *ee* and 60% conversion with 98% *ee*, respectively [15]. The DKR of a large number of structurally diverse secondary alcohols was carried

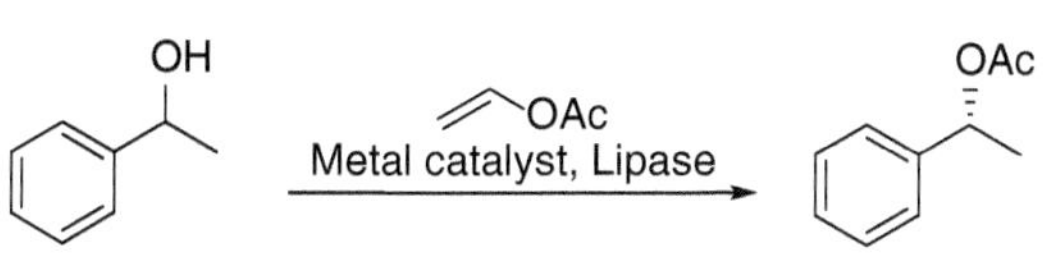

Scheme 4.11 Chemoenzymatic DKR of phenethyl alcohol using vinyl acetate as acyl donor.

out using isopropenyl acetate as the acyl donor in the presence of dicarbonylchlorido(pentabenzylcyclopentadienyl)ruthenium and CALB (Novozym 435) in toluene, and the racemic alcohols were converted into the corresponding (*R*)-acetates in excellent yields and *ee* values (Scheme 4.12) [16].

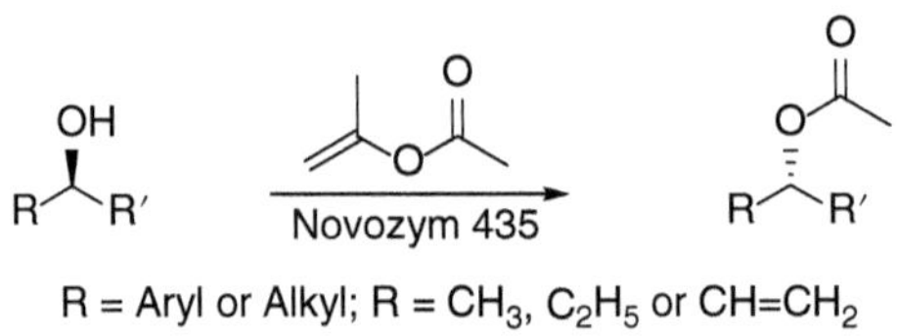

Scheme 4.12 Chemoenzymatic DKR of chiral alcohol using isopropenyl acetate as acyl donor. Source: Adapted from Mari et al. [16].

Some other acyl donors such as trifluoroethyl acetate, *p*-chlorophenyl acetate or the corresponding butyrate, and other carboxylates have also been used in the DKR of chiral alcohols. The acyl donor plays an important role for the outcome of the DKR process based on metal-catalyzed racemization of alcohols. When the ruthenium-catalyzed racemization of phenethyl alcohol was combined with CALB (Novozym 435)-catalyzed transesterification, the acylated (*R*)-phenethyl alcohol, phenethyl alcohol, and acetophenone were obtained in different ratios for different acyl donors, as shown in Scheme 4.13 [17].

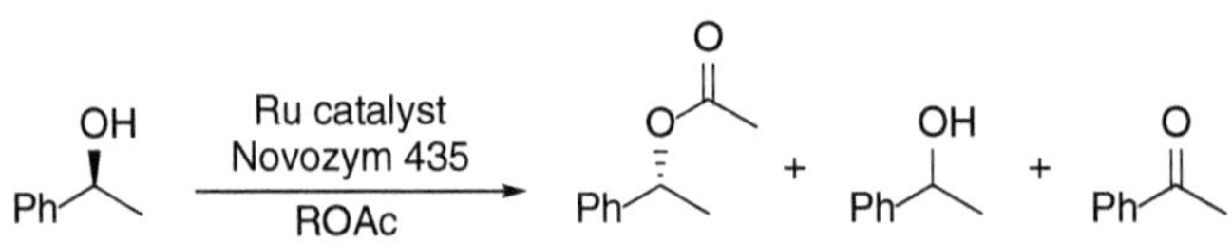

R = vinyl	50	0	50
R = isopropenyl	72	0	28
R = *p*-chlorophenyl	88	12	0

Scheme 4.13 Effects of acyl donor on the chemoenzymatic DKR of phenethyl alcohol. Source: Adapted from Larsson et al. [17].

In the case of alkenyl acetate, the lipase-catalyzed acyl transfer led to the formation of aldehyde, which can act as a hydrogen acceptor and oxidize the substrate alcohol to ketone. In some other cases, the alcohol formed in the acyl transfer reaction can compete with the substrate for the enzyme and/or racemization catalyst, complicating the DKR process. In this context, aryl esters such as *p*-chlorophenyl acetate gained attention and have been proved to be good acyl donors for the DKR of different chiral alcohols. Phenyl acetate is not an active acyl donor for the lipase-catalyzed transesterification reaction, and the number of halogen atoms on the phenyl ring exerts an effect on the transesterification rate. Therefore, by varying the number of halogen atoms, it is possible that effective DKR of a given alcohol can be achieved by adjusting the rate of the transesterification to match that of the metal-catalyzed racemization.

The racemization of secondary alcohols can be realized by the transition-metal-catalyzed hydrogen transfer reaction, which involves metal hydride as the key intermediate. It is generally assumed that there are two different hydridic pathways: a metal monohydride mechanism and a metal dihydride mechanism (Scheme 4.14), and the mechanism in action is dictated by the metal catalyst. For example, the metal dihydride mechanism operates for ruthenium dihalide catalyst precursors, whereas the metal monohydride mechanism applies to rhodium, iridium, and most non-dihalide ruthenium complexes [18].

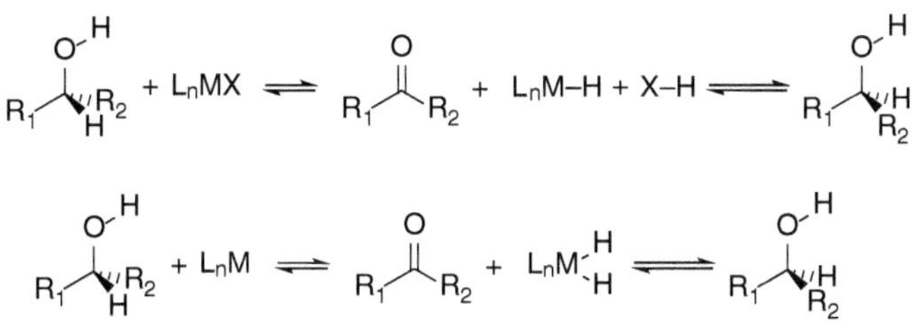

Scheme 4.14 Two different hydridic pathways for metal-catalyzed racemization of secondary alcohols.

An early efficient system for the DKR of secondary alcohols was developed by Bäckvall and coworkers, in which *p*-chlorophenyl acetate was used as the acyl donor and the robust ruthenium catalyst **1** was employed for the racemization [17, 19]. A variety of racemic aromatic and aliphatic secondary alcohols were effectively resolved by combining CALB-catalyzed transesterification with the ruthenium-catalyzed racemization to give the corresponding (*R*)-acetates in 78–92% yields and >99% *ee* (Scheme 4.15). It is worth noting that addition of an external base or the corresponding ketone is not needed for the racemization. When *p*-chlorophenyl acetate was replaced by commercially available isopropenyl acetate [17], similar results were obtained for 1-arylethanols to furnish the corresponding

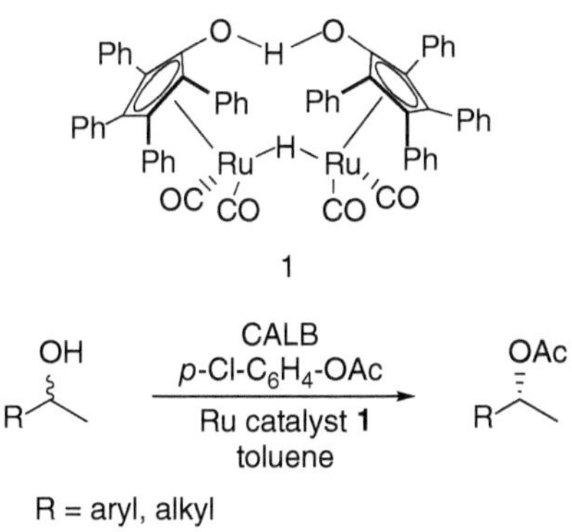

Scheme 4.15 DKR of secondary alcohols using *p*-chlorophenyl acetate as acyl donor.

(*R*)-acetates in 83–95% yields and >99% *ee* (Scheme 4.16). In this case, an appropriate hydrogen source has to be used to prevent the ketone formation; otherwise, the yield would decrease.

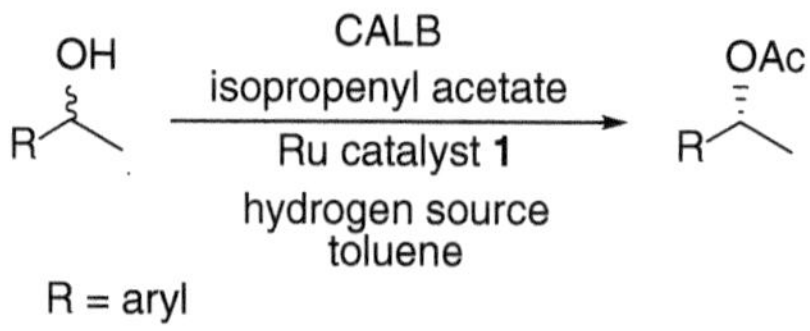

Scheme 4.16 DKR of secondary alcohols using isopropenyl acetate as acyl donor.

This first-generation Bäckvall-type of system, employing CALB, the Shvo catalyst, and 4-chlorophenyl acetate was also efficient for the DKR of a range of substituted heterocyclic racemic ethanols, hence extending the toolbox to access this class of valuable chiral heterocyclic alcohols with high yields and stereoselectivity [20].

$RuCl(\eta^5\text{-indenyl})(PPh_3)_2$ is also a good catalyst for the racemization of 1-phenylethanol in the presence of triethylamine and molecular oxygen (5 mol %) as the oxidant [21]; the combination of this racemization protocol with lipase-catalyzed transesterification afforded (*R*)-phenylethyl acetate in 85% yield and 96% *ee*. The DKR of 1-phenylethanol was also carried out by using $RuCl_2(PPh_3)_3$ or $Ru(TsN(CH_2)_2NH)$(*p*-cymene) in the presence of 2,2,6,6-tetramethylpiperidin-1-oxyl (TEMPO) as the racemization catalyst, and good yields and excellent enantioselectivity have been achieved [22].

While the DKR of chiral alcohols by the combination of metal and enzyme catalysts was performed at elevated temperature in the early studies, a series of metal catalysts have been developed recently for the racemization of secondary alcohols at room temperature. For example, aminocyclopentadienyl ruthenium chloride $RuCl[\eta^5\text{-}C_5Ph_4NHCH(CH_3)_2](CO)_2$, $RuCl(\eta^5\text{-}C_5Ph_5)(CO)_2$ and $RuBr(\eta^5\text{-}C_5Ph_5)(CO)_2$ catalyze the racemization of 1-phenylethanol in the presence of potassium *t*-butyloxide and sodium carbonate, enabling efficient DKR of secondary alcohols by combining with the CALB-catalyzed transesterification using isopropenyl acetate as the acyl donor (Scheme 4.17) [23–25].

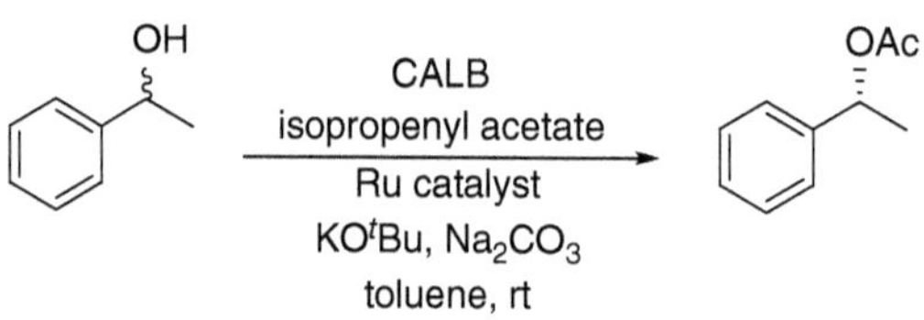

Scheme 4.17 DKR of secondary alcohols at room temperature. Source: Choi et al. [23]; Martín-Matute et al. [24]; Martín-Matute et al. [25].

The substrate scope of dicarbonylchlorido (pentabenzylcyclopentadienyl)-ruthenium and CALB system in the DKR of secondary alcohols has been examined. For 31 structurally different alcohols, the corresponding (*R*)-acetates were obtained in excellent yields and >99% *ee*, in most cases, by using isopropenyl acetate as acyl donor. These results suggested that the ruthenium complex catalyzed the racemization in a highly compatible manner with the enzymatic kinetic resolution of secondary alcohol [16].

The above-discussed transition-metal racemization catalysts require anaerobic conditions for high efficiency, due to their air sensitivity, which reduced the yield of (*R*)-1-phenylethyl acetate to around 60% [26]. In this context, air-stable Ru complexes $[\eta^5\text{-}C_5Ph_4(OCH_2Ph)]Ru(CO)_2X$ (X = Cl, Br, and I) were developed for the racemization of 1-phenylethanol. Notably, potassium phosphate could be used as base instead of potassium *tert*-butoxide, which decomposed the ruthenium complexes. By combining the ruthenium chloride complex with lipase as the catalyst system, the DKR of racemic 1-phenylethanol proceeded efficiently at room temperature in the air, giving (*R*)-1-phenylethyl acetate in >99% yield and >99% *ee* [26]. Furthermore, the ruthenium chloride complex was attached to polyester, forming a polymer-supported recyclable catalyst, which could be reused three times for the aerobic DKR of racemic 1-phenylethanol. Both the ruthenium chloride complex and the polyester-bound one were highly efficient catalysts for the DKR of a diversity of secondary alcohols [26]. The $\eta^5\text{-}C_5Ph_4(OCH_2Ph)$ ligand could be replaced with $\eta^5\text{-}C_5Ph_4(OCOPh)$ to give the complex $[\eta^5\text{-}C_5Ph_4(OCOPh)]Ru(CO)_2Cl$. The substituent on the phenyl rings of $[\eta^5\text{-}C_5Ph_4(OCOPh)]Ru(CO)_2Cl$ derivatives exerted effect on the racemization and DKR of secondary alcohols. The ruthenium complexes, $[\eta^5\text{-}C_5Ar_4(OCOAr)]Ru(CO)_2Cl$ with electron-donating substituents on the phenyl rings, showed better activities than those with electron-withdrawing groups in the racemization of 1-phenylethanol [27].

A carbonyl ligand in the cyclopentadienylruthenium dicarbonyl complexes was also substituted with PPh_3, producing a new class of catalysts for alcohol racemization in the presence of silver oxide at room temperature under aerobic conditions. This alcohol racemization was found to be compatible with the lipase-catalyzed esterification of alcohols by isopropenyl acetate, resulting in effective DKR of secondary alcohols under ambient conditions [28]. The neutral (η^5-phenylindenyl)ruthenium$(PPh_3)_2Cl$ complex also catalyzed the racemization of sec-alcohols in the presence of KO^tBu. By adding sodium tetrakis[3,5-bis(trifluoromethyl)phenyl]borate ($NaBAr_F$) into the dichloromethane solution of (η^5-phenylindenyl)ruthenium$(PPh_3)_2Cl$, a cationic ruthenium complex was obtained in quantitative yield, which was a highly active catalyst for the racemization of enantiomerically pure secondary alcohols in the presence of a weak base such as K_2CO_3. This complex was found to be compatible with CALB, enabling an efficient DKR of secondary alcohols without the action of a strong base [29].

The reaction of $Ru_3(CO)_{12}$ with pentaphenylcyclopenta-2,4-dienol in toluene gave a cyclopentadienyl benzoyl ruthenium(II) complex (Scheme 4.18), which was stable in air. In the presence of 5% ruthenium complex and one equivalent of potassium phosphate, (*S*)-1-phenylethanol was racemized at room temperature within six hours, and 90% of the catalyst could be recovered with the same activity. Efficient DKR of various types of secondary alcohols have been achieved by using this ruthenium(II) complex and CALB with isopropenyl acetate as the acyl donor [30].

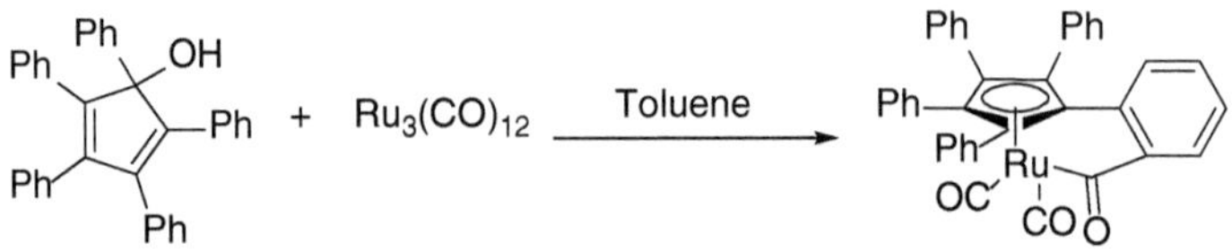

Scheme 4.18 Preparation of cyclopentadienyl benzoyl ruthenium(II) complex.

The aminomethylpyridine (ampy) ruthenium complexes were a class of readily accessible and active catalysts for hydrogen transfer reactions. When these ampy-based ruthenium complexes were tested as alcohol racemization catalysts, the *ortho*-metalated-ampy Ru complex **2** was found to be a highly active racemization catalyst, which catalyzed the racemization of 1-phenylethanol at 70 °C with 100% conversion in 10 min. When the Ru-catalyzed alcohol racemization was combined with lipase-catalyzed esterification, the cycloruthenated amine **3** was most active for the DKR of 1-phenylethanol, generating the (*R*)-1-phenylethylacetate in 86% yield and >99% *ee* [31].

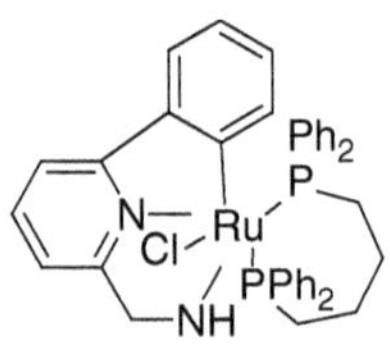

2

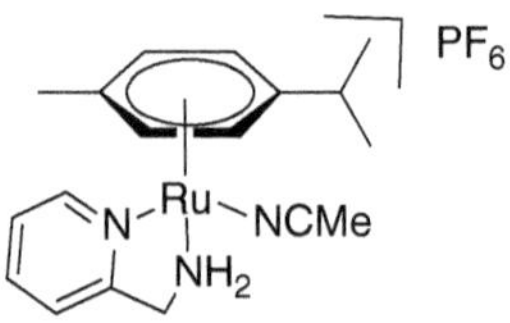

3

A dinuclear ruthenium complex bearing tetrafluorosuccinate and (*rac*)-BINAP ligands was reported to catalyze the racemization of secondary alcohols. It was successfully combined with immobilized CALB in the DKR of aliphatic and aromatic secondary alcohols with isopropyl butyrate as the acyl donor. The catalyst performed optimally at 70 °C and only 0.1 mol % of racemization catalyst relative to the substrate was required. Although addition of the ketone corresponding to the substrate was not required, it stabilized the active Ru complex, and thus was beneficial for the DKR reaction [32].

In addition to Ru complexes, other metal complexes could also serve as the racemization catalysts based on the reversible transfer hydrogenation strategy and have been successfully applied in DKR of secondary alcohols. The pincer-type iron complexes effectively catalyzed the complete racemization of a wide range of benzylic alcohols under mild reaction conditions. An air- and moisture-sensitive iron hydride complex **4** was directly used in combination with CALB for the DKR of benzylic and aliphatic secondary alcohols with *p*-chlorophenyl acetate as acyl donor, giving the corresponding (*R*)-acetates in good yields and with excellent *ee* values [33]. It had been reported that the iron complex **4** could be *in situ* generated from the air- and moisture-stable precursor **5** (Scheme 4.19). This readily accessible iron complex was then employed together with CALB in the metalloenzymatic DKR of secondary alcohols. When phenyl acetate was used as the acyl donor, various benzylic alcohols were effectively converted to enantiomerically pure products in good to high yields [34].

In the presence of molecular hydrogen at 1 atm as hydrogen donor, iron hydride complex **4** in combination with CALB catalyzed the transformation of a wide range of benzylic, aliphatic, and (hetero)aromatic ketones, as well as diketones, into the corresponding acetates with high yields and enantioselectivities, via chemoenzymatic hydrogenation–racemization and kinetic resolution cascade [35].

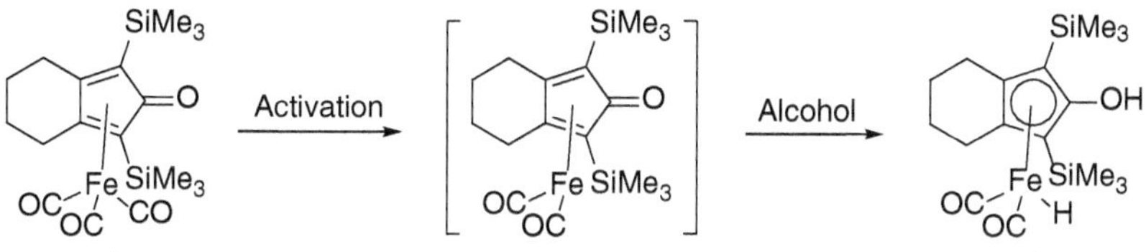

Scheme 4.19 Preparation of iron hydride complex **4**.

A class of bifunctional amidoiridium complexes, Cp*Ir[κ^2(N,C)-(NHCR$_2$-2-C$_6$H$_4$)] (Cp* = 1,2,3,4,5-pentamethylcyclopentadienyl; R = CH_3 and C_6H_5), have been reported to be highly efficient for the transfer hydrogenation of ketones and aerobic oxidation of alcohols. These bifunctional C–N chelate Ir complexes were coupled with lipase CALB for the DKR of racemic secondary alcohols using phenyl acetate as the acyl donor. The chemoenzymatic DKR was carried out at ambient temperature under neutral conditions. A series of racemic aryl alcohols were transformed into the corresponding (*R*)-acetates in 72–93% isolated yields and >99% *ee* [36].

Dinuclear Al complexes and Al alkoxides catalyzed the Meerwein–Ponndorf–Verley–Oppenauer (MPVO) reaction, in which Oppenauer oxidation of the alcohol was followed by non-stereoselective reduction of the resulting ketone, leading to the racemization of alcohol. The reaction of $AlMe_3$ with the bidentate ligands 1,1′-bi-2-naphthol or 2,2′-biphenol in 1 : 1 ratio generated a very effective catalyst, 10 mol% of which racemized benzylic alcohol completely at room temperature within three hours. This Al-catalyzed alcohol racemization was combined with lipase-catalyzed alcohol esterification using phenylvinyl acetate as acyl donor, resulting in the DKR of both aromatic and aliphatic secondary alcohols to give the corresponding (*R*)-acetates in high yield and optical purity [37].

Not only have a number of homogeneous catalysts, including Ru, Ir, Rh, and Al complexes, been demonstrated as effective catalysts for racemization of secondary alcohols under mild reaction conditions and successfully incorporated with enzymatic kinetic resolution of chiral alcohols, but different heterogeneous alcohol racemization catalysts have also been developed and shown to be compatible with enzymatic esterification, resulting in effective DKR of chiral alcohols. Zeolites constitute a large class of heterogeneous catalysts that catalyze a diversity of chemical reactions. In this context, the racemization of 1-phenylethanol has been investigated over a number of zeolites. Their catalytic activity varied as the Si/Al ratio changed. The low racemization activity of the heterogeneous catalysts and the incompatibility of the kinetic resolution and the racemization procedure usually limit the development of efficient DKR processes. As such, Wuyts et al. developed a biphasic system for the DKR of benzylic alcohols. An acidic H-beta zeolite-catalyzed alcohol racemization in the aqueous phase was combined with the lipase-catalyzed kinetic resolution in an organic phase system, affording enantiomerically pure esters in high yield. The drawback of this conceptually elegant system was the unproductive hydrolysis of the acyl donor, and thus a large excess of acyl donor was needed to reach reaction completion [38, 39].

In order to perform the racemization of benzylic alcohols in organic solvents, beta zeolites were doped with low concentrations of other metals to generate active sites. The resulting hydrophobic Zr-beta and Al-beta with a high Si content were found to be good catalysts for the racemization of 1-phenylethanol. By combining them with an immobilized CALB, the DKR of 1-phenylethanol was achieved in toluene at 60 °C to give the (*R*)-phenylethyl ester. Instead of the Meerwein–Ponndorf–Verley–Oppenauer oxidation–reduction mechanism, the racemization of 1-phenylethanol over these Zr-beta or Al-beta catalysts proceeded through an acid-catalyzed hydration–dehydration mechanism, resulting in styrene formation and thus potentially low yield of (*R*)-phenylethyl ester. The *ee* value of the product ester was affected by the acyl donor. For vinyl butyrate or vinyl octanoate as the acyl donor, 92% or 98% *ee* was obtained, respectively, while only 67% *ee* was achieved when isopropenyl acetate, a more common acyl donor, was used. The lower *ee* value of the product ester was attributed to the racemization of the product catalyzed by the zeolite catalyst, because the acetate was less sterically hindered to enter the pore space (active sites) of the zeolite catalyst [40]. It was postulated that the zeolite catalysts might offer shape selectivity for the racemization of different substrates. As such, beta–silicalite-1 core–shell microcomposites with controllable shell thickness were synthesized and used as catalysts for the racemization of 1-phenylethanol. The inert silicalite-1 shell covered the external acidic sites of the beta zeolite core, allowing the alcohols to penetrate the silicalite-1 shell to access the acidic sites at the core beta for racemization, but excluding the product (*R*)-esters to access the acidic sites at the core beta owing to their larger size. Therefore, by using the optimized core–shell catalyst, small and readily available acyl donors such as isopropenyl acetate could be employed in the DKR of various secondary alcohols [41].

In addition to zeolites, various transition-metal-based heterogeneous catalysts have shown great potential for the one-pot DKRs of secondary alcohols, due to the advantages of heterogeneous catalysis, such as the easier workup protocols, simple product isolation, and catalyst recycling [42]. The Shvo catalyst and $[RuCl_2(p\text{-cymene})]_2$ were immobilized on phosphotungstic acid modified g-alumina, and the resulting heterogeneous catalysts were combined with CALB in the DKR of 1-phenylethanol in *n*-octane using vinyl octanoate as the acyl donor, affording the product ester in up to 89% yield and >99% *ee* [43]. The covalently immobilized $[Ph_4(\eta^4\text{-}C_4CO)]Ru(CO)_3$ complex on the functionalized polystyrene served as good racemization catalyst in combination with CALB for DKR of several benzylic and aliphatic alcohols. The polystyrene-immobilized Ru complex and CALB could be successfully used for three runs with 95–99% yields and 99% *ee* for the DKR of 1-phenylethanol [26]. This process was then applied as a key step in the enantioselective synthesis of rivastigmine from commercially available 3′-hydroxyacetophenone in five steps with overall 57% yield (Scheme 4.20) [44].

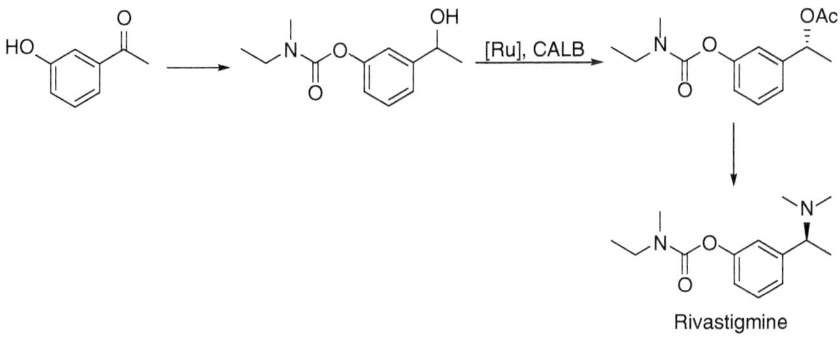

Scheme 4.20 Chemoenzymatic synthesis of rivastigmine via DKR as the key step. Source: Adapted from Han et al. [44].

It had been reported that vanadyl sulfate ($VOSO_4$) and vanadium oxide (V_2O_4) catalyzed the racemization of benzylic alcohols in *n*-octane at 80 °C. The reusable and heterogeneous racemization catalyst $VOSO_4$ was combined with an immobilized CALB in the DKR of benzylic alcohol to give (*R*)-1-phenylethyl octanoate in good yield and high optical purity [45]. ^{18}O-Labeling experiments showed that the V-catalyzed racemization proceeded through formation of a carbenium ion intermediate as the key step.

The combination of $VOSO_4$ with an immobilized CALB was also successfully applied in the DKR of 1-(2,6-dichloro-3-fluorophenyl)ethanol, giving (*R*)-1-(2,6-dichloro-3-fluorophenyl)ethyl acetate (Scheme 4.21), an important intermediate for the synthesis of the anticancer agent crizotinib for the treatment of non-small cell lung carcinoma. In the continuous-flow process of DKR the (*R*)-ester was obtained in 57% conversion and 98% *ee* [46].

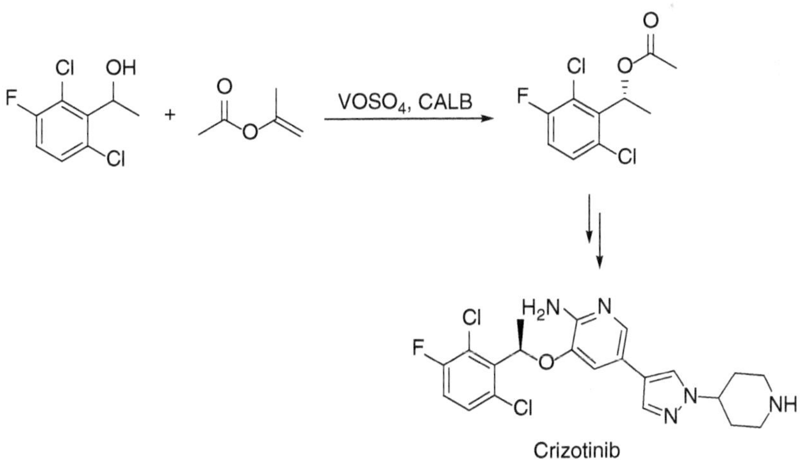

Scheme 4.21 Chemoenzymatic synthesis of the intermediate of crizotinib.

It has been known that the oxovanadium (V) compounds, $[VO(OR)_3]$, catalyze the interconversion of allylic alcohols through an allyl vanadate intermediate, affording a thermodynamic mixture of two regioisomers (Scheme 4.22). A new DKR method was established by employing this vanadium-catalyzed isomerization in combination with the lipase-catalyzed esterification. By using the stable compound $VO(OSiPh_3)_3$ as the racemization catalyst, and ethoxyvinyl acetate or vinyl acetate as the acyl donor, this DKR protocol was successfully applied to various racemic allylic alcohols, even with the tertiary alcohol functional group, to produce the corresponding (*R*)-esters with 91–99% *ee* in 80–96% yields (Scheme 4.23) [47].

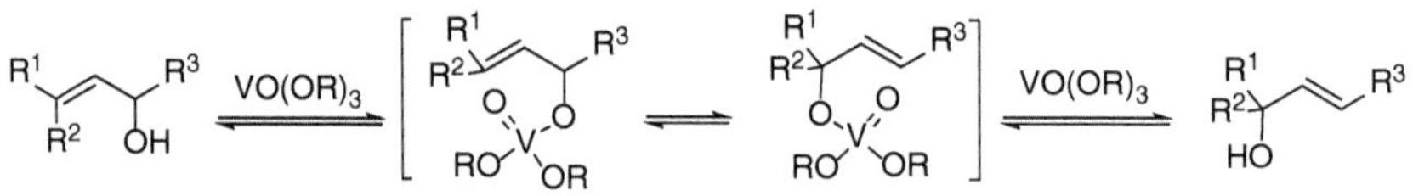

Scheme 4.22 V-catalyzed racemization of allylic alcohols.

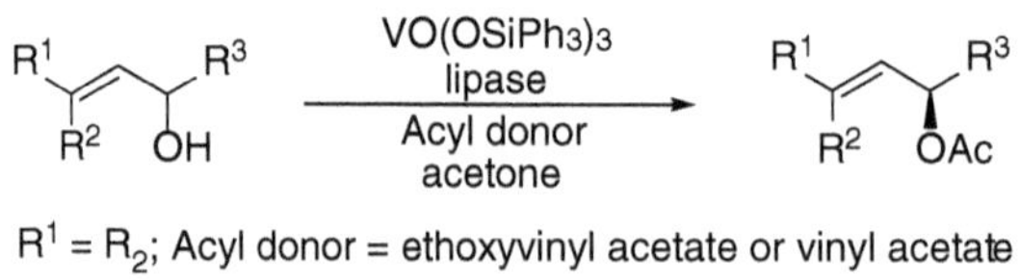

Scheme 4.23 Chemoenzymatic DKR of allylic alcohols. Source: Adapted from Akai et al. [47].

However, for the unsymmetrically substituted allyl alcohols (such as (*E*)-4-phenyl-3-buten-2-ol) where R^1 and R^2 are different, the racemization was very slow and not compatible with the lipase-catalyzed kinetic resolution, making the protocol not effective. In order to address this problem, the solvent, the reaction temperature, and the vanadium catalyst were studied to search for an effective racemization method using (*E*)-4-phenyl-3-buten-2-ol as the substrate. Under the catalysis of a polymer-bound vanadyl phosphate or $VO(OSiPh_3)_3$, the racemization of (*E*)-4-phenyl-3-buten-2-ol was significantly improved in CH_3CN at 35 °C. These vanadyl compounds were then combined with immobilized CALB in the transformation of (*E*)-4-(4′-methoxyphenyl)-3-buten-2-ol in CH_3CN at 35 °C, generating (*R*,*E*)-4-(4′-methoxyphenyl)-3-buten-2-yl acetate in 97% isolated yield and >99% *ee* by using $VO(OSiPh_3)_3$ as the racemization catalyst. Various racemic allyl alcohols were converted into the corresponding enantiomerically enriched allyl acetates by this DKR protocol, which involved the vanadium-catalyzed racemization of the alcohols via the migration of the hydroxyl group, followed by lipase-catalyzed chemo and enantioselective acetylation (Scheme 4.24) [48]. This method was also effective for the dynamic kinetic transformation of 1-phenyl-2,4-hexadien-1-ol into (*R*,3*E*,5*E*)-6-phenyl-3,5-hexadien-2-yl acetate with 88% yield and 98% *ee* via hydroxyl migration across two C=C bonds (Scheme 4.25).

The oxovanadium compound was later immobilized inside mesoporous silica (MPS, pore diameter 2.7 nm), resulting in a novel heterogeneous racemization

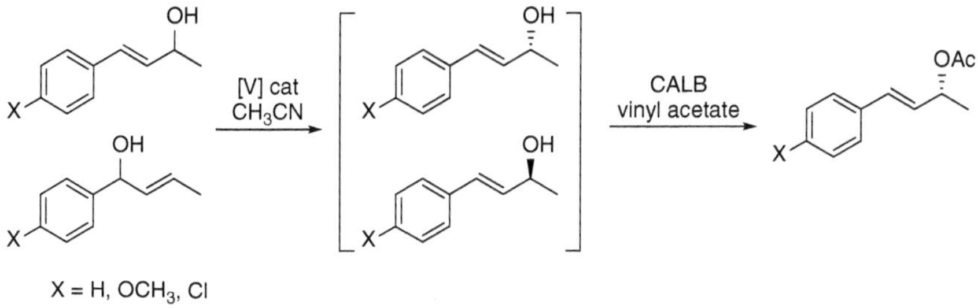

Scheme 4.24 Chemoenzymatic dynamic kinetic transformation of allylic alcohols. Source: Adapted from Akai et al. [48].

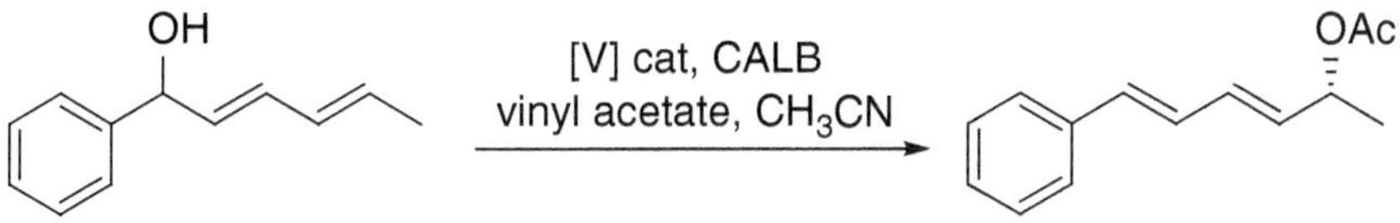

Scheme 4.25 Chemoenzymatic dynamic kinetic transformation of dienols.

catalyst (V-MPS3). The V-MPS3 catalyst showed excellent compatibility with the commercially available immobilized CALB, and the DKRs of a wide range of racemic allyl alcohols were achieved to give the corresponding (*R*)-esters with 64–99% yields and 92–>99% *ee* (Scheme 4.26). Furthermore, the combined lipase–V-MPS3 catalyst was reusable and the high catalytic performance was sustained up to the sixth run, as demonstrated in the DKR of 1-phenyl-2-buten-1-ol, affording the acetate of (*R*)-4-phenyl-3-buten-2-ol in quantitative chemical and optical yields [49].

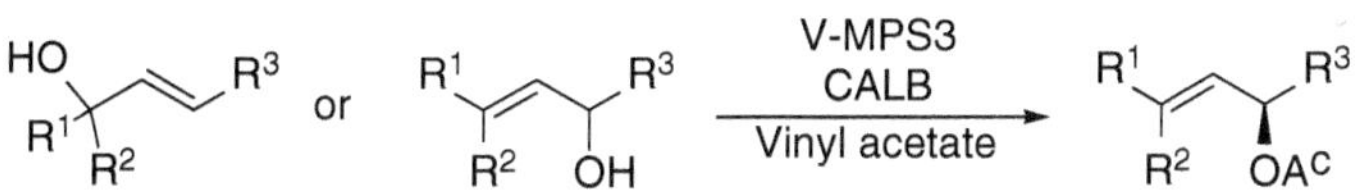

Scheme 4.26 Chemoenzymatic dynamic kinetic resolution (transformation) of allylic alcohols with immobilized oxovanadium catalyst.

The V-MPS3 catalyst showed much higher activity than that of the previously used $O = V(OSiPh_3)_3$ and the polymer-supported oxovanadium compound. V-MPS2 and V-MPS4 were also prepared from the MPS with pore sizes of approximately 2 and 4 nm, respectively, in a similar manner as for the preparation of V-MPS3. They were combined with lipase CALB in the dynamic kinetic transformation of 1-phenyl-2-buten-1-ol, and the acetate of (*R*)-4-phenyl-3-buten-2-ol was obtained with 98% or 99% *ee* in all cases, but the reaction using V-MPS2 was slower. By choosing suitable V-MPS catalyst, a diversity of allyl alcohols and benzylic alcohols were effectively converted into the corresponding acetates in 76–99% isolated yields and 94–>99% *ee* [50].

The combination of oxovanadium catalyst V-MPS3 with CALB effectively catalyzed DKR of racemic 4-(cyclohex-1-en-1-yl)but-3-en-2-ol with 2,2,2-trifluoroethyl propiolate as acyl donor, affording the (*R*)-ester, which underwent intramolecular Diels–Alder reaction in toluene with 5.0 mol% 3,5-di-*tert*-butyl-4-hydroxytoluene (BHT) to give (3*R*,3α*S*,8α*S*)-3-methyl-3α,5,6,7,8,8α-hexahydronaphtho[2,3-c]furan-1(3*H*)-one (Scheme 4.27). Addition of QuadraPure AMPA (1.0 w/w) into the DKR step significantly facilitated the reaction, and the final product could be obtained in up to 81% yield and 98% *ee*. By using (*Z*)-3-(phenylsulfonyl)acrylate as the acyl donor, the resulting ester underwent intramolecular Diels–Alder reaction under the same reaction conditions, affording (3*R*,3α*S*,8α*R*,9*S*,9α*S*)-3-methyl-9-(phenylsulfonyl)-3α,5,6,7,8,8α,9,9α-octahydronaphtho[2,3-c]furan-1(3*H*)-one and its diastereomer in 4 : 1 ratio in a one-pot process. By treatment with 1,8-diazabicyclo[5.4.0]undec-7-ene (DBU) in dichloromethane, this diastereomeric mixture was quantitatively converted to (3*R*,3α*S*,8α*S*)-3-methyl-3α,5,6,7,8,8α-hexahydronaphtho[2,3-c]furan-1 (3*H*)-one with 98% *ee*, which was used as the precursor for the total synthesis of (−)-himbacine (Scheme 4.28), a tetracyclic piperidine alkaloid of pharmaceutical importance isolated from the Australian pine tree of *Galbulimima* species [51].

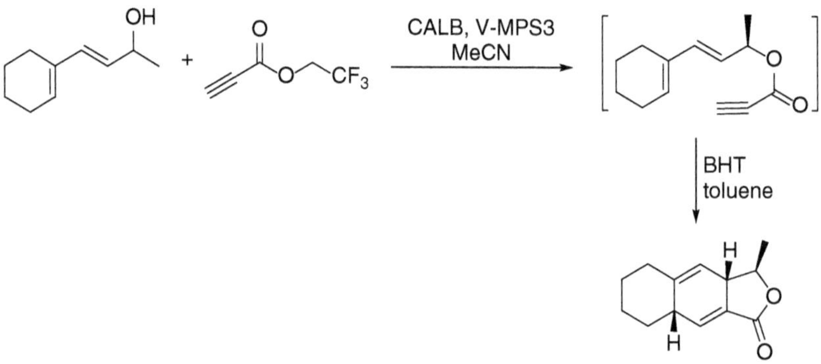

Scheme 4.27 Lipase/oxovanadium co-catalyzed DKR of racemic 4-(cyclohex-1-en-1-yl) but-3-en-2-ol followed by intramolecular Diels–Alder reaction.

The combination of immobilized oxovanadium catalyst V-MPS with CALB also effectively catalyzed DKR of *tert*-allylic alcohols. By using V-MPS4 as the racemization catalyst and vinyl acetate as acyl donor, racemic 1-substituted cyclohex-2-enols were converted to the corresponding (*R*)-3-substituted cyclohex-2-enyl acetate with high yields and optical purity. The (*R*)-esters obtained underwent Ireland–Claisen rearrangement to furnish 1-substituted (*S*)-cyclohex-2-enylacetic acids bearing an all-carbon quaternary stereogenic center (Scheme 4.29). This method was applied to the total synthesis of (−)-crinane, a core structure of a class of natural alkaloids, from cyclohex-2-enone in seven steps with an overall yield of 39% and 98% *ee* (Scheme 4.30) [52].

The V-MPS4/lipase-co-catalyzed DKR of racemic propargyl alcohols was also studied. However, it has been known that the oxovanadium moiety reacted with propargyl alcohol, resulting in Meyer–Schuster rearrangement to form an enal. This irreversible process competed with the racemization of allyl alcohols,

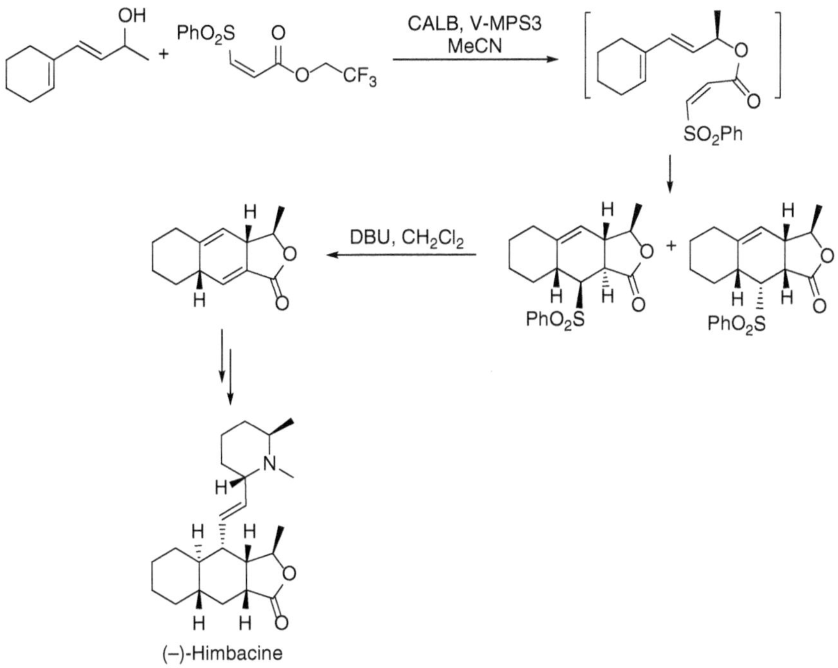

Scheme 4.28 One-pot chemoenzymatic DKR of racemic 4-(cyclohex-1-en-1-yl) but-3-en-2-ol and intramolecular Diels–Alder reaction.

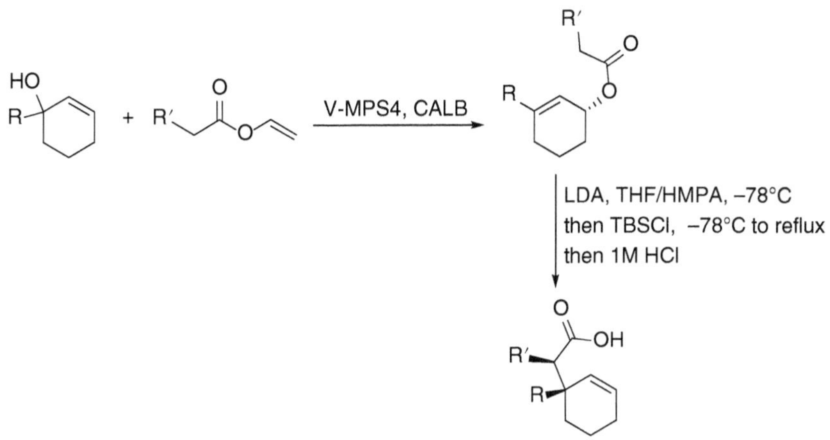

Scheme 4.29 Chemoenzymatic DKR of racemic 1-substituted cyclohex-2-enols followed by Ireland–Claisen rearrangement.

leading to a significant decrease in the yields of DKR product (Scheme 4.31). (Trifluoromethyl)benzene as a solvent was found to accelerate the racemization of propargyl alcohols and suppressed the oxovanadium-catalyzed Meyer–Schuster rearrangement. Therefore, various racemic aryl propargyl alcohols were transformed into the corresponding propargyl esters with up to 99% yields and 99% *ee* by

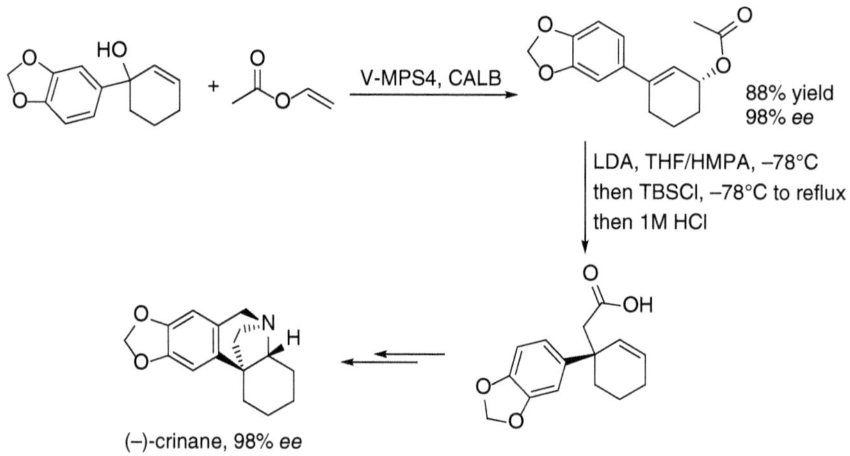

Scheme 4.30 Chemoenzymatic synthesis of (–)-crinane. Source: Adapted from Kawanishi et al. [52].

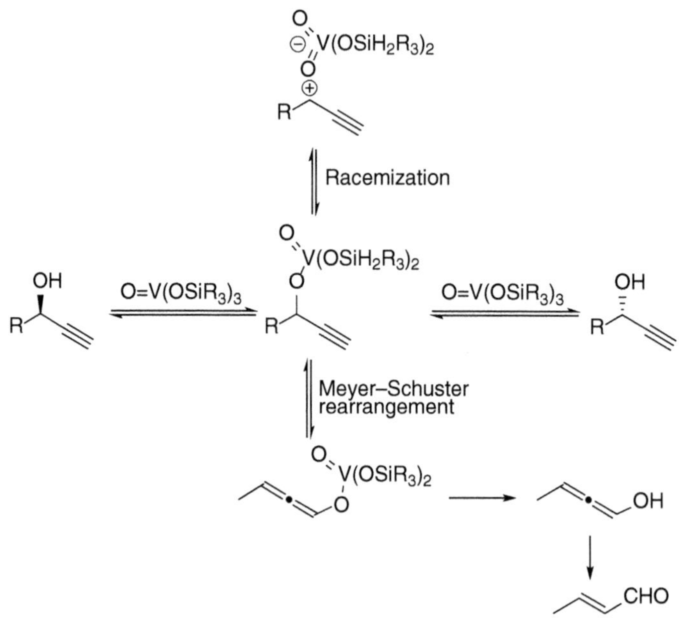

Scheme 4.31 Oxovanadium-catalyzed Meyer–Schuster rearrangement and racemization of propargyl alcohols.

conducting the lipase/oxovanadium-co-catalyzed DKR in (trifluoromethyl)benzene and use of vinyl decanoate in (trifluoromethyl)benzene or vinyl butyrate in acetonitrile (Scheme 4.32) [53].

The racemization of allylic alcohols with Ru catalysts such as Shvo's catalyst proceeded via a conjugated enone intermediate (non-coordinated or coordinated), followed by its reduction at the carbonyl position by a ruthenium hydride. At the

Scheme 4.32 Lipase/oxovanadium-co-catalyzed DKR of racemic aryl propargyl alcohols. Source: Adapted from Kawanishi et al. [53].

same time, competing hydride addition could occur at the olefin bond, resulting in the isomerization of the starting material to the saturated ketone. The α,β-unsaturated carbonyl intermediate might escape reduction as an oxidized side product. These undesired reactions could be minimized by performing the racemization with Shvo's catalyst in dry toluene and the kinetic resolution by lipase PS from *Burkholderia cepacia* immobilized on Celite (PS-IM) with *p*-chlorophenyl acetate as acyl donor. Under these conditions, the DKR of 3-phenylcyclohex-2-enol afforded (*R*)-3-phenylcyclohex-2-enyl acetate in 97% *ee* and 91% yield, offering a new approach to access a range of optically active functionalized allylic alcohols in good yields and high *ee*s (Scheme 4.33) [54].

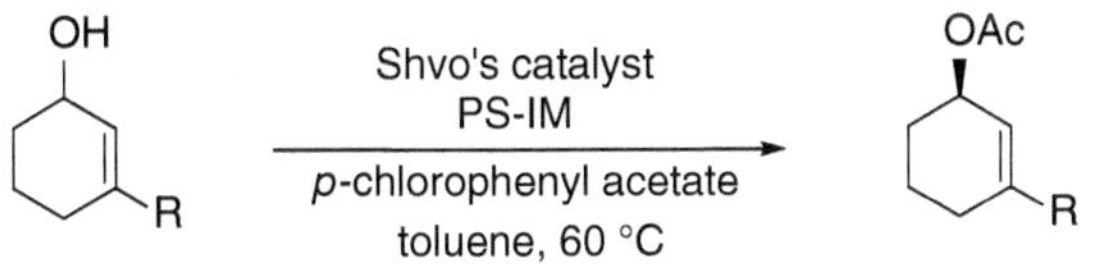

Scheme 4.33 DKR of 3-substituted cyclohex-2-enols. Source: Adapted from Lihammar et al. [54].

However, for hydroxy-ketones such as 4-hydroxy-2-methylcyclopent-2-en-1-one, the Shvo's catalyst triggered an intramolecular redox reaction, resulting in the formation of 2-methylcyclopent-2-en-1,4-dione. On the other hand, $[RuCl_2(cymene)]_2$ complex catalyzed the efficient racemization of 4-hydroxy-2-methylcyclopent-2-en-1-one in the presence of *N*,*N*,*N'*,*N'*-tetramethylpropanediamine in 1,2-dichloroethane, but low enantioselectivity of lipase-catalyzed kinetic resolution under these conditions made the DKR disappointing. Therefore, 4-hydroxy-2-methylcyclopent-2-en-1-one was converted to 4-hydroxy-2-methylcyclopent-2-en-1-one *O*-trityloxime. The DKR of this *O*-trityloxime using $[RuCl_2(cymene)]_2$ and CALB in the presence of trimethylamine with isopropenyl acetate as acyl donor afforded the (*S*)-acetate in up to 88% yield and 98% *ee* (Scheme 4.34) [56].

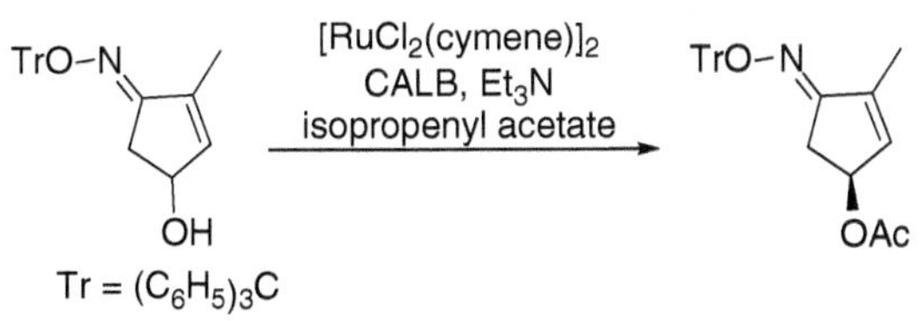

Scheme 4.34 DKR of 4-hydroxy-2-methylcyclopent-2-en-1-one *O*-trityloxime. Source: Adapted from Michalak et al. [55].

Metallic palladium nanoparticles (NPs) with a size of approximately 2.7 nm was immobilized on a MPS, and the resulting catalyst (Pd@SBA-15) showed highly efficient racemization activity on secondary alcohols. The racemization was combined with lipase-catalyzed kinetic resolution in a one-pot process, giving the corresponding (*R*)-acetate with good yields and excellent *ee* values. The immobilized lipase and Pd catalysts could be recovered and reused for several runs without losing the catalytic efficiency. Microwave heating was applied to accelerate the racemization, thus greatly shortening the reaction time required for the complete DKR [55]. Household fluorescent light was also utilized to facilitate the metal-catalyzed racemization of secondary alcohols under ambient conditions. The diruthenium complex was inactive for the racemization of (*S*)-1-phenylethanol at 30 and 50 °C in the dark. When the reaction solution was illuminated with fluorescent light (30 W) at 30 °C, (*S*)-1-phenylethanol was racemized within 10 min. The photoinduced Ru-catalyzed alcohol racemization was then applied in combination with CALB-catalyzed esterification in the chemoenzymatic DKR of racemic secondary alcohols to give optically pure acetates with high yields and excellent optical purity under mild conditions (Scheme 4.35) [57].

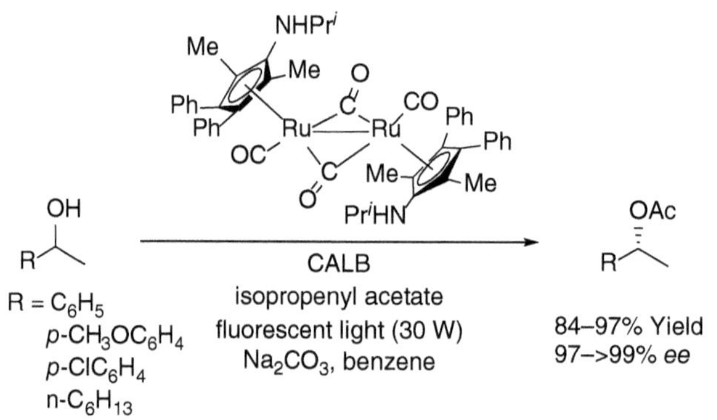

Scheme 4.35 DKR of chiral alcohols via photoinduced Ru-catalyzed racemization. Source: Adapted from Do et al. [57].

Supercritical carbon dioxide ($scCO_2$) was conceptually considered as an environmentally benign reaction medium to replace conventional and potentially hazardous solvents in a wide range of synthetic processes [58]. Accordingly, using an immobilized lipase PS CI from *B. cepacia* as catalyst and vinyl acetate as acyl donor, the (dynamic) kinetic resolution of racemic 1-phenylethanol gave (*R*)-phenylethyl acetate with 48–49% yield and 98–99% *ee* in $scCO_2$, higher yield than that in hexane (30–35% yield, 98–99% *ee*) under comparable conditions. By combining the lipase-catalyzed KR with the racemization of the unreacted alcohol catalyzed by either $[Ru(p\text{-cymene})Cl_2]_2$ or the acid catalyst Nafion SAC in $scCO_2$, the yield of the (*R*)-phenylethyl acetate was increased to 70% or 85%, respectively. The *ee* values of the products in the DKR were higher in $scCO_2$ (96% Ru-catalyst, 85% Nafion) than those in hexane (91%, 81%). These results demonstrated that a viable DKR of secondary alcohols could be established in $scCO_2$ [59].

As shown above, the commercially available CALB was used for the DKR of secondary alcohols in most cases. The limited accessibility to highly active and enantioselective enzymes for the kinetic resolution of racemic alcohols would hamper the wide application of these DKR processes in organic synthesis. In this context, a highly active enzyme was prepared by coating *B. cepacia* lipase with an ionic surfactant to overcome the low activity problem of the commercially available *B. cepacia* lipase (BCL, brand names: Lipase PS and Lipase PS-C) in the DKR of chiral alcohols. The most active ionic-surfactant-coating *B. cepacia* lipase (ISCBCL) (18% protein) preparation displayed three orders of magnitude higher activity than the crude enzyme Lipase PS, and more active than CALB (20% protein). In combination with aryl-substituted cyclopentadienyl ruthenium complexes as the racemization catalyst, the ISCBCL showed excellent performance in the DKR of a wide range of secondary alcohols (RCH(OH)Ar), with isopropenyl acetate as the acyl donor in toluene. In most cases, excellent yields and high enantiopurities were obtained, and the shape of the aliphatic chain (*R*) might result in the configuration switching of product acetates in the DKR [60]. The substrate scope and enantioselectivity of ionic surfactant-coated ISCBCL in kinetic and DKR were further investigated with more structurally diverse substrates, including 6 boron-containing alcohols, 24 chiral propargyl alcohols, and 14 diarylmethanols. ISCBCL catalyzed the KR of most tested substrates with high enantioselectivity, especially the sterically demanding secondary alcohols with two bulky substituents at the hydroxymethine center. In combination with a ruthenium-based racemization catalyst at 25–60 °C, about half of the tested alcohols were converted into the corresponding acetates with high yields (>80%) and excellent enantiopurity (>90% *ee*) [61].

The lipase/metal-catalyzed DKR reactions of secondary alcohols usually afforded the ester products of (*R*)-configuration. The subtilisin-catalyzed kinetic resolution of secondary alcohols generated the (*S*)-configurated ester products, offering a complementary stereoselectivity to the lipase-based KRs (Scheme 4.36). However, the application of subtilisin in kinetic resolution of chiral alcohols has been limited due to the low activity and stability of subtilisin in nonaqueous biocatalysis.

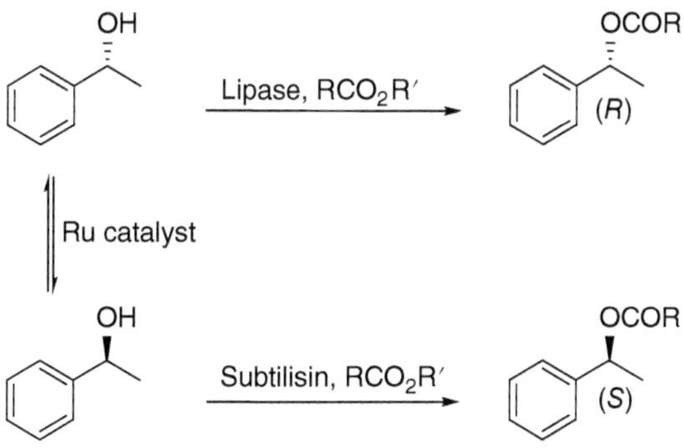

Scheme 4.36 Lipase- and subtilisin-catalyzed kinetic resolution of chiral alcohols.

In order to solve these problems, the commercially available serine protease, subtilisin Carlsberg, was treated with a nonionic surfactant, polyoxyethylene (10) cetyl ether ($C_{16}H_{33}(OCH_2CH_2)_nOH$, $n = 10$). The surfactant-treated subtilisin (STS) exhibited significantly enhanced activity and stability in organic media such as tetrahydrofuran (THF) than its untreated counterpart. The STS was applied in combination with an aminocyclopentadienylruthenium complex in the DKR of a wide range of racemic secondary alcohols in THF with trifluoroethyl butyrate as the acyl donor, transforming them into the corresponding (*S*)-configurated butyrates in high yields and *ee* values [62].

Although STS offered a useful enantiocomplementary counterpart of (*R*)-selective lipase for the DKR of secondary alcohols, the (*S*)-selective DKRs required very long reaction time (three to four days) due to poor compatibility between the enzyme and the ruthenium catalyst. As such, the more active racemization catalyst, (η^5-C_5Ph_5)Ru$(CO)_2$Cl, was employed to facilitate the racemization of alcohols rapidly at room temperature, thus allowing the use of a non-thermostable enzyme. After the effects of subtilisin treatment with surfactants and the acyl donor on the activity and enantioselectivity of the catalytic system toward the kinetic resolution of 1-phenylethanol were investigated, the best results were obtained by using isopropenyl valerate as the acyl donor, and with the enzyme treated with surfactants, octyl β-D-glycopyranoside and polyoxyethylene(10) cetyl ether, in a ratio of 4 : 1 : 1. The combination of catalyst (η^5-C_5Ph_5)Ru$(CO)_2$Cl and the specially treated subtilisin Carlsberg was about 4–5 times faster than the previously reported (*S*)-selective DKRs. The DKR of various racemic alcohols proceeded smoothly under very mild reaction conditions to give the corresponding esters in high yields with excellent enantioselectivities [62, 63].

A subtilisin from *Bacillus licheniformis* was treated with an ionic surfactant, and when the resulting ionic-surfactant-coating *B. licheniformis* subtilisin (ISCBLS) was used for the DKR of secondary alcohols, ISCBCL displayed an extremely enhanced activity in the transesterification of trifluoroethyl butyrate with 1-phenylethanol in THF, and high enantioselectivity for the kinetic resolution of most tested secondary alcohols, such as *m*- or *p*-substituted 1-phenylethanols. The DKRs of these secondary alcohols by the combination of ISCBLS and a ruthenium-based racemization catalyst provided the (*S*)-configuration products with good results (80–94% yield, 90–99% *ee*) [64].

Optically pure diarylmethanols and 1,2-diarylethanols are useful building blocks for the synthesis of pharmaceutically important compounds such as antihistaminic, antiarrhythmic, anticholinergic, and anticancer agents. While the enzyme/metal-catalyzed DKR offered a novel approach to access the desirable single enantiomer of these sterically bulky alcohols from their racemic forms, the most used enzyme in the previous DKR of secondary alcohols was CALB, which usually exhibited high enantioselectivity toward the kinetic resolution of secondary alcohols carrying one small and one significantly larger substituent at the hydroxymethine center. The enzymes applicable to chiral alcohols with two bulky substituents at the hydroxymethine center, such as 1,2-diarylethanols and diarylmethanol, were relatively rare. It was recently found that a commercially available *Pseudomonas stutzeri* lipase (PSL) was highly enantioselective ($E > 200$) toward the transesterification 1,2-diarylethanols with isopropenyl acetate as acyl donor in toluene at room temperature. The enzymatic enantioselectivity was dramatically reduced toward 1,3-diphenyl-1-propanol (E = 38) and 1,4-diphenyl-2-butanol (E = 8). A practical procedure was thus developed for the DKR of 1,2-diarylethanols by employing the highly enantioselective lipase PSL as the resolution catalyst and a ruthenium complex as the racemization catalyst, converting the racemic 1,2-diarylethanols into the corresponding acetates with good yields (95–97%) and high enantiomeric excesses (96–99%) [65].

When a lipoprotein lipase (LPL) from *Burkholderia* species was coated with both dextrin (D) and an ionic surfactant via lyophilization, the activated lipoprotein lipase (LPL-D1) was about 3000-fold more active than its native counterpart toward the kinetic resolution of diarylmethanols with isopropenyl acetate as acyl donor in toluene at 40 °C. The LPL-D1-catalyzed kinetic resolution was combined with Ru-catalyzed racemization, leading to effective DKR of various diarylmethanols. Twenty four out of 31 tested substrates were transformed to the corresponding acetates with satisfactory yields (71–96%) and high enantiopurities (90–99% *ee*). The DKR reaction of phenyl-(*p*-trimethylsilylphenyl)methanol provided the (*R*)-acetate in 82% yield and 96% *ee*, which was then converted to *L*-cloperastine, an antitussive drug, in 57% overall yield (Scheme 4.37). This demonstrated the synthetic application of the DKR procedure [66].

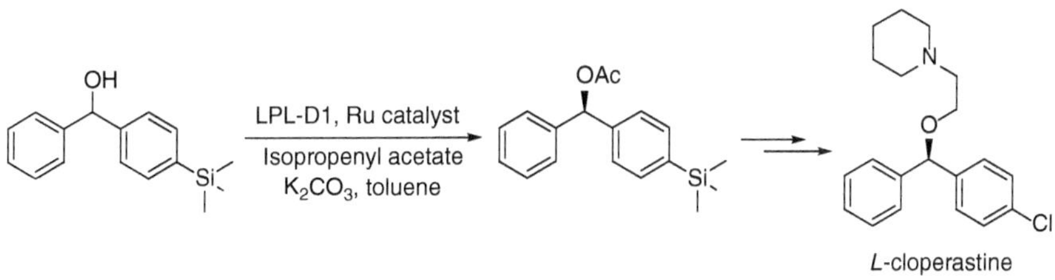

Scheme 4.37 DKR of phenyl-(*p*-trimethylsilylphenyl)methanol.

This highly active LPL-D1 also displayed perfect enantioselectivity (E = >200) toward the esterification of alkyl aryl carbinols carrying a trimethylsilyl group. The LPL-D1-catalyzed esterification was then successfully applied in combination with a Ru-based alcohol racemization in the DKR of these substrates, affording single enantiomeric products (>99% *ee*) in good yields. The resulting products were readily desilylated or halodesilylated to yield enantiopure alkyl aryl carbinols, establishing a particularly useful method for the preparation of alkyl aryl carbinols carrying a long alkyl chain (C6–C10, Scheme 4.38). It was, otherwise, pretty challenging to access these important alcohols in optically pure form [67].

Recently, by employing a bulky substituted cyclopentadienyl ruthenium complex **6** as the racemization catalyst and the anionic surfactant-activated lipoprotein lipase LPL-D1 as the acyl transfer catalyst, the DKR of various secondary alcohols was achieved using isopropenyl acetate as acyl donor under base-free conditions at room temperature, affording the corresponding (*R*)-acetates in 92–99% yields and 91–99% *ee* values. Furthermore, several α-arylallyl alcohols were also transformed into the (*R*)-acetates in 87–91% yields and 91–99% *ee* values (Scheme 4.39) [68].

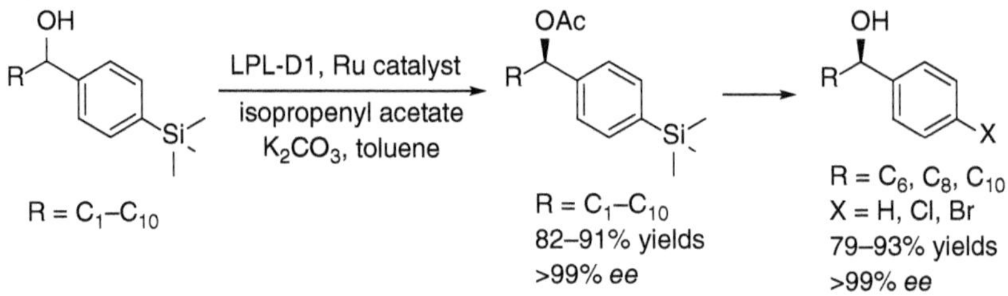

Scheme 4.38 Preparation of alkyl aryl carbinols carrying a long alkyl chain.

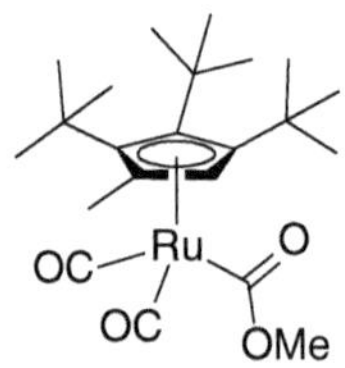

6

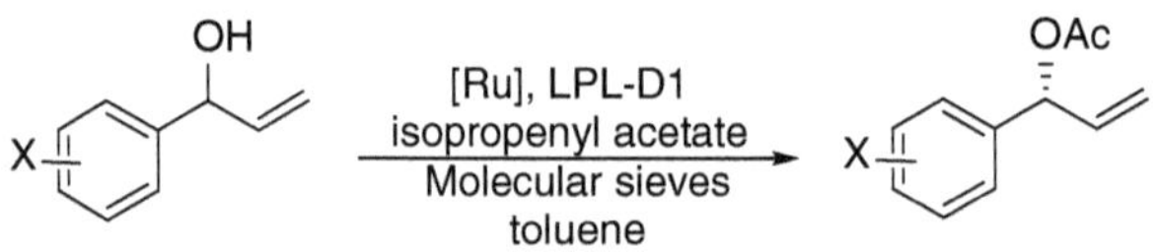

Scheme 4.39 DKR of α-arylallyl alcohols under base-free conditions. Source: Adapted from Yun et al. [68].

In the enzyme/metal-catalyzed DKR of secondary alcohols, the metal-catalyzed *in situ* racemization of the remaining substrate enantiomer proceeded via a reversible oxidation/reduction through a prochiral ketone intermediate. However, the racemization of primary alcohols with an unfunctionalized stereogenic center in the β-position differed, leading to the possibility that the primary alcohol might have been oxidized to aldehyde under the action of a metal catalyst, and that the aldehyde could undergo enolization, resulting in an indirect racemization of primary alcohol. Enolization of the aldehyde could be facilitated by elevated temperature or acid/base catalysis. Shvo's catalyst was found to catalyze the racemization of 2-phenylpropanol through the following three steps: (i) ruthenium-catalyzed dehydrogenation of the alcohol, (ii) enolization of the resulting aldehyde, and (iii) ruthenium-catalyzed reduction of the aldehyde. A lipase from *B. cepacia* was reported to catalyze a highly efficient kinetic resolution of various primary alcohols in the presence of vinyl 3-phenylpropanoates as acyl donors. Based on these results, an efficient enzyme- and metal-catalyzed DKR of primary alcohols was established by employing Amano Lipase PS-D I and Shvo's catalyst as the catalysts and 4-nitrophenyl 3-[4-(trifluoromethyl)phenyl]propanoates as acyl donors (Scheme 4.40). Various β-racemic primary alcohols were converted to the corresponding alcohol esters in good to high yields and optical purity (up to 99% *ee*) [69].

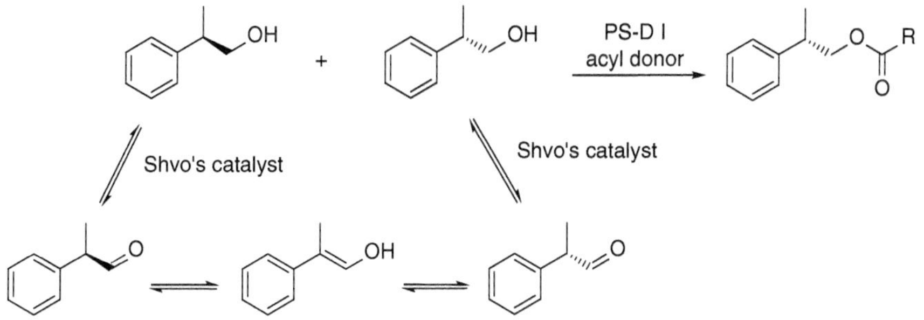

Scheme 4.40 Chemoenzymatic DKR of β-racemic primary alcohols.

Optically pure chlorohydrins are versatile precursors for the synthesis of chiral epoxides, β-aminoalcohols, pyrrolidines, and functionalized cyclopropane derivatives. DKR protocols involving a ruthenium catalyst for the racemization and a lipase for enzymatic resolution had been successfully used in the preparation of many enantiomerically pure alcohols. However, the most used CALB showed low activity toward chlorohydrins. By employing lipase PS-C "Amano" II and $(\eta^5\text{-}C_5Ph_5)Ru(CO)_2Cl$ as catalysts and isopropenyl acetate as acyl donor in dry toluene, a diversity of racemic aromatic chlorohydrins were transformed into the corresponding (*S*)-acetate with high yields and *ee* values [70]. The ruthenium- and enzyme-catalyzed DKR was then applied as the key step in the enantioselective synthesis of (*R*)-bufuralol (Scheme 4.41), the most studied benzofuran derivative and a nonselective β-adrenoceptor blocking agent [71].

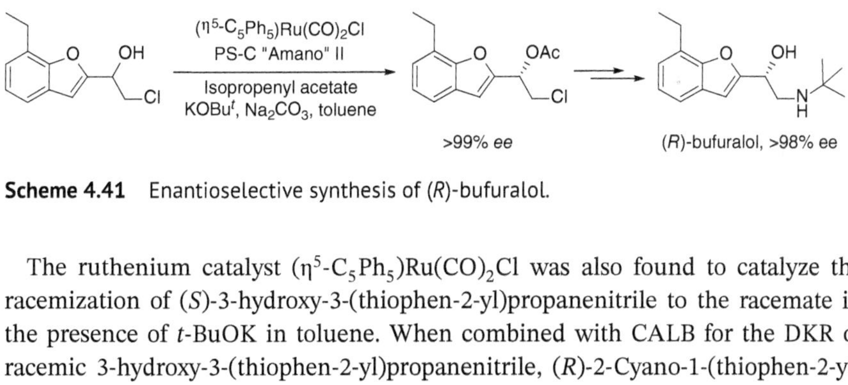

Scheme 4.41 Enantioselective synthesis of (*R*)-bufuralol.

The ruthenium catalyst $(\eta^5\text{-}C_5Ph_5)Ru(CO)_2Cl$ was also found to catalyze the racemization of (*S*)-3-hydroxy-3-(thiophen-2-yl)propanenitrile to the racemate in the presence of *t*-BuOK in toluene. When combined with CALB for the DKR of racemic 3-hydroxy-3-(thiophen-2-yl)propanenitrile, (*R*)-2-Cyano-1-(thiophen-2-yl) ethyl acetate was obtained using isopropenyl acetate as the acyl donor in 87% yield and 98% *ee* (Scheme 4.42). The (*R*)-β-cyano acetate was then used in the synthesis of (*R*)-duloxetine, an antidepressant drug targeting the presynaptic cell [72].

Chiral β-amino alcohols are important structural moieties in biologically active compounds, and serve as chiral ligands for asymmetric catalysis. In addition to α-halo alcohols, β-azido alcohols also are useful precursors for the preparation of chiral β-amino alcohols via azido-reduction. Therefore, DKR of β-azido alcohols was explored by combining lipase-catalyzed resolution with ruthenium-catalyzed alcohol racemization. A variety of racemic β-azido alcohols were efficiently

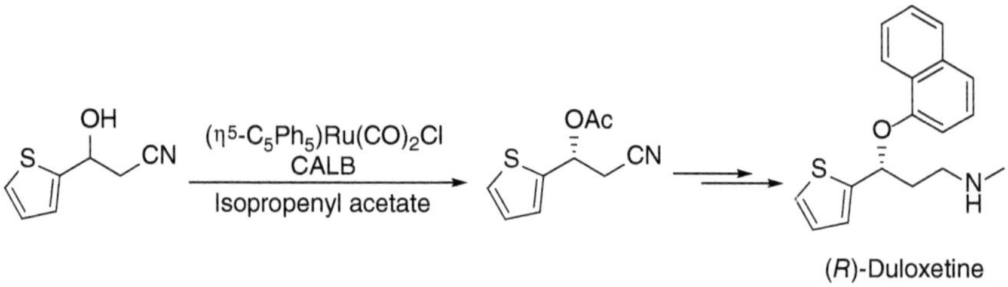

Scheme 4.42 DKR of racemic β-hydroxynitrile.

transformed by employing CALB and Shvo's catalyst in the DKR into the corresponding optically pure acetates (up to 98% conversion and 99% *ee*). Among them, (*S*)-1-azido-2-acetoxy-3-(1-naphthoxy)propane was used to synthesize the antihypertensive drug (*S*)-propanolol (Scheme 4.43) [73].

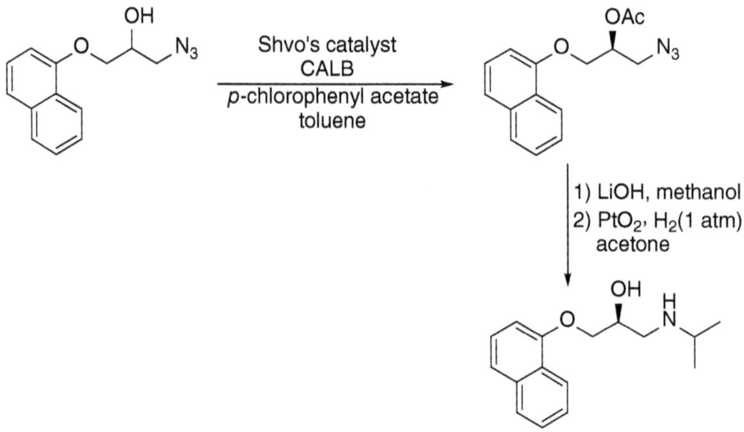

Scheme 4.43 DKR of racemic β-azido alcohols. Source: Adapted from Pàmies et al. [73].

DKR of hydroxyl carboxylic acid esters with the hydroxyl group at different positions had been achieved via enzymatic resolution in combination with ruthenium-catalyzed racemization. Several α-hydroxy esters were converted into the corresponding α-acetoxy acid esters in good yields and excellent *ee*s by using lipase PS-C, Shvo's ruthenium catalyst, and 4-chlorophenyl acetate as acyl donor in cyclohexane (Scheme 4.44) [74]. DKR of β-hydroxy esters was also realized under similar conditions. The racemic β-hydroxy esters were synthesized by the aldol reaction of aldehyde and ethyl acetate. It was worth noting that the aldol reaction was incorporated with the DKR, and a one-pot procedure was developed to access

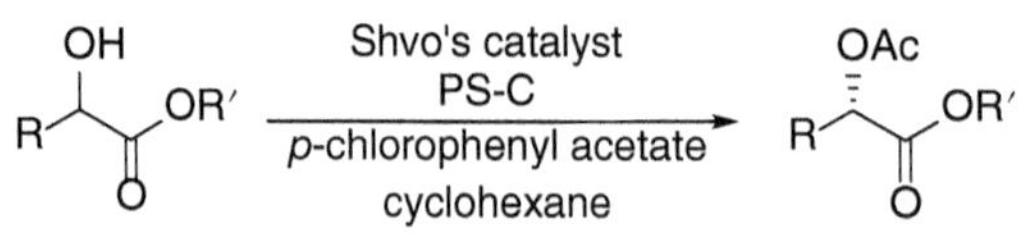

Scheme 4.44 DKR of racemic α-hydroxy esters. Source: Adapted from Huerta et al. [74].

the β-hydroxy ester derivatives with high enantiomeric purity (up to 99% *ee*) from simple starting materials (Scheme 4.45) [75].

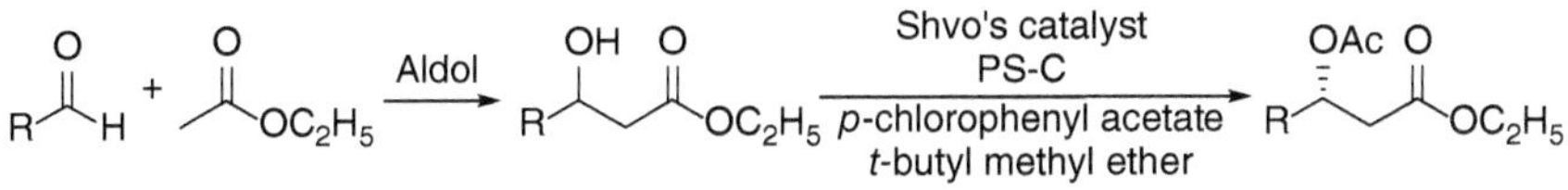

Scheme 4.45 DKR of racemic β-hydroxy esters. Source: Adapted from Huerta et al. [75].

While CALB showed low enantioselectivity toward the kinetic resolution of racemic γ-hydroxy amides, lipases PS-C and *P. fluorescens* lipase (PF) catalyzed highly enantioselective transesterification of these substrates. The enzyme PS-C tolerated both variation in the chain length and different functionalities, giving good to high enantioselectivity (*E* values of up to >250). When PS-C-catalyzed kinetic resolution was combined with Shvo's ruthenium-catalyzed racemization, the corresponding γ-acetoxy amides were obtained in high enantiomeric purity (up to 98% *ee*). However, a small amount of ketone was detected in the DKR (Scheme 4.46). 2,4-Dimethyl-3-pentanol (DMP) could be used as a hydrogen source to suppress the oxidation of the hydroxyl group, leading to higher yields of γ-acetoxy amides [76].

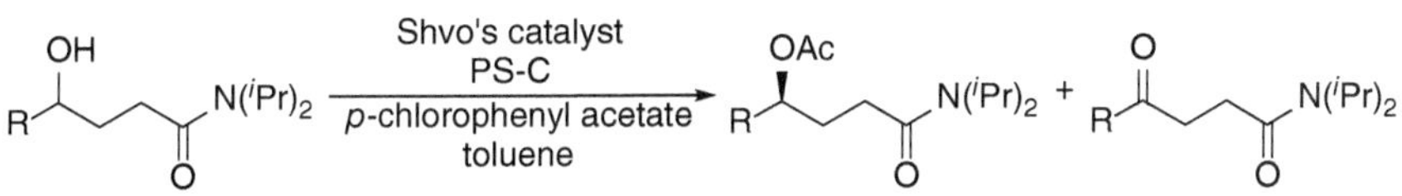

Scheme 4.46 DKR of racemic γ-hydroxy amides.

Lipase PS-C also showed excellent enantioselectivity toward kinetic resolution of racemic δ-hydroxy esters (*E* value up to 360). The combination of lipase-catalyzed transesterification with Shvo's ruthenium-catalyzed alcohol racemization resulted in an efficient DKR in up to 92% conversion and up to 99% *ee*. The (*R*)-δ-acetoxy hexanoic acid *t*-butyl ester and heptanoic acid *t*-butyl ester obtained were converted to δ-lactones (*R*)-6-methyl- and (*R*)-6-ethyl-tetrahydropyran-2-one (Scheme 4.47),

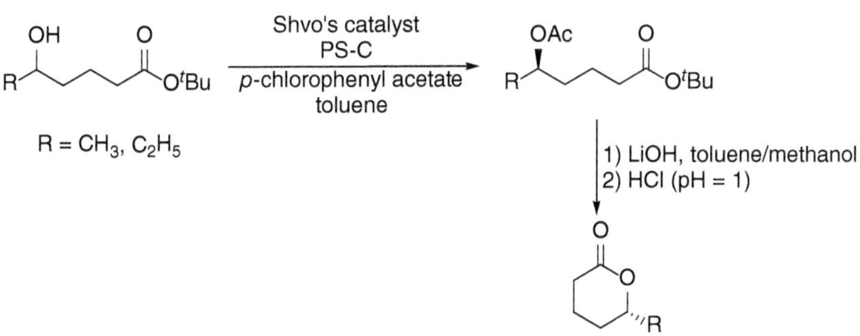

Scheme 4.47 DKR of racemic δ-hydroxy esters.

respectively, the important building blocks for natural products and biologically active compounds. The (*S*)-δ-hydroxy heptanoic acid *t*-butyl ester obtained via kinetic resolution was transformed into (*S*)-5-(*tert*-butyldimethylsiloxy)heptanal (Scheme 4.48), a key intermediate in the synthesis of the widely used commercial insecticide Spinosyn A [77].

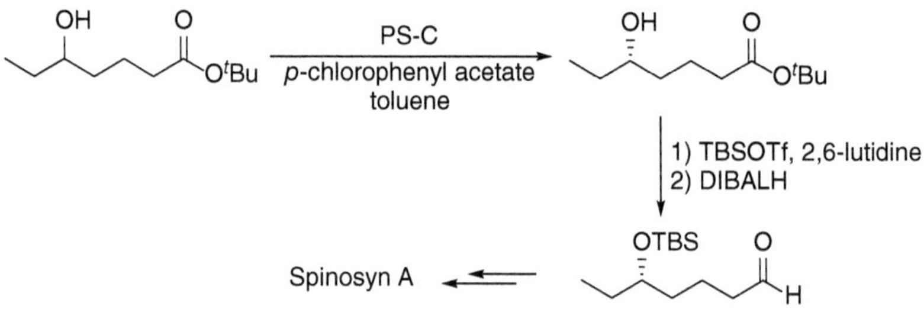

Scheme 4.48 Chemoenzymatic synthesis of (*S*)-5-(*tert*-butyldimethylsiloxy) heptanal.

Optically pure C_2-symmetric diols are important sources of chiral auxiliaries and ligands and serve as precursors in the preparation of other chiral compounds such as *trans*-2,5-disubstituted pyrrolidines. It was highly desirable to convert the commercial mixture of diols (*dl*/*meso* ≈ 1 : 1) into only a single enantiomer. The lipase-catalyzed kinetic resolution offered a tool to obtain one enantiomer, but it would give a maximum theoretical yield of 25%. The combination of enzymatic acylation of the symmetrical diols with ruthenium-catalyzed isomerization of the diol would provide a way to access only one stereoisomer in high yield. Indeed, the *meso*/*dl* mixtures of various diols were transformed into the corresponding enantiomerically enriched (*R*,*R*)-diacetates, by employing Shvo's ruthenium complex for the *in situ* isomerization, and CALB for the acetylation of diols with 4-chlorophenyl acetate as the acyl donor in toluene at 70 °C. The yield and stereoselectivity were dependent on the structure of a substrate diol. For example, 2,4-pentanediol and 2,5-hexanediol were transformed into the corresponding diacetates with the ratios of (*R*,*R*): *meso* being 38 : 62 and 86 : 14, respectively (Scheme 4.49) [78].

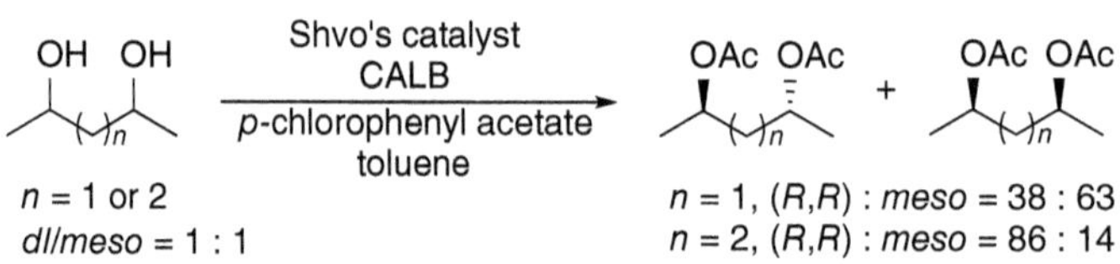

Scheme 4.49 DKR of 2,4-pentanediol and 2,5-hexanediol. Source: Adapted from Persson et al. [78].

The enzyme-catalyzed acylations of 2,5-hexanediol and its monoacetates and the intermediates produced under the action of the ruthenium catalyst were investigated, leading to a better understanding of the dynamic kinetic asymmetric transformation (DYKAT) (Scheme 4.50). The enantiopure diacetates of 2,5-hexanediol were made accessible by combining a fast ruthenium-catalyzed epimerization with a highly selective enzymatic transesterification. The more active ruthenium complex (η^5-C_5Ph_5)Ru$(CO)_2$Cl was coupled with CALB in the presence of isopropenyl acetate under the optimized conditions, producing the (*R*,*R*)-diacetate in 99% yield and >99 % *ee*, with the (*R*,*R*): *meso* ratio being 94 : 6. Similarly, the (*R*,*R*)-diacetate of 2,4-pentanediol was obtained in 96% yield and >99 % *ee*, with the (*R*,*R*): *meso* ratio being 97 : 3 (Scheme 4.51) [79].

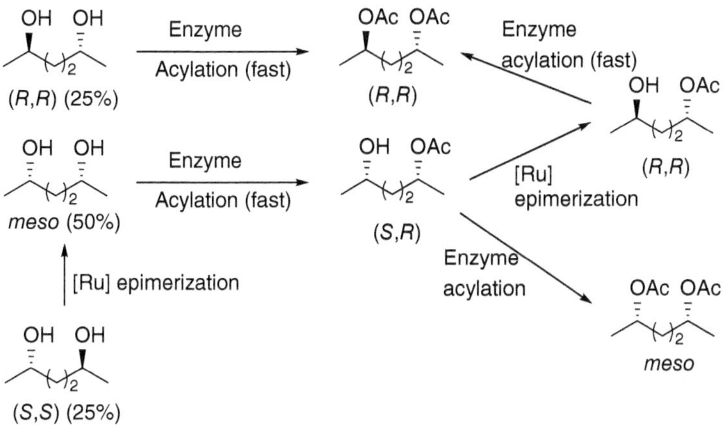

Scheme 4.50 Mechanism details of DKR of 2,5-hexanediol.

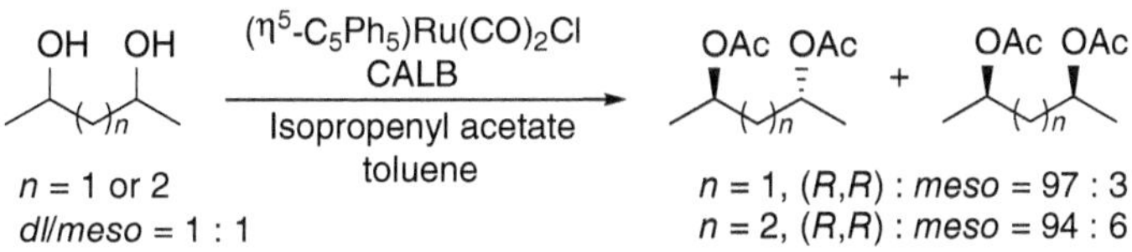

Scheme 4.51 DKR of 2,4-pentanediol and 2,5-hexanediol under optimized conditions. Source: Adapted from Martín-Matute et al. [79].

By employing the same DYKAT system, some other unsymmetrical 1,4-diols including a β-chloro-substituted 1,4-diol were also converted into the corresponding (*R*,*R*)-diacetates with high yield and stereoselectivity. The optically pure 1,4-diacetates are useful precursors for the synthesis of enantiopure 2,5-disubstituted pyrrolidines (Scheme 4.52) [80].

The combined lipase and ruthenium catalysis was also successfully applied to the DYKAT of both symmetrical and unsymmetrical 1,5-diols (Scheme 4.53). The corresponding enantiomerically pure diacetates were prepared in high diastereoselectivity and demonstrated as intermediates for the synthesis of natural products (+)-Solenopsin A, enantiopure 2,6-disubstituted piperidine, and 3,5-disubstituted morpholine (Scheme 4.54) [81, 82].

For an unsymmetrical 1,4-diol with one large group such as 1-phenyl-1,4-pentanediol, only the less sterically hindered alcohol group could be accessible to

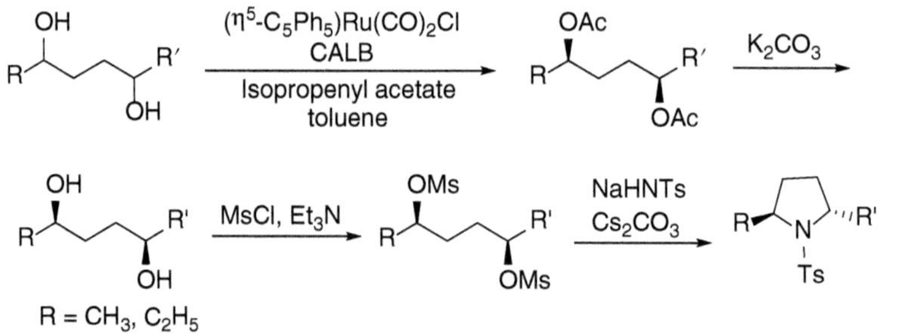

Scheme 4.52 Chemoenzymatic synthesis of enantiopure 2,5-disubstituted pyrrolidines. Source: Adapted from Borén et al. [80].

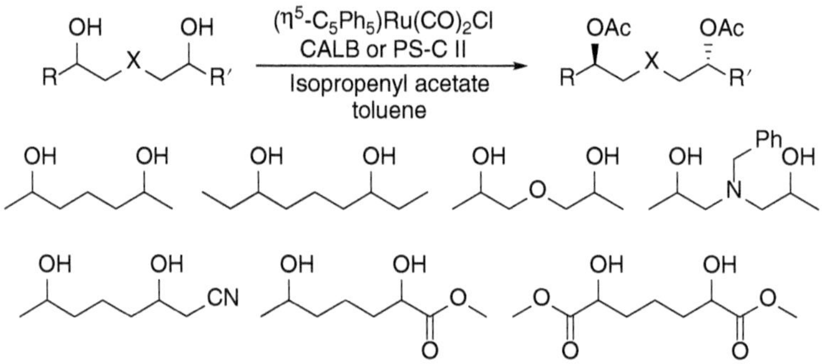

Scheme 4.53 DYKAT of symmetrical and unsymmetrical 1,5-diols.

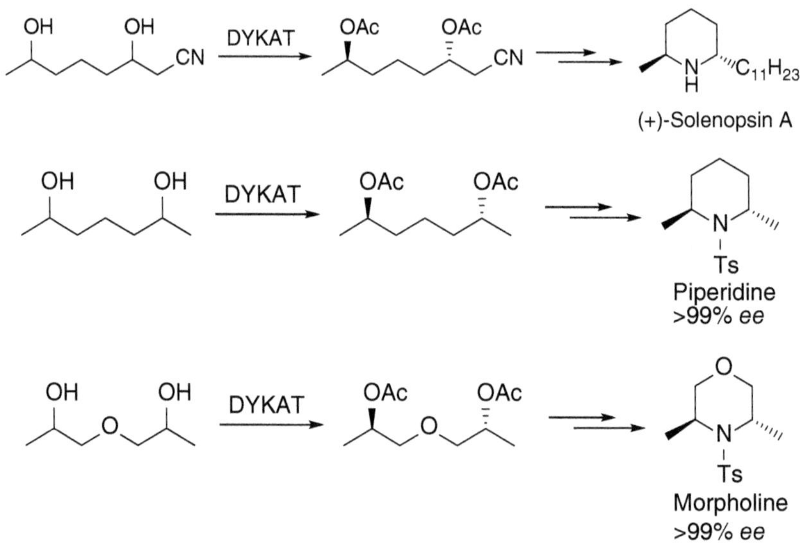

Scheme 4.54 Chemoenzymatic synthesis of (+)-Solenopsin A, enantiopure 2,6-disubstituted piperidine, and 3,5-disubstituted morpholine. Source: Leijondahl et al. [81]; Leijondahl et al. [82].

the enzyme. The lipase-catalyzed kinetic resolution in combination with a ruthenium-catalyzed hydrogen transfer process thus resulted in a DYKAT of the least hindered alcohol group. At the same time, the other hydroxy group was oxidized to ketone under the reaction conditions. Therefore, the corresponding γ-acetoxy ketone was obtained in high enantiomeric purity with concomitant formation of small amounts of diketones (Scheme 4.55) [83].

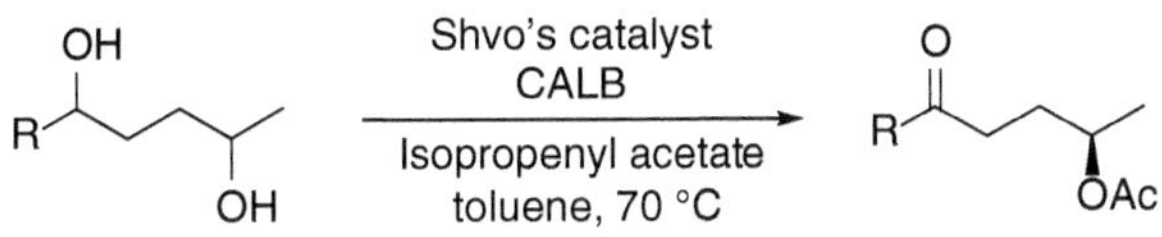

Scheme 4.55 DYKAT of unsymmetrical 1,4-diol with one large group. Source: Adapted from Martín-Matute et al. [83].

cis-1,3-Cyclohexanediol was effectively desymmetrized via CALB-catalyzed transesterification with *p*-chlorophenyl acetate as the acyl donor in several organic solvents to give (1*S*,3*R*)-3-(acetoxy)-1-cyclohexanol in high yield (up to 93%) and excellent enantioselectivity (up to >99.5% *ee*). The following DYKAT of (1*S*,3*R*)-3-acetoxy-1-cyclohexanol employing CALB and (η^5-C_5Ph_5)Ru$(CO)_2$Cl catalysts at room temperature afforded the (*R*,*R*)-diacetate in a high *trans/cis* ratio (91 : 9) and in excellent enantioselectivity of >99% (Scheme 4.56). The DYKAT of racemic *cis*/*trans*-1,3-cyclohexanediol proceeded effectively by using lipase PS-C and Shvo's ruthenium complex, giving a high ratio of *cis*-diacetate (*cis*/*trans* = 97 : 3, Scheme 4.57) [84].

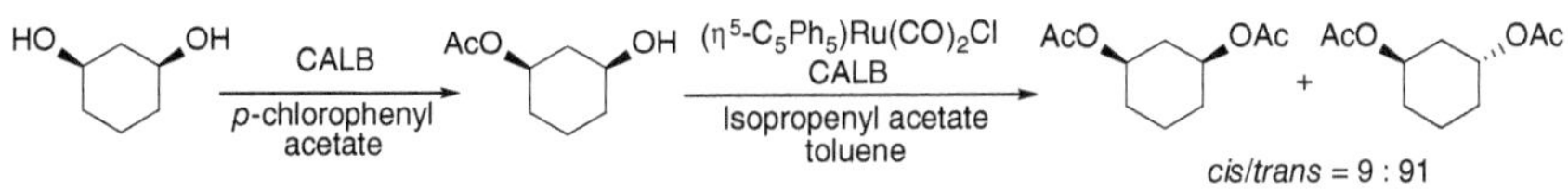

Scheme 4.56 DYKAT of *cis*-1,3-cyclohexanediol.

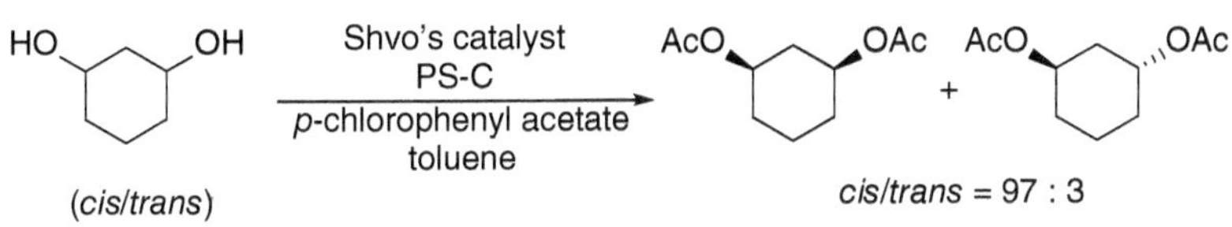

Scheme 4.57 DYKAT of racemic *cis/trans*-1,3-cyclohexanediol. Source: Adapted from Fransson et al. [84].

From the above discussion, we can see that the DKR of racemic alcohols involves lipases or esterases. For racemic β-haloalcohols, a haloalcohol dehalogenase (HheC) also catalyzed its kinetic resolution by converting β-haloalcohols into epoxides. At the same time, it was found that iridacyclic complex **7** catalyzed the racemization of β-haloalcohols. A chemoenzymatic DKR of racemic β-haloalcohols was then established by employing this iridacycle and a mutant Cys153Ser/Trp249Phe of haloalcohol dehalogenase HheC, yielding the corresponding (*R*)-epoxides in up to 90% conversions and 98% *ee* values (Scheme 4.58) [85].

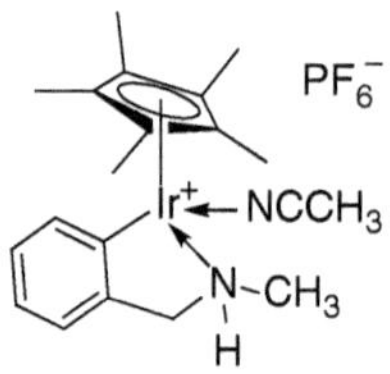

7

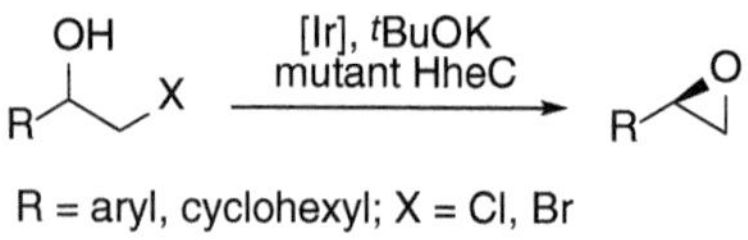

Scheme 4.58 DKR of racemic β-haloalcohols involving an iridacycle and a haloalcohol dehalogenase. Source: Adapted from Haak et al. [85].

4.5 DKR of Chiral Amines

Similar to DKR of racemic alcohols, combination of the enzymatic kinetic resolution of racemic amine with a chemical-catalyzed amine racemization leads to the conversion of a racemic mixture of amine into its single enantiomer in theoretical 100% yield. It is essential that the racemization catalyst should be efficient and compatible with the enzyme catalyst of kinetic resolution. The DKR of chiral amine was first reported by Reetz and Schimossek in 1996. The combination of thermally stable CALB and palladium on charcoal (Pd/C) was applied to the DKR of 1-phenylethylamine with ethyl acetate as the acyl donor in trimethylamine; the acetylated (*R*)-1-phenylethylamine was obtained with 64% yield and 99% *ee* in eight days [86]. Since then, different metal catalysts have been developed for the racemization and combined with enzymes for the efficient DKR of chiral amines. DKR has become one of the practical routes to prepare enantiomerically pure amines, an important class of building blocks for the asymmetric synthesis of more complex compounds, such as pharmaceuticals, natural products, and other bioactive molecules [87].

The first metal catalyst used for the chemoenzymatic DKR of amines is Pd/C. In the presence of molecular hydrogen, Pd/C also catalyzed the reduction of prochiral ketoximes to racemic amines. Therefore, eight different ketoximes were transformed into the corresponding acetylated amines in high yields (70–89%) and optical purities (94–99% *ee*), by employing CALB as the resolution catalyst, Pd/C as the reduction/racemization catalyst, and ethyl acetate as the acyl donor. The reactions were performed with diisopropylethylamine in toluene at 60 °C under hydrogen gas (1 atm) for five days (Scheme 4.59) [88].

Palladium can also be deposited on other supporting materials to prepare the catalysts for amine racemization. Pd on alkaline earth supports catalyzed the racemization of (*S*)-1-phenylethylamine with higher activity and better selectivity than Pd/C. The main side product was ethylbenzene with less than 10% for Pd/$BaSO_4$ and Pd/$CaCO_3$ as the catalyst. The racemization was proposed to proceed via

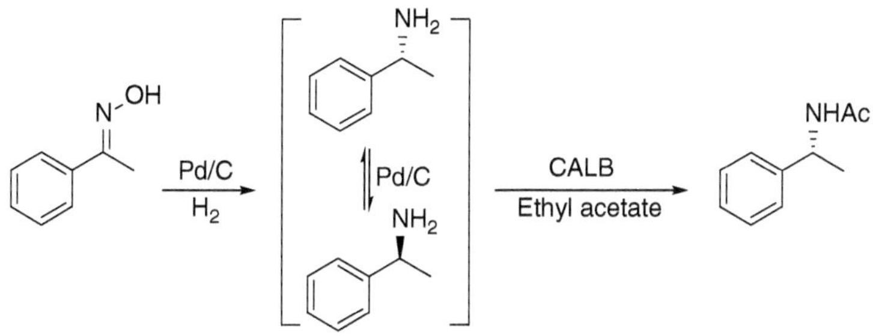

Scheme 4.59 Dynamic asymmetric reduction of prochiral ketoxime9 to optically active acetylated amine. Source: Adapted from Choi et al. [88].

dehydrogenation, followed by hydrogenation of the imine formed. Condensation of the amine with the imine intermediate generated bis(1-phenylethyl)amine, which was hydrogenated to give ethylbenzene and 1-phenylethylamine (Scheme 4.60). The Pd/$BaSO_4$ was thus successfully applied in combination with CALB under molecular hydrogen (10 kPa) to the DKR of benzylic amines; the corresponding (*R*)-amides were obtained with high yields and enantioselectivities [89, 90]. Moreover, ammonium formate could be used as an alternative hydrogen source in this DKR process [91].

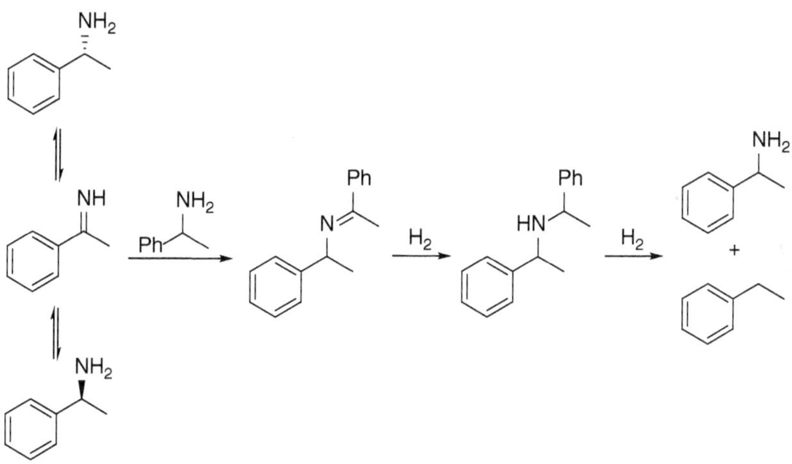

Scheme 4.60 Pd-catalyzed racemization of 1-phenylethylamine and the side reactions.

Palladium NPs entrapped in aluminum hydroxide, Pd/AlO(OH) (average diameter of 2.34 nm), showed much higher activity toward the racemization of optically active (*S*)-1-phenylethylamine than the commercially available Pd/Al_2O_3. By coupling Pd/AlO(OH) with lipase CALB in the one-pot DKR process using ethyl acetate or ethyl methoxyacetate as acyl donor in toluene at 70 °C, a wide range of benzylic and aliphatic primary amines were transformed into the corresponding (*R*)-amides in 84–99% yields and 97–99% *ee* [92]. By changing the preparation procedure, palladium NPs Pd/AlO(OH) with an average diameter of 1.73 nm were obtained. Faster DKR of benzylic amines was realized by coupling the smaller palladium NP catalyst with CALB under the same reaction conditions [93]. The DKR of racemic amino acid amides and β-amino esters to optically active amino acid derivatives had also been accomplished by using the heterogeneous Pd racemization catalyst and CALB enzyme [94, 95]. This DKR protocol had been successfully applied as the key step to the synthesis of rasagiline mesylate [96] and a potent calcimimetic (+)-NPS R-568 (Scheme 4.61) [97].

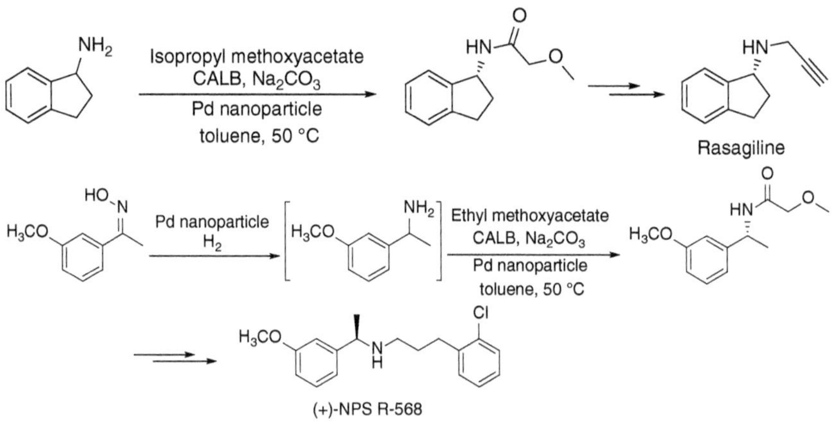

Scheme 4.61 Synthesis of rasagiline mesylate and potent calcimimetic (+)-NPS R-568 via DKR of the amine intermediates. Source: Adapted from Han et al. [97].

Pd catalysts on amine-functionalized silica and layered double hydroxides (LDHs) supports were also examined for racemization and DKR of chiral benzylic amines. The performance of Pd catalysts was greatly affected by the supports, and selectivity for the racemization of (*S*)-1-phenylethylamine decreased in the following order: Pd/amine-functionalized silica, Pd/alkaline earth supports, Pd/LDH, Pd/SiO_2, and Pd/C. Pd catalysts on 3-aminopropyl functionalized silica or 3-(1-piperazino)propyl functionalized silica were used in combination with CALB for the one-pot DKR of various benzylic amines, and the corresponding (*R*)-amides were obtained in over 90% yields with 99% *ee* [98].

The DKR of racemic benzylic amines had also been successfully achieved by using palladium NPs supported on amino-functionalized siliceous mesocellular foam (Pd-AmP-MCF) as the racemization catalyst, in combination with CALB-catalyzed acylation using ethyl methoxyacetate as the acyl donor. The one-pot process was carried out in toluene at 70 °C in the presence of molecular hydrogen, affording the corresponding amides in high yields and excellent *ee*s. Pd-AmP-MCF showed good performance for the racemization of 1-phenylethylamine at temperatures below 70 °C, thus enabling an enzyme that was less thermostable than CALB to be used for the DKR of chiral amines. The commercially available Amano Lipase PS-C1 that exhibited low stability at temperatures over 60 °C was chosen and combined with Pd-AmP-MCF for the DKR of 1-phenylethylamine in toluene at 50 °C. As a result, the primary amine was converted into (*R*)-2-methoxy-*N*-(1-phenylethyl)acetamide in 82% isolated yield and 99% *ee*. Furthermore, the heterogeneous Pd-AmP-MCF could be reused up to five runs [99]. The effects of various alkalic salts, which were introduced into the pores of Pd/MCF by the incipient wetness impregnation method, on their catalytic activity and selectivity for the racemization of (*S*)-1-phenylethylamine were studied. Basic alkalic salts greatly enhanced the selectivity of Pd catalysts, although the catalytic activity was slightly decreased. Compared to the unmodified Pd/MCF, which showed 69% selectivity for 1-phenylethylamine, the selectivities of alkalic potassium salt-modified K_3PO_4–Pd/MCF, K_2CO_3–Pd/MCF, and CH_3COOK–Pd/MCF were significantly improved to 97%, 96%, and 91%, respectively. In combination with lipase CALB, the modified Pd catalysts efficiently catalyzed the DKR of 1-phenylethylamine to give the (*R*)-amide with up to 97% yield and 99% *ee* [100].

Pd NPs were encapsulated inside the cages of both the bare and ethylenediamine-grafted MIL-101 via a facile incipient wetting procedure followed by reduction with hydrogen. The resulting catalysts (Pd/MIL-101 and Pd/ED-MIL-101) catalyzed the racemization of 1-phenylethylamine. The 1.5%-Pd/MIL-101 and 1.5%-Pd/ED-MIL-101 were then combined with lipase CALB in the one-pot DKR of racemic 1-phenylethylamine using ethyl methoxyacetate as the acyl donor, giving optically pure (*R*)-amide (>99% *ee*). Especially for 1.5%-Pd/ED-MIL-101 as the racemization catalyst, excellent conversion (up to 99%) and selectivity (93%) were achieved. The amine modification of MIL-101 could likely change the basic property of the cage surface and help confine Pd NPs in the cages, thus enhancing the catalyst performance. The combined heterogeneous chemoenzymatic catalysts could be recovered and reused eight times without significant loss of efficiency [101].

In the above studies, lipases and palladium catalysts were immobilized on different supports and catalyzed the reactions in separate heterogeneous phases. New hybrid biocatalysts had also been developed by having a covalently bound enzyme joined to a metallic catalyst support. This facilitated the racemization and acetylation steps occurring in close proximity. Lipase CALB was immobilized on functionalized Pd-SiO_2 NPs, and the resulting hybrid biocatalyst was applied to the DKR of 1-phenylethylamine. The (*R*)-amide was obtained in 82% conversion and >99% *ee* by using the hybrid catalyst with only 1% of Pd, reaching a higher productivity (2.21 mg of product h^{-1} mg of support^{-1}) than that by commercially available lipase N435® (0.76 mg of product h^{-1} mg of support^{-1}) [102].

Although platinum was known as an efficient catalyst for hydrogenation and dehydrogenation procedure, resulting in racemization, it had been rarely applied in the chemoenzymatic DKR process. This could be due to its inhibitory effect on the enzyme activity in the acylation reaction, leading to enzyme deactivation and probably poor racemization. As such, platinum NPs were encapsulated in zeolitic microcapsule (ZMC) with an intact thin shell of silicalite-1. The resulting catalyst was successfully used in combination with the immobilized CALB in the DKR of phenylethylamine. The existence of the silicalite-1 shell effectively prevented the deactivation of both enzyme and Pt by locating them in separate regions in the reaction system, and high conversion (80%) and selectivity (95%) were achieved [103].

Raney metals such as Raney nickel and cobalt also catalyzed the racemization of various chiral amines with high selectivity in the presence of hydrogen. For benzylic primary amines, the formation of side products, e.g. secondary amines, could be suppressed by varying the hydrogen pressure. Very high selectivity was achieved in the racemization of aliphatic amines over Raney catalysts, and DKR of racemic 2-hexylamine by using immobilized CALB and Raney nickel in one pot gave the acetylated amide in 95% yield and 97 % *ee* [104].

The ruthenium complexes are efficient alcohol racemization catalysts and have been widely used in combination with lipases for the DKR of various secondary alcohols. These complexes usually have modest activities for the racemization of primary amines. A modified Shvo's catalyst with *p*-MeO substituent showed high activity for the racemization of benzyl amines. The combination of this ruthenium complex with CALB effectively catalyzed the DKR of a variety of benzyl primary amines in toluene at 90 °C for three days by using isopropyl acetate as the acetylating agent, giving the corresponding (*R*)-amides in high yield and high enantioselectivity [105]. By using isopropyl 2-methoxyacetate as acyl donor and adding DMP as hydrogen donor to suppress the side product formation, this DKR proceeded smoothly at 100 °C, and complete conversion was achieved for most substrates within 26 hours [106]. Dibenzyl carbonate could also serve as the acyl donor for the CALB-catalyzed kinetic resolution of 1-phenylethylamine with an *E* value of 156. By replacing the acyl donor with dibenzyl carbonate, the above chemoenzymatic DKR process worked well with *p*-substituted phenylethylamines and aliphatic amines, affording the corresponding carbamates in good to high yields and high *ee* [107]. Since the benzyloxy could be removed by either mild Pd catalysis or weak acid catalysis, this prevented release of the free amine from the carbamate products under harsh conditions. *C. antarctica* lipase A (CALA) showed high enantioselectivity toward β-amino esters, but it was not stable under the conditions for the Ru-catalyzed racemization of chiral amines. When it was immobilized in functionalized mesocellular foam in the presence of sucrose, the resulting enzyme showed dramatically improved enantioselectivity and thermostability. The immobilized CALA was successfully combined with the ruthenium racemization catalyst to enable the DKR of ethyl 3-amino-3-phenylpropanoate, giving the (*S*)-amide in 85% conversion and 89% *ee* [108].

The above-discussed metal-catalyzed amine racemization procedures were carried out at temperatures higher than 70 °C, and the thermally stable CALB was usually used as the biocatalyst for the kinetic resolution step, leading only to (*R*)-selective chemoenzymatic DKR of primary amines. It was desirable to develop amine racemization processes at low temperature, enabling the synthesis of either (*R*)- or (*S*)-amides by changing the enzyme catalyst. In this context, the radical racemization procedure mediated by sulfanyl radicals was examined. Efficient racemization of amines could be achieved at low temperature by initiating the formation of sulfanyl radical with irradiation at 350 nm in the presence of 2,2′-azobis(2-methylpropionitrile) (AIBN). By employing either CALB or alkaline protease as the biocatalyst for the kinetic resolution, (*R*)- and (*S*)-selective chemoenzymatic DKRs of amines were achieved at temperatures below 40 °C, allowing the synthesis of the corresponding (*R*)- or (*S*)-amides in good yield and high *ee* (Scheme 4.62) [109, 110].

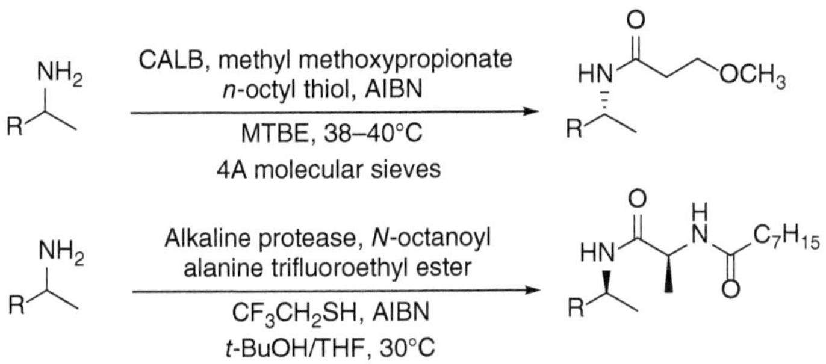

Scheme 4.62 DKR of chiral amines via radical racemization. Source: Poulhès et al. [109]; El Blidi et al. [110].

4.6 DKRs of Other Compounds

In the above sections, the metal-catalyzed racemization of alcohols or amines was combined with lipase-catalyzed acylation of alcohol or amine functional group to achieve the DKR, giving the corresponding optically enriched chiral alcohols or amines in up to 100% yield. The metal-catalyzed racemization of a target substrate could also be coupled with other enzymatic stereoselective reactions to accomplish the DKR of the target compound. For example, the coupling of rhodium-catalyzed racemization at the allylic position of various *para*-substituted 3-aryl-4-pentenoic acids and enzymatic esterification of the acid functionality with an alkoxyl donor resulted in DKR of these chiral unsaturated carboxylic acids to afford the corresponding enantiomerically enriched esters with very high isolated yields (Scheme 4.63) [111]. Similarly, the kinetic resolution based on irreversible enzymatic esterification of carboxylic acids with triethyl orthobenzoate $PhC(OEt)_3$ was combined with Rh-catalyzed racemization for the DKR of 3-hydroxy-3-(4-nitrophenyl)propanoic acid, and both (*R*)- and (*S*)-enantiomers of ethyl ester were obtained with high yields by employing immobilized CALB or lipase Amano AK, respectively (Scheme 4.64) [112].

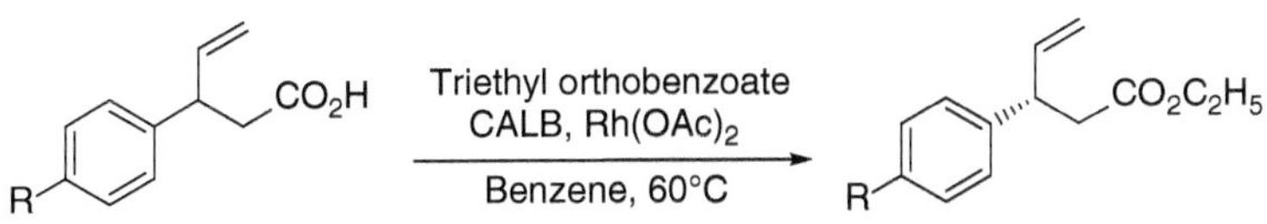

Scheme 4.63 DKR of 3-aryl-4-pentenoic acids. Source: Adapted from Koszelewski et al. [111].

It has been known that haloalkane dehalogenases catalyze the kinetic resolution of α-haloesters and α-haloamides by converting one enantiomer into the

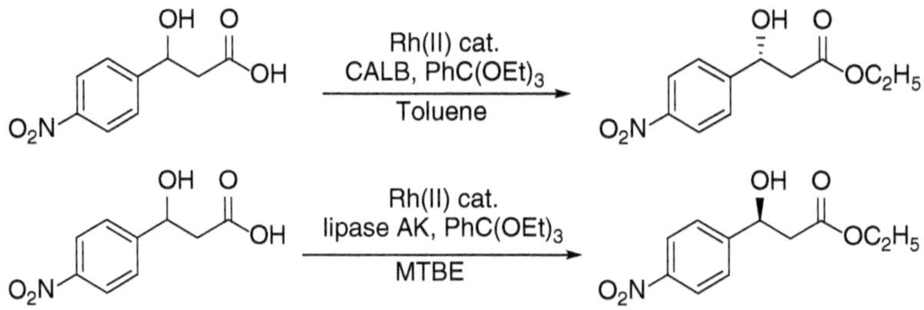

Scheme 4.64 DKR of 3-hydroxy-3-(4-nitrophenyl)propanoic acid. Source: Adapted from Koszelewski et al. [112].

α-hydroxy esters or α-hydroxy amides. A polymer-based phosphonium bromide (cross-linked $PhCH_2CPh_3^+$ Br^-) was found to be an efficient racemization agent for α-bromoesters and α-bromoamides. Combination of this racemization with haloalkane dehalogenase-catalyzed kinetic resolution was then successfully applied to the DKR of *N*-phenyl-2-bromopropionamide using a bacterial haloalkane dehalogenase DbjA, affording (*S*)-*N*-phenyl-2-hydroxypropionamide in high yield and *ee* (Scheme 4.65) [113].

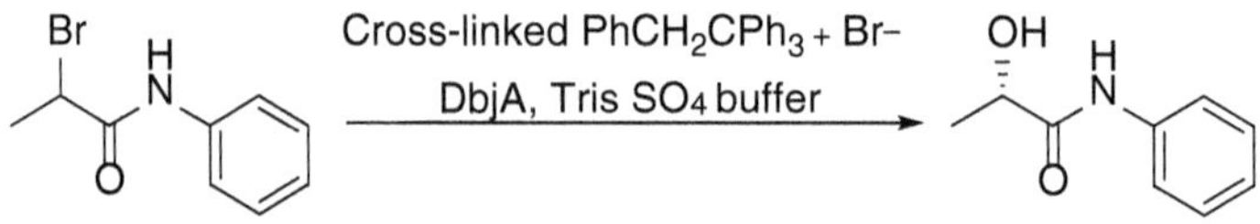

Scheme 4.65 DKR of *N*-phenyl-2-bromopropionamide. Source: Adapted from Westerbeek et al. [113].

In the above discussion, a catalyst was required in the racemization of chiral compounds. It is possible that the racemization can take place simultaneously under the reaction conditions for the stereoselective enzymatic reaction, thus resulting in the DKR of the chiral compounds. Two paradigmatic examples are the DKR of cyanohydrins and the compounds with activated hydrogen at the chiral center, in which the racemization occurred via the reversible cyanohydrin formation from the corresponding aldehydes and hydrogen cyanide, and the reversible H^+-exchange at the chiral center, respectively. Enantiopure 2-hydroxy acids could be synthesized by DKR of racemic cyanohydrin via nitrilase-catalyzed enantioselective hydrolysis and reversible decomposition/formation of cyanohydrin under the slightly alkaline conditions. (*R*)-Mandelic acid was industrially produced on a multi-ton scale by this methodology. By screening a large set of unique nitrilases from environmental DNA libraries, highly active and enantioselective nitrilases were obtained and applied to the synthesis of a wide array of (*R*)-mandelic acid analogs with high yields and enantioselectivities (95–99% *ee*) via DKR (Scheme 4.66) [114].

For compounds such as aldehydes, ketones, acids, and their derivatives with a chiral center at the α-position, the α-H is activated and reversible H^+-exchange can occur under basic conditions, resulting in the racemization of the chiral center. This reversible process can be combined with an irreversible reaction and applied

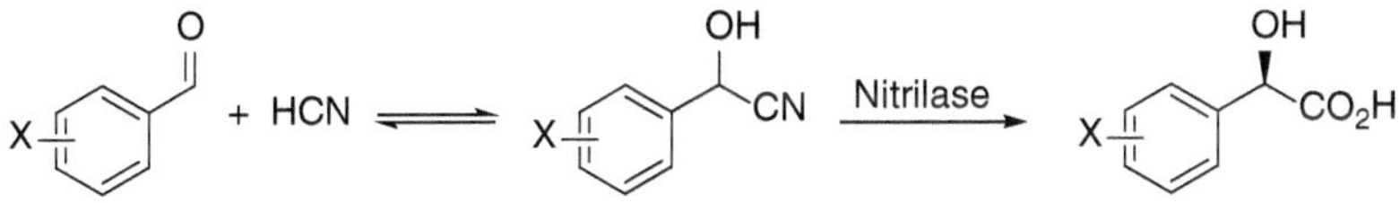

Scheme 4.66 DKR of mandelonitrile analogs. Source: Adapted from DeSantis et al. [114].

in the DKR of these chiral compounds. The most studied was the dynamic reductive kinetic resolution (DYRKR) of β-keto acid derivatives with an α-substituent [115]. For the α-substituted β-keto acid derivatives, next to the carbonyl group is a chiral center bearing an acidic proton. Under the reaction conditions, the stereocenter could rapidly epimerize through an enol(ate) intermediate. The highly stereoselective carbonyl reductase catalyzed the reduction of one of the two enantiomers to give one enantiomer of the β-hydroxyacid derivative. The other non-active enantiomer of the α-substituted β-keto acid derivative was converted simultaneously to the active enantiomer, thus leading to an effective DYRKR to afford one of the four possible stereoisomeric products with formation of two stereocenters in one transformation. One of the most important DYRKR transformations was the reduction of β-keto esters. Many microbial strains harboring alcohol dehydrogenases (ADHs) were evaluated for the reduction of β-keto esters and some successful examples had been reported. For example, the strain *Geotrichum candidum* catalyzed the DYRKR of ethyl 2-methyl-3-oxobutanoate to give ethyl (2*S*,3*S*)- 2-methyl-3-hydroxybutanoate in 80% yield with 98% *de* and 98% *ee* [116]. The stereospecific reduction of ethyl 2-oxocycloheptane carboxylates by *Kloeckera magna* gave the *cis*-(1*R*,2*S*)-hydroxyester in 80% yield with 99% *de* and 94% *ee* [117]. While these highly selective DYRKR successfully delivered one of four possible stereoisomeric products, often the reduction was less selective and hence undesired isomers were formed. For the microbial transformations, this could be due to the existence of multiple active enzymes or the enzyme was not selective. To mitigate this problem, a straightforward approach was to employ individual carbonyl reductases with high stereoselectivity so that the DYRKR could be achieved to afford the desired stereoisomeric product. The reduction of a series of α-substituted β-ketoesters was evaluated with several purified ADHs or crude preparations of overexpressed ADH in *Escherichia coli*. The corresponding β-hydroxyesters were obtained with excellent enantio- and diastereoselectivities in many cases. ADH-A from *Rhodococcus ruber*, CPADH from *Candida parapsilosis*, and TesADH from *Thermoanaerobacter ethanolicus* afforded *syn*-(2*R*,3*S*) α-substituted β-hydroxyesters with 90–99% *de* and >99% *ee* (Scheme 4.67). With *anti*-Prelog ADHs such as alcohol dehydrogenase from *Lactobacillus brevis* (LBADH) from *Lactobacillus brevis*, *syn*-(2*S*,3*R*) products were obtained for α-acetyl-γ-butyrolactone with 90% *de* and >99% *ee* (Scheme 4.68). RasADH from *Ralstonia* sp. catalyzed the reduction of isopropyl 2-benzyl-3-oxobutanoate with high diastereoselectivity, leading to *anti*-(2*S*,3*S*) β-hydroxyester with 90% *de* and >99% *ee* (Scheme 4.69) [118]. The enzymatic DYRKR of β-ketoesters with the α-substituent being chloro, alkoxy, and amino groups had also been achieved by employing suitable biocatalysts (Scheme 4.70) [119–121]. It should be noted that the selectivities of reduction were enzyme-dependent and highly affected by the structure of the α-substituted β-ketoesters.

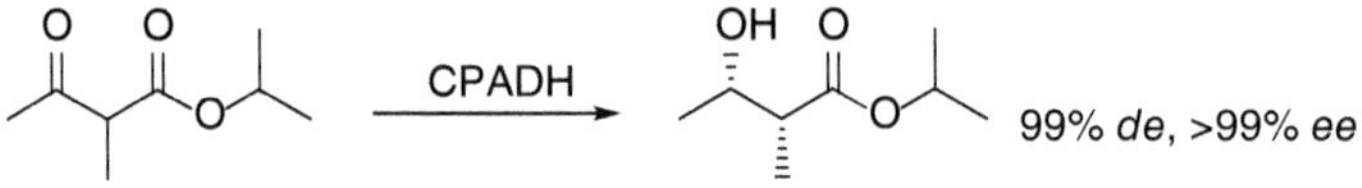

Scheme 4.67 Synthesis of *syn*-(2*R*,3*S*) α-substituted β-hydroxyesters.

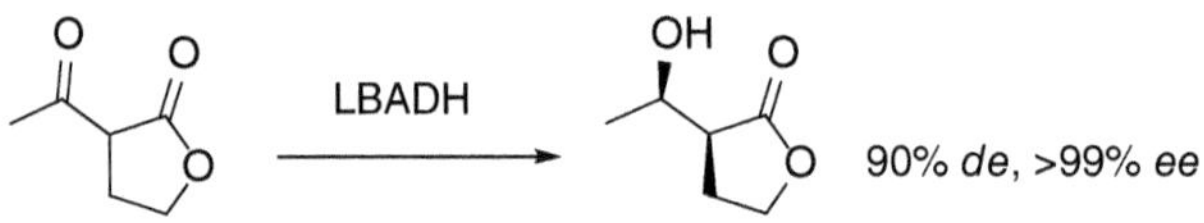

Scheme 4.68 DYRKR of α-acetyl-γ-butyrolactone.

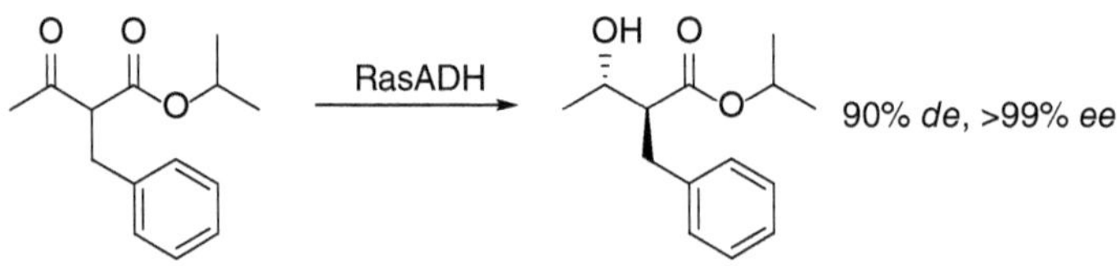

Scheme 4.69 DYRKR of isopropyl 2-benzyl-3-oxobutanoate. Source: Adapted from Cuetos et al. [118].

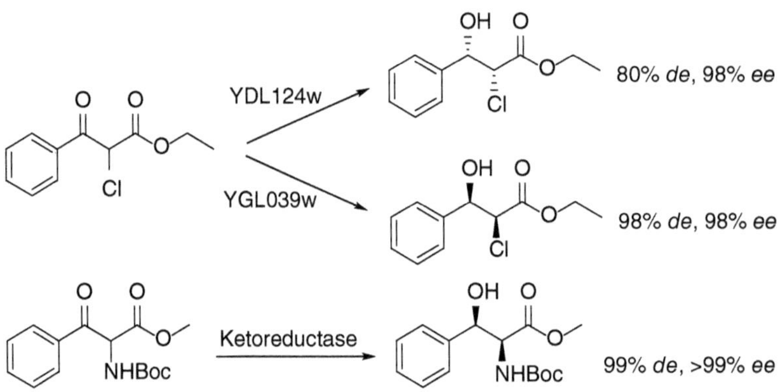

Scheme 4.70 DYRKR of β-ketoesters with the α-substituent of chloro or amino group. Source:Feske et al. [119]; Perrone et al. [120]; Xu et al. [121].

The ADH-catalyzed DYRKR of other β-keto compounds such as cyclic β-keto nitriles [122], dimethyl α-substituted-β-keto phosphonate [123, 124], and cyclic β-keto sulfones [125] had also been realized by using Baker's yeast or the recombinant enzymes derived from Baker's yeast. With the growing cells of *Saccharomyces montanus* strain as the biocatalyst, 2-oxocyclopentanecarbonitrile was reduced under optimized reaction conditions, giving (1*S*,2*S*)-2-hydroxycyclopentanecarbonitrile in 89% yield and 97% *ee* (Scheme 4.71). Similarly, 2-oxocyclohexanecarbonitrile was reduced via DKR to furnish (1*S*,2*S*)-2-hydroxycyclohexanecarbonitrile with 85% yield and 93% *ee* by using the resting cells of *S. montanus*. The resulting cyclic hydroxynitriles were further transformed into the corresponding optically active cyclic hydroxy amines or hydroxy methylamines [122].

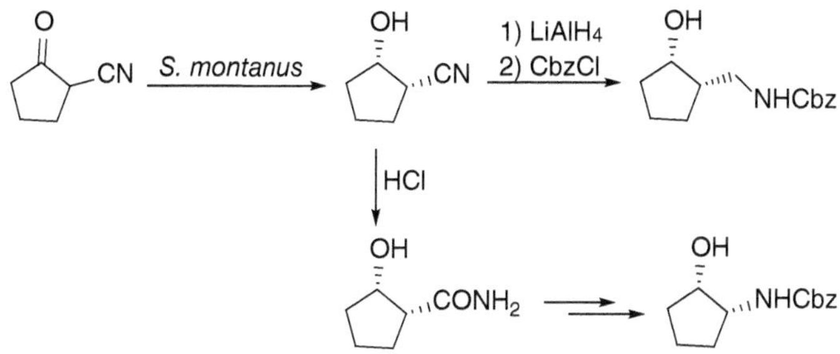

Scheme 4.71 DYRKR of 2-oxocyclopentanecarbonitrile.

After screening the recombinant reductases from *Saccharomyces cerevisiae*, YBR149w and YDR368w were identified to catalyze the reduction of dimethyl (1-chloro-2-oxopropyl)phosphonate to afford dimethyl *syn*-(1*R*,2*S*)-(1-chloro-2-hydroxypropyl)phosphonate with 95% *de* and 99% *ee*, while YHR104w catalyzed the reduction of this keto phosphate to give the *anti*-(1*S*,2*S*)-product with 92% *de* and 99% *ee* through DKR (Scheme 4.72). Dimethyl *syn*-(1*R*,2*S*)-(1-chloro-2-hydroxypropyl)phosphonate served as a precursor for the synthesis of (−)-fosfomycin [123]. Very recently, a series of β-keto arylphosphonates with different α-substituents were reduced to the corresponding *syn*- or *trans*-α-substituted-β-hydroxyphosphonates with high de and *ee* values by choosing an appropriate carbonyl reductase [124].

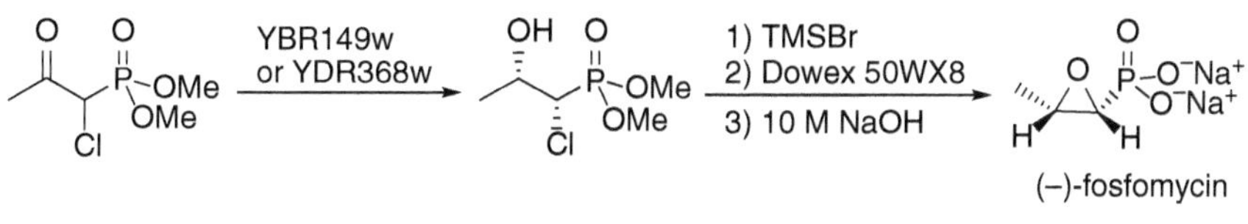

Scheme 4.72 Enzymatic dynamic kinetic reduction of dimethyl (1-chloro-2-oxopropyl) phosphonate.

The DKRs of 2-benzenesulfonylcyclopentanone and 2-benzenesulfonylcyclohexanone were achieved with Baker's yeast, generating the (1*S*,2*R*)-2-benzenesulfonylcyclopentanone and (1*S*,2*R*)-2-benzenesulfonylcyclohexanone with 98 : 2 dr and >95% *ee*, respectively (Scheme 4.73) [126].

Scheme 4.73 Enzymatic dynamic kinetic reduction of 2-benzenesulfonylcycloalkanones. Source: Adapted from Maguire et al. [126].

In addition to the β-keto systems, aldehydes or ketones with an adjacent chiral carbon but less acidic α-hydrogen could also perform the DYRKR transformation under the action of an effective ADH in the suitable reaction environment. Racemic α-arylaldehydes were converted to the corresponding (*R*)- or (*S*)-chiral alcohols via this DYRKR approach by using metal or enzyme catalyst for the asymmetric aldehyde reduction. The ADHs, HLADH from horse liver and SsADH from *Sulfolobus solfataricus*, catalyzed the asymmetric reduction of arylpropanals to give (*S*)-arylpropanols. When the reduction was carried out under slightly basic conditions, a diversity of (*S*)-arylpropanols were prepared by the reduction of racemic arylpropanals with high yield and *ee* (Scheme 4.74) [127, 128]. Recently, both (*R*)-and (*S*)-enantiomers of various arylpropanols were synthesized similarly by using the (*R*)- and (*S*)-selective mutants of a thermophilic ADH TbADH from *Thermoanaerobacter brockii* (Scheme 4.74) [129].

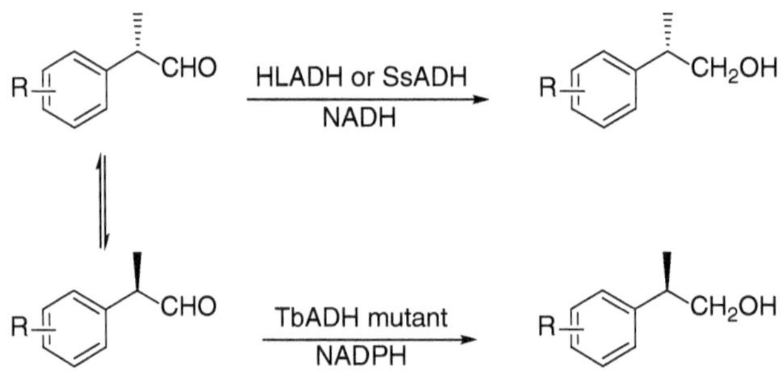

Scheme 4.74 DYRKR of arylpropanals. Source: Giacomini et al. [127]; Galletti et al. [128].

An ADH isozyme 10 from *S. solfataricus* (ADH-10) was identified to catalyze the DYRKR of Profen-type 2-arylpropanal, furnishing (*S*)-profenols corresponding to the NSAIDs (nonsteroidal anti-inflammatory drugs) such as naproxen, ibuprofen, flurbiprofen, ketoprofen, and fenoprofen, in up to 96% yields and up to 99% values [130].

The alcohol dehydrogenase (ADH-A) from *R. ruber* DSM 44541 catalyzed the reduction of racemic α-substituted benzylketone in ~38% conversion with high diastereoselectivity (99% *de*) and enantioselectivity (99% *ee*), and the ketone was recovered in 60% *ee*. Addition of ion exchange resins or triethylamine into the reaction system facilitated the epimerization of the remaining ketone, leading to the *in situ* racemization. Under the basic conditions, DYRKRs of various racemic 3-arylalkanones had been achieved by using ADH-A or commercially available enzymes with an opposite *anti*-Prelog selectivity, to give the corresponding (*S,S*)-substituted propan-2-ols or (*R,R*)-substituted propan-2-ols with good conversions and both diastereoselectivity and stereoselectivity (Scheme 4.75) [131, 132].

The enzymatic reduction of ketones was usually carried out at pH < 8 and ambient temperature, but for the ketones with less acidic α-H, the epimerization of the α-stereogenic center required higher pH and temperature. For example, fast racemization of *tert*-butyl (1-oxo-1-phenylhex-5-yn-2-yl)carbamate could only be realized at elevated temperature (≥45 °C) and high pH (≥pH 10). A commercially

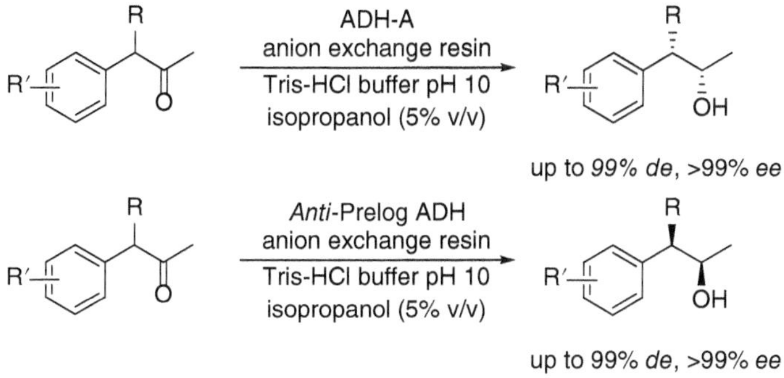

Scheme 4.75 DYRKR of 3-arylalkanones. Source: Mangas-Sánchez et al. [131]; Méndez-Sánchez et al. [132].

available ketoreductase (KRED) catalyzed its reduction at pH 7 and 30 °C, giving the desired (*R*,*R*) alcohol with high diastereoselectivity (16 : 1 d.r.) and excellent enantioselectivity (>99% *ee*). Mutation of this enzyme resulted in a variant that was stable at 45 °C and pH 10, under which conditions the *S*-enantiomer of *tert*-butyl (1-oxo-1-phenylhex-5-yn-2-yl) carbamate was rapidly epimerized. By using this mutant enzyme, DYRKR of *tert*-butyl (1-oxo-1-phenylhex-5-yn-2-yl)carbamate proceeded smoothly to afford the desired (*R*,*R*) alcohol with >100 : 1 d.r. and >99% *ee* (Scheme 4.76), enabling an efficient stereoselective synthesis of vibegron, a potent and selective β_3-adrenergic receptor agonist which has recently emerged as a promising new class of agents for the treatment of overactive bladder syndrome (OAB) [133].

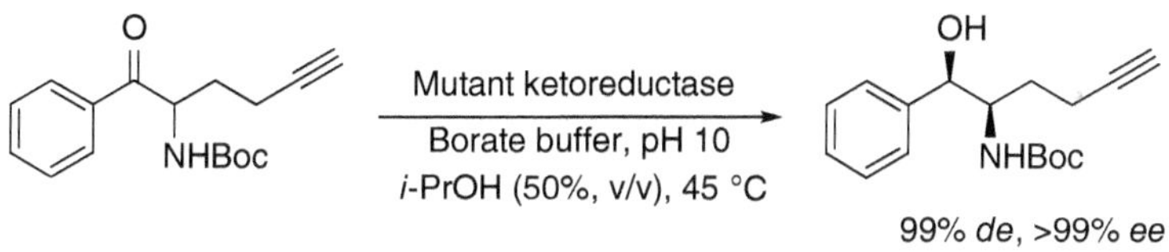

Scheme 4.76 DYRKR of *tert*-butyl (1-oxo-1-phenylhex-5-yn-2-yl)carbamate.

For unprotected α-aminoketones such as 3-amino-4-phenylbutan-2-one, racemization could not occur in borate buffer up to pH 9.5, at which the KREDs generally functioned. Pyridoxal-5-phosphate (PLP) was found to effectively catalyze racemization of (*S*)-3-amino-4-phenylbutan-2-one in MES buffer at pH 6.5 in the presence of catalytic amount of $NiSO_4$. α-Aminoketone could dimerize to give dihydropyrazine, which was irreversibly transformed to pyrazine upon oxidation although dimerization was reversible under aerobic reaction conditions. Combination of PLP-catalyzed racemization and enzymatic ketone reduction by a Codexis KRED P2-D11 under N_2 atmosphere resulted in effective DYRKR of 3-amino-4-phenylbutan-2-one affording (2*S*,3*S*)-3-amino-4-phenylbutan-2-ol in 85% yield with 11 : 1 *dr* and 98% *ee*. By employing Codexis KRED P1-B05 as biocatalyst, a diversity of aromatic and aliphatic α-aminoketones were transformed

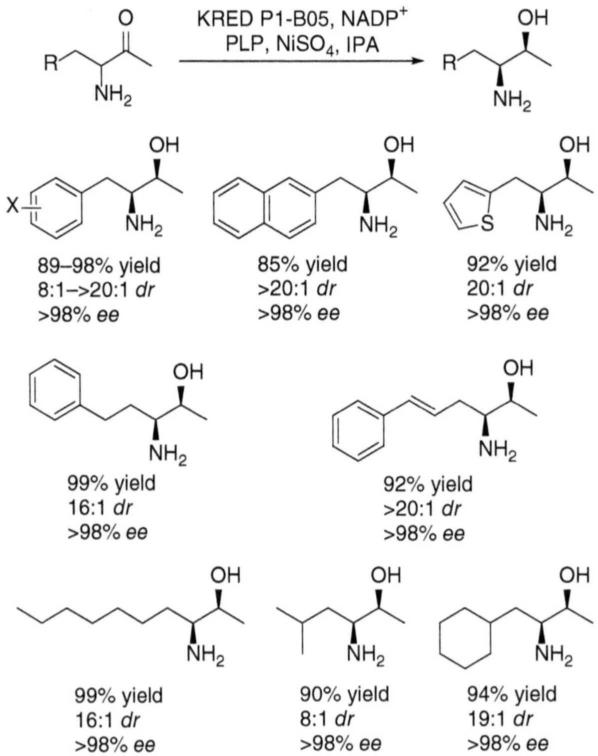

Scheme 4.77 DYRKR of aromatic and aliphatic α-aminoketones.

into the corresponding optically active 1,2-amino alcohols in excellent yields and up to >20 : 1 *dr* and >98% *ee* (Scheme 4.77). By choosing appropriate KREDs, all four possible stereoisomers of 2-amino-1-phenylpentan-3-ol were prepared in high yield, diastereoselectivity, and excellent enantioselectivity (Scheme 4.78). PLP-mediated chiral amine racemization pairing with ketone bioreduction thus enabled rapid access to the desired 1,2-amino alcohol stereoisomer, a prevalent motif in pharmaceutical and agrochemical molecules such as HIV antiviral drugs Atazanavir and Darunavir (Scheme 4.79) [134].

In addition to carbonyl reduction, this class of ketone compounds can undergo other enzymatic transformations. Other enzymes such as transaminases can also recognize the α-chiral center, leading to the kinetic resolution of these compounds. For example, transaminases recognizing the α-chiral center adjacent to an aldehyde moiety were identified to promote DKR amination of a few γ-oxo-β-alkyl esters, giving access to both (*R*)- and (*S*)-enantiomers of γ-amino-β-alkyl esters in high optical purity (Scheme 4.80), the important building blocks for antiepileptic drugs Brivaracetam and Pregabalin [135].

Transaminase also catalyzed the amination of the carbonyl group of α-alkyl-β-keto esters to generate α-alkylated β-amino esters by using isopropylamine as an amino donor in aqueous medium. Although low diastereoselectivities were observed for

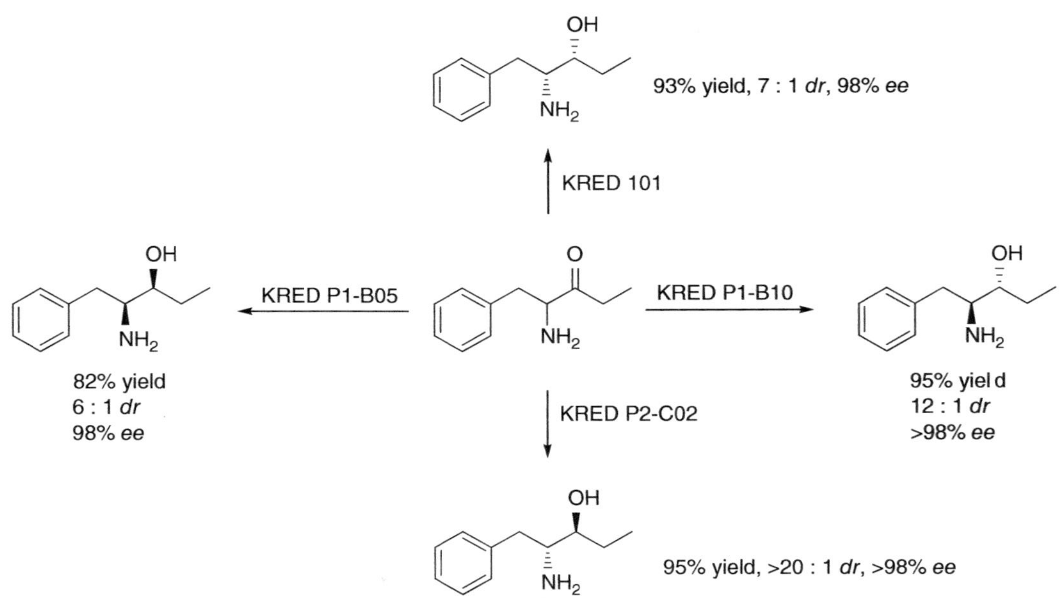

Scheme 4.78 Preparation of four stereoisomers of 2-amino-1-phenylpentan-3-ol via chemoenzymatic DYRKR.

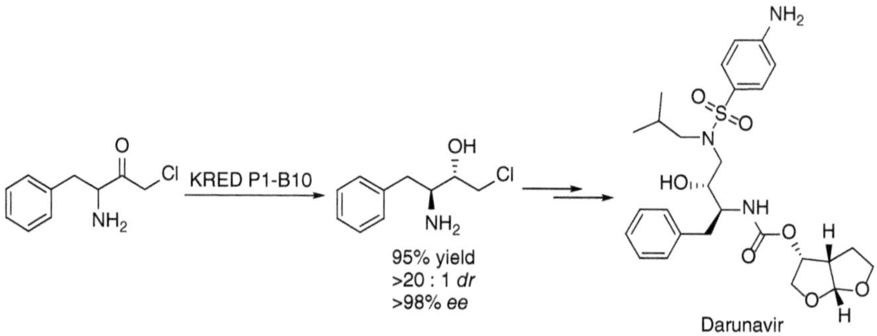

Scheme 4.79 Preparation of key intermediate for HIV antiviral drugs Darunavir via chemoenzymatic DYRKR. Source: Adapted from Cao et al. [134].

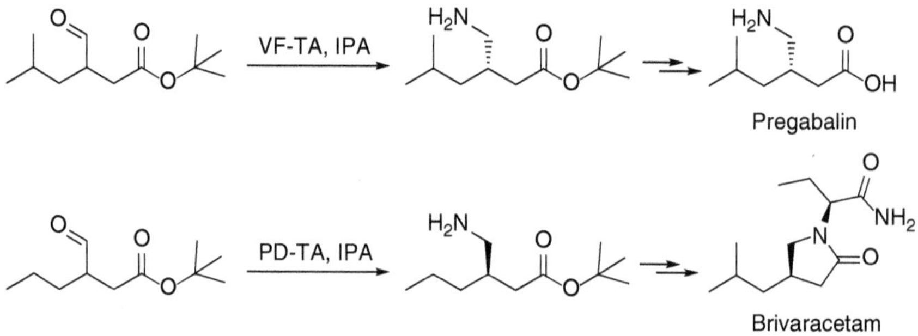

Scheme 4.80 Dynamic kinetic resolution transamination of aldehydes with α-chiral center.

acyclic substrates with the tested transaminases, the dynamic amination kinetic resolution of ethyl 2-oxocyclopentanecarboxylate was realized using isopropylamine as the amino donor to give ethyl (1*S*,2*S*)-2-aminocyclopentanecarboxylate in 58% yield with 98% *de* and >99% *ee* (Scheme 4.81) [136].

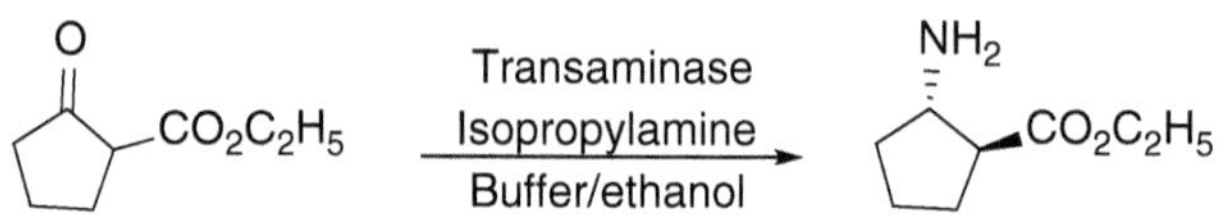

Scheme 4.81 Dynamic kinetic resolution transamination of ethyl 2-oxocyclopentanecarboxylate. Source: Adapted from Cuetos et al. [136].

This DKR strategy was also applied to the synthesis of various optically active α-alkyl-β-amino amides from racemic α-alkyl-β-keto amides. A series of commercially available (*S*)- and (*R*)-selective transaminases were evaluated toward DKR transamination of *N*-benzyl-2-methyl-3-oxobutanamide using isopropylamine as the amine donor. In most cases, *anti*-*N*-benzyl-3-amino-2-methyl-butanamide was obtained as the major product with low to moderate *dr* and excellent *ee*. When the DKR reactions with (*R*)-selective transaminases were performed at pH 10, an improvement of *dr* values favoring *anti*-*N*-benzyl-3-amino-2-methyl-butanamide was observed, whereas it was not the case for (*S*)-selective transaminases. A set of α-alkyl-β-keto amides were then assessed for the DKR transamination, and the *anti*-diastereoisomer was preferred in all cases except for *N*-benzyl-2-benzyl-3-oxobutanamide. Both *anti*- and *syn*-diastereoisomers had excellent *ee* values, but the *anti*:*syn* ratios were greatly dependent on the enzyme, substrate structure, and reaction pH. Several diastereoenriched α-substituted β-amino amides were prepared with varied *dr* and >99% *ee* by this procedure (Scheme 4.82) [137].

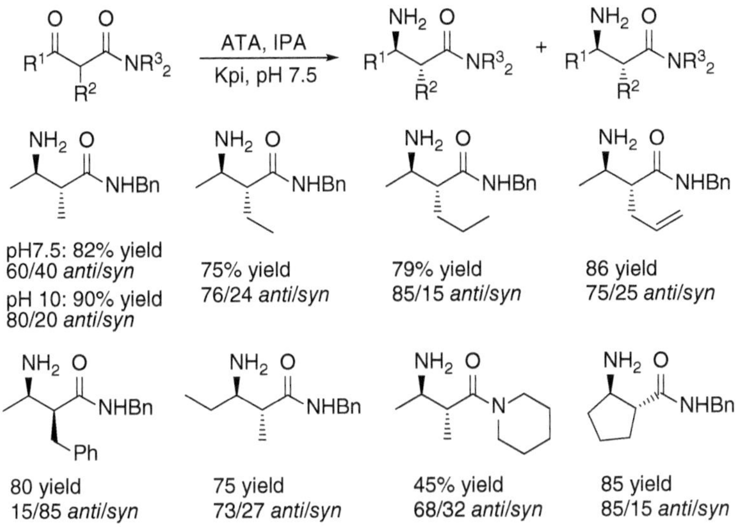

Scheme 4.82 Dynamic kinetic resolution transamination of α-alkyl-β-keto amides. Source: Adapted from Mourelle-Insua et al. [137].

Baeyer–Villiger monooxygenases (BVMOs) catalyze the asymmetric Baeyer–Villiger oxidations and oxygenation of heteroatom-containing compounds (S, Se, P and N). The possibility that these enzymes could discriminate the α-chiral center of ketone substrates, thus making it useful in the DKR to yield the corresponding esters with high conversion and optical purity, was explored. As such, 4-hydroxyacetophenone monooxygenase (HAPMO) from *P. fluorescens* ACB was employed for the DKRs of racemic benzyl ketones in the presence of a weak anion exchange resin, producing the (*S*)-benzyl esters with 58–83% *ee* in 46–88% conversions (Scheme 4.83) [138].

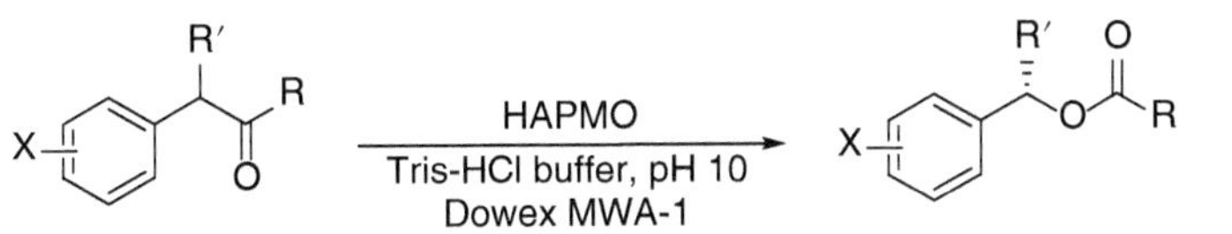

Scheme 4.83 HAPMO-catalyzed DKR of benzyl ketones. Source: Adapted from Rodríguez et al. [138].

Forró et al. examined if *in situ* base-catalyzed racemization could occur in carboxylic acid esters with an adjacent chiral center that has an acidic proton, and if lipase-catalyzed ester hydrolysis could lead to the DKR of these compounds. Indeed, CALB catalyzed the hydrolysis of ethyl 1,2,3,4-tetrahydro-β-carboline-1-carboxylate in NH_4OAc buffer (pH 8.0, 30 °C), affording (*R*)-1,2,3,4-tetrahydro-β-carboline-1-carboxylic acid with 98% *ee* and 90% yield. When alcalase was used as the biocatalyst for the hydrolysis in borate buffer (pH 8.0, 30 °C), the (*S*)-enantiomer was obtained with 60% *ee* and 66% yield (Scheme 4.84) [139]. Similarly, (*R*)-6-hydroxy-1,2,3,4-tetrahydroisoquinoline-1-carboxylic acid and its analog (*R*)-6-methoxy-1,2,3,4-tetrahydroisoquinoline-1-carboxylic acid were synthesized through CALB-catalyzed dynamic kinetic hydrolysis of the corresponding 1,2,3,4-tetrahydroisoquinoline-1-carboxylic esters in both organic solvents and NH_4OAc buffer at pH 8.5 in >87% yields and >99% *ee* [140].

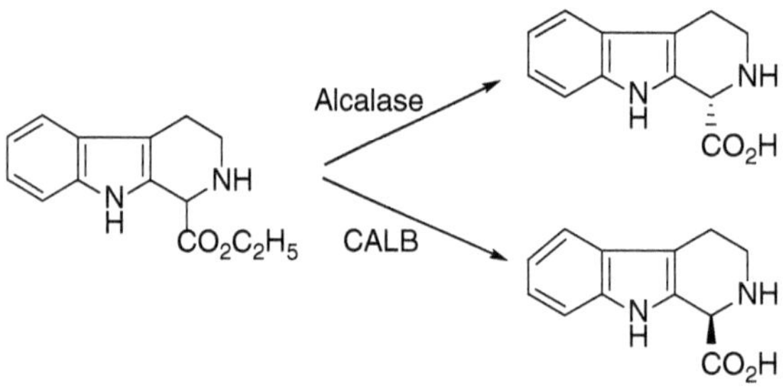

Scheme 4.84 DKR of ethyl 1,2,3,4-tetrahydro-β-carboline-1-carboxylate. Source: Adapted from Megyesi et al. [139].

The racemization of dimethyl (1,3-dihydro-2*H*-isoindol-1-yl)phosphonate also proceeded without external auxiliaries under mild reaction conditions. The DKR of this phosphonate was achieved via lipase-catalyzed alkoxycarbonylation reaction by using *C. antarctica* lipase A (CAL-A) as the biocatalyst and allyl 3-methoxyphenyl carbonate as the acyl donor in toluene at 30 °C, yielding the (*R*)-allyl carbamate in 58% isolated yield and 96% *ee* (Scheme 4.85) [141].

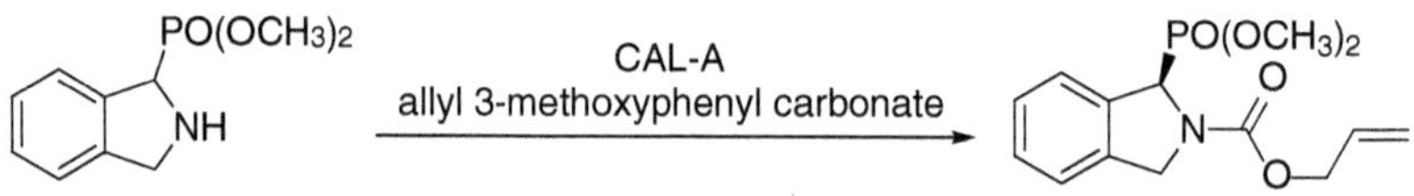

Scheme 4.85 DKR of dimethyl (1,3-dihydro-2*H*-isoindol-1-yl) phosphonate. Source: Adapted from López-Iglesias et al. [141].

4.7 Summary and Outlook

Since the mid-1990s chemoenzymatic DKR has become an emerging methodology for asymmetric organic synthesis by converting a racemate into a single enantiomer of the target compound, especially for chiral alcohols and amines.

Various homogeneous and heterogeneous metal catalysts have been developed for the racemization of chiral alcohol and amines via different reaction mechanisms depending on the metal catalyst and substrate structure. Their combination with lipases in the DKR has been widely applied to the synthesis of optically pure alcohols and amines of high values. Another major racemization approach is based on the activated hydrogen at the chiral carbon center adjoining an electron-withdrawing group such as ketone, which undergoes an enzymatic reduction or amination under action of an ADH or transaminase, favorably resulting in one of the four possible stereoisomers of the product. In spite of the tremendous advances in this research area, the basic requirements in reaction dynamics for an efficient DKR and the compatibility of a chemical racemization and biotransformation often present challenges for developing new DKR processes. However, the improved availability of biocompatible metal and organocatalysts for racemization, and enzyme catalysts for KR of chiral molecules makes more chemical space to address these issues and explore the application of chemoenzymatic DKR in the preparation of single stereoisomers of high-value chiral compounds. It can be expected that more and more delicately designed and efficient chemoenzymatic DKR processes would appear in the coming years.

References

1 Robinson, D.E.J.E. and Bull, S.D. (2003). Kinetic resolution strategies using non-enzymatic catalysts. *Tetrahedron: Asymmetry* 14 (11): 1407–1446.

2 Keith, J.M., Larrow, J.F., and Jacobsen, E.N. (2001). Practical considerations in kinetic resolution reactions. *Advanced Synthesis & Catalysis* 343 (1): 5–26.

3 Ahmed, M., Kelly, T., and Ghanem, A. (2012). Applications of enzymatic and non-enzymatic methods to access enantiomerically pure compounds using kinetic resolution and racemisation. *Tetrahedron* 68 (34): 6781–6802.

4 Chen, C.S., Fujimoto, Y., Girdaukas, G., and Sih, C.J. (1982). Quantitative analyses of biochemical kinetic resolutions of enantiomers. *Journal of the American Chemical Society* 104 (25): 7294–7299.

5 Pàmies, O. and Bäckvall, J.-E. (2003). Combination of enzymes and metal catalysts. A powerful approach in asymmetric catalysis. *Chemical Reviews* 103 (8): 3247–3262.

6 Seddigi, Z.S., Malik, M.S., Ahmed, S.A. et al. (2017). Lipases in asymmetric transformations: recent advances in classical kinetic resolution and lipase–metal combinations for dynamic processes. *Coordination Chemistry Reviews* 348: 54–70.

7 Kazlauskas, R.J., Weissfloch, A.N.E., Rappaport, A.T., and Cuccia, L.A. (1991). A rule to predict which enantiomer of a secondary alcohol reacts faster in reactions catalyzed by cholesterol esterase, lipase from *Pseudomonas cepacia*, and lipase from *Candida rugosa*. *The Journal of Organic Chemistry* 56 (8): 2656–2665.

8 Kitamura, M., Ohkuma, T., Tokunaga, M., and Noyori, R. (1990). Dynamic kinetic resolution in BINAP—ruthenium(II) catalyzed hydrogenation of 2-substituted 3-oxo carboxylic esters. *Tetrahedron: Asymmetry* 1 (1): 1–4.

9 Allen, J.V. and Williams, J.M.J. (1996). Dynamic kinetic resolution with enzyme and palladium combinations. *Tetrahedron Letters* 37 (11): 1859–1862.

10 Ebbers, E.J., Ariaans, G.J.A., Houbiers, J.P.M. et al. (1997). Controlled racemization of optically active organic compounds: prospects for asymmetric transformation. *Tetrahedron* 53 (28): 9417–9476.

11 Felfer, U., Goriup, M., Koegl, M. .F. et al. (2005). The substrate spectrum of mandelate racemase: minimum structural requirements for substrates and substrate model. *Advanced Synthesis & Catalysis* 347 (7-8): 951–961.

12 Verho, O. and Bäckvall, J.-E. (2015). Chemoenzymatic dynamic kinetic resolution: a powerful tool for the preparation of enantiomerically pure alcohols and amines. *Journal of the American Chemical Society* 137 (12): 3996–4009.

13 de Miranda, A.S., Miranda, L.S.M., and de Souza, R.O.M.A. (2015). Lipases: valuable catalysts for dynamic kinetic resolutions. *Biotechnology Advances* 33 (5): 372–393.

14 Lee, J.H., Han, K., Kim, M.-J., and Park, J. (2010). Chemoenzymatic dynamic kinetic resolution of alcohols and amines. *European Journal of Organic Chemistry* 2010 (6): 999–1015.

15 Dinh, P.M., Howarth, J.A., Hudnott, A.R. et al. (1996). Catalytic racemisation of alcohols: applications to enzymatic resolution reactions. *Tetrahedron Letters* 37 (42): 7623–7626.

16 Mari, P., Denys, M., Reko, L., and Kanerva, L.T. (2011). Dynamic kinetic resolution of a wide range of secondary alcohols: cooperation of dicarbonylchlorido(pentabenzylcyclopentadienyl)ruthenium and CAL-B. *European Journal of Organic Chemistry* 2011 (8): 1452–1457.

17 Larsson, A.L.E., Persson, B.A., and Bäckvall, J.E. (1997). Enzymatic resolution of alcohols coupled with ruthenium-catalyzed racemization of the substrate alcohol. *Angewandte Chemie International Edition in English* 36 (11): 1211–1212.

18 Oscar, P. and Bäckvall, J.-E. (2001). Studies on the mechanism of metal-catalyzed hydrogen transfer from alcohols to ketones. *Chemistry – A European Journal* 7 (23): 5052–5058.

19 Persson, B.A., Larsson, A.L.E., Le Ray, M., and Bäckvall, J.-E. (1999). Ruthenium- and enzyme-catalyzed dynamic kinetic resolution of secondary alcohols. *Journal of the American Chemical Society* 121 (8): 1645–1650.

20 Vallin, K.S.A., Wensbo Posaric, D., Hameršak, Z. et al. (2009). Efficient chemoenzymatic dynamic kinetic resolution of 1-heteroaryl ethanols. *The Journal of Organic Chemistry* 74 (24): 9328–9336.

21 Koh, J.H., Jung, H.M., Kim, M.-J., and Park, J. (1999). Enzymatic resolution of secondary alcohols coupled with ruthenium-catalyzed racemization without hydrogen mediator. *Tetrahedron Letters* 40 (34): 6281–6284.

22 Dijksman, A., Elzinga, J.M., Li, Y.-X. et al. (2002). Efficient ruthenium-catalyzed racemization of secondary alcohols: application to dynamic kinetic resolution. *Tetrahedron: Asymmetry* 13 (8): 879–884.

23 Choi, J.H., Kim, Y.H., Nam, S.H. et al. (2002). Aminocyclopentadienyl ruthenium chloride: catalytic racemization and dynamic kinetic resolution of alcohols at ambient temperature. *Angewandte Chemie International Edition* 41 (13): 2373–2376.

24 Martín-Matute, B., Edin, M., Bogár, K. et al. (2005). Combined ruthenium(II) and lipase catalysis for efficient dynamic kinetic resolution of secondary alcohols. Insight into the racemization mechanism. *Journal of the American Chemical Society* 127 (24): 8817–8825.

25 Martín-Matute, B., Edin, M., Bogár, K., and Bäckvall, J.-E. (2004). Highly compatible metal and enzyme catalysts for efficient dynamic kinetic resolution of alcohols at ambient temperature. *Angewandte Chemie International Edition* 43 (47): 6535–6539.

26 Kim, N., Ko, S.-B., Kwon, M.S. et al. (2005). Air-stable racemization catalyst for dynamic kinetic resolution of secondary alcohols at room temperature. *Organic Letters* 7 (20): 4523–4526.

27 Lee, J.H., Kim, N., Kim, M.-J., and Park, J. (2011). Substituent effect on catalytic activities of [{η^5-Ar4C4COC(=O)Ar}Ru(CO)2Cl] in racemization and DKR of secondary alcohols. *ChemCatChem* 3 (2): 354–359.

28 Ko, S.-B., Baburaj, B., Kim, M.-J., and Park, J. (2007). Air-stable racemization catalysts for the dynamic kinetic resolution of secondary alcohols. *The Journal of Organic Chemistry* 72 (18): 6860–6864.

29 Fernández-Salas, J.A., Manzini, S., and Nolan, S.P. (2014). A cationic ruthenium complex for the dynamic kinetic resolution of secondary alcohols. *Chemistry – A European Journal* 20 (41): 13132–13135.

30 Chen, Q. and Yuan, C. (2008). An unexpected ruthenium complex and its unique behavior as catalyst in dynamic kinetic resolution of secondary alcohols. *Chemical Communications* 42: 5333–5335.

31 Eckert, M., Brethon, A., Li, Y.-X. et al. (2007). Study of the efficiency of amino-functionalized ruthenium and ruthenacycle complexes as racemization catalysts in the dynamic kinetic resolution of 1-phenylethanol. *Advanced Synthesis & Catalysis* 349 (17-18): 2603–2609.

32 van Nispen, S.F.G.M., van Buijtenen, J., Vekemans, J.A.J.M. et al. (2006). Efficient dynamic kinetic resolution of secondary alcohols with a novel tetrafluorosuccinato ruthenium complex. *Tetrahedron: Asymmetry* 17 (15): 2299–2305.

33 El-Sepelgy, O., Alandini, N., and Rueping, M. (2016). Merging iron catalysis and biocatalysis—iron carbonyl complexes as efficient hydrogen autotransfer catalysts in dynamic kinetic resolutions. *Angewandte Chemie International Edition* 55 (43): 13602–13605.

34 Gustafson, K.P.J., Guðmundsson, A., Lewis, K., and Bäckvall, J.-E. (2017). Chemoenzymatic dynamic kinetic resolution of secondary alcohols using an air- and moisture-stable iron racemization catalyst. *Chemistry – A European Journal* 23 (5): 1048–1051.

35 El-Sepelgy, O., Brzozowska, A., and Rueping, M. (2017). Asymmetric chemoenzymatic reductive acylation of ketones by a combined iron-catalyzed hydrogenation–racemization and enzymatic resolution cascade. *ChemSusChem* 10 (8): 1664–1668.

36 Sato, Y., Kayaki, Y., and Ikariya, T. (2012). Efficient dynamic kinetic resolution of racemic secondary alcohols by a chemoenzymatic system using bifunctional iridium complexes with C–N chelate amido ligands. *Chemical Communications* 48 (30): 3635–3637.

37 Berkessel, A., Sebastian-Ibarz, M.L., and Müller, T.N. (2006). Lipase/aluminum-catalyzed dynamic kinetic resolution of secondary alcohols. *Angewandte Chemie International Edition* 45 (39): 6567–6570.

38 Wuyts, S., De Temmerman, K., De Vos, D., and Jacobs, P. (2003). A zeolite–enzyme combination for biphasic dynamic kinetic resolution of benzylic alcohols. *Chemical Communications* 15: 1928–1929.

39 Wuyts, S., De Temmerman, K., De Vos, D.E., and Jacobs, P.A. (2005). Acid zeolites as alcohol racemization catalysts: screening and application in biphasic dynamic kinetic resolution. *Chemistry – A European Journal* 11 (1): 386–397.

40 Zhu, Y., Fow, K.-L., Chuah, G.-K., and Jaenicke, S. (2007). Dynamic kinetic resolution of secondary alcohols combining enzyme-catalyzed transesterification and zeolite-catalyzed racemization. *Chemistry – A European Journal* 13 (2): 541–547.

41 Wang, J., Do, D.-M., Chuah, G.-K., and Jaenicke, S. (2013). Core–shell composite as the racemization catalyst in the dynamic kinetic resolution of secondary alcohols. *ChemCatChem* 5 (1): 247–254.

42 Långvik, O., Saloranta, T., Murzin, D.Y., and Leino, R. (2015). Heterogeneous chemoenzymatic catalyst combinations for one-pot dynamic kinetic resolution applications. *ChemCatChem* 7 (24): 4004–4015.

43 Im, J.-S., Ahn, S.-H., and Park, Y.-H. (2013). Reaction characteristics of the dynamic kinetic resolution of 1-phenylethanol over Ru complexes immobilized on PTA-modified γ-alumina and Novozym 435®. *Chemical Engineering Journal* 234: 49–56.

44 Han, K., Kim, C., Park, J., and Kim, M.-J. (2010). Chemoenzymatic synthesis of rivastigmine via dynamic kinetic resolution as a key step. *The Journal of Organic Chemistry* 75 (9): 3105–3108.

45 Wuyts, S., Wahlen, J., Jacobs, P.A., and De Vos, D.E. (2007). Heterogeneous vanadium catalysts for racemization and chemoenzymatic dynamic kinetic resolution of benzylic alcohols. *Green Chemistry* 9 (10): 1104–1108.

46 de França, A.d.S., Silva, M.V.M., Neves, R.V. et al. (2018). Studies on the dynamic resolution of crizotinib intermediate. *Bioorganic & Medicinal Chemistry* 26 (7): 1333–1337.

47 Akai, S., Tanimoto, K., Kanao, Y. et al. (2006). A dynamic kinetic resolution of allyl alcohols by the combined use of lipases and [VO(OSiPh3)3]. *Angewandte Chemie International Edition* 45 (16): 2592–2595.

48 Akai, S., Hanada, R., Fujiwara, N. et al. (2010). One-pot synthesis of optically active allyl esters via lipase–vanadium combo catalysis. *Organic Letters* 12 (21): 4900–4903.

49 Egi, M., Sugiyama, K., Saneto, M. et al. (2013). A mesoporous-silica-immobilized oxovanadium cocatalyst for the lipase-catalyzed dynamic kinetic resolution of racemic alcohols. *Angewandte Chemie International Edition* 52 (13): 3654–3658.

50 Sugiyama, K., Oki, Y., Kawanishi, S. et al. (2016). Spatial effects of oxovanadium-immobilized mesoporous silica on racemization of alcohols and application in lipase-catalyzed dynamic kinetic resolution. *Catalysis Science & Technology* 6 (13): 5023–5030.

51 Sugiyama, K., Kawanishi, S., Oki, Y. et al. (2018). Lipase-catalyzed asymmetric synthesis of naphtho[2,3-c]furan-1(3*H*)-one derivatives by a one-pot dynamic kinetic resolution/intramolecular Diels–Alder reaction: total synthesis of (−)-himbacine. *Bioorganic & Medicinal Chemistry* 26 (7): 1378–1386.

52 Kawanishi, S., Sugiyama, K., Oki, Y. et al. (2017). Preparation of optically active cycloalkenes bearing all-carbon quaternary stereogenic centres via lipase–oxovanadium combo-catalysed dynamic kinetic resolution. *Green Chemistry* 19 (2): 411–417.

53 Kawanishi, S., Oki, S., Kundu, D., and Akai, S. (2019). Lipase/oxovanadium co-catalyzed dynamic kinetic resolution of propargyl alcohols: competition between racemization and rearrangement. *Organic Letters* 21 (9): 2978–2982.

54 Lihammar, R., Millet, R., and Bäckvall, J.-E. (2013). Enzyme- and ruthenium-catalyzed dynamic kinetic resolution of functionalized cyclic allylic alcohols. *The Journal of Organic Chemistry* 78 (23): 12114–12120.

55 Xu, Y., Wang, M., Feng, B. et al. (2017). Dynamic kinetic resolution of aromatic sec-alcohols by using a heterogeneous palladium racemization catalyst and lipase. *Catalysis Science & Technology* 7 (24): 5838–5842.

56 Michalak, K., Wicha, J., and Wójcik, J. (2016). Studies towards dynamic kinetic resolution of 4-hydroxy-2-methylcyclopent-2-en-1-one and its E-O-trityloxime. *Tetrahedron* 72 (32): 4813–4820.

57 Do, Y., Hwang, I.-C., Kim, M.-J., and Park, J. (2010). Photoactivated racemization catalyst for dynamic kinetic resolution of secondary alcohols. *The Journal of Organic Chemistry* 75 (16): 5740–5742.

58 Leitner, W. (2002). Supercritical carbon dioxide as a green reaction medium for catalysis. *Accounts of Chemical Research* 35 (9): 746–756.

59 Benaissi, K., Poliakoff, M., and Thomas, N.R. (2009). Dynamic kinetic resolution of rac-1-phenylethanol in supercritical carbon dioxide. *Green Chemistry* 11 (5): 617–621.

60 Kim, H., Choi, Y.K., Lee, J. et al. (2011). Ionic-surfactant-coated Burkholderia cepacia lipase as a highly active and enantioselective catalyst for the dynamic kinetic resolution of secondary alcohols. *Angewandte Chemie International Edition* 50 (46): 10944–10948.

61 Kim, C., Lee, J., Cho, J. et al. (2013). Kinetic and dynamic kinetic resolution of secondary alcohols with ionic-surfactant-coated Burkholderia cepacia lipase:

substrate scope and enantioselectivity. *The Journal of Organic Chemistry* 78 (6): 2571–2578.

62 Kim, M.-J., Chung, Y.I., Choi, Y.K. et al. (2003). (S)-Selective dynamic kinetic resolution of secondary alcohols by the combination of subtilisin and an aminocyclopentadienylruthenium complex as the catalysts. *Journal of the American Chemical Society* 125 (38): 11494–11495.

63 Borén, L., Martín-Matute, B., Xu, Y. et al. (2006). (S)-Selective kinetic resolution and chemoenzymatic dynamic kinetic resolution of secondary alcohols. *Chemistry – A European Journal* 12 (1): 225–232.

64 Kim, K., Lee, E., Kim, C. et al. (2017). Ionic-surfactant-coated subtilisin: activity, enantioselectivity, and application to dynamic kinetic resolution of secondary alcohols. *Organic & Biomolecular Chemistry* 15 (41): 8836–8844.

65 Kim, M.-J., Choi, Y.K., Kim, S. et al. (2008). Highly enantioselective dynamic kinetic resolution of 1,2–diarylethanols by a lipase–ruthenium couple. *Organic Letters* 10 (6): 1295–1298.

66 Lee, J., Oh, Y., Choi, Y.K. et al. (2015). Dynamic kinetic resolution of diarylmethanols with an activated lipoprotein lipase. *ACS Catalysis* 5 (2): 683–689.

67 Cho, J., Lee, J., Park, J., and Kim, M.-J. (2015). Highly enantioselective dynamic kinetic resolution of alkyl aryl carbinols carrying a trimethylsilyl group with a highly active lipoprotein lipase preparation. *Tetrahedron: Asymmetry* 26 (15): 840–845.

68 Yun, I., Park, J.Y., Park, J., and Kim, M.-J. (2019). Base-free dynamic kinetic resolution of secondary alcohols with a ruthenium–lipase couple. *The Journal of Organic Chemistry* 84 (24): 16293–16298.

69 Strübing, D., Krumlinde, P., Piera, J., and Bäckvall, J.-E. (2007). Dynamic kinetic resolution of primary alcohols with an unfunctionalized stereogenic center in the β-position. *Advanced Synthesis & Catalysis* 349 (10): 1577–1581.

70 Träff, A., Bogár, K., Warner, M., and Bäckvall, J.-E. (2008). Highly efficient route for enantioselective preparation of chlorohydrins via dynamic kinetic resolution. *Organic Letters* 10 (21): 4807–4810.

71 Johnston, E.V., Bogár, K., and Bäckvall, J.-E. (2010). Enantioselective synthesis of (R)-bufuralol via dynamic kinetic resolution in the key step. *The Journal of Organic Chemistry* 75 (13): 4596–4599.

72 Träff, A., Lihammar, R., and Bäckvall, J.-E. (2011). A chemoenzymatic dynamic kinetic resolution approach to enantiomerically pure (R)- and (S)-duloxetine. *The Journal of Organic Chemistry* 76 (10): 3917–3921.

73 Pàmies, O. and Bäckvall, J.-E. (2001). Dynamic kinetic resolution of β-azido alcohols. An efficient route to chiral aziridines and β-amino alcohols. *The Journal of Organic Chemistry* 66 (11): 4022–4025.

74 Huerta, F.F., Laxmi, Y.R.S., and Bäckvall, J.-E. (2000). Dynamic kinetic resolution of α-hydroxy acid esters. *Organic Letters* 2 (8): 1037–1040.

75 Huerta, F.F. and Bäckvall, J.-E. (2001). Enantioselective synthesis of β-hydroxy acid derivatives via a one-pot aldol reaction–dynamic kinetic resolution. *Organic Letters* 3 (8): 1209–1212.

76 Fransson, A.-B.L., Borén, L., Pàmies, O., and Bäckvall, J.-E. (2005). Kinetic resolution and chemoenzymatic dynamic kinetic resolution of functionalized γ-hydroxy amides. *The Journal of Organic Chemistry* 70 (7): 2582–2587.

77 Pàmies, O. and Bäckvall, J.-E. (2002). Enzymatic kinetic resolution and chemoenzymatic dynamic kinetic resolution of δ-hydroxy esters. An efficient route to chiral δ-lactones. *The Journal of Organic Chemistry* 67 (4): 1261–1265.

78 Persson, B.A., Huerta, F.F., and Bäckvall, J.-E. (1999). Dynamic kinetic resolution of secondary diols via coupled ruthenium and enzyme catalysis. *The Journal of Organic Chemistry* 64 (14): 5237–5240.

79 Martín-Matute, B., Edin, M., and Bäckvall, J.-E. (2006). Highly efficient synthesis of enantiopure diacetylated C2-symmetric diols by ruthenium- and enzyme-catalyzed dynamic kinetic asymmetric transformation (DYKAT). *Chemistry – A European Journal* 12 (23): 6053–6061.

80 Borén, L., Leijondahl, K., and Bäckvall, J.-E. (2009). Dynamic kinetic asymmetric transformation of 1,4-diols and the preparation of trans-2,5-disubstituted pyrrolidines. *Tetrahedron Letters* 50 (26): 3237–3240.

81 Leijondahl, K., Borén, L., Braun, R., and Bäckvall, J.-E. (2009). Enzyme- and ruthenium-catalyzed dynamic kinetic asymmetric transformation of 1,5-diols. Application to the synthesis of (+)-solenopsin A. *The Journal of Organic Chemistry* 74 (5): 1988–1993.

82 Leijondahl, K., Borén, L., Braun, R., and Bäckvall, J.-E. (2008). Enantiopure 1,5-diols from dynamic kinetic asymmetric transformation. Useful synthetic intermediates for the preparation of chiral heterocycles. *Organic Letters* 10 (10): 2027–2030.

83 Martín-Matute, B. and Bäckvall, J.-E. (2004). Ruthenium- and enzyme-catalyzed dynamic kinetic asymmetric transformation of 1,4-diols: synthesis of γ-hydroxy ketones. *The Journal of Organic Chemistry* 69 (26): 9191–9195.

84 Fransson, A.-B.L., Xu, Y., Leijondahl, K., and Bäckvall, J.-E. (2006). Enzymatic resolution, desymmetrization, and dynamic kinetic asymmetric transformation of 1,3-cycloalkanediols. *The Journal of Organic Chemistry* 71 (17): 6309–6316.

85 Haak, R.M., Berthiol, F., Jerphagnon, T. et al. (2008). Dynamic kinetic resolution of racemic β-haloalcohols: direct access to enantioenriched epoxides. *Journal of the American Chemical Society* 130 (41): 13508–13509.

86 Reetz, M.T. and Schimossek, K. (1996). Lipase-catalyzed dynamic kinetic resolution of chiral amines: use of palladium as the racemization catalyst. *CHIMIA* 50 (12): 668–669.

87 Kim, Y., Park, J., and Kim, M.J. (2011). Dynamic kinetic resolution of amines and amino acids by enzyme–metal cocatalysis. *ChemCatChem* 3 (2): 271–277.

88 Choi, Y.K., Kim, M.J., Ahn, Y., and Kim, M.-J. (2001). Lipase/palladium-catalyzed asymmetric transformations of ketoximes to optically active amines. *Organic Letters* 3 (25): 4099–4101.

89 Parvulescu, A., Vos, D.D., and Jacobs, P. (2005). Efficient dynamic kinetic resolution of secondary amines with Pd on alkaline earth salts and a lipase. *Chemical Communications* 42: 5307–5309.

90 Andrade, L.H., Silva, A.V., and Pedrozo, E.C. (2009). First dynamic kinetic resolution of selenium-containing chiral amines catalyzed by palladium (Pd/BaSO4) and *Candida antartica* lipase (CAL-B). *Tetrahedron Letters* 50 (30): 4331–4334.

91 de Miranda, A.S., de Souza, R.O.M.A., and Miranda, L.S.M. (2014). Ammonium formate as a green hydrogen source for clean semi-continuous enzymatic dynamic kinetic resolution of (+/−)-α-methylbenzylamine. *RSC Advances* 4 (26): 13620–13625.

92 Kim, M.-J., Kim, W.-H., Han, K. et al. (2007). Dynamic kinetic resolution of primary amines with a recyclable Pd nanocatalyst for racemization. *Organic Letters* 9 (6): 1157–1159.

93 Kim, Y., Park, J., and Kim, M.-J. (2010). Fast racemization and dynamic kinetic resolution of primary benzyl amines. *Tetrahedron Letters* 51 (42): 5581–5584.

94 Choi, Y.K., Kim, Y., Han, K. et al. (2009). Synthesis of optically active amino acid derivatives via dynamic kinetic resolution. *The Journal of Organic Chemistry* 74 (24): 9543–9545.

95 Engström, K., Shakeri, M., and Bäckvall, J.-E. (2011). Dynamic kinetic resolution of β-amino esters by a heterogeneous system of a palladium nanocatalyst and *Candida antarctica* lipase A. *European Journal of Organic Chemistry* 2011 (10): 1827–1830.

96 Ma, G., Xu, Z., Zhang, P. et al. (2014). A novel synthesis of rasagiline via a chemoenzymatic dynamic kinetic resolution. *Organic Process Research & Development* 18 (10): 1169–1174.

97 Han, K., Kim, Y., Park, J., and Kim, M.-J. (2010). Chemoenzymatic synthesis of the calcimimetics (+)-NPS R-568 via asymmetric reductive acylation of ketoxime intermediate. *Tetrahedron Letters* 51 (27): 3536–3537.

98 Parvulescu, A.N., Jacobs, P.A., and De Vos, D.E. (2009). Support influences in the Pd-catalyzed racemization and dynamic kinetic resolution of chiral benzylic amines. *Applied Catalysis A: General* 368 (1): 9–16.

99 Gustafson, K.P.J., Lihammar, R., Verho, O. et al. (2014). Chemoenzymatic dynamic kinetic resolution of primary amines using a recyclable palladium nanoparticle catalyst together with lipases. *The Journal of Organic Chemistry* 79 (9): 3747–3751.

100 Jin, Q., Jia, G., Zhang, Y., and Li, C. (2014). Modification of supported Pd catalysts by alkalic salts in the selective racemization and dynamic kinetic resolution of primary amines. *Catalysis Science & Technology* 4 (2): 464–471.

101 Xu, S., Wang, M., Feng, B. et al. (2018). Dynamic kinetic resolution of amines by using palladium nanoparticles confined inside the cages of amine-modified MIL-101 and lipase. *Journal of Catalysis* 363: 9–17.

102 de Souza, S.P., Leão, R.A.C., Bassut, J.F. et al. (2017). New biosilified Pd-lipase hybrid biocatalysts for dynamic resolution of amines. *Tetrahedron Letters* 58 (52): 4849–4854.

103 Shi, J., Li, X., Wang, Q. et al. (2012). Platinum-encapsulated zeolitically microcapsular catalyst for one-pot dynamic kinetic resolution of phenylethylamine. *Journal of Catalysis* 291: 87–94.

104 Parvulescu, A.N., Jacobs, P.A., and De Vos, D.E. (2008). Heterogeneous raney nickel and cobalt catalysts for racemization and dynamic kinetic resolution of amines. *Advanced Synthesis & Catalysis* 350 (1): 113–121.

105 Paetzold, J. and Bäckvall, J.E. (2005). Chemoenzymatic dynamic kinetic resolution of primary amines. *Journal of the American Chemical Society* 127 (50): 17620–17621.

106 Veld, M.A.J., Hult, K., Palmans, A.R.A., and Meijer, E.W. (2007). Fast DKR of amines using isopropyl 2-methoxyacetate as acyl donor. *European Journal of Organic Chemistry* 2007 (32): 5416–5421.

107 Hoben, C.E., Kanupp, L., and Bäckvall, J.-E. (2008). Practical chemoenzymatic dynamic kinetic resolution of primary amines via transfer of a readily removable benzyloxycarbonyl group. *Tetrahedron Letters* 49 (6): 977–979.

108 Shakeri, M., Engström, K., Sandström, A.G., and Bäckvall, J.-E. (2010). Highly enantioselective resolution of β-amino esters by Candida antarctica lipase A immobilized in mesocellular foam: application to dynamic kinetic resolution. *ChemCatChem* 2 (5): 534–538.

109 Poulhès, F., Vanthuyne, N., Bertrand, M.P. et al. (2011). Chemoenzymatic dynamic kinetic resolution of primary amines catalyzed by CAL-B at 38–40°C. *The Journal of Organic Chemistry* 76 (17): 7281–7286.

110 El Blidi, L., Vanthuyne, N., Siri, D. et al. (2010). Switching from (R)- to (S)-selective chemoenzymatic DKR of amines involving sulfanyl radical-mediated racemization. *Organic & Biomolecular Chemistry* 8 (18): 4165–4168.

111 Koszelewski, D., Brodzka, A., Żądło, A. et al. (2016). Dynamic kinetic resolution of 3-Aryl-4-pentenoic acids. *ACS Catalysis* 6 (5): 3287–3292.

112 Koszelewski, D., Zysk, M., Brodzka, A. et al. (2015). Evaluation of a new protocol for enzymatic dynamic kinetic resolution of 3-hydroxy-3-(aryl)propanoic acids. *Organic & Biomolecular Chemistry* 13 (45): 11014–11020.

113 Westerbeek, A., Szymański, W., Feringa, B.L., and Janssen, D.B. (2011). Dynamic kinetic resolution process employing haloalkane dehalogenase. *ACS Catalysis* 1 (12): 1654–1660.

114 DeSantis, G., Zhu, Z., Greenberg, W.A. et al. (2002). An enzyme library approach to biocatalysis: development of nitrilases for enantioselective production of carboxylic acid derivatives. *Journal of the American Chemical Society* 124 (31): 9024–9025.

115 Applegate, G.A. and Berkowitz, D.B. (2015). Exploiting enzymatic dynamic reductive kinetic resolution (DYRKR) in stereocontrolled synthesis. *Advanced Synthesis & Catalysis* 357 (8): 1619–1632.

116 Buisson, D., Azerad, R., Sanner, C., and Larcheveque, M. (1990). Stereocontrolled reduction of alpha-methyl beta-ketoesters by *Geotrichum candidum*. *Biocatalysis* 3 (1-2): 85–93.

117 Danchet, S., Bigot, C., Buisson, D., and Azerad, R. (1997). Dynamic kinetic resolution in the microbial reduction of α-monosubstituted β-oxoesters: the reduction of 2-carbethoxycycloheptanone and 2-carbethoxy-cyclooctanone. *Tetrahedron: Asymmetry* 8 (11): 1735–1739.

118 Cuetos, A., Rioz-Martínez, A., Bisogno, F.R. et al. (2012). Access to enantiopure α-alkyl-β-hydroxy esters through dynamic kinetic resolutions employing purified/overexpressed alcohol dehydrogenases. *Advanced Synthesis & Catalysis* 354 (9): 1743–1749.

119 Feske, B.D., Kaluzna, I.A., and Stewart, J.D. (2005). Enantiodivergent, biocatalytic routes to both taxol side chain antipodes. *The Journal of Organic Chemistry* 70 (23): 9654–9657.

120 Perrone, M.G., Santandrea, E., Scilimati, A. et al. (2004). Baker's yeast-mediated reduction of ethyl 2-(4-chlorophenoxy)-3-oxoalkanoates intermediates for potential PPARα ligands. *Tetrahedron: Asymmetry* 15 (22): 3501–3510.

121 Xu, F., Chung, J.Y.L., Moore, J.C. et al. (2013). Asymmetric synthesis of cis-2,5-disubstituted pyrrolidine, the core scaffold of β3-AR agonists. *Organic Letters* 15 (6): 1342–1345.

122 Dehli, J.R. and Gotor, V. (2002). Dynamic kinetic resolution of 2-oxocycloalkanecarbonitriles: chemoenzymatic syntheses of optically active cyclic β- and γ-amino alcohols. *The Journal of Organic Chemistry* 67 (19): 6816–6819.

123 Marocco, C.P., Davis, E.V., Finnell, J.E. et al. (2011). Asymmetric synthesis of (−)-fosfomycin and its trans-(1S,2S)-diastereomer using a biocatalytic reduction as the key step. *Tetrahedron: Asymmetry* 22 (18): 1784–1789.

124 Wang, Z., Zeng, Y., Wu, X. et al. (2020). Access to chiral α-substituted-β-hydroxy arylphosphonates enabled by biocatalytic dynamic reductive kinetic resolution. *Organic & Biomolecular Chemistry* 18 (14): 2672–2677.

125 Deasy, R.E. and Maguire, A.R. (2014). Baker's-yeast-mediated reduction of sulfur-containing compounds. *European Journal of Organic Chemistry* 2014 (18): 3737–3756.

126 Maguire, A.R. and O'Riordan, N. (1999). Dynamic kinetic resolution in the baker's yeast mediated reduction of 2-benzenesulfonylcycloalkanones. *Tetrahedron Letters* 40 (52): 9285–9288.

127 Giacomini, D., Galletti, P., Quintavalla, A. et al. (2007). Highly efficient asymmetric reduction of arylpropionic aldehydes by Horse Liver Alcohol Dehydrogenase through dynamic kinetic resolution. *Chemical Communications* 39: 4038–4040.

128 Galletti, P., Emer, E., Gucciardo, G. et al. (2010). Chemoenzymatic synthesis of (2S)-2-arylpropanols through a dynamic kinetic resolution of 2-arylpropanals with alcohol dehydrogenases. *Organic & Biomolecular Chemistry* 8 (18): 4117–4123.

129 Dong, Y., Yao, P., Cui, Y. et al. (2018). Manipulating the stereoselectivity of a thermostable alcohol dehydrogenase by directed evolution for efficient asymmetric synthesis of arylpropanols. *Biological Chemistry* 400 (3): 313–321.

130 Friest, J.A., Maezato, Y., Broussy, S. et al. (2010). Use of a robust dehydrogenase from an archael hyperthermophile in asymmetric catalysis–dynamic reductive kinetic resolution entry into (S)-profens. *Journal of the American Chemical Society* 132 (17): 5930–5931.

131 Mangas-Sánchez, J., Busto, E., Gotor, V., and Gotor-Fernández, V. (2013). One-pot synthesis of enantiopure 3,4-dihydroisocoumarins through dynamic reductive kinetic resolution processes. *Organic Letters* 15 (15): 3872–3875.

132 Méndez-Sánchez, D., Mangas-Sánchez, J., Busto, E. et al. (2016). Dynamic reductive kinetic resolution of benzyl ketones using alcohol dehydrogenases and anion exchange resins. *Advanced Synthesis & Catalysis* 358 (1): 122–131.

133 Xu, F., Kosjek, B., Cabirol, F.L. et al. (2018). Synthesis of vibegron enabled by a ketoreductase rationally designed for high pH dynamic kinetic reduction. *Angewandte Chemie International Edition* 57 (23): 6863–6867.

134 Cao, J. and Hyster, T.K. (2020). Pyridoxal-catalyzed racemization of α-aminoketones enables the stereodivergent synthesis of 1,2-amino alcohols using ketoreductases. *ACS Catalysis* 10 (11): 6171–6175.

135 Fuchs, C.S., Farnberger, J.E., Steinkellner, G. et al. (2018). Asymmetric amination of α-chiral aliphatic aldehydes via dynamic kinetic resolution to access stereocomplementary brivaracetam and pregabalin precursors. *Advanced Synthesis & Catalysis* 360 (4): 768–778.

136 Cuetos, A., Lavandera, I., and Gotor, V. (2013). Expanding dynamic kinetic protocols: transaminase-catalyzed synthesis of α-substituted β-amino ester derivatives. *Chemical Communications* 49 (91): 10688–10690.

137 Mourelle-Insua, Á., Méndez-Sánchez, D., Galman, J.L. et al. (2019). Efficient synthesis of α-alkyl-β-amino amides by transaminase-mediated dynamic kinetic resolutions. *Catalysis Science & Technology* 9 (15): 4083–4090.

138 Rodríguez, C., de Gonzalo, G., Rioz-Martínez, A. et al. (2010). BVMO-catalysed dynamic kinetic resolution of racemic benzyl ketones in the presence of anion exchange resins. *Organic & Biomolecular Chemistry* 8 (5): 1121–1125.

139 Megyesi, R., Mándi, A., Kurtán, T. et al. (2017). Dynamic kinetic resolution of ethyl 1,2,3,4-tetrahydro-β-carboline-1-carboxylate: use of different hydrolases for stereocomplementary processes. *European Journal of Organic Chemistry* 2017 (32): 4713–4718.

140 Forró, E., Megyesi, R., Paál, T.A., and Fülöp, F. (2016). Efficient dynamic kinetic resolution method for the synthesis of enantiopure 6-hydroxy- and 6-methoxy-1,2,3,4-tetrahydroisoquinoline-1-carboxylic acid. *Tetrahedron: Asymmetry* 27 (24): 1213–1216.

141 López-Iglesias, M., Arizpe, A., Sayago, F.J. et al. (2016). Lipase-catalyzed dynamic kinetic resolution of dimethyl (1,3-dihydro-2H-isoindol-1-yl) phosphonate. *Tetrahedron* 72 (46): 7311–7316.

5

Chemoenzymatic Concurrent Deracemization

In the previous chapter, dynamic kinetic resolution was discussed, in which a racemic substrate (**A**) is completely transformed into one single enantiomer of the product (**B**). Deracemization is the complete transformation of a racemic compound into its one single enantiomer of identical structure (Scheme 5.1). Deracemization is different from dynamic kinetic resolution since no new product is produced in this process and the additional steps to remove the resolving agents from the products are not needed. As such, it is a more efficient method when the substrate and the desired product possess an identical chemical structure.

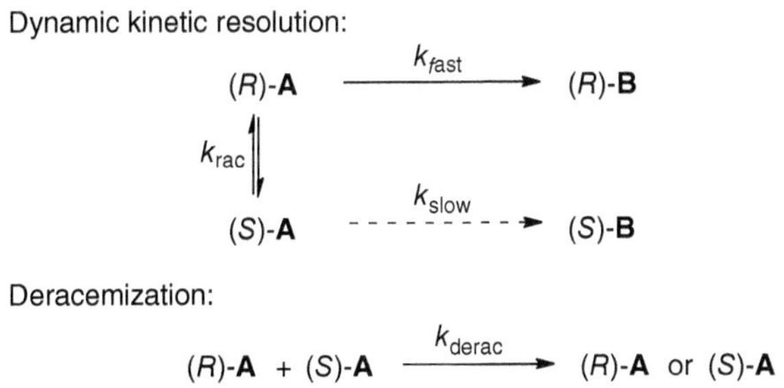

Scheme 5.1 Dynamic kinetic resolution and deracemization.

Deracemization can proceed in different modes (Scheme 5.2). One mode is the linear deracemization in which both enantiomers of a racemic compound (**A**) are first converted into an intermediate (**C**), which is transformed back to one of its enantiomers via an enantiospecific reaction. Another approach is that one enantiomer reacts enantioselectively to give the intermediate, which is converted nonselectively into its racemic mixture or enantioselectively into the unreacted enantiomer. For the nonselective case, the process is also referred to as cyclic deracemization, while the enantioselective process is often called stereoinversion. A deracemization process comprises two half-reactions in which the chirality of the stereogenic center is destructed and regenerated, and at least one of them has to operate enantioselectively.

Destruction and regeneration of the chirality of a stereogenic center have often been achieved by the oxidation–reduction process. These processes can be the

Chemo-Enzymatic Cascade Reactions, First Edition. Dunming Zhu.

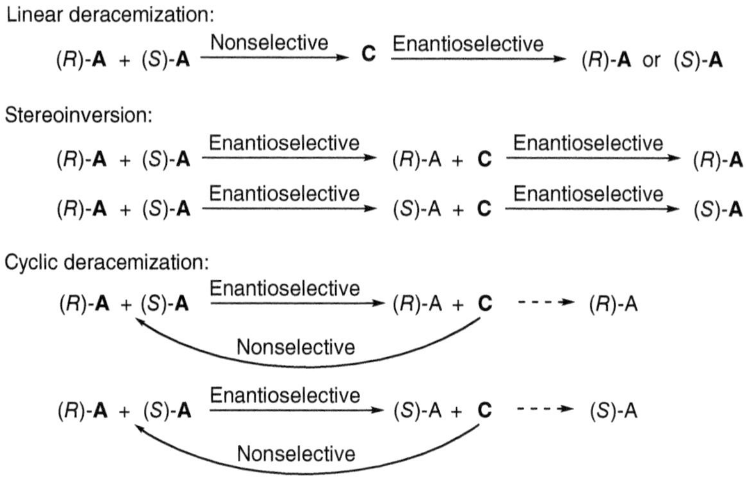

Scheme 5.2 Three deracemization modes.

combination of chemical–chemical, chemical–enzymatic, or enzymatic–enzymatic reactions. For the chemical–chemical process, the main challenge is that the oxidant and reductant can easily and directly quench each other in a single reactor. Therefore, a purely chemical-driven concurrent deracemization is very rare to date, and sequential operation or physical isolation of oxidation and reduction has been employed to overcome this problem [1]. The deracemization techniques are currently mainly focused on the use of biocatalyst systems. For the enzymatic–enzymatic process, microbial strains and recombinant enzymes have been employed and several review articles have appeared [2–5]. The advances in chemoenzymatic concurrent deracemization of amines, alcohols, and other compounds will be the topics of this chapter [5, 6].

5.1 Deracemization of Amino Acids and Amines

Deracemization of compounds bearing a chiral amino group can be achieved in a "one-pot single-step" process, including an enzyme-mediated enantioselective oxidation of a racemic amine and the nonselective chemical reduction of the resulting prochiral imine. An early example of such chemoenzymatic deracemization is the synthesis of L-proline from the racemate by using a D-amino acid oxidase and sodium borohydride system (Scheme 5.3). D-Amino acid oxidase catalyzes the enantiospecific oxidation of D-proline in the racemate to form Δ1-pyrroline-2-carboxylic acid, which is immediately reduced to D- and L-proline by sodium borohydride. Racemic proline is eventually converted to optically pure L-proline with 98% yield by cyclic repetition of this oxidation–reduction sequence [7]. By employing this chemoenzymatic deracemization procedure, optically pure L-pipecolic acid had also been prepared from the racemic pipecolic acid [8].

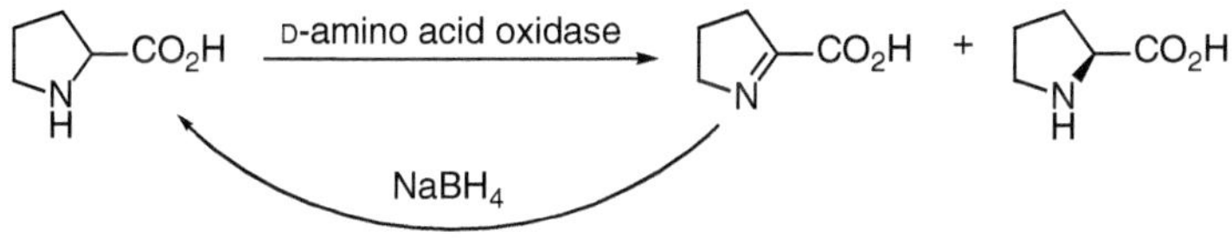

Scheme 5.3 Chemoenzymatic deracemization of proline by using a D-amino acid oxidase and sodium borohydride system.

Since sodium borohydride reacts with water, a large excess of reducing agent (up to 500 equiv.) is required. Thus, a milder and water-stable hydride reducing agent, $NaCNBH_3$, is used instead of $NaBH_4$ for the deracemization of racemic amino acids. In this case, less amount of reducing agent (as little as 3 equiv.) was used in the deracemization of proline, and L-proline was obtained in >99% yield and >99% ee. This improved methodology was then applied to the deracemization of DL-piperazine-2-carboxylic acid, to give the L-enantiomer in 86% yield and 99% ee (Scheme 5.4), which is a component of the HIV-protease inhibitor Crixivan [9]. However, replacement of $NaBH_4$ by $NaCNBH_3$ resulted in a decrease in the reduction rate of the imine intermediate so that it became the rate-limiting step of the reaction. By using porcine kidney D-amino acid oxidase (DAAO) and $NaBH_4$, deracemization of acyclic DL-α-amino acids also proceeded smoothly to give the corresponding L-amino acids in high yields (75–90%) and excellent ee (>98%, Scheme 5.5).

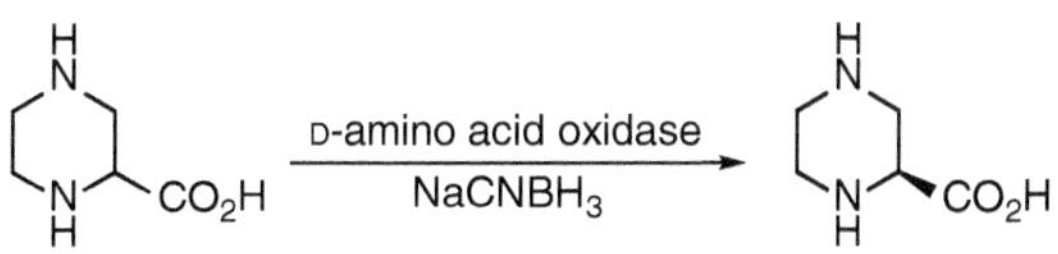

Scheme 5.4 Chemoenzymatic deracemization of DL-piperazine-2-carboxylic acid.

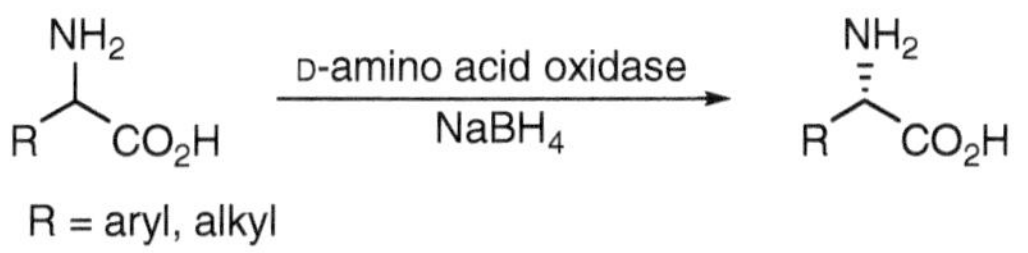

Scheme 5.5 Chemoenzymatic deracemization of acyclic DL-α-amino acids.

For β-substituted α-amino acids, the amino acid oxidases usually do not recognize the chirality at the β-carbon. As such, (2*RS*,3*RS*)-2-amino-3-methylhexanoic acid and (2*RS*,3*RS*)-β-methyl-phenylalanine, which were prepared as a mixture of diastereomers from the corresponding racemic aldehydes via the Strecker reaction, were converted into the corresponding mixture of 2*S*-stereoisomers with >99% ee for each stereoisomer using D-amino acid oxidase (TvDAAO) from *Trigonopsis variabilis* and either $NaBH_4$ or $NaCNBH_3$ as the reducing agent (Scheme 5.6) [10]. Celite-immobilized TvDAAO was combined with ammonia borane in the deracemization of racemic 2-amino-3-(6-*o*-tolylpyridin-3-yl)propanoic acid to prepare its (*S*)-enantiomer with 76–79% yield and >99.9% ee (Scheme 5.7), which is a key

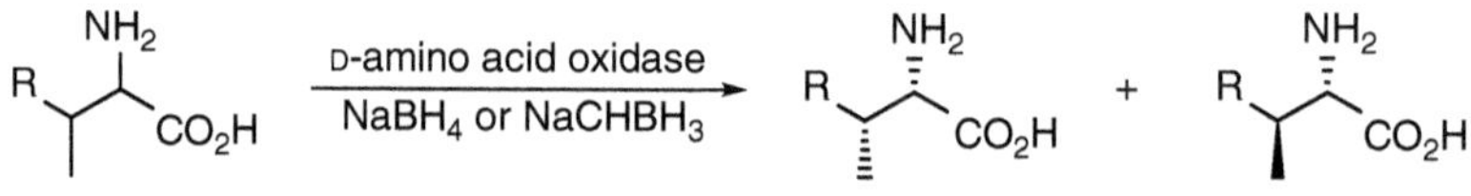

Scheme 5.6 Chemoenzymatic deracemization of (2*RS*,3*RS*)-2-amino-3-methylhexanoic acid and (2*RS*,3*RS*)-β-methyl-phenylalanine. Source: Based on Enright et al. [10].

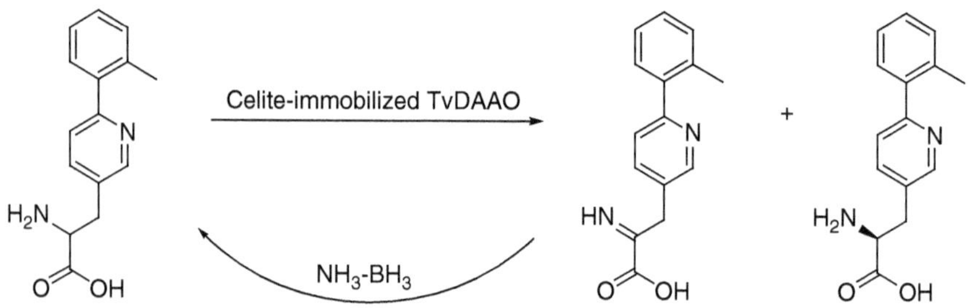

Scheme 5.7 Chemoenzymatic deracemization of 2-amino-3-(6-*o*-tolylpyridin-3-yl) propanoic acid.

intermediate for the synthesis of glucagon-like peptide-1 (GLP-1) mimics or GLP-1 receptor modulators potentially useful for the treatment of type II diabetes [11].

When L-amino acid oxidase was used in the deracemization of DL-α-amino acids, enantiomerically pure D-amino acids should be obtained. Indeed, by using L-amino acid oxidase from *Proteus myxofaciens* and ammonia boranes as chemical reducing agent, a range of D-amino acids were produced in up to 90% yields and >99% ee from their racemates (Scheme 5.8). Amine–boranes are stable in water at neutral or basic pH, soluble and unreactive toward a wide range of protic and aprotic solvents. In addition, the reducing ability was tunable by the complexing amine; thus amine–boranes served as a class of reducing agents with high versatility and selectivity. Pd/C/HCO_2NH_4 could also be used as reducing agent in combination with amino acid oxidases for the deracemization of α-amino acids, although the reduction was slower than amine–boranes [12].

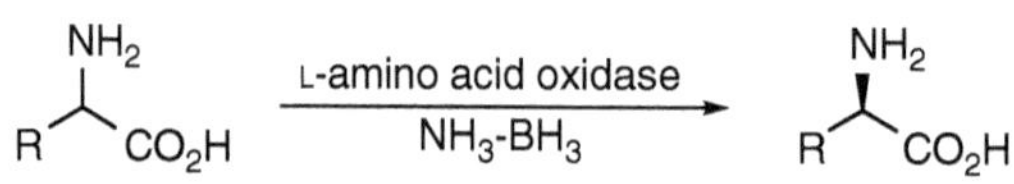

Scheme 5.8 Chemoenzymatic deracemization of α-amino acids to D-amino acids.

In the cyclic deracemization with enantioselective amino acid oxidase and nonselective chemical reducing agent, one enantiomer is transformed to the other one. This technique can also be used in the interconversion of the two enantiomers. By employing porcine kidney D-amino acid oxidase and $NaCNBH_3$, D-proline was stereoinverted to L-proline with >99% yield and >99% ee [9].

(2*R*,3*R*)-β-Methylphenylalanine was successfully converted to its (2*S*,3*R*)-diastereoisomer by using D-AAO from porcine kidney and ammonia borane. Instead, L-AAO from snake venom resulted in the stereoinversion of (2*S*,3*R*)- to (2*R*,3*R*)-β-methylphenylalanine. Similarly, L-isoleucine was transformed to D-*allo*-isoleucine in 87% yield and 99% de in the presence of L-AAO from *Proteus myxofaciens* (Scheme 5.9) [10].

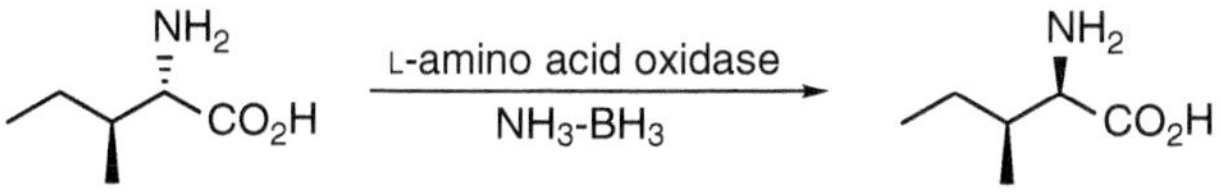

Scheme 5.9 Chemoenzymatic stereoinversion of -isoleucine to D-*allo*-isoleucine. Source: Based on Enright et al. [10].

(2*R*,3*S*) and (2*S*,3*R*)-Diastereoisomers of a series of β-methyl-β-phenylalanine analogs were prepared via asymmetric hydrogenation of the corresponding α,β-unsaturated α-acetamido acid methyl esters by using either [Rh(*R*,*R*)-Et-DuPhos (COD)]BF_4 or [Rh(*S*,*S*)-Et-DuPhos(COD)]BF_4 as the catalyst, respectively, followed by hydrolysis with HCl. These β,β-disubstituted didehydroamino acid esters could be synthesized from L-threonine methyl ester by a few chemical steps. Subsequently, D-AAO from *Trigonopsis variabilis* and snake venom L-AAO were utilized in combination with ammonia borane complex to enable the stereoinversion of (2*R*,3*S*) and (2*S*,3*R*)-diastereoisomers, furnishing the (2*S*,3*S*) and (2*R*,3*R*)-diastereomers, respectively (Scheme 5.10) [13].

Recently, a specific L-amino acid oxidase (RebO) from the actinomycete *Lechevalieria aerocolonigenes* was found to enantioselectively catalyze the oxidation of L-tryptophan to the imine intermediate. It was thus applied to the stereoinversion of various halogenated L-tryptophan derivatives into the corresponding D-tryptophan derivatives via the cyclic deracemization approach. In the presence of an ammonia borane complex D-configured halotryptophans were obtained in ~70% yields and >98% ee under optimized reaction conditions. The halogenated L-tryptophan derivatives could be prepared via enzymatic halogenation of L-tryptophan catalyzed by tryptophan halogenase. A sequential one-pot reaction cascade was then developed by coupling the dynamic stereoinversion with enzymatic halogenation for the synthesis of D-5-bromotryptophan and D-7-bromotryptophan with about 90% conversion and >92% ee (Scheme 5.11) [14].

In addition to chemical reduction, an unselective electrochemical reduction, in combination with selective enzymatic oxidation, led to deracemization and stereoinversion of α-amino acids. L-Leucine was prepared from racemic leucine or D-leucine with D-amino acid oxidase from *Trigonopsis variabilis* as enzyme catalyst in coupling with the electrochemical reduction (Scheme 5.12) [15].

Optically active 1,2,3,4-tetrahydroisoquinoline carboxylic acids are important building blocks in organic synthesis. Recently, a D-amino acid oxidase from *Fusarium solani* M-0718 (FsDAAO) was applied in combination with ammonia borane complex for the deracemization of racemic carboxyl-substituted tetrahydroisoquinolines (Scheme 5.13). Optically pure (*S*)-1,2,3,4-tetrahydroisoquinoline-1-carboxylic

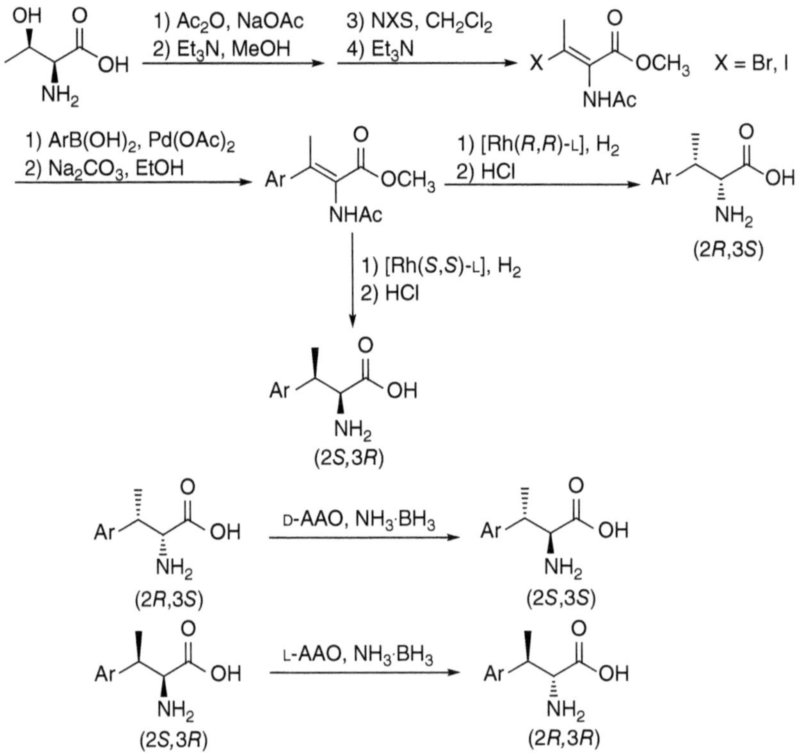

Scheme 5.10 Chemoenzymatic synthesis of four diastereomers of β-methyl-β-phenylalanine analogs. Source: Based on Roff et al. [13].

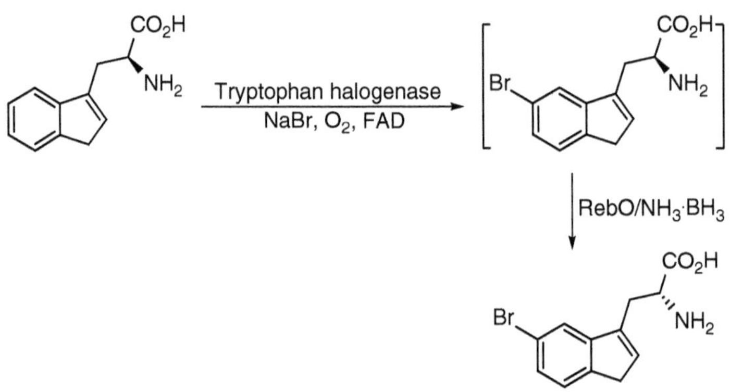

Scheme 5.11 Chemoenzymatic synthesis of D-5-bromotryptophan. Source: Based on Schnepel et al. [14].

acid and (S)-1,2,3,4-tetrahydroisoquinoline-3-carboxylic acid were isolated with 82% and 73% yield, respectively [16]. This D-amino acid oxidase also showed high activity and enantioselectivity toward other D-amino acids such as D-phenylglycine and D-phenylalanine, but no activity toward 1,2,3,4-tetrahydroisoquinoline and 1-methyl-1,2,3,4-tetrahydroisoquinoline.

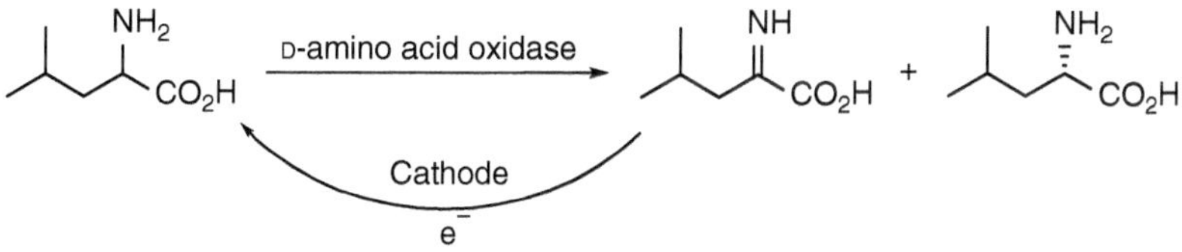

Scheme 5.12 Electrochemoenzymatic deracemization of leucine. Source: Based on Märkle et al. [15].

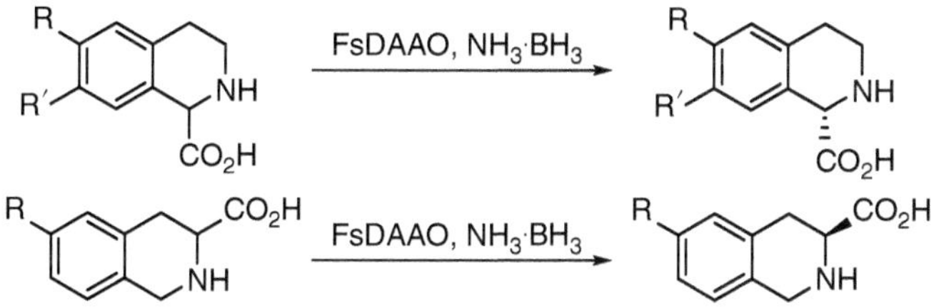

Scheme 5.13 Chemoenzymatic deracemization of 1,2,3,4-tetrahydroisoquinoline-1-carboxylic acids and 1,2,3,4-tetrahydroisoquinoline-3-carboxylic acids.

L-Amino acid deaminase, a membrane-associated and flavin adenine dinucleotide (FAD)-containing enzyme, also catalyzed the stereoselective deamination of L-amino acids to the imino acids, which was spontaneously hydrolyzed to yield the corresponding α-keto acids and ammonia. Recently, L-amino acid deaminase from *Proteus myxofaciens* (PmaLAAD) was produced as soluble protein with a stably bounded flavin cofactor in *Escherichia coli*, and the purified enzyme catalyzed the deamination of a number of natural and synthetic L-amino acids to give the corresponding α-keto acids. Starting from a racemic mixture, D,L-1-naphthylalanine was kinetically resolved into the D-enantiomer with this deaminase. Moreover, L-4-nitrophenylalanine was completely stereoinverted into the D-enantiomer by using PmaLAAD and a small molar excess of borane *tert*-butylamine complex for the *in situ* reduction of the imino acid intermediate (Scheme 5.14) [17].

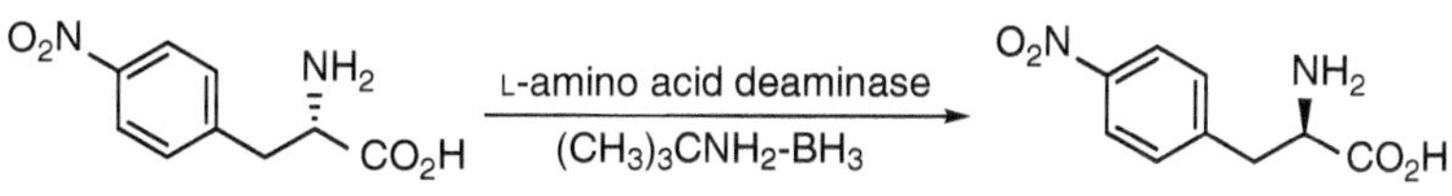

Scheme 5.14 Chemoenzymatic stereoinversion of L-4-nitrophenylalanine. Source: Based on Rosini et al. [17].

This deracemization strategy of concurrent enantioselective oxidation and nonselective reduction has been then extended to the deracemization of simple amines by employing a mutant of the Type II monoamine oxidase from *Aspergillus niger* (MAO-N). The mutant (Asn336Ser) was obtained by a high-throughput colorimetric screen of a random mutagenesis library of the wild-type enzyme with (*S*)-α-methylbenzylamine as the target substrate. After screening a range of reducing agents such as sodium borohydride, Pd/C/HCO_2NH_4, amine–borane complexes, and ammonia borane complex, ammonia borane complex was identified as the optimal reducing agent. Deracemization of racemic α-methylbenzylamine was performed to give (*R*)-α-methylbenzyamine in 77% yield and 93% ee (Scheme 5.15) [18].

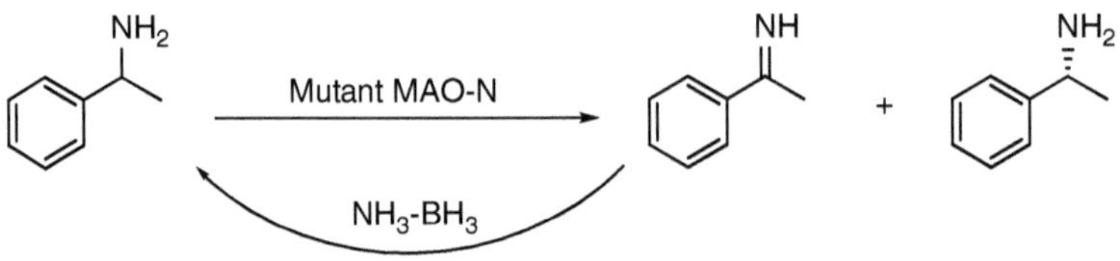

Scheme 5.15 Chemoenzymatic deracemization of α-methylbenzylamine. Source: Based on Alexeeva et al. [18].

Substrate profile studies showed that the variant enzyme possessed broad substrate specificity and high *S*-enantioselectivity toward the primary amines with the amino group flanked by a methyl group and a bulky alkyl/aryl group. However, it exhibited low activity toward chiral secondary amines such as 1-methyltetrahydroisoquinoline, not permitting efficient preparative deracemization reactions [19]. Further random mutagenesis and library screening using 1-methyltetrahydroisoquinoline as the substrate led to the identification of a variant enzyme (Asn336Ser/Ile246Met) with improved activity and high enantioselectivity. This new variant was applied to the deracemization of racemic 1-methyltetrahydroisoquinoline on a preparative scale to yield (*R*)-enantiomer in 71% isolated yield and 99% ee (Scheme 5.16) [20]. When the immobilized enzyme was used, the product (*R*)-1-methyltetrahydroisoquinoline could be isolated in 95% yield. The preparative deracemization of racemic 2-phenylpyrroldine was also achieved at a substrate concentration of 100 mm (14.7 g/l) with the immobilized enzyme and ammonia borane to give (*R*)-2-phenylpyrrolidine in 80% yield and 98% ee (Scheme 5.17).

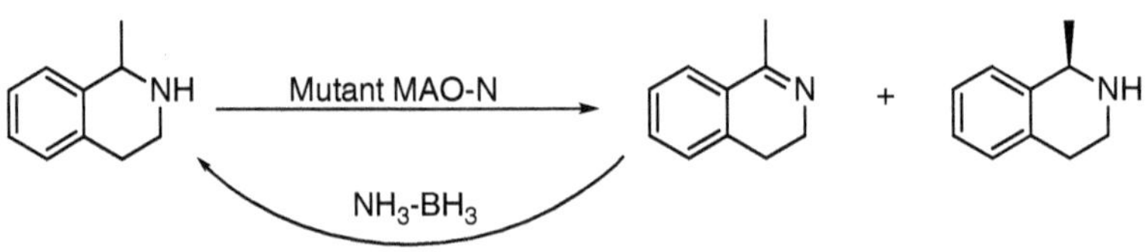

Scheme 5.16 Chemoenzymatic deracemization of 1-methyltetrahydroisoquinoline. Source: Based on Carr et al. [20].

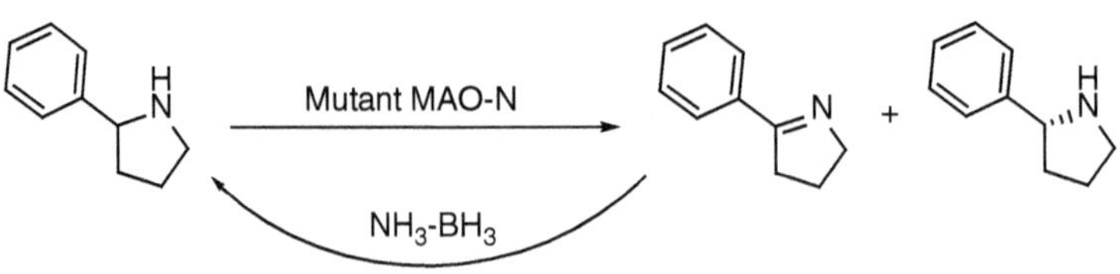

Scheme 5.17 Chemoenzymatic deracemization of 2-phenylpyrroldine.

One of these variants, I246M/N336S/M348K/T384N/D385S (MAO-N-D5), also showed high activity toward some chiral tertiary amines, especially those containing a pyrrolidine ring flanked by bulky aryl groups. The racemate of *N*-methyl-2-phenylpyrrolidine was deracemized by coupling this mutant amine oxidase with ammonia borane in a one-pot process, yielding (*R*)-*N*-methyl-2-phenylpyrrolidine in 75% isolated yield and 99% ee (Scheme 5.18) [21]. The MAO-N-D5 variant was also able to mediate the efficient deracemization of (±)-2-propylpiperidine (coniine) to furnish the (*R*)-enantiomer in 90% ee (Scheme 5.19) [22]. The piperidine alkaloid coniine is a potent neurotoxin isolated from the plant *Conium maculatum* (poison hemlock) and the 2-substituted piperidine is a common structural motif found in many alkaloid natural products.

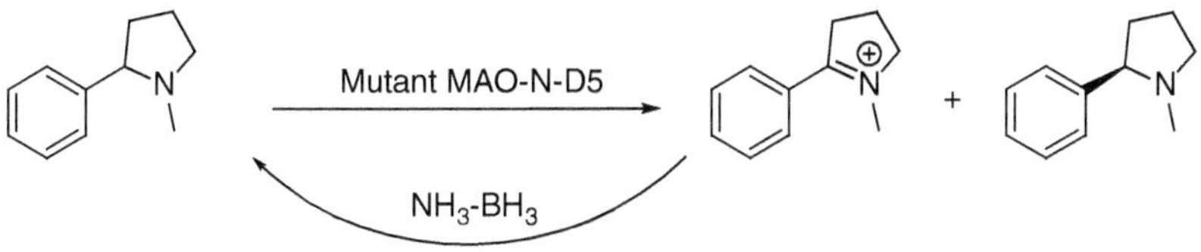

Scheme 5.18 Chemoenzymatic deracemization of *N*-methyl-2-phenylpyrroldine. Source: Based on Dunsmore et al. [21].

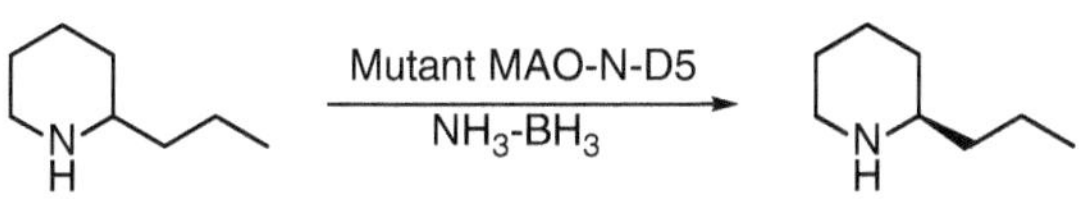

Scheme 5.19 Chemoenzymatic deracemization of (±)-2-propylpiperidine. Source: From Ghislieri et al. [22].

The MAO-N-D5 variant also showed good enantioselectivity toward 2-aryl azepanes and was applied to the kinetic resolution affording the corresponding (*R*)-2-aryl azepanes with up to 99% ee. However, the deracemization of 2-aryl azepanes using MAO-N mutant enzymes in combination with ammonia borane complex failed, due to the ring opening of the imine intermediates generating the amino ketones, which were reduced to the amino alcohols (Scheme 5.20). But 1-methyl-2,3,4,5-tetrahydro-1*H*-benzo[c]azepine can be deracemized to give the

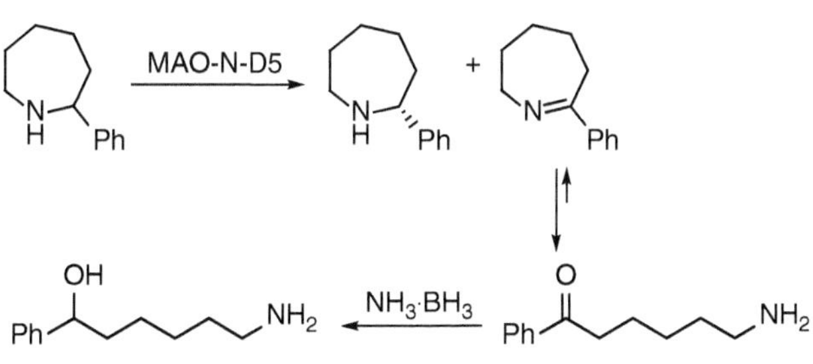

Scheme 5.20 Chemoenzymatic reactions of phenyl azepane with MAO-N mutant and ammonia borane.

(*R*)-enantiomer in 70% yield and 94% ee or 90% yield and 98% ee by utilizing MAO-N mutant D9 or D11, respectively (Scheme 5.21) [23].

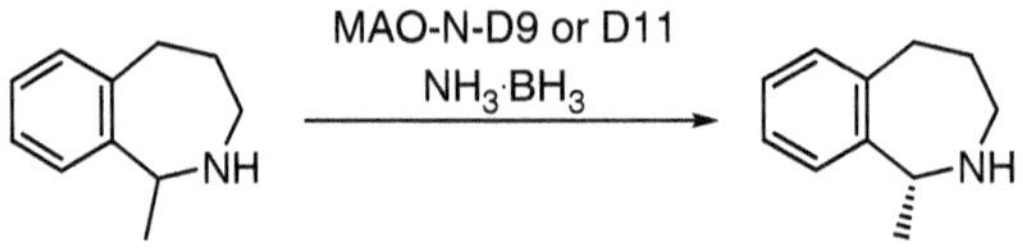

Scheme 5.21 Chemoenzymatic deracemization of 1-methyl-2,3,4,5-tetrahydro-1*H*-benzo[c]azepine. Source: From Zawodny et al. [23].

The alkaloid (±)-crispine A and one of its deoxygenated analogs are also good substrates for the variant amine oxidase MAO-N-D5, and their deracemization reactions produced the biologically active (*R*)-stereoisomer with 97% ee [24]. However, the deracemization of (±)-crispine A required 40 hours to proceed to completion. Modification of the active site of MAO-N-D5 resulted in a new mutant MAO-N-D9 (F210L, L213T, M242Q, M246T, W430H) with about 990-fold higher specific activity than that of the parent MAO-N-D5 variant, allowing the complete deracemization of the alkaloid crispine A within two hours to give (*R*)-crispine A in >97% ee (Scheme 5.22) [25].

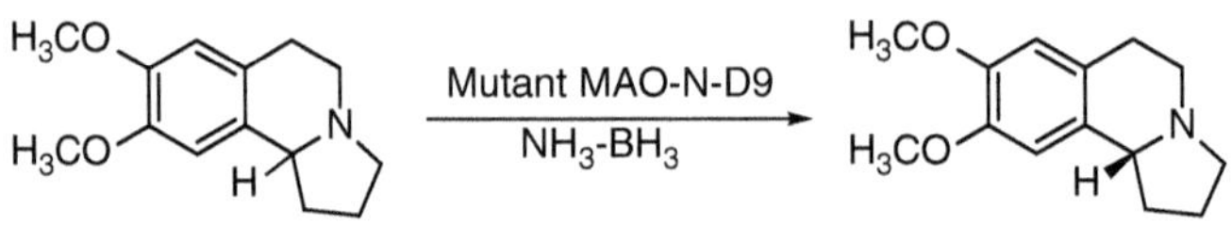

Scheme 5.22 Chemoenzymatic deracemization of (±)-crispine A. Source: Based on Rowles et al. [25].

When W430H in MAO-N-D9 variant was replaced with W430G, the new variant (MAO-N-D11) showed high activity and enantioselectivity for the oxidation of (*S*)-4-chlorobenzhydrylamine. Interestingly, the MAO-N-D11 variant showed high (*R*)-selectivity toward 1-phenyltetrahydroisoquinoline, which was in contrast to the (*S*)-selectivity observed for all previously reported substrates for MAO-N. By using MAO-N-D11 and NH_3-BH_3, deracemization of racemic 4-chlorobenzhydrylamine resulted in the formation of (*R*)-enantiomer, a precursor for the synthesis of Levocetirizine, in 45% isolated yield and 97% ee (Scheme 5.23). The low isolated yield was due to partial hydrolysis of the intermediate imine to the ketone during the deracemization process. Much higher yield (90%) was

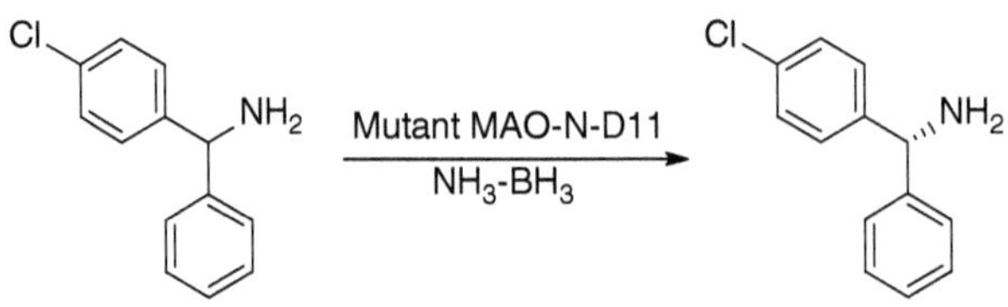

Scheme 5.23 Chemoenzymatic deracemization of 4-chlorobenzhydrylamine.

obtained in the deracemization of racemic 1-phenyltetrahydroisoquinoline to the (*S*)-enantiomer (98% ee), an important intermediate for the synthesis of Solifenacin (Scheme 5.24) [22].

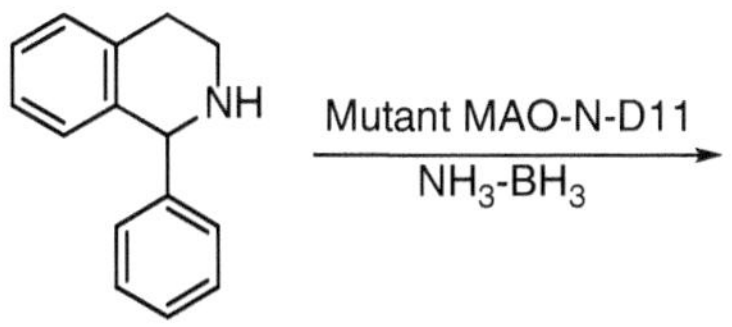

Scheme 5.24 Chemoenzymatic deracemization of 1-phenyltetrahydroisoquinoline. Source: Based on Ghislieri et al. [22].

The stereochemical switch to preference for the (*R*)-enantiomer also applied to the substrate benzylisoquinolines. MAO-N-D11 was found to oxidize the (*R*)-enantiomer of all the tested benzylisoquinoline substrates, and the substitution pattern on the isoquinoline ring exerted some effects on the enantioselectivity. The (*S*)-enantiomers of some 1-benzylisoquinolines were prepared from the corresponding racemates in high isolated yield (up to 85%) and excellent optical purity (>97% ee) via the one-pot deracemization protocol using MAO-N-D11 and ammonia borane (Scheme 5.25) [26].

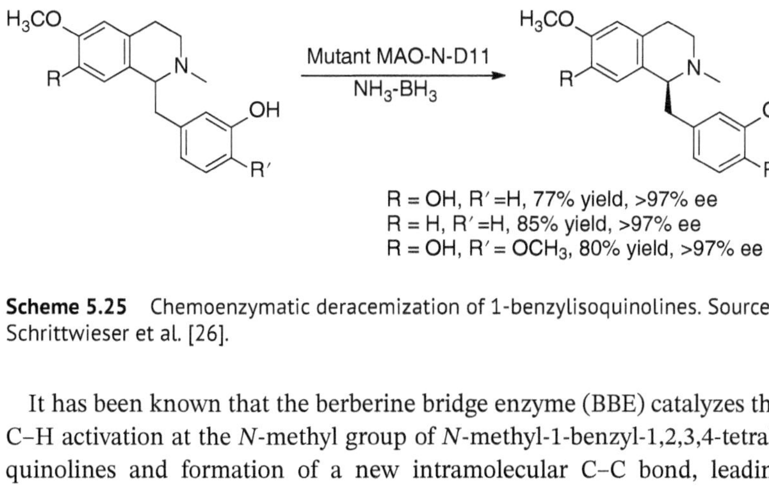

Scheme 5.25 Chemoenzymatic deracemization of 1-benzylisoquinolines. Source: Based on Schrittwieser et al. [26].

It has been known that the berberine bridge enzyme (BBE) catalyzes the aerobic C–H activation at the *N*-methyl group of *N*-methyl-1-benzyl-1,2,3,4-tetrahydroisoquinolines and formation of a new intramolecular C–C bond, leading to the kinetic resolution of *N*-methyl-1-benzyl-1,2,3,4-tetrahydroisoquinolines to yield (*S*)-berbines and the unreacted (*R*)-*N*-methyl-1-benzyl-1,2,3,4-tetrahydroisoquinolines. The deracemization of benzylisoquinolines can be coupled with the BBE-catalyzed cyclization to synthesize the optically pure (*S*)-berbines. Since BBE is inhibited by the ammonia borane complex, *t*-$BuNH_2$-BH_3, Me_3N-BH_3, and morpholine-BH_3 were tested as alternative reductants, and morpholine-BH_3 worked best for the deracemization and was compatible with BBE. Therefore, (*R*)-*N*-methyl-1-benzyl-1,2,3,4-tetrahydroisoquinolines were effectively stereoinverted to the (*S*)-stereomer by employing MAO-N-D11 and morpholine-BH_3. Two optically pure (*S*)-berbines were produced at preparative scale via the one-pot processes by incorporating the stereoinversion reaction with the BBE-catalyzed cyclization in either sequence or concurrent manner (Scheme 5.26) [27].

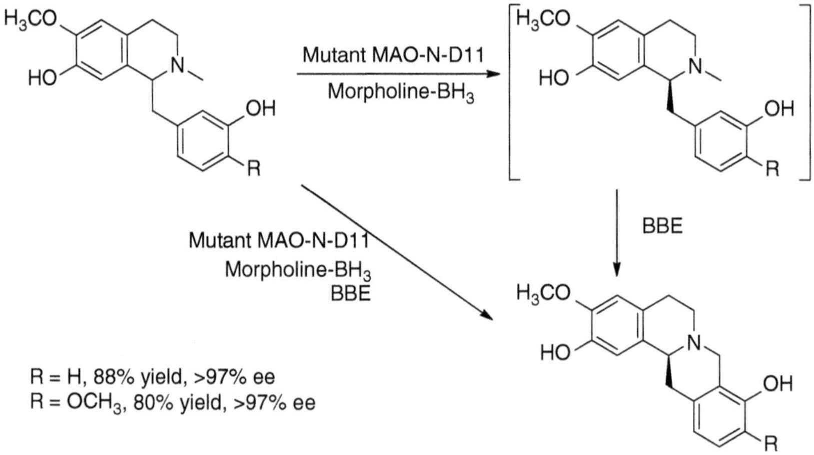

Scheme 5.26 One-pot sequence or concurrent synthesis of optically pure (*S*)-berbines. Source: Based on Schrittwieser et al. [27].

The tetrahydro-β-carboline (THBC) ring system is an important structural motif of bioactive alkaloid natural products. The enantiomerically pure β-carbolines could be prepared via deracemization of the racemates employing the D9 or D11 variants of monoamine oxidase MAO-N in combination with a nonselective chemical reducing agent. Eleagnine is an alkaloid isolated from *Chrysophyllum albidum* with potent analgesic, anti-inflammatory, and weak antioxidant properties. Leptaflorin has been isolated from *Peganum harmala* and from *Leptactina densiflora*, which is a known psychedelic. Both (*R*)-Eleagnine and (*R*)-Leptaflorin were synthesized in 99% ee by employing MAO-N-D9 in coupling with ammonia borane complex for deracemization of the corresponding racemates (Scheme 5.27). Interestingly, a switch from (*S*) to (*R*) in enantioselectivity of the MAO-N variants D9 and D11 was observed as the C-1 substituent of the THBC was varied from methyl and ethyl group to a larger alkyl or phenyl group (Scheme 5.28) [28].

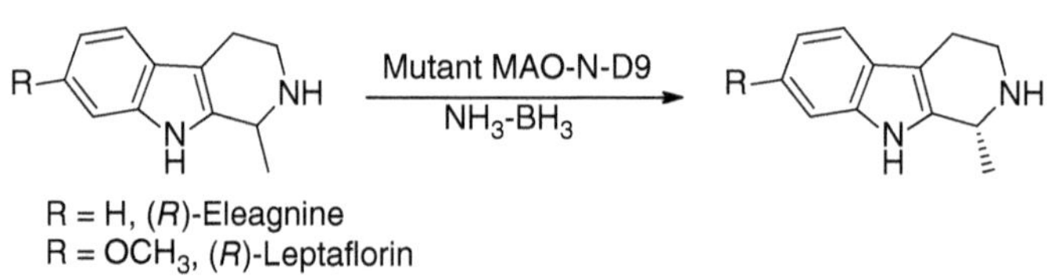

Scheme 5.27 One-pot synthesis of (*R*)-Eleagnine and (*R*)-Leptaflorin via deracemization.

Harmicine is an alkaloid natural product isolated from the Malaysian plant *Kopsia griffithii* with strong anti-*Leishmania* activity. The naturally occurring (*R*)-harmicine was prepared by two routes involving the deracemization of the racemic harmicine

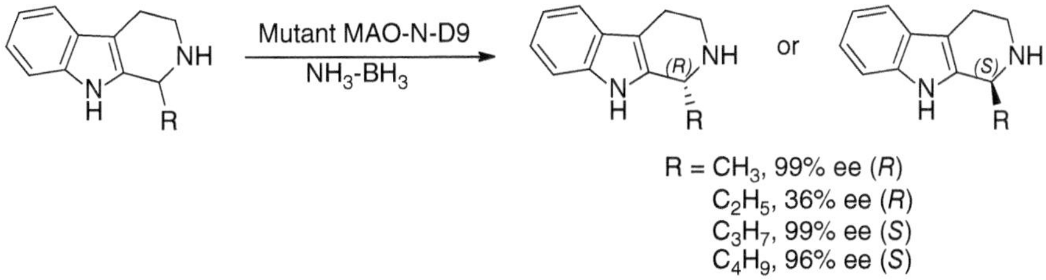

Scheme 5.28 Deracemization of C-1 substituted tetrahydro-β-carbolines. Source: From Ghislieri et al. [28].

by MAO-N-D9/ammonia borane system as the key step. The overall transformation from pyrrolidine to (*R*)-harmicine was the shortest route reported back in 2013 and it represented an oxidative asymmetric Pictet–Spengler reaction under aqueous conditions using molecular oxygen as the stoichiometric oxidant (Scheme 5.29) [22].

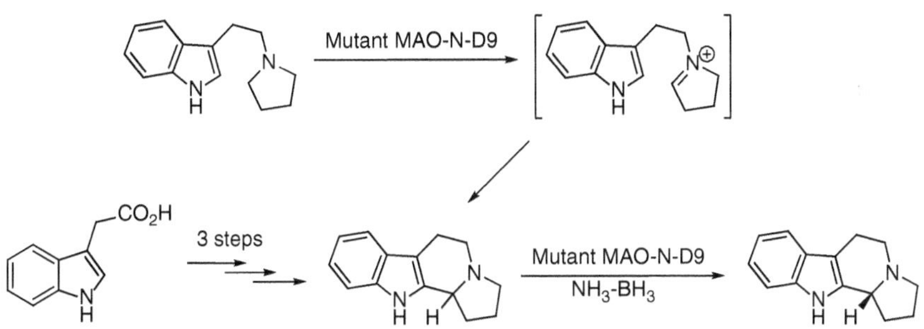

Scheme 5.29 Chemoenzymatic synthesis of (*R*)-harmicine via deracemization as the key step. Source: Based on Ghislieri et al. [22].

The above discussion demonstrated that monoamine oxidase variants from MAO-N displayed remarkable substrate scope and tolerance for sterically demanding motifs and served as an effective toolbox for the production of enantiomerically pure chiral amines. Recently, 132 racemic chiral amines of diverse structures, including α-substituted methylbenzylamines, benzhydrylamines, 1,2,3,4-tetrahydronaphthylamines (THNs), indanylamines, allylic and homoallylic amines, and propargyl amines, were screened against the MAO-N variants D5, D9, and D11. Among these variant enzymes, MAO-N-D9 exhibited the highest activity for most of the tested substrates. The deracemizations of a set of selected primary amines were performed with MAO-N-D9 in combination with ammonia borane complex, yielding the corresponding (*R*)-amines in excellent enantioselectivity (>99% ee) with moderate to good yields (55–80%) [29].

As an alternative to the fungal-sourced enzyme (MAO-N), a cloned cyclohexylamine oxidase (CHAO) from *Brevibacterium oxydans* IH-35A was tested for the selective oxidation of structurally different amines including cycloalkyl primary amines, alkyl aryl amines, and α-carbon-substituted aliphatic amines, all of which were found to be suitable substrates. The deracemizations of several amines were carried out by either recombinant whole cells or crude enzyme extracts in combination with a borane–ammonia complex as reducing agent, and the corresponding (*R*)-amines were obtained with excellent enantiomeric ratios (>99 : 1) and good isolated yields (62–75%) [30].

The substrate profiles of CHAO and five of its single-amino acid substitution mutants (L199A, M226A, Y321A, Y321F, and L353M) were further investigated with 38 structurally diverse amines. The wild-type (wt) CHAO and variant L353M when applied in combination with ammonia borane complex as reducing agent to the deracemization of 1-aminotetraline, gave the (*R*)-enantiomer, a precursor of an antidepressant drug Norsertraline, in good yield (73–76%, Scheme 5.30) [31].

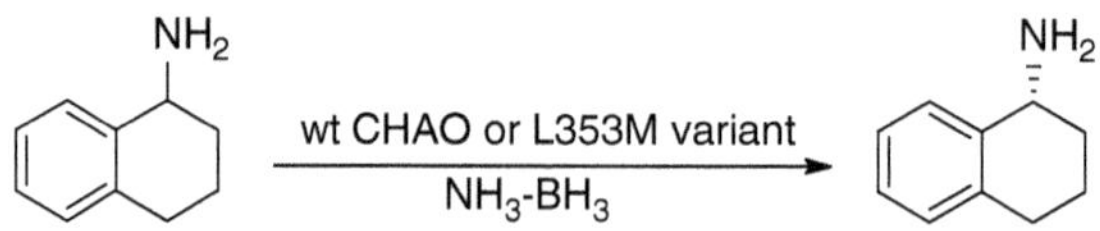

Scheme 5.30 Deracemization of 1-aminotetraline. Source: Based on Li et al. [31].

Although the wt CHAO exhibited high substrate specificity toward alicyclic amines and *sec*-alkylamines, it had virtually no activity toward secondary amines. Rational structure-guided engineering of the CHAO protein with iterative saturation mutagenesis was thus performed, and functional screening of the resulting libraries using 2-methyl-1,2,3,4-tetrahydroquinoline (2-methyl-THQ) as a substrate led to several mutants that exhibited high activity toward 2-substituted THQs with high enantioselectivity. Deracemization of 2-methyl-THQ was achieved by using *E. coli* whole cells of the triple mutant T198FL199SM226F in combination with $NH_3{\cdot}BH_3$, yielding (*R*)-2-methyl-THQ in 76% isolated yield with 98% ee (Scheme 5.31) [32]. Some other 2-substituted-THQs were also deracemized by employing L225A, Q233A, or Y459T variant to give the corresponding chiral 2-substituted-THQ derivatives with high ee values (up to 99%) and isolated yield in the range of 58–92%. Interestingly, the L225A mutants reversed the enantiopreference for 2-isopropyl-THQ and 2-cyclopropyl-THQ (Scheme 5.32) [33].

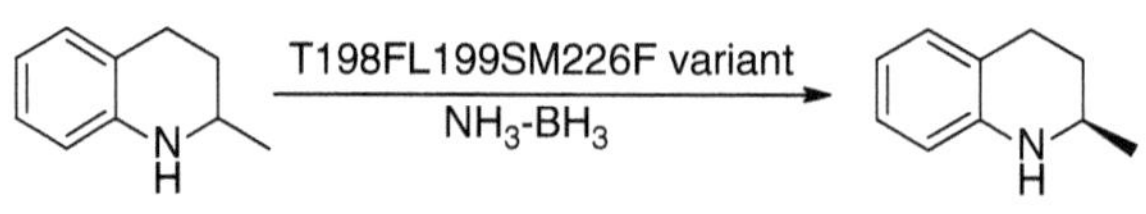

Scheme 5.31 Deracemization of 2-methyl-THQ. Source: Based on Li et al. [32].

Phenyl-substituted 2-(2-phenylethyl)THQ derivatives could be synthesized through a convergent one-pot Rh(I)-catalyzed addition/condensation sequence of alkyl vinyl ketones and aminophenylboronic acids. The racemic mixture was deracemized by MAO-N-D9 variant/ammonium borane system, delivering the corresponding *R*-enantiomers in 50–99% ee (Scheme 5.33) [34]. The deracemization of racemic 2-methyl-THQ derivatives with substituent on the benzene ring was achieved by employing a monoamine oxidase (MAO5) from *Pseudomonas monteilii*

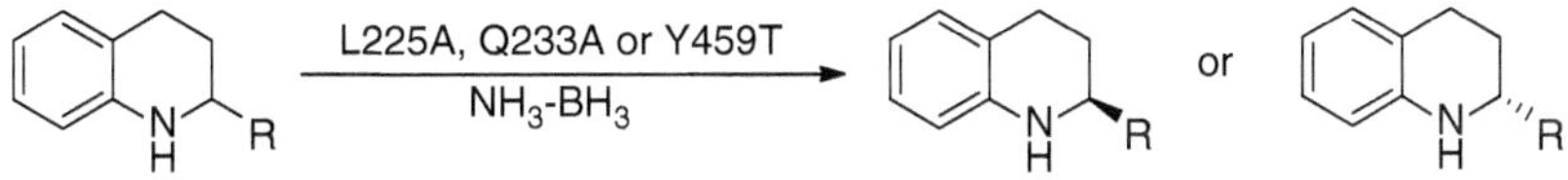

R =				
	isopropyl	L225		7.2% ee
		Q233A	95.6% ee	
	cyclopropyl	L225A		91.3% ee
		Q233A	>99% ee	
	allyl	L225A	93.8% ee	
	pentyl	L225A	>99% ee	
	benzyl	L225A	98.8% ee	
	2-(*m*,*p*-dimethoxyphenylethyl)	Y459T	94.2% ee	

Scheme 5.32 Deracemization of 2-substituted-THQ derivatives. Source: Based on Yao et al. [33].

ZMU-T01 as the biocatalyst to prepare the corresponding *R*-enantiomer with up to >99 % ee (Scheme 5.34) [35].

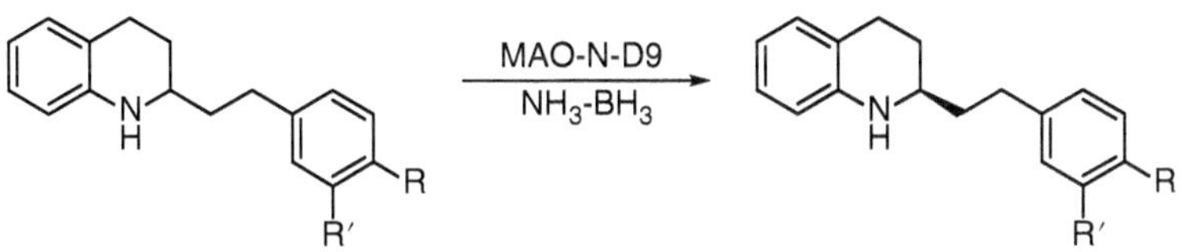

R = R′ = H; R = R′ = OCH_3; R = OCH_3, R′ = OH; R = OH, R′ = OCH_3

Scheme 5.33 Deracemization of phenyl-substituted 2-(2-phenylethyl)THQ derivatives. Source: Based on Cosgrove et al. [34].

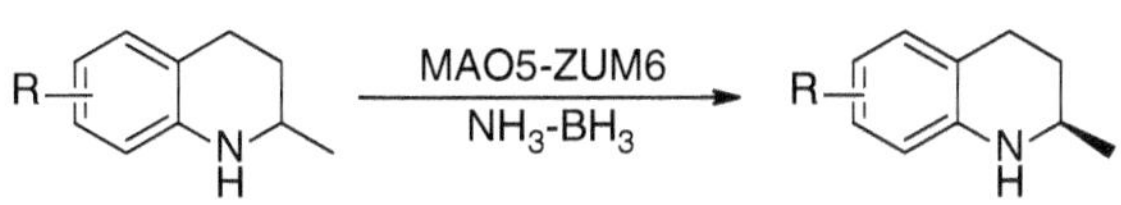

Scheme 5.34 Deracemization of 2-methyl-1,2,3,4-tetrahdroquinoline derivatives. Source: Based on Deng et al. [35].

More CHAO variants suitable for the deracemization of secondary amines have been identified by orthogonally assaying designed mutants with a library of structurally diverse substrates. Particularly, mutants Y321I and Y321I/M226T displayed high catalytic efficiency toward 1-(4-methoxybenzyl)-1, 2, 3, 4, 5, 6, 7, 8-octahydroisoquinoline and (*S*)-*N*-(prop-2-yn-1-yl)-2, 3-dihydro-1*H*-inden-1-amine. Their deracemizations were thus carried out with crude enzyme extracts of the respective mutants, affording (*R*)-*N*-(prop-2-yn-1-yl)-2, 3-dihydro-1*H*-inden-1-amine in 51% isolated yield and 93% ee, and (*S*)-1-(4-methoxybenzyl)-1, 2, 3, 4, 5, 6, 7, 8-octahydroisoquinoline in 78% isolated yields and 99% ee, respectively (Schemes 5.35 and 5.36) [36]. The CHAO variant Y321I/M226T was also applied to the chemoenzymatic deracemization of racemic valine ethyl ester or inversion of L-valine ethyl ester. In both cases D-valine ethyl ester was prepared in high yield (up to 95 %) with excellent optical purity (>99 % ee, Scheme 5.37). Interestingly, CHAO and its variants showed opposite enantiopreference for valine ethyl ester and phenylalanine ethyl ester [37].

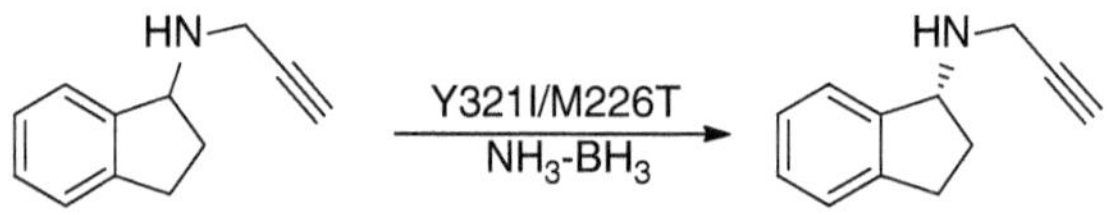

Scheme 5.35 Deracemization of *N*-(prop-2-yn-1-yl)-2, 3-dihydro-1*H*-inden-1-amine. Source: Based on Li et al. [36].

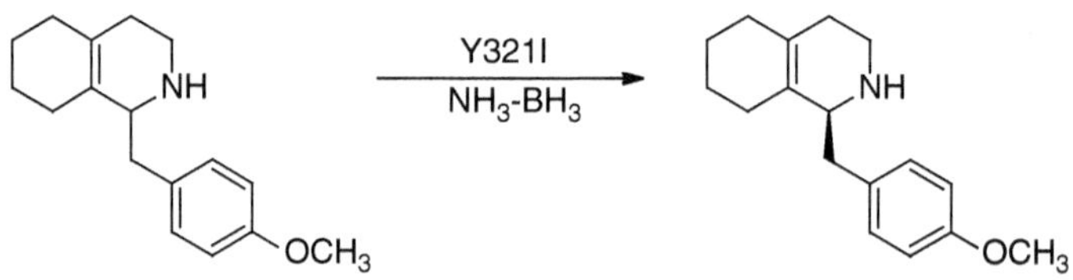

Scheme 5.36 Deracemization of 1-(4-methoxybenzyl)-1, 2, 3, 4, 5, 6, 7, 8-octahydroisoquinoline. Source: Based on Li et al. [36].

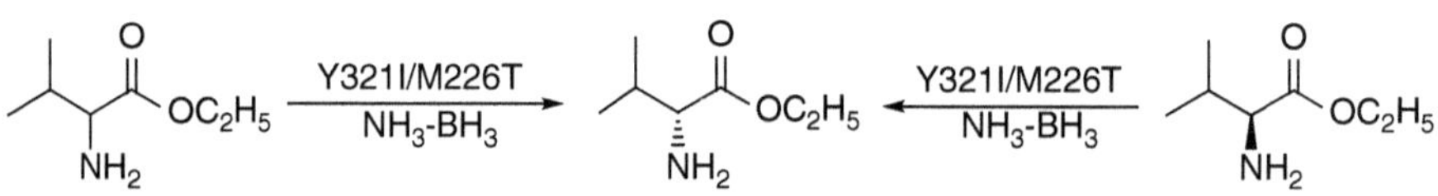

Scheme 5.37 Deracemization of racemic valine ethyl ester or inversion of L-valine ethyl ester.

Recently, a new CHAO enzyme from *Erythrobacteraceae bacterium* CCH12-C2 was reported to carry out chemoenzymatic deracemization of 1-benzyl-3,4,5,6,7,8-hexahydroisoquinoline (1-benzyl-HHIQ) derivatives, affording the corresponding (*S*)-configurated 1-benzyl-HHIQ derivatives in up to 91% yields and 99% ee at semipreparative scale (Scheme 5.38) [38].

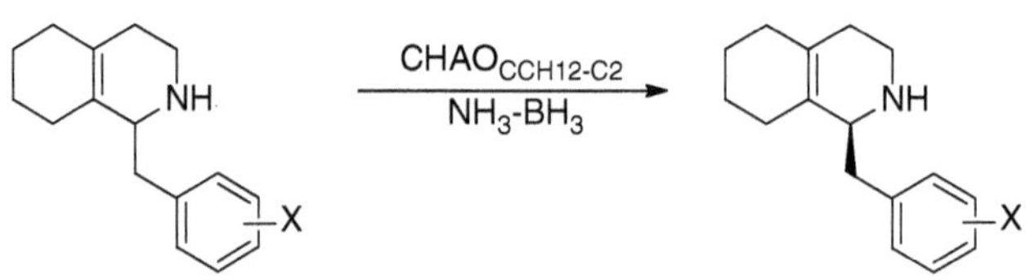

Scheme 5.38 Deracemization of 1-benzyl-3,4,5,6,7,8-hexahydroisoquinoline derivatives. Source: Based on Wu et al. [38].

The above monoamine oxidases (MAO-N, CHAO, and MAO5) and their mutants have been demonstrated to be useful biocatalysts for the synthesis of enantiomerically pure primary, secondary, and tertiary chiral amines via the Turner chemoenzymatic deracemization strategy. However, these enzymes were highly *S*-selective although some variants showed *R*-selectivity toward certain secondary amines. An enantiocomplementary *R*-selective amine oxidase was developed by engineering a double mutant of 6-hydroxy-D-nicotine oxidase (6-HDNO). The mutant, 6-HDNO E350L/E352D enzyme, exhibited broadened substrate scope and high enantioselectivity, and when applied to the deracemization of a range of racemic amines such as 2-substituted piperidines, *S*-configured products were obtained in high ee (Scheme 5.39) [39].

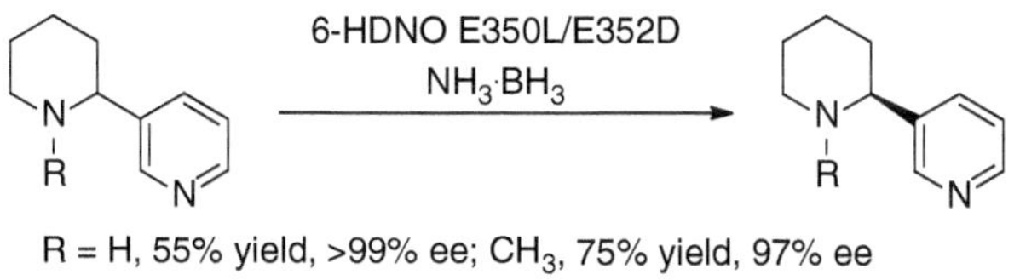

Scheme 5.39 Deracemization of 2-substituted piperidines. Source: Based on Heath et al. [39].

Porcine kidney D-amino acid oxidase (pkDAO) had also been evolved into *R*-stereoselective amine oxidase by directed evolution. The resulting variants gave a significantly altered substrate profile toward *R*-chiral amines. Racemic α-methylbenzylamine was deracemized into the (*S*)-enantiomer [40]. Based on the X-ray crystallographic analyses of pkDAO and its variants, a new double mutant was rationally created for the oxidation of (*S*)-chiral amines with bulky substituents. The variant enzyme (I230A/R283G) catalyzed the deracemization of racemic 4-Cl-benzhydrylamine (CBHA) in the presence of reducing agents such as $NaBH_4$ in water, yielding (*R*)-CBHA with 96 % ee (Scheme 5.40) [41].

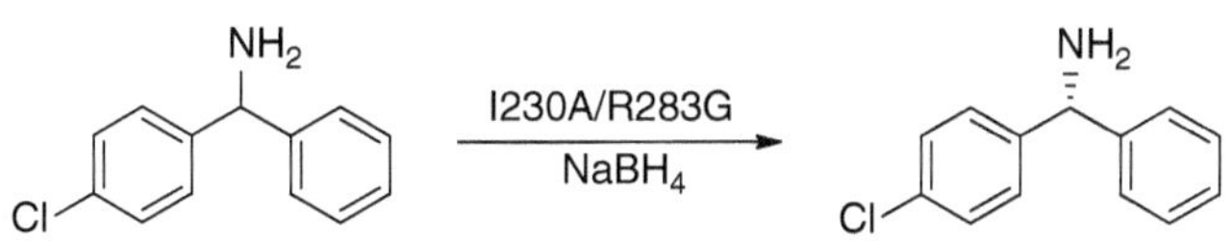

Scheme 5.40 Deracemization of 4-Cl-benzhydrylamine. Source: Based on Yasukawa et al. [41].

Recently, an NADPH-dependent reductive aminase from *Aspergillus oryzae* (AspRedAm) was identified to catalyze the reductive amination of ketones generating the corresponding *R*-configured chiral amines. RedAms belong to the imine reductase enzyme family and also catalyze the reverse oxidation of amines to imines. AspRedAm could catalyze the kinetic resolution of various racemic secondary and primary amines to give *S*-configured amines with up to 99% ee in the presence of an NADPH oxidase (NOX) for recycling of the cofactor. The AspRedAm-NOX system was thus applied in combination with ammonia borane to enable the deracemization of racemic amines. A number of racemic cyclic and acyclic amines including rasagiline and salsolidine were deracemized to (*S*)-configurated amines with up to >98% conversions and >99% ee (Scheme 5.41). An AspRedAm mutant W210A was obtained by enzyme engineering with the opposite enantiopreference, enabling deracemization to give the (*R*)-enantiomers [42].

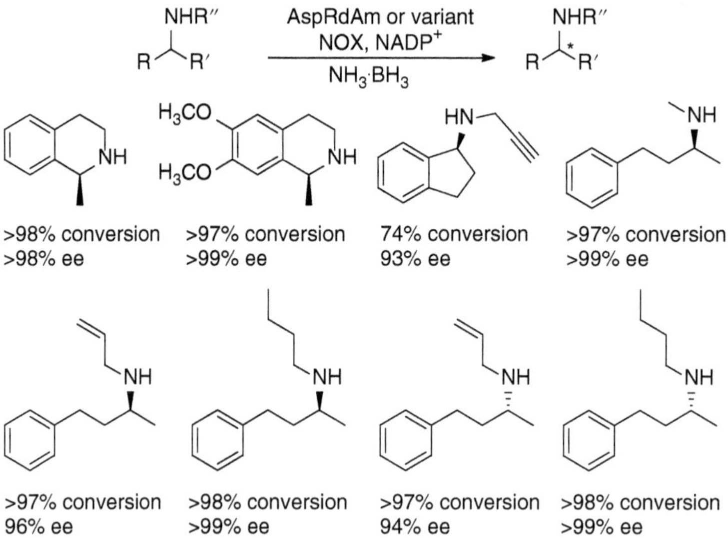

Scheme 5.41 AspRedAm-NOX system in combination with ammonia borane for the deracemization of racemic amines.

In addition to the chemical imine reduction, visible-light-driven reduction of imines has also been incorporated with amine oxidase-catalyzed enantioselective oxidation of amines for the deracemization of racemic cyclic amines. Excitation of water-soluble photosensitizer trisodium *fac*-tris[2-(5′-sulfonatophenyl)pyridine] iridate $Na_3[Ir(sppy)_3]$ [sppy = 2-(5′-sulfonatophenyl)pyridine] promoted electron transfer to a cyclic imine, generating a highly reactive α-amino alkyl radical that was transformed into the corresponding amines by hydrogen atom transfer (HAT) from ascorbic acid. The resulting racemic amines were then converted to the enantioenriched amines by the cyclic deracemization comprising of the monoamine oxidase (MAO-N-D9)-catalyzed oxidation of amine and the photoreduction of imine (Scheme 5.42) [43].

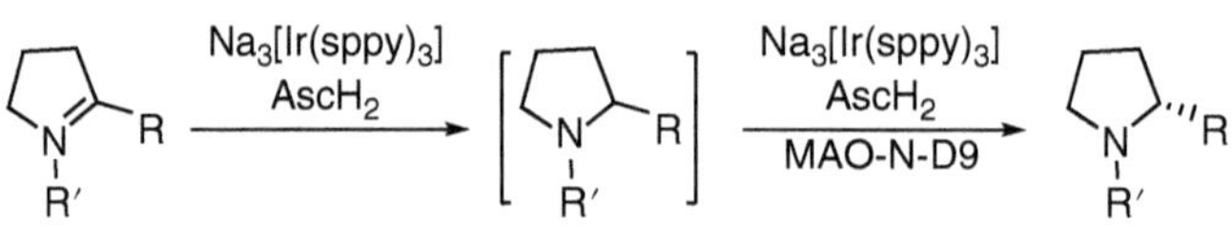

R′ = H, R = cyclohexyl, 92% yield, >99% ee
R′ = H, R = *n*-butyl, >95% yield, 98% ee
R′ = H, R = benzyl, >95% yield, 8% ee
R′ = methyl, R = cyclohexyl, >95% yield, >99% ee

Scheme 5.42 One-pot concurrent photochemoenzymatic asymmetric synthesis of chiral amines. Source: Based on Guo et al. [43].

5.2 Deracemization of Hydroxy Acids and Alcohols

The above cyclic deracemization strategy has also been used to convert racemic lactate or L-lactate into D-lactate. In this case, coupling of L-specific oxidation of the racemate with L-lactate oxidase and non-enantiospecific reduction of pyruvate to DL-lactate with sodium borohydride enabled the "one-pot single-step" deracemization of lactate to furnish D-lactate in 99% yield with more than 99% ee (Scheme 5.43). DL-α-Hydroxybutyrate was converted to D-α-hydroxybutyrate in the same way, though at a slower rate [44]. L-Lactic acid was also inverted into D-lactic acid via oxidation of L-lactate by L-lactate dehydrogenase and the electrochemical reduction of the pyruvate formed with simultaneous cofactor-recycling (Scheme 5.44) [45].

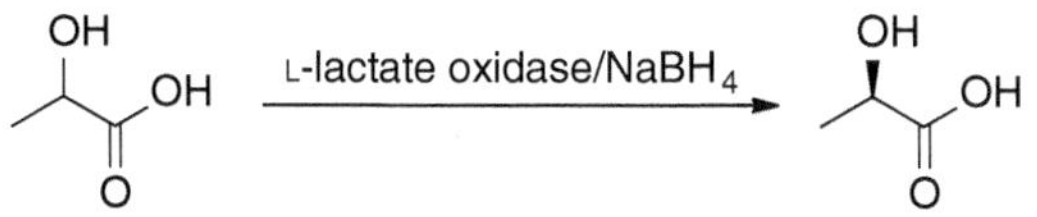

Scheme 5.43 Deracemization of lactic acid.

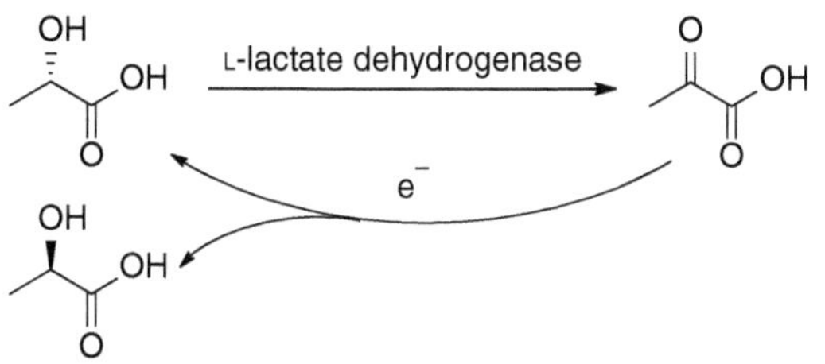

Scheme 5.44 Electrochemoenzymatic stereoinversion of L-lactic acid. Source: Based on Biade et al. [45].

The immobilized cells of *Geotrichum candidum* on water-absorbing polymer BL-100 catalyzed the enantioselective oxidation of secondary alcohols in an ionic liquid [emim][BF_4], yielding the optically active (*R*)-configurated alcohols. The *Geotrichum candidum*-catalyzed kinetic resolution of racemic secondary alcohols was combined with $NaBH_4$ reduction of the *in situ*-generated ketones, and the reactions were carried out in the mixture of 2-(*N*-morpholino)ethanesulfonic acid (MES) buffer solution and ionic liquid, resulting in deracemization of the racemic alcohols to give the corresponding (*R*)-alcohols in up to 96% yields and 99% ee (Scheme 5.45) [46].

Similarly, the plant *Gardenia jasminoides* cells also stereoselectively oxidized the (*S*)-isomers of secondary alcohols in Tris–HCl buffer. This kinetic resolution was applied to a chemoenzymatic deracemization of racemic alcohols using the immobilized *G. jasminoides* cells in calcium alginate beads in combination with $NaBH_4$ for the reduction of the resulting ketone intermediates. The sequential repetition of biocatalytic oxidation and chemical reduction in one-pot afforded the corresponding (*R*)-alcohols in 82–90% yields and 71–96% ee [47].

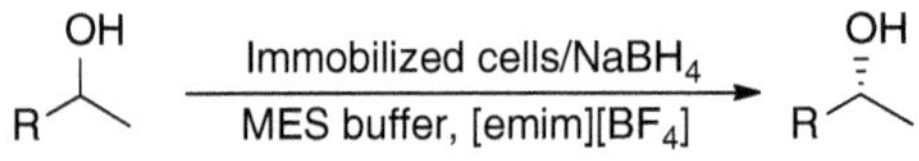

R = Aryl, Alkyl

Scheme 5.45 Deracemization of racemic secondary alcohols. Source: Based on Tanaka et al. [46].

Recently, the *E. coli* cells expressing the heme acquisition system A (HasA)-type hemophore from *Pseudomonas fluorescens* (HasApf) was found to catalyze the enantioselective oxidation of the *R*-enantiomer of 1-(6-methoxynaphthalen-2-yl)ethanol using dioxygen as the oxidant, resulting in the kinetic resolution of the racemate to give the *S*-enantiomer and the ketone. This enantioselective enzymatic oxidation was then combined with nonselective reduction of the *in situ*-formed ketone using $NaBH_4$ in aqueous medium. Racemic 1-(6-methoxynaphthalen-2-yl)ethanol was thus transformed into (*S*)-1-(6-methoxynaphthalen-2-yl)ethanol in >90% yield and >99% ee (Scheme 5.46) [48]. HasApf was immobilized on porous ceramic particles (ImHasApf), and evaluated for the kinetic resolution of racemic 1-(6-methoxynaphthalen-2-yl)ethanol and other secondary alcohols such as 1-(2-naphthyl)ethanol and substituted phenyl ethanols. The corresponding (*S*)-isomers of 1-(6-methoxynaphthalen-2-yl)ethanol and 1-(6-methoxynaphthalen-2-yl) ethanol were prepared in >90% yields and >99% ee from the racemates via one-pot chemoenzymatic deracemization using ImHasApf and $NaBH_4$ [49].

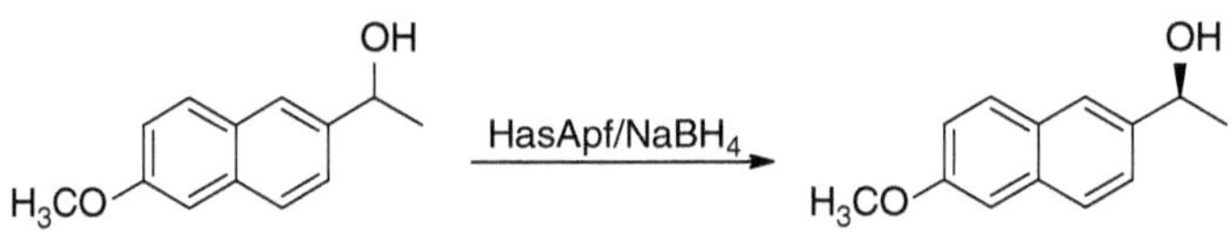

Scheme 5.46 Deracemization of 1-(6-methoxynaphthalen-2-yl)ethanol. Source: Based on Nagaoka [48].

The other approach to deracemization of racemic alcohols is the linear deracemization, which involves the combination of non-stereoselective oxidation of chiral alcohols with the stereoselective reduction of the resulting ketones. In a recent report, an iridium-catalyzed oxidation of chlorohydrins was coupled concurrently with the asymmetric bioreduction of the *in situ*-generated ketones using the alcohol dehydrogenase (ADH-A) from *Rhodococcus ruber* in one-pot manner, leading to the deracemization of racemates into the corresponding enantioenriched chlorohydrins. The iridacyclic complex catalyzed the hydrogen transfer from chlorohydrins to the hydrogen acceptors such as 6,6-dimethyl-2-chlorocyclohexanone in a two-phase system of Tris–HCl buffer (50 mM, pH 7.5) and toluene in a ratio of 3 : 7, resulting in the formation of the α-chloroketones, which could be spontaneously reduced by enzyme ADH-A employing formate dehydrogenase (FDH) and formate for NADH recycling. The racemic chlorohydrins were thus transformed into enantioenriched

(*R*)-chlorohydrins (Scheme 5.47). However, the optical purity of the products was limited because the desired (*R*)-enantiomer was oxidized unnecessarily and the Ir-catalyst also catalyzed the alcohol racemization [50].

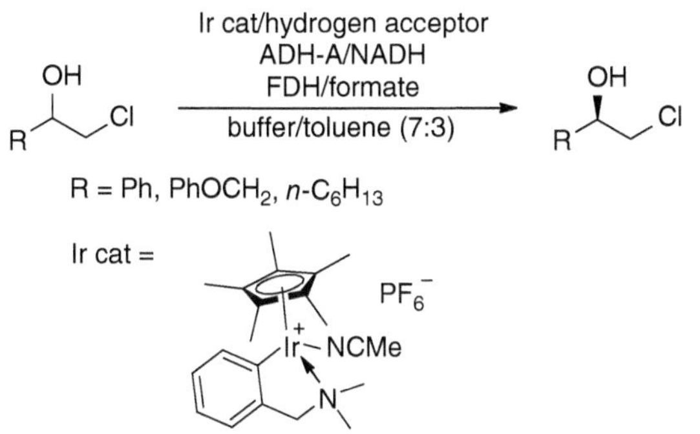

Scheme 5.47 Chemoenzymatic deracemization of chlorohydrins via one-pot concurrent oxidation and reduction.

This strategy could also be carried out in a sequential manner. For example, racemic 1-arylethanols and lineal aliphatic alcohols were first oxidized using 2,2,6,6-tetramethylpiperidin-1-oxyl and iodine. The oxidation reaction was facilitated by sonication and the corresponding ketone intermediates were quantitatively formed after just one hour. After simple destruction of iodine in the same pot, either Prelog or anti-Prelog alcohol dehydrogenases were added into the reaction mixture to reduce the ketones (Scheme 5.48). This one-pot stepwise chemical oxidation and bioreduction processes yielded enantiopure 1-arylethanols and lineal aliphatic alcohols with excellent yields [51].

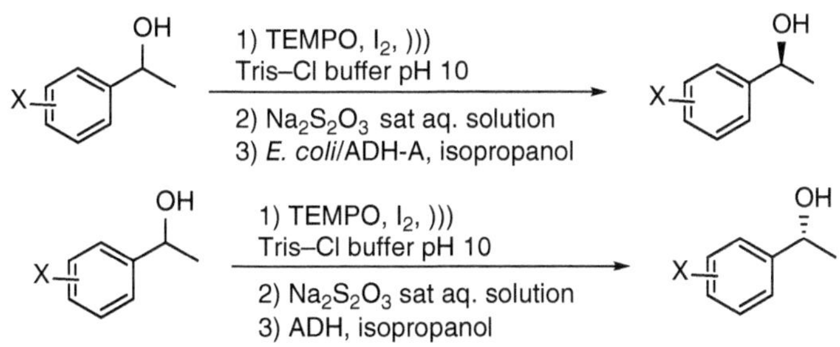

Scheme 5.48 Chemoenzymatic deracemization of alcohols via one-pot sequential oxidation and reduction.

A laccase from *Trametes versicolor* had been paired with (2,2,6,6-tetramethylpiperidin-1-yl)oxyl (TEMPO) to serve as an efficient catalyst system for the oxidation of β,β-dihalogenated secondary alcohols employing bubbling oxygen in a NaOAc buffer and methyl *t*-butyl ether biphasic medium, cleanly providing the corresponding ketones. By using this catalytic system, 2,2-dichloro-1-phenylethanol was oxidized in 95% conversion. The resulting 2,2-dichloro-acetophenone was then reduced by an alcohol dehydrogenase from *Rhodococcus ruber* (ADH-A) with 2-propanol as co-substrate for cofactor regeneration in a "coupled-substrate" mode. This chemoenzymatic deracemization of 2,2-dichloro-1-phenylethanol afforded the (*R*)-enantiomer in 97% conversion and 97% ee (Scheme 5.49) [52].

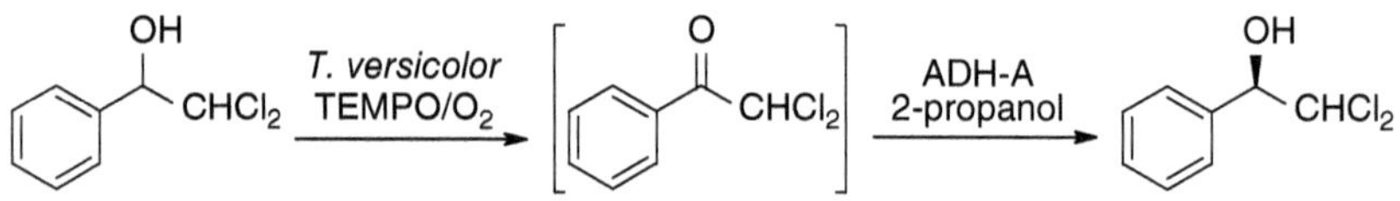

Scheme 5.49 One-pot deracemization of 2,2-dichloro-1-phenylethanol. Source: Based on Kedziora et al. [52].

This mild one-pot deracemization methodology had also been applied to racemic 2-phenyl-1-propanol. The nonselective oxidation of 2-phenyl-1-propanol mediated by laccase/TEMPO generated racemic 2-phenylpropionaldehyde. Various alcohol dehydrogenases were tested toward the enantioselective reduction of the racemic aldehyde. The commercially available HLADH and Evo-1.1.200 catalyzed the dynamic kinetic reduction of 2-phenylpropionaldehyde, furnishing the (*S*)- and (*R*)-2-phenyl-1-propanol in up to 94% and 97% ee, respectively, at pH 8–10. The coupling of the oxidation and reduction enabled the deracemization of 2-phenyl-1-propanol to afford the (*S*)- and (*R*)-2-phenyl-1-propanol with 71 and 72% isolated yields and 82% and 86% ee, respectively (Scheme 5.50) [53].

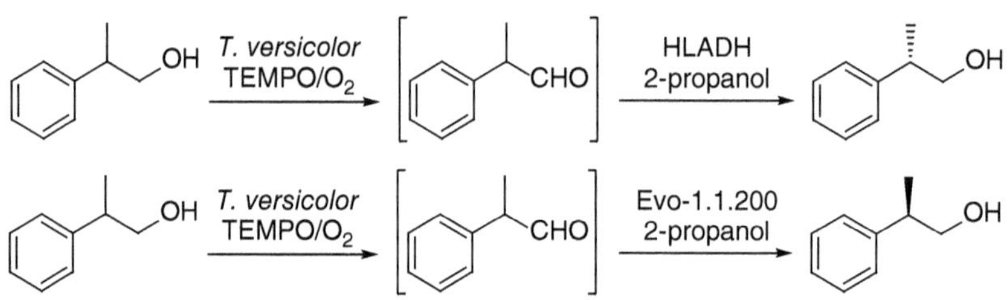

Scheme 5.50 One-pot deracemization of 2-phenyl-1-propanol. Source: Based on Diaz-Rodriguez et al. [53].

Sodium hypochlorite (NaOCl) and 2-azaadamantane *N*-oxyl (AZADO) mediate efficient nonselective oxidation of racemic secondary alcohols to the corresponding ketones. This was recently coupled with a highly stereoselective enzymatic reduction of the intermediate ketone, enabling the one-pot deracemization of a large group of secondary alcohols and diols with diverse structural features. A set of synthetically important enantiopure alcohols (>99% ee) were prepared with up to 98 % isolated yields. The stereoisomeric rac/meso mixture of diols 1,1′-[(1,1′-biphenyl)-3,3′-diyl]diethanol was converted into a single stereoisomer (1*R*,1′*R*)-1,1′-[(1,1′-biphenyl)-3,3′-diyl]diethanol in 95% yield with >99% ee and >99 : <1 dr (Scheme 5.51) [54].

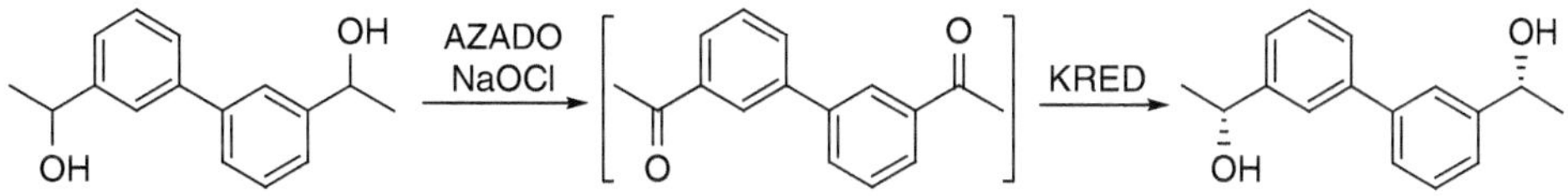

Scheme 5.51 One-pot deracemization of 1,1′-[(1,1′-biphenyl)-3,3′-diyl]diethanol. Source: Based on Liardo et al. [54].

5.3 Deracemization of Chiral Sulfoxides

Chiral sulfoxide is an important component of biologically active compounds, such as the marketed pharmaceutical omeprazole. A wealth of metal-based, organocatalytic, and enzymatic methods have been reported for the asymmetric synthesis of these chiral compounds. Recently, a sequential deracemization of racemic sulfoxides was developed for the preparation of asymmetric sulfoxides. In this two-step protocol, *E. coli* cells catalyzed the reduction of (*R*)-sulfoxide to the thioether, which was then oxidized using a heterogeneous Ta_2O_5–SiO_2 catalyst to the racemic sulfoxide. (*S*)-Methyl tolyl sulfoxide was obtained in 97.5% ee and 56% yield after three deracemization cycles of racemate [55]. Had the enzymatic reduction and the chemical oxidation been compatible, a "one-pot" chemoenzymatic deracemization of chiral sulfoxides could be achieved to access enantiomerically pure sulfoxides. More recently, this was realized by a pair of compatible enzyme and organocatalyst. The methionine sulfoxide reductase A (MsrA) catalyzed the highly enantioselective reduction of chiral sulfoxides, resulting in kinetic resolution of sulfoxides with the formation of sulfides. This reaction was combined with the oxidation of sulfides by an oxaziridine-type oxidant in a biphasic system, leading to the efficient deracemization of chiral sulfoxides (Scheme 5.52). A wide range of sulfoxides were converted into the corresponding enantiomerically pure sulfoxides with high yields [56].

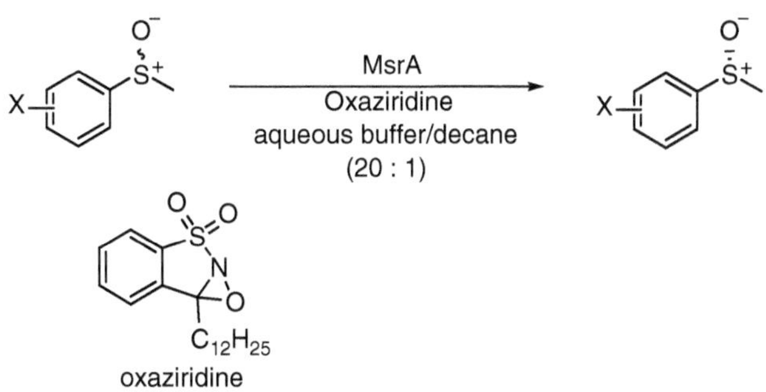

Scheme 5.52 One-pot deracemization of chiral sulfoxides.

5.4 Summary and Outlook

Deracemization transforms a racemic compound into a single enantiomer of identical structure, offering an attractive methodology for the preparation of enantiomerically enriched chiral compounds, especially in the cases where the racemic mixture can be readily synthesized. Deracemization of a racemic compound usually proceeds via a two-step process of destruction and re-construction of the chiral center either chemically or biochemically. While the one-pot concurrent chemical–chemical process has been rarely reported due to the difficulty in implementing chemical destruction and re-construction of a chiral center under same reaction conditions [57, 58], the one-pot chemoenzymatic deracemization has been successfully performed for the compounds with chiral amine or alcohol moiety as discussed above. The biocatalytic–biocatalytic deracemization or stereoinversion of chiral amines and alcohols has also been achieved in one-pot concurrent manner by coupling two redox enzymes with properly manipulated cofactor recycling systems [5]. The ever-increasing variety of enzyme catalysts with different activity and better performance will allow design of novel chemoenzymatic deracemization by combining a suitable enzyme with a biocompatible chemical reaction in a harmonic manner, expanding the application scope beyond chiral amines and alcohols. For example, a dehydrogenation coupling with ene-reductase-catalyzed reduction might lead to deracemization of a chiral molecule with the carbon stereocenter not connecting to a nitrogen or oxygen atom. Therefore, more and more new state-of-art chemoenzymatic systems for deracemization of various classes of chiral compounds will be developed and widely applied to the sustainable production of enantiomerically pure compounds that are in high demand in the coming years.

References

1 Ji, Y., Shi, L., Chen, M.-W. et al. (2015). Concise redox deracemization of secondary and tertiary amines with a tetrahydroisoquinoline core via a nonenzymatic process. *Journal of the American Chemical Society* 137 (33): 10496–10499.

2 Gruber, C.C., Lavandera, I., Faber, K., and Kroutil, W. (2006). From a racemate to a single enantiomer: deracemization by stereoinversion. *Advanced Synthesis & Catalysis* 348 (14): 1789–1805.

3 Musa, M.M., Hollmann, F., and Mutti, F.G. (2019). Synthesis of enantiomerically pure alcohols and amines via biocatalytic deracemisation methods. *Catalysis Science & Technology* 9 (20): 5487–5503.

4 Turner, N.J. (2010). Deracemisation methods. *Current Opinion in Chemical Biology* 14 (2): 115–121.

5 Aranda, C., Oksdath-Mansilla, G., Bisogno, F.R., and de Gonzalo, G. (2020). Deracemisation processes employing organocatalysis and enzyme catalysis. *Advanced Synthesis & Catalysis* 362 (6): 1233–1257.

6 Batista, V.F., Galman, J.L., Pinto, D.C.G.A. et al. (2018). Monoamine oxidase: tunable activity for amine resolution and functionalization. *ACS Catalysis* 8 (12): 11889–11907.

7 Huh, J.W., Yokoigawa, K., Esaki, N., and Soda, K. (1992). Synthesis of l-proline from the racemate by coupling of enzymatic enantiospecific oxidation and

chemical non-enantiospecific reducation. *Journal of Fermentation and Bioengineering* 74 (3): 189–190.
8 Soda, K., Oikawa, T., and Yokoigawa, K. (2001). One-pot chemo-enzymatic enantiomerization of racemates. *Journal of Molecular Catalysis B: Enzymatic* 11 (4): 149–153.
9 Beard, T.M. and Turner, N.J. (2002). Deracemisation and stereoinversion of α-amino acids using D-amino acid oxidase and hydride reducing agents. *Chemical Communications* 3: 246–247.
10 Enright, A., Alexandre, F.-R., Roff, G. et al. (2003). Stereoinversion of β- and γ-substituted α-amino acids using a chemo-enzymatic oxidation–reduction procedure. *Chemical Communications* 20: 2636–2637.
11 Chen, Y., Goldberg, S.L., Hanson, R.L. et al. (2011). Enzymatic preparation of an (S)-amino acid from a racemic amino acid. *Organic Process Research & Development* 15 (1): 241–248.
12 Alexandre, F.-R., Pantaleone, D.P., Taylor, P.P. et al. (2002). Amine–boranes: effective reducing agents for the deracemisation of dl-amino acids using l-amino acid oxidase from *Proteus myxofaciens*. *Tetrahedron Letters* 43 (4): 707–710.
13 Roff, G.J., Lloyd, R.C., and Turner, N.J. (2004). A versatile chemo-enzymatic route to enantiomerically pure β-branched α-amino acids. *Journal of the American Chemical Society* 126 (13): 4098–4099.
14 Schnepel, C., Kemker, I., and Sewald, N. (2019). One-pot synthesis of d-halotryptophans by dynamic stereoinversion using a specific l-amino acid oxidase. *ACS Catalysis* 9 (2): 1149–1158.
15 Märkle, W. and Lütz, S. (2008). Electroenzymatic strategies for deracemization, stereoinversion and asymmetric synthesis of amino acids. *Electrochimica Acta* 53 (7): 3175–3180.
16 Ju, S., Qian, M., Xu, G. et al. (2019). Chemoenzymatic approach to (S)-1,2,3,4-tetrahydroisoquinoline carboxylic acids employing D-amino acid oxidase. *Advanced Synthesis & Catalysis* 361 (13): 3191–3199.
17 Rosini, E., Melis, R., Molla, G. et al. (2017). Deracemization and stereoinversion of α-amino acids by l-amino acid deaminase. *Advanced Synthesis & Catalysis* 359 (21): 3773–3781.
18 Alexeeva, M., Enright, A., Dawson, M.J. et al. (2002). Deracemization of α-methylbenzylamine using an enzyme obtained by in vitro evolution. *Angewandte Chemie International Edition* 41 (17): 3177–3180.
19 Carr, R., Alexeeva, M., Enright, A. et al. (2003). Directed evolution of an amine oxidase possessing both broad substrate specificity and high enantioselectivity. *Angewandte Chemie International Edition* 42 (39): 4807–4810.
20 Carr, R., Alexeeva, M., Dawson, M.J. et al. (2005). Directed evolution of an amine oxidase for the preparative deracemisation of cyclic secondary amines. *ChemBioChem* 6 (4): 637–639.
21 Dunsmore, C.J., Carr, R., Fleming, T., and Turner, N.J. (2006). A chemo-enzymatic route to enantiomerically pure cyclic tertiary amines. *Journal of the American Chemical Society* 128 (7): 2224–2225.

22 Ghislieri, D., Green, A.P., Pontini, M. et al. (2013). Engineering an enantioselective amine oxidase for the synthesis of pharmaceutical building blocks and alkaloid natural products. *Journal of the American Chemical Society* 135 (29): 10863–10869.

23 Zawodny, W., Montgomery, S.L., Marshall, J.R. et al. (2018). Chemoenzymatic synthesis of substituted azepanes by sequential biocatalytic reduction and organolithium-mediated rearrangement. *Journal of the American Chemical Society* 140 (51): 17872–17877.

24 Bailey, K.R., Ellis, A.J., Reiss, R. et al. (2007). A template-based mnemonic for monoamine oxidase (MAO-N) catalyzed reactions and its application to the chemo-enzymatic deracemisation of the alkaloid (+/-)-crispine A. *Chemical Communications* 35: 3640–3642.

25 Rowles, I., Malone, K.J., Etchells, L.L. et al. (2012). Directed evolution of the enzyme monoamine oxidase (MAO-N): highly efficient chemo-enzymatic deracemisation of the alkaloid (±)-crispine A. *ChemCatChem* 4 (9): 1259–1261.

26 Schrittwieser, J.H., Groenendaal, B., Willies, S.C. et al. (2014). Deracemisation of benzylisoquinoline alkaloids employing monoamine oxidase variants. *Catalysis Science & Technology* 4 (10): 3657–3664.

27 Schrittwieser, J.H., Groenendaal, B., Resch, V. et al. (2014). Deracemization by simultaneous bio-oxidative kinetic resolution and stereoinversion. *Angewandte Chemie International Edition* 53 (14): 3731–3734.

28 Ghislieri, D., Houghton, D., Green, A.P. et al. (2013). Monoamine oxidase (MAO-N) catalyzed deracemization of tetrahydro-β-carbolines: substrate dependent switch in enantioselectivity. *ACS Catalysis* 3 (12): 2869–2872.

29 Herter, S., Medina, F., Wagschal, S. et al. (2018). Mapping the substrate scope of monoamine oxidase (MAO-N) as a synthetic tool for the enantioselective synthesis of chiral amines. *Bioorganic & Medicinal Chemistry* 26 (7): 1338–1346.

30 Leisch, H., Grosse, S., Iwaki, H. et al. (2011). Cyclohexylamine oxidase as a useful biocatalyst for the kinetic resolution and dereacemization of amines. *Canadian Journal of Chemistry* 90 (1): 39–45.

31 Li, G., Ren, J., Iwaki, H. et al. (2014). Substrate profiling of cyclohexylamine oxidase and its mutants reveals new biocatalytic potential in deracemization of racemic amines. *Applied Microbiology and Biotechnology* 98 (4): 1681–1689.

32 Li, G., Ren, J., Yao, P. et al. (2014). Deracemization of 2-methyl-1,2,3,4-tetrahydroquinoline using mutant cyclohexylamine oxidase obtained by iterative saturation mutagenesis. *ACS Catalysis* 4 (3): 903–908.

33 Yao, P., Cong, P., Gong, R. et al. (2018). Biocatalytic route to chiral 2-substituted-1,2,3,4-tetrahydroquinolines using cyclohexylamine oxidase muteins. *ACS Catalysis* 8 (3): 1648–1652.

34 Cosgrove, S.C., Hussain, S., Turner, N.J., and Marsden, S.P. (2018). Synergistic chemo/biocatalytic synthesis of alkaloidal tetrahydroquinolines. *ACS Catalysis* 8 (6): 5570–5573.

35 Deng, G., Wan, N., Qin, L. et al. (2018). Deracemization of phenyl-substituted 2-methyl-1,2,3,4-tetrahydroquinolines by a recombinant monoamine oxidase from *Pseudomonas monteilii* ZMU-T01. *ChemCatChem* 10 (11): 2374–2377.

36 Li, G., Yao, P., Cong, P. et al. (2016). New recombinant cyclohexylamine oxidase variants for deracemization of secondary amines by orthogonally assaying designed mutants with structurally diverse substrates. *Scientific Reports* 6: 24973.

37 Gong, R., Yao, P., Chen, X. et al. (2018). Accessing d-valine synthesis by improved variants of bacterial cyclohexylamine oxidase. *ChemCatChem* 10 (2): 387–390.

38 Wu, X., Huang, Z., Wang, Z. et al. (2020). Asymmetric synthesis of a key dextromethorphan intermediate and its analogues enabled by a new cyclohexylamine oxidase: enzyme discovery, reaction development, and mechanistic insight. *The Journal of Organic Chemistry* 85 (8): 5598–5614.

39 Heath, R.S., Pontini, M., Bechi, B., and Turner, N.J. (2014). Development of an R-selective amine oxidase with broad substrate specificity and high enantioselectivity. *ChemCatChem* 6 (4): 996–1002.

40 Yasukawa, K., Nakano, S., and Asano, Y. (2014). Tailoring D-amino acid oxidase from the pig kidney to R-stereoselective amine oxidase and its use in the deracemization of α-methylbenzylamine. *Angewandte Chemie International Edition* 53 (17): 4428–4431.

41 Yasukawa, K., Motojima, F., Ono, A., and Asano, Y. (2018). Expansion of the substrate specificity of porcine kidney D-amino acid oxidase for S-stereoselective oxidation of 4-Cl-benzhydrylamine. *ChemCatChem* 10 (16): 3500–3505.

42 Aleku, G.A., Mangas-Sanchez, J., Citoler, J. et al. (2018). Kinetic resolution and deracemization of racemic amines using a reductive aminase. *ChemCatChem* 10 (3): 515–519.

43 Guo, X., Okamoto, Y., Schreier, M.R. et al. (2018). Enantioselective synthesis of amines by combining photoredox and enzymatic catalysis in a cyclic reaction network. *Chemical Science* 9 (22): 5052–5056.

44 Oikawa, T., Mukoyama, S., and Soda, K. (2001). Chemo-enzymatic D-enantiomerization of DL-lactate. *Biotechnology and Bioengineering* 73 (1): 80–82.

45 Biade, A.E., Bourdillon, C., Laval, J.M. et al. (1992). Complete conversion of L-lactate into D-lactate. A generic approach involving enzymic catalysis, electrochemical oxidation of NADH and electrochemical reduction of pyruvate. *Journal of the American Chemical Society* 114 (3): 893–897.

46 Tanaka, T., Iwai, N., Matsuda, T., and Kitazume, T. (2009). Utility of ionic liquid for *Geotrichum candidum*-catalyzed synthesis of optically active alcohols. *Journal of Molecular Catalysis B: Enzymatic* 57 (1): 317–320.

47 Magallanes-Noguera, C., Ferrari, M.M., Kurina-Sanz, M., and Orden, A.A. (2012). Deracemization of secondary alcohols by chemo-enzymatic sequence with plant cells. *Journal of Biotechnology* 160 (3): 189–194.

48 Nagaoka, H. (2016). Heterogeneous asymmetric oxidation catalysis using hemophore HasApf. Application in the chemoenzymatic deracemization of sec-alcohols with sodium borohydride. *Catalysts* 6 (3): 38.

49 Nagaoka, H. (2016). Continuous reusability using immobilized HasApf in chemoenzymatic deracemization: a new heterogeneous enzyme catalysis. *Biomolecules* 6 (4): 41.

50 Mutti, F.G., Orthaber, A., Schrittwieser, J.H. et al. (2010). Simultaneous iridium catalysed oxidation and enzymatic reduction employing orthogonal reagents. *Chemical Communications* 46 (42): 8046–8048.

51 Méndez-Sánchez, D., Mangas-Sánchez, J., Lavandera, I. et al. (2015). Chemoenzymatic deracemization of secondary alcohols by using a TEMPO–iodine–alcohol dehydrogenase system. *ChemCatChem* 7 (24): 4016–4020.

52 Kedziora, K., Diaz-Rodriguez, A., Lavandera, I. et al. (2014). Laccase/TEMPO-mediated system for the thermodynamically disfavored oxidation of 2,2-dihalo-1-phenylethanol derivatives. *Green Chemistry* 16 (5): 2448–2453.

53 Diaz-Rodriguez, A., Rios-Lombardia, N., Sattler, J.H. et al. (2015). Deracemisation of profenol core by combining laccase/TEMPO-mediated oxidation and alcohol dehydrogenase-catalysed dynamic kinetic resolution. *Catalysis Science & Technology* 5 (3): 1443–1446.

54 Liardo, E., Ríos-Lombardía, N., Morís, F. et al. (2018). A straightforward deracemization of sec-alcohols combining organocatalytic oxidation and biocatalytic reduction. *European Journal of Organic Chemistry* 2018 (23): 3031–3035.

55 Tudorache, M., Nica, S., Bartha, E. et al. (2012). Sequential deracemization of sulfoxides via whole-cell resolution and heterogeneous oxidation. *Applied Catalysis A: General* 441: 42–46.

56 Nosek, V. and Míšek, J. (2018). Chemoenzymatic deracemization of chiral sulfoxides. *Angewandte Chemie International Edition* 57 (31): 9849–9852.

57 Lackner, A.D., Samant, A.V., and Toste, F.D. (2013). Single-operation deracemization of 3H-indolines and tetrahydroquinolines enabled by phase separation. *Journal of the American Chemical Society* 135 (38): 14090–14093.

58 Qu, P., Kuepfert, M., Jockusch, S., and Weck, M. (2019). Compartmentalized nanoreactors for one-pot redox-driven transformations. *ACS Catalysis* 9 (4): 2701–2706.

6

One-pot Concurrent Chemoenzymatic Reactions

For "one-pot concurrent chemoenzymatic reactions," the chemical and biocatalytic reactions occur simultaneously in one-pot without high accumulation of the intermediates. This is the most efficient mode of chemoenzymatic cascade, in which multiple reactions can be realized in one single process operation, presenting huge operational and economic advantages and serving as an important tool for more sustainable practice of the chemical industry. The one-pot concurrent processes are particularly beneficial for the situations where the reaction intermediates are unstable or have inhibitory effects on the catalysts due to the fast consumption and low concentration of the intermediates. At the same time, there are many challenges to be overcome for the implementation of such one-pot concurrent processes. The reaction conditions of chemical reaction and biotransformation must be completely compatible. The inhibition or inactivation of (bio)catalysts has to be addressed, which often occurs due to the large number of components in the complicated reaction system. Usually, the metal- and organo-catalyzed reactions are carried out in organic solvents, while the biocatalytic transformations require aqueous buffer as the reaction media. In order to solve this incompatibility problem, scientists and engineers are designing chemical catalysts for use in aqueous reaction medium and evolving enzyme catalysts with tolerance to organic solvents, so that these catalysts can work under the same reaction conditions. Reaction engineering has also been carried out for developing efficient and economical chemoenzymatic one-pot one-step processes.

The one-pot concurrent chemoenzymatic reactions were explored in the early 1980s. An excellent example involving cooperation of a bio- and a chemocatalyst is the one-pot chemoenzymatic conversion of a mixture of D-fructose and D-glucose to D-mannitol. In this concurrent one-pot process, D-glucose isomerase catalyzes the interconversion of D-glucose and D-fructose, and platinum metal simultaneously catalyzes the selective hydrogenation of D-fructose to give D-mannitol in 46 % yield in water (pH 7–8, 60 °C, 20 atm hydrogen, Scheme 6.1) [1]. In order to obtain high yield of D-mannitol, there were some challenges to this process: (a) the hydrogenation of D-fructose to D-mannitol should be highly preferred to hydrogenation of D-glucose, (b) the interconversion of D-glucose to D-fructose should be fast, and (c) the mutual interference of the enzyme and metal catalysts should be minimal. As such, different immobilized D-glucose isomerases and

Chemo-Enzymatic Cascade Reactions, First Edition. Dunming Zhu.

metal catalysts were investigated with the aim to improve the yield of D-mannitol. By employing D-glucose isomerase immobilized on silica in combination with a Cu/silica hydrogenation catalyst, 62–66% yields of D-mannitol were obtained from D-glucose or a 1 : 1 mixture of D-glucose and D-fructose [2].

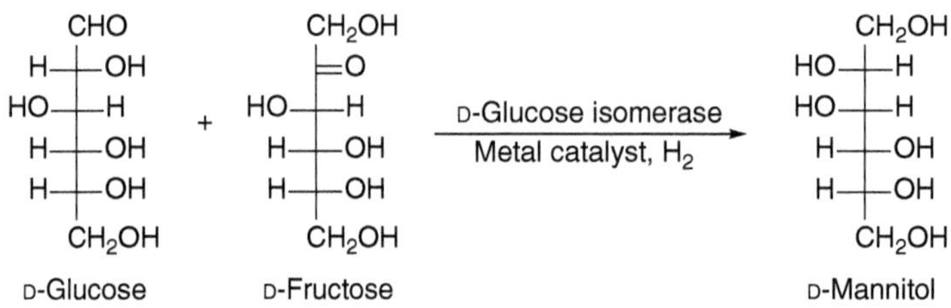

Scheme 6.1 One-pot chemoenzymatic conversion of a mixture of D-fructose and D-glucose to D-mannitol. Source: Based on Makkee et al. [1].

After more than a decade of "silence" and it was only until the mid-1990s, chemoenzymatic dynamic kinetic resolution (DKR) was developed, whereby chemocatalyzed racemization was combined with enzymatic kinetic resolution to convert a racemate into the corresponding enantiomerically pure product. The powerfulness of this method in the synthesis of optically pure alcohols and amines led to the revival of the field of chemoenzymatic reactions. Another widely studied one-pot concurrent chemoenzymatic reaction is the chemoenzymatic deracemization of a racemate into one of its enantiomers, in which one enantiomer is converted into the other by a chemoenzymatic process. The chemoenzymatic DKR and deracemization of racemates represent two paradigms of integrating chemical and biocatalytic reactions in a one-pot concurrent process. These two research areas have attracted considerable interest, and rapid development has taken place in the past years as they were discussed in the previous two chapters. In this chapter, the one-pot concurrent reactions other than DKR and deracemization of racemates will be described.

In addition to the abovementioned, well-studied chemoenzymatic DKR and deracemization of racemates, there is increasing number of other true one-pot concurrent chemoenzymatic cascades that have been reported in the literature, but examples and scope of this elegant one-pot one-step process are still limited. This is largely due to the incompatibility of the different catalysts and their respective reaction conditions. To overcome this incompatibility, one possible approach was the spatial separation of the two reactions by controlling the microenvironment by compartmentalization. Advances in microbial engineering and synthetic biology have enabled the manipulation of microorganisms to produce target chemicals from glucose. The combination of microbial synthesis with chemical transformations offers a tremendous opportunity for the direct production of important fine chemicals for various industrial applications from renewable feedstocks. Recently, some biocompatible chemical reactions have been developed and integrated into microbial fermentation for the synthesis of some value-added chemicals. Herein, the true one-pot concurrent chemoenzymatic cascades including those of microbial fermentation with biocompatible chemical reactions will be discussed first, followed by those realized via compartmentalization.

6.1 One-pot Concurrent Chemoenzymatic Cascades

In the true one-pot concurrent chemoenzymatic cascades, the chemical reaction can occur with or without chemical catalyst and before or after the biocatalytic reaction, as shown in Scheme 6.2. In the first case, the product (B) of the chemical reaction serves as the substrate for the biocatalyst and is concurrently converted to product (C), or B can serve as a cofactor for the enzymatic reaction (i.e. the chemical reaction replaces the biocatalytic cofactor recycling in the biological systems). In the second situation, the biocatalytic reaction produces a product that is spontaneously transformed to product (C) with or without the action of a chemical catalyst, or it can serve as a reagent for the chemical reaction to convert compound (D) to product (C).

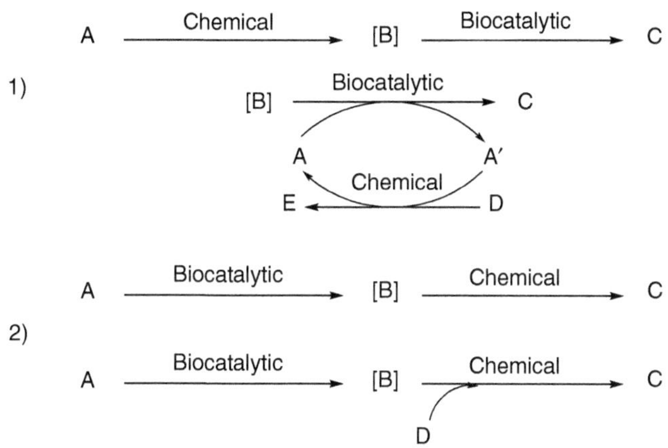

Scheme 6.2 Different modes of one-pot concurrent chemoenzymatic cascades.

Up to now, various families of enzymes have been widely studied as catalysts for organic synthesis and they have been explored for applicability in one-pot concurrent chemoenzymatic cascades. Various advances in this research area are discussed here and the following presentation is organized according to the enzyme types.

6.1.1 Lipases

Lipases catalyze the stereoselective acylation of chiral alcohols and amines to form esters or amides, respectively. They have been widely used for the (dynamic) kinetic resolution of racemic alcohols and amines. This has been discussed in Chapter 4. The lipase-catalyzed acylation can also be combined with the reactions of alcohol or amine formation. In this scheme, the *in situ*-generated chiral alcohols or amines can be stereoselectively acylated to the desired optically pure esters or amides.

In the metal and enzyme-co-catalyzed DKR of racemic alcohols and amines, the metal catalyst promotes the racemization of alcohols and amines via the reversible dehydrogenation of the alcohol/amine and hydrogenation of the resulting ketone/imine intermediate. These metal catalysts can often mediate the reduction of ketones to afford alcohols. Therefore, a metal catalyst may simultaneously promote ketone reduction and alcohol racemization, and be combined with lipase-catalyzed acylation of alcohol, leading to the asymmetric reductive acylation of ketones in a one-pot process. This was demonstrated by employing ruthenium complex **1** as a procatalyst for the ketone reduction and alcohol racemization and *Candida antarctica* lipase B (CALB) for the alcohol acylation. When these multistep reactions were carried out in toluene at 70 °C utilizing 2,6-dimethylheptan-4-ol as the hydrogen donor and 4-chlorophenyl acetate as the acyl donor, a set of aromatic and aliphatic ketones were transformed into the corresponding (*R*)-acetate in 82–100% yields and 95–99% ee (Scheme 6.3) [3].

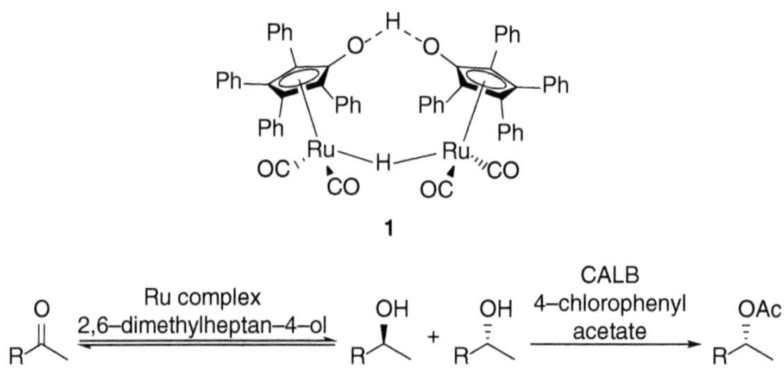

Scheme 6.3 One-pot chemoenzymatic reductive acylation of ketones. Source: Based on Jung et al. [3].

Lipase catalyzes the acyl transfer of enol acetate to form enol, which can isomerize into ketone. The enol acetates of these ketones were thus converted to the chiral acetates in 80–95% yields and 79–99% ee under the same reaction conditions using 2,6-dimethylheptan-4-ol of higher purity than 90%, but without the added acyl donor, 4-chlorophenyl acetate (Scheme 6.4). Formations of ketones (1–16%) and alcohols (1–9%) were detected in the reaction, and their amounts were dependent on the substrates and affected by the purity of the hydrogen donor. This process represented a highly atom-economic transformation since no external acyl donor was added.

In the above asymmetric reductive acylation of ketones, the chiral acetates were produced in high yields and optical purities. However, the produced 2,6-dimethylheptan-4-one and unreacted 4-chlorophenyl acetate in product mixtures made product separation and purification difficult, thus hindering the practical applications of this one-pot process. In order to solve the product separation problems, molecular hydrogen and formic acid were tested as the hydrogen

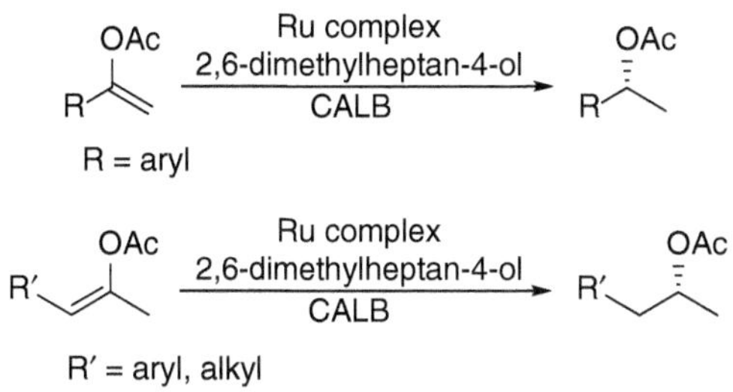

Scheme 6.4 One-pot chemoenzymatic concerted conversion of enol acetates to chiral acetates.

donor. Molecular hydrogen was effective for these transformations using ethyl acetate as the acyl donor. A variety of ketones were reductively acylated under 1 atm of hydrogen gas in ethyl acetate, furnishing the corresponding (*R*)-acetates in 71–89% yields and up to 99% ee (Scheme 6.5). The transformation of enol acetates was conducted with the same Ru catalyst and lipase under 1 atm of hydrogen in toluene, affording the chiral acetates in 74–91% yields and up to 99% ee (Scheme 6.6) [4].

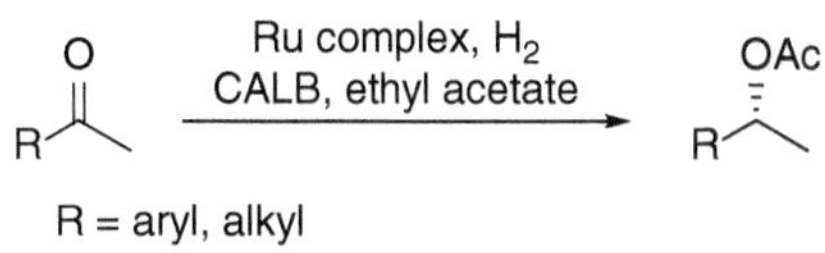

Scheme 6.5 One-pot chemoenzymatic reductive acylation of ketones.

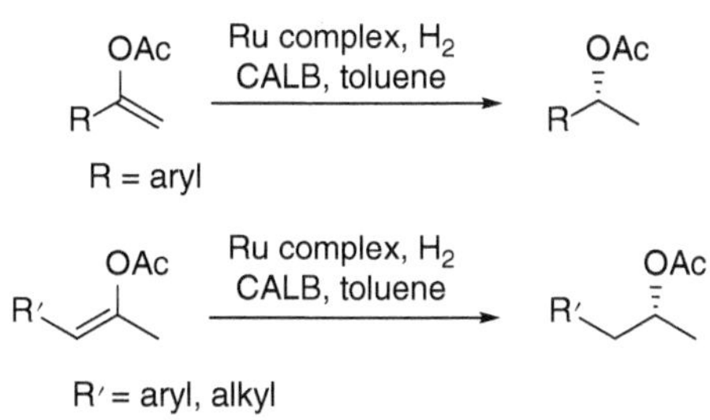

Scheme 6.6 One-pot chemoenzymatic concerted conversion of enol acetates to chiral acetates. Source: Based on Jung et al. [4].

This one-pot multi-reaction process had also been applied to ketones with acyloxy functional group in the same molecule serving as the acyl donor. By using the same Ru and lipase catalysts, a group of acyloxyphenyl ketones were transformed into the corresponding optically active esters of hydroxyphenyl alcohols with 92–96% yields and 96–98% ee under 1 atm hydrogen in toluene (Scheme 6.7). The asymmetric transformation involved the Ru-catalyzed reversible ketone reduction and lipase-catalyzed kinetic resolution of the *in situ*-generated alcohol via intramolecular acyl transfer [5].

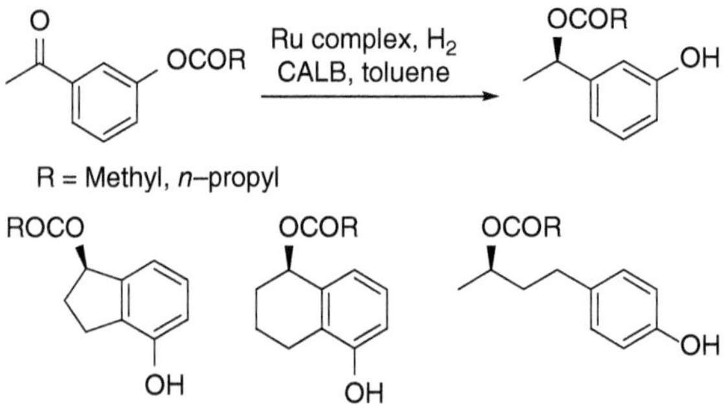

Scheme 6.7 One-pot chemoenzymatic reductive intramolecular acyl transfer of acyloxyphenyl ketones.

Acyclic chiral 1,3-aminoalcohols with two chiral centers are useful building blocks in the asymmetric synthesis of many biologically active compounds. The above ruthenium complex and lipase CALB were also combined for the reduction of β-aminoketones and DKR of the resulting 1,3-aminoalcohols. A range of Boc-protected β-aminoketones were prepared with >98% ee by the enantioselective L- or D-proline-catalyzed Mannich reaction of imines with acetone. When the dynamic kinetic asymmetric transformation of the enantiopure β-aminoketones was performed using *p*-chlorophenyl acetate as acyl donor under an atmosphere of H_2 in toluene, both *syn* and *anti* diastereomers of acetylated 1,3-aminoalcohols were obtained in 73–93% yields, 92/8–98/2 dr, and 99% ee (Scheme 6.8) [6].

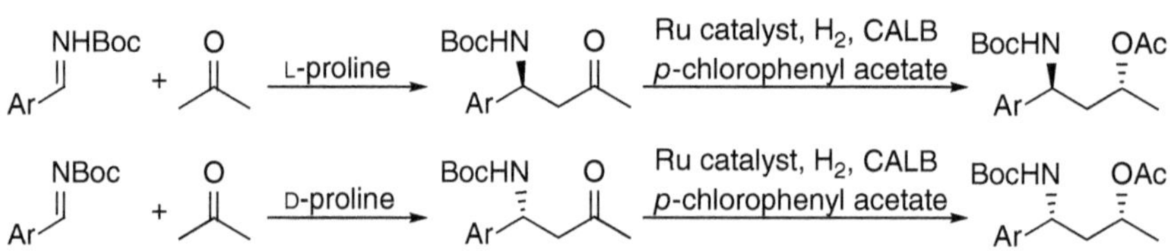

Scheme 6.8 Dynamic kinetic asymmetric transformation of β-aminoketones. Source: Based on Millet et al. [6].

An iron carbonyl complex **2** had also been reported to catalyze the hydrogenation of prochiral ketones and dehydrogenation of the resulting chiral alcohols. The iron-catalyzed redox reactions would effect the racemization of chiral alcohols, thus enabling the DKR of racemic alcohols by incorporating with enantioselective enzymatic acylation. Therefore, by employing iron carbonyl complex **2** and the lipase B from *C. antarctica* (CALB) as the catalysts, molecular hydrogen as hydrogen donor and ethyl acetate as acyl donor, a wide range of benzylic, aliphatic, (hetero)aromatic ketones, and diketones were transformed into the corresponding (*R*)-acetates in 72–98% yields and 90–99% ee via one-pot chemoenzymatic reductive acylation (Scheme 6.9) [7].

In the above case, the hydrogenation of ketone functional group and the racemization of alcohol group were promoted by one metal catalyst. These two

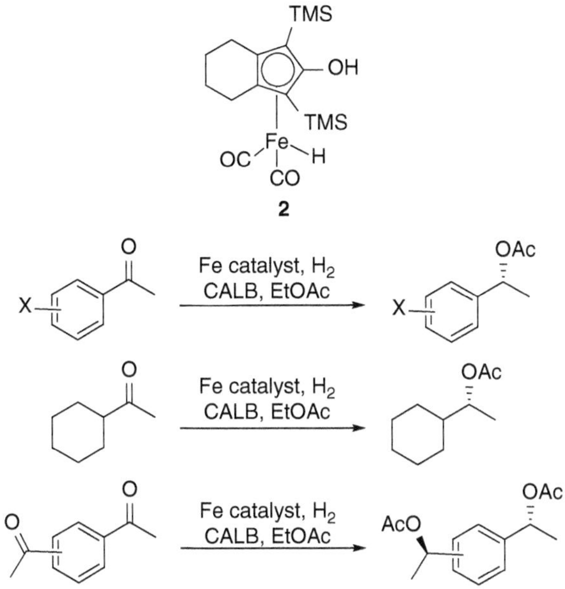

Scheme 6.9 One-pot chemoenzymatic reductive acylation of chiral alcohols and diols. Source: Based on El-Sepelgy et al. [7].

transformations could also be mediated by separate heterogeneous metal catalysts. Chiral 2-hydroxy-1-indanone was transformed into *cis*-1-amino-2-indanol; both are useful building blocks for pharmaceuticals and chiral catalysts/agents in asymmetric synthesis. It was found that Pd/Al_2O_3 catalyzed the regioselective hydrogenation of 1,2-indanedione to give racemic 2-hydroxy-1-indanone in 99% regioselectivity at up to 56% conversion. This metal-catalyzed regioselective hydrogenation was then combined with the chemoenzymatic DKR of the resulting racemic 2-hydroxy-1-indanone to access this chiral hydroxyketone in optically pure form. By using a heterogeneous Ru catalyst for racemization and lipase AK for acylation of 2-hydroxy-1-indanone with trifluoroethyl acetate as the acyl donor, the hydrogenation and DKR were conducted in ethyl acetate in one-pot concurrent operation, and (*R*)-2-acetoxy-1-indanone was obtained in 86–92% ee and 39% isolated yield (Scheme 6.10) [8].

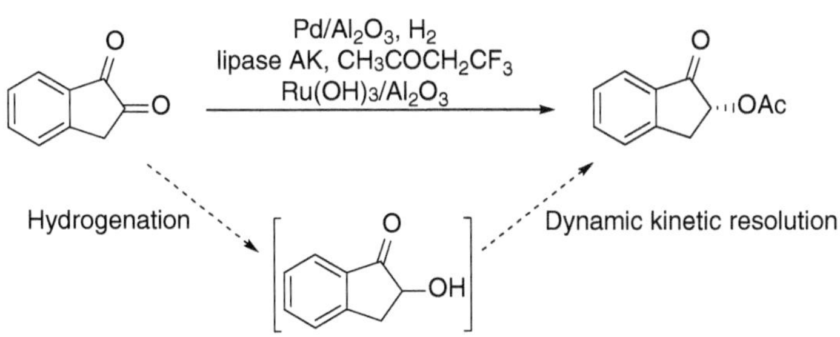

Scheme 6.10 One-pot concurrent synthesis of (*R*)-2-acetoxy-1-indanone. Source: Based on Langvik et al. [8].

The Pd/Al_2O_3 catalyst was applied as catalyst for ketone reduction and alcohol racemization in the one-pot conversion of acetophenone into (*R*)-1-phenylethyl acetate. However, the transformation was complicated by the dehydration of the intermediate 1-phenylethanol to styrene, which was instantaneously hydrogenated to give ethyl benzene (Scheme 6.11). The Pd catalysts on different supports were examined for their ability to prevent the side reactions, and the slightly acidic $Pd–Al_2O_3$ was found to show the best selectivity for one-pot synthesis of (*R*)-1-phenylethyl acetate from acetophenone [9, 10].

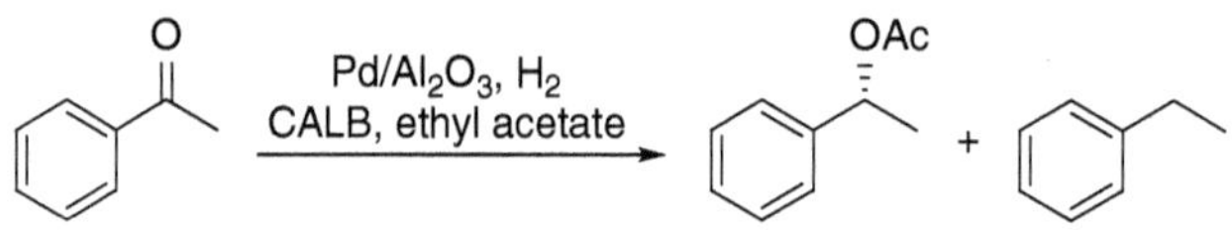

Scheme 6.11 One-pot chemoenzymatic transformation of acetophenone into (*R*)-1-phenylethyl acetate.

In the nitroaldol (Henry) reaction, a nitroalkane was reversibly added to a carbonyl compound under mild basic conditions to form a β-nitroalcohol, which could be resolved by the lipase-mediated acylation. Indeed, equimolar amounts of 4-trifluoromethylbenzaldehyde, 2-fluorobenzaldehyde, 3-nitrobenzaldehyde, 2-chlorobenzaldehyde, and 2,4-dichlorobenzaldehyde were mixed with 1 equivalent of 2-nitropropane in the presence of trimethylamine to form a set of nitroaldol adducts, from which the (*R*)-enantiomer of the adduct ester from 3-nitrobenzaldehyde and 2-nitropropane was obtained as the major product with 99% ee by treating the dynamic system with lipase PS-C I from *Pseudomonas cepacia* as the biocatalyst and *p*-chlorophenyl acetate as the acyl donor, although the relative concentration of this nitroaldol adduct was among the lowest in the system in the absence of the enzyme. A minor product from the addition of 2-nitropropane to 4-trifluoromethylbenzaldehyde was produced with 98% ee (Scheme 6.12) [11].

This one-pot concurrent process of nitroaldol (Henry) reaction and lipase-catalyzed transesterification had been further demonstrated with various benzaldehydes and 2-nitropropane, whereby the corresponding β-nitroalkanol derivatives were produced in high yield and enantiomeric purity (up to 92% yield and 99% ee, Scheme 6.13) [12]. The combination of the intramolecular nitroaldol reaction of 6-nitroheptanal with lipase-catalyzed resolution of the resulting nitroaldol adduct 2-nitrocyclohexanol had also been investigated, but the one-pot sequential process gave a mixture of the products (Scheme 6.14) [13].

Zhang et al. established a double parallel dynamic system via nitroaldol reaction and hemithioacetal formation of three aromatic aldehydes (3-nitrobenzaldehyde, 2-chlorobenzaldehyde, and 2,4-dichlorobenzaldehyde) with 2-nitropropane and 1-butanethiol in the presence of triethylamine. Lipase-catalyzed transesterification reactions toward the resulting nitroaldol and hemithioacetal intermediates were examined. Under the action of *Burkholderia cepacia* lipase (BCL) with phenyl acetate as the acyl donor, the nitroaldol and hemithioacetal intermediates

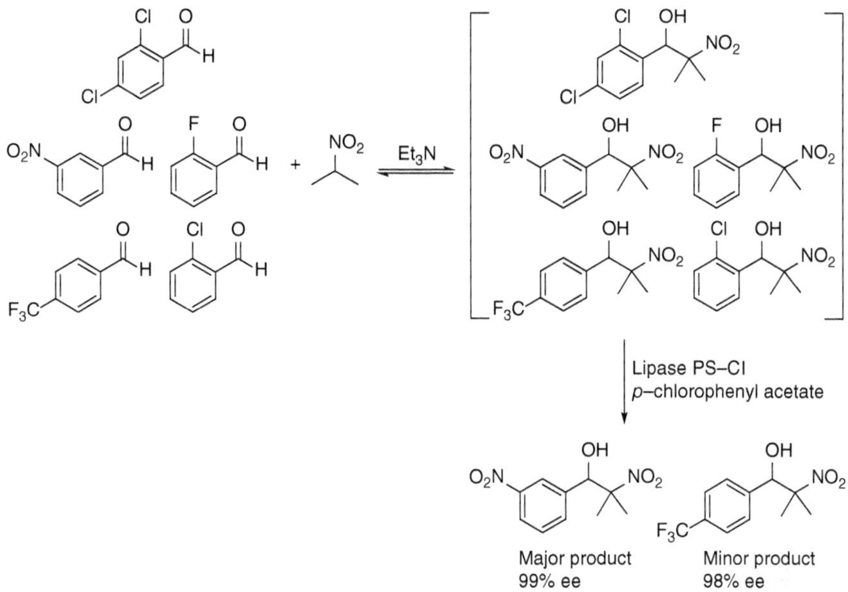

Scheme 6.12 One-pot lipase-catalyzed kinetic resolution of dynamic nitroalcohol library created by nitroaldol (Henry) reaction. Source: Based on Vongvilai et al. [11].

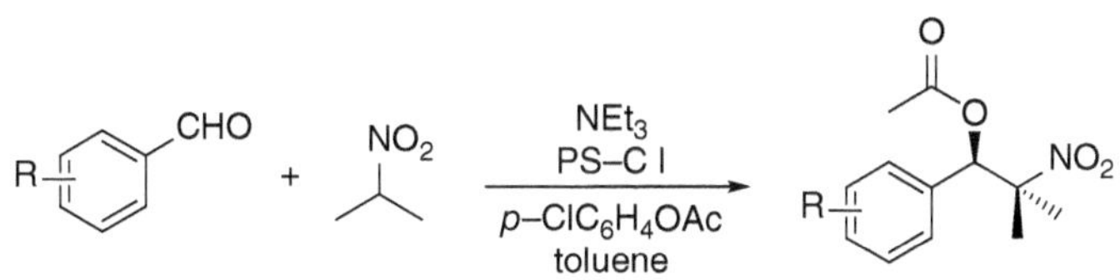

Scheme 6.13 One-pot concurrent synthesis of acetylated β-nitroalkanols. Source: Based on Vongvilai et al. [12].

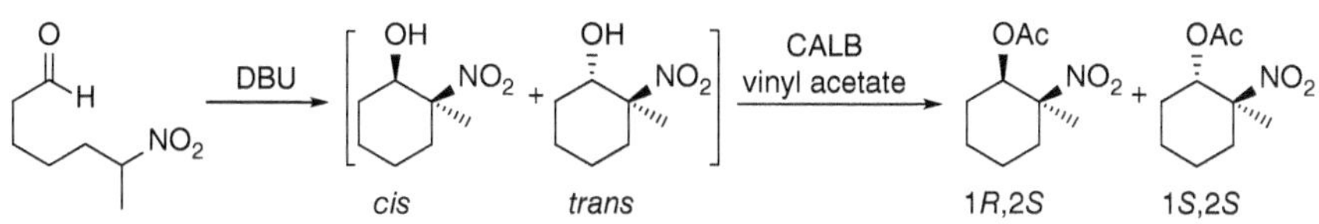

Scheme 6.14 One-pot sequential intramolecular nitroaldol reaction and lipase-catalyzed resolution. Source: Based on Foley et al. [13].

of 3-nitrobenzaldehyde were acetylated to give the products in different ratios depending on the solvent (Scheme 6.15) [14].

Under the action of ZnI_2 and 1,1,3,3-tetramethylguanidine (TMG), (*E*)-1-(2, 4-dimethoxyphenyl)-2-nitropropene, (*E*)-1-(3,4-dimethoxyphenyl)-2-nitropropene, and 1,4-dithiane-2,5-diol underwent reversible domino thia-Michael–Henry reaction to form a dynamic system including different stereoisomers of substituted thiolanes. By employing BCL as biocatalyst and isopropenyl acetate as acyl donor for the stereoselective acylation of thiolanes, the (*3R*,*4S*,*5R*)-stereoisomer of

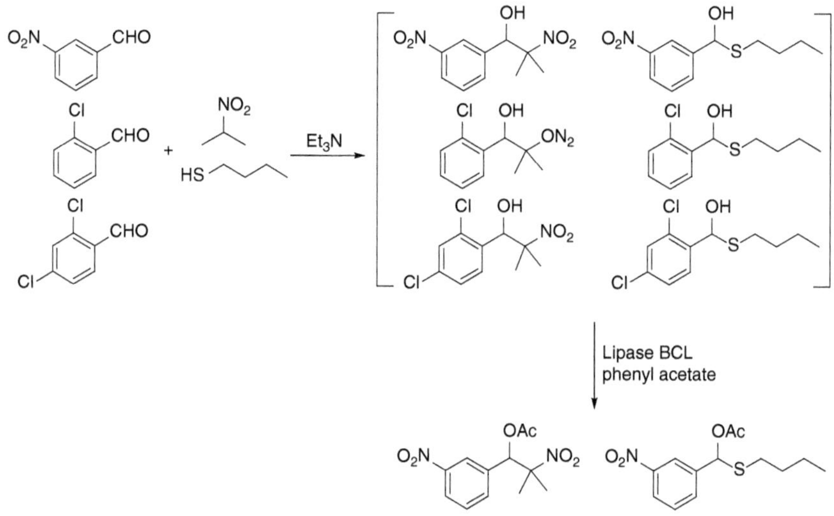

Scheme 6.15 One-pot lipase-catalyzed kinetic resolution of a double parallel dynamic library. Source: Based on Zhang et al. [14].

3,4-dimethoxyphenyl-substituted thiolane was obtained as the major product with 98% ee (Scheme 6.16) [15].

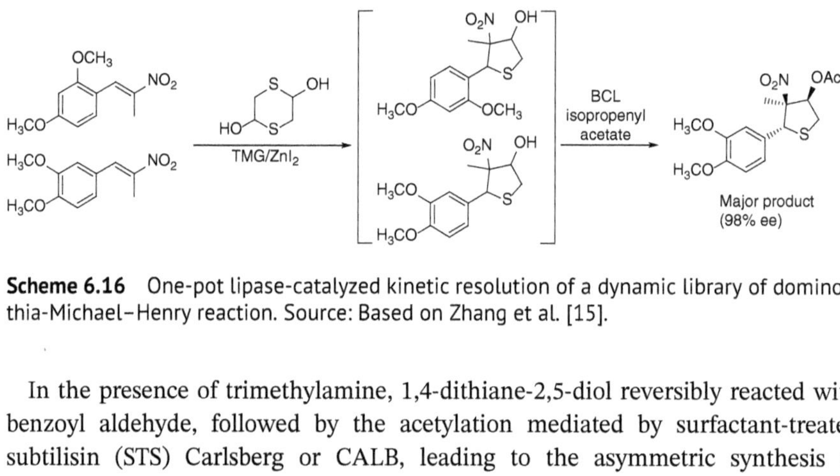

Scheme 6.16 One-pot lipase-catalyzed kinetic resolution of a dynamic library of domino thia-Michael–Henry reaction. Source: Based on Zhang et al. [15].

In the presence of trimethylamine, 1,4-dithiane-2,5-diol reversibly reacted with benzoyl aldehyde, followed by the acetylation mediated by surfactant-treated subtilisin (STS) Carlsberg or CALB, leading to the asymmetric synthesis of ((2*R*)-5-acetoxy-1,3-oxathiolan-2-yl)methyl benzoate or ((2*S*)-5-acetoxy-1,3-oxathiolan-2-yl)methyl benzoate (Scheme 6.17), respectively, which are important intermediates for a range of useful pharmaceuticals [16].

Thiols such as methyl 2-sulfanylacetate reacted with various aliphatic and aromatic aldehydes in the presence of triethylamine to form γ-hydroxyesters, which subsequently underwent intramolecular lactonization under the action of CALB to give 1,3-oxathiolan-5-one products in good conversions with moderate to good ee values (Scheme 6.18) [17, 18]. Similarly, the addition of methyl 2-sulfanylacetate to nitrones under basic conditions resulted in the reversible formation of γ-aminoesters

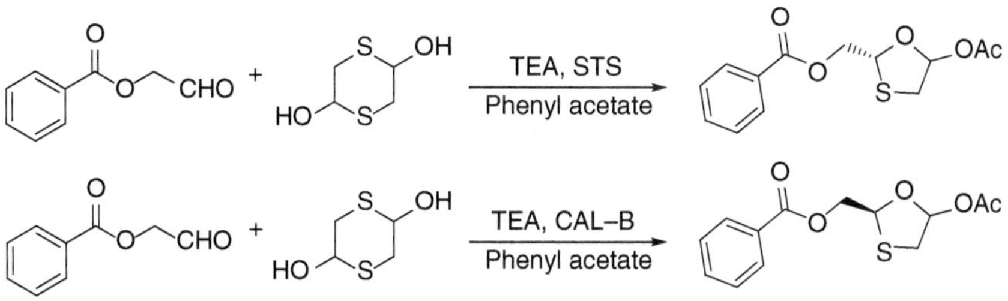

Scheme 6.17 One-pot concurrent asymmetric synthesis of both antipodes of (5-acetoxy-1,3-oxathiolan-2-yl)methyl benzoate.

that underwent a lipase-catalyzed cyclization in the presence of an acyl donor. By using CALB as catalyst and isopropenyl acetate as the acyl donor, this domino addition–lactonization process proceeded smoothly for some nitrones, leading to efficient formation of oxathiazinanones with high ee (Scheme 6.19) [19].

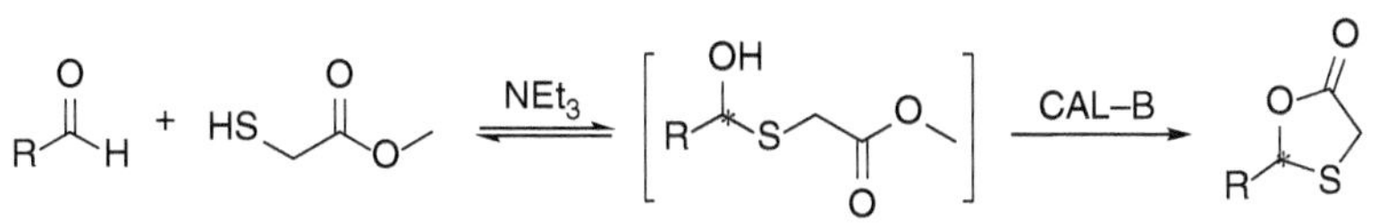

Scheme 6.18 One-pot concurrent asymmetric synthesis of 1,3-oxathiolan-5-one. Sources: Based on Sakulsombat et al. [17]; Zhang et al. [18].

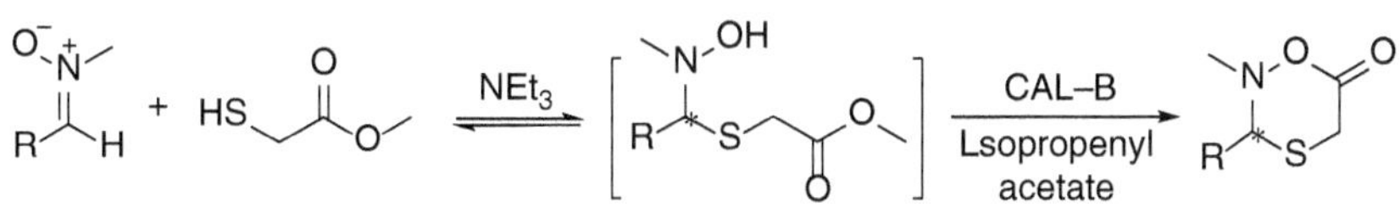

Scheme 6.19 One-pot concurrent asymmetric synthesis of oxathiazinanones. Source: Based on Zhang et al. [18].

Under the action of amines such as trimethylamine or 4-diazabicyclo[2.2.2]octane (DABCO), 6-*tert*-butoxy-2*H*-pyran-3(6*H*)-one went through rearrangement to produce (±)-*trans*-4-*tert*-butoxy-5-hydroxycyclopent-2-enone, which could be kinetically resolved to give (+)-*trans*-4-*tert*-butoxy-5-hydroxycyclopent-2-enone and the corresponding (−)-acetylated cyclopentenone (Scheme 6.20). When the rearrangement and the enzymatic resolution were carried out within the same step in one-pot, the (−)-acetylated cyclopentenone was obtained in 64% yield with 77% ee. The slower rate of rearrangement reaction under the enzymatic resolution reaction conditions indicated that the conditions for the rearrangement and the lipase-catalyzed acylation were partially incompatible with each other. Indeed, when the DABCO-mediated rearrangement and the lipase-catalyzed resolution were performed sequentially, the acetylated (−)-*trans*-4-*tert*-butoxy-5-hydroxycyclopent-2-enone was prepared in 81% yield and 96% ee [20].

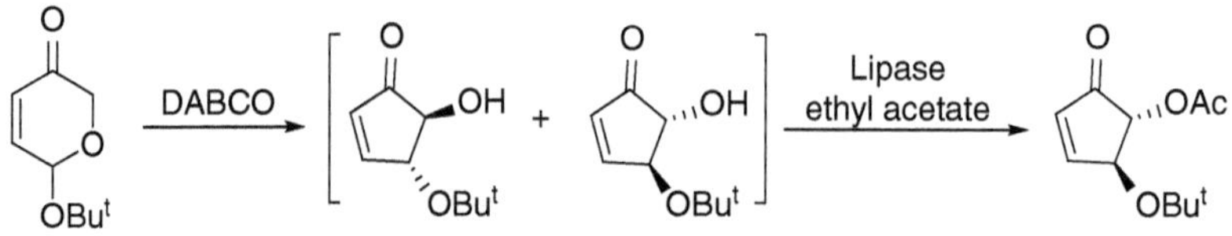

Scheme 6.20 One-pot chemoenzymatic transformation of 6-*tert*-butoxy-2*H*-pyran-3(6*H*)-one to acetylated (−)-*trans*-4-*tert*-butoxy-5-hydroxycyclopent-2-enone.

Lipase also catalyzes the enantioselective acylation of chiral amines. This enzymatic acyl transfer reaction when incorporated with chemical formation of amines led to one-pot concerted chemoenzymatic process for the asymmetric synthesis of chiral amines in optically pure form. For example, a Pd nanocatalyst Pd/AlO(OH)$_6$ catalyzed the reduction of the ketoxime of 3′-methoxyacetophenone to methyl (3′-methoxybenzyl)amine, which could be acylated by using CALB as the catalyst and ethyl methoxyacetate as the acyl donor. The reduction and acylation were carried out in a one-pot procedure, yielding the *R*-amide. The asymmetric reductive acylation of the ketoxime of 3′-methoxyacetophenone was thus achieved in a one-pot manner as shown in Scheme 6.21, and applied to the synthesis of a potent calcimimetic (+)-NPS R-568 from commercially available 3′-methoxyacetophenone [21].

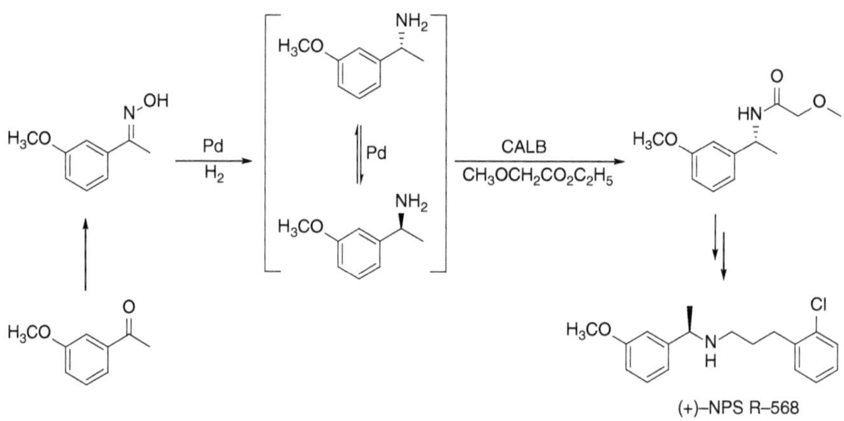

Scheme 6.21 Synthesis of a potent calcimimetic (+)-NPS R-568 involving one-pot chemoenzymatic reductive acylation of ketoxime.

In the above discussion, lipase-catalyzed acyl transfer reaction is combined with chemical transformation of generating racemic alcohol in one-pot concurrent process, leading to asymmetric synthesis of optically active acylated alcohols. In addition, lipase-catalyzed hydrolysis has also been coupled with chemical reaction in one-pot concurrent manner to access a pool of chiral alcohol derivatives. Wen et al. investigated the use of a titanium-complex-mediated acetylcyanation of (1-naphthyloxy)acetaldehyde with acetyl cyanide in CH_2Cl_2

to generate racemic 1-cyano-2-(naphthalen-1-yloxy)ethyl acetate. CALB catalyzed the selective hydrolysis of the (*R*)-acetylated cyanohydrin furnishing (*S*)-1-cyano-2-(naphthalen-1-yloxy)ethyl acetate. The resulting cyanohydrin decomposed into (1-naphthyloxy)acetaldehyde, which could react *in situ* with acetyl cyanide to give racemic 1-cyano-2-(naphthalen-1-yloxy)ethyl acetate, while achieving recycling of the *R*-enantiomer. The lipase-catalyzed kinetic resolution and the *R*-enantiomer recycling were performed in one-pot two-phase system, resulting in the full conversion of (1-naphthyloxy)acetaldehyde into (*S*)-1-cyano-2-(naphthalen-1-yloxy)ethyl acetate, which was further transformed into (*S*)-propanolol (Scheme 6.22). In the same manner, (*R*)-dichloroisoproterenol and (*R*)-pronethalol were prepared starting from 3,4-dichlorobenzaldehyde and 2-naphthylaldehyde, respectively. Thus, a novel strategy of recycling the undesired enantiomer in lipase-catalyzed kinetic resolution in a one-pot process was established and applied to the synthesis of the precursors for β-adrenergic antagonists such as (*S*)-propanolol, (*R*)-dichloroisoproterenol, and (*R*)-pronethalol [22].

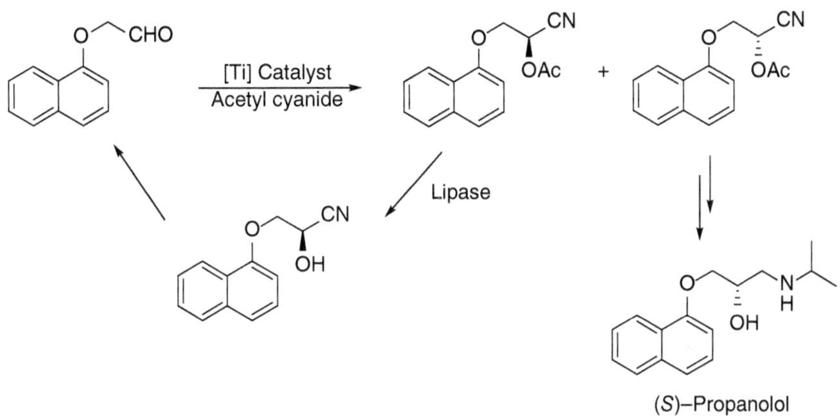

Scheme 6.22 Synthesis of (*S*)-propanolol involving one-pot chemoenzymatic conversion of (1-naphthyloxy)acetaldehyde into (*S*)-1-cyano-2-(naphthalen-1-yloxy)ethyl acetate.

Supramolecular encapsulation of organometallic complexes prevents their diffusion into the bulk solution. This would be helpful for the development of chemoenzymatic one-pot reactions, because it prohibits the interaction of metal complexes with the enzymes that potentially compromise their activity. The tetrahedral Ga_4L_6 [L = N,N'-bis(2,3-dihydroxybenzoyl)-1,5-diaminonaphthalene] cluster is stable at neutral pH in water and provides a well-defined cavity for reaction. Me_3PAu^+ was encapsulated in this cluster, resulting in a gold(I)–Ga_4L_6 host–guest complex with an eight-fold rate enhancement relative to that of the free gold complex in the hydroalkoxylation of allenes. Lipases and esterases were tested as partners with the gold(I)–Ga_4L_6 supramolecular cluster in the tandem process starting from allenic acetates. The enzymatic hydrolysis of allenic acetate generated the alcohol intermediate, which was cyclized to give the substituted tetrahydrofuran product under gold catalysis (Scheme 6.23). For the chiral allenic acetates, lipases or esterases showed enantioselectivity for the hydrolysis. The racemic allenic acetates were kinetically resolved to give (*R*)-alcohols and (*S*)-allenic acetates. The (*R*)-alcohols were transformed to the chiral substituted tetrahydrofuran products in the following gold-catalyzed intramolecular hydroalkoxylation of allenes (Scheme 6.24). These observations underlined the advantages of the supramolecular approach and suggested that encapsulation of reactive complexes may provide a general strategy for carrying out classic organic reactions in the presence of biocatalysts [23].

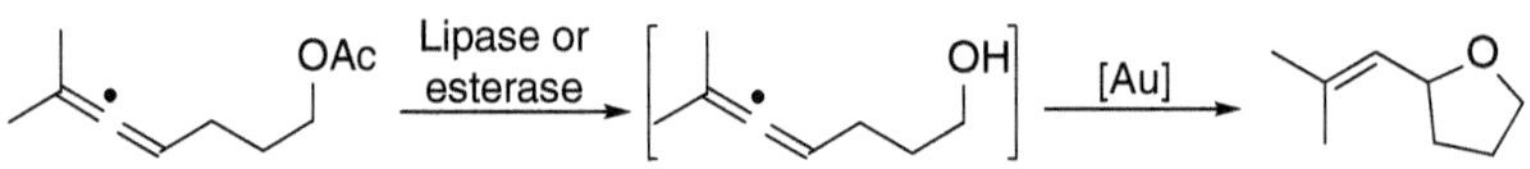

Scheme 6.23 One-pot chemoenzymatic transformation of allenic acetate to substituted tetrahydrofuran.

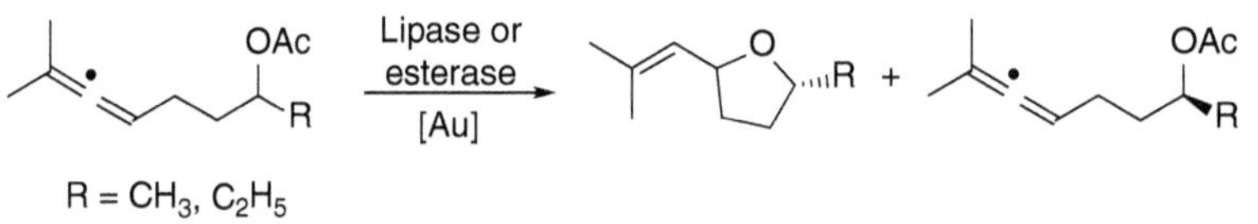

Scheme 6.24 One-pot chemoenzymatic asymmetric transformation of allenic acetate to substituted tetrahydrofuran.

Passerini reaction is a multicomponent reaction involving an aldehyde (or ketone), a carboxylic acid, and an isocyanide to generate an α-acyloxy amide. Lipase catalyzed the hydrolysis of alkenyl esters to give a carboxylic acid and an alkenyl alcohol, and the latter spontaneously isomerizes to an aldehyde or ketone. When this enzymatic hydrolysis was carried out under suitable conditions in the presence of isocyanide, the Passerini reaction occurred simultaneously in a one-pot manner. Indeed, this lipase-promoted tandem reaction proceeded smoothly in dichloromethane with 0.5% (v/v) H_2O. The chemoenzymatic sequence was followed by lipase-catalyzed kinetic resolution of the Passerini product, producing the corresponding α-hydroxy carboxamides in high enantiomeric purity (up to 99% ee, Scheme 6.25). This

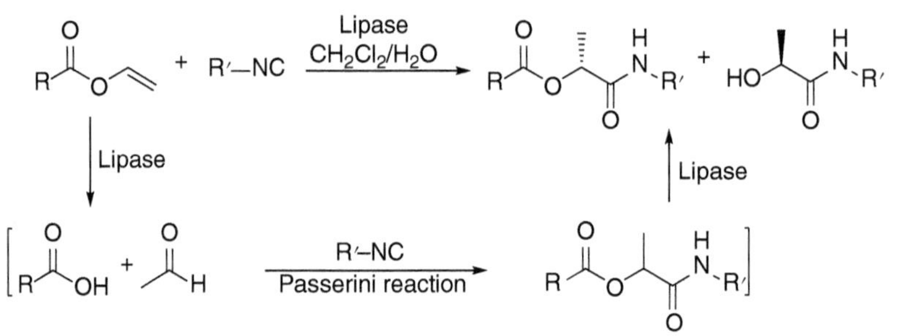

Scheme 6.25 One-pot chemoenzymatic process involves two lipase-catalyzed hydrolysis and Passerini reactions.

one-pot, three-step chemoenzymatic process involved two lipase-catalyzed hydrolysis and Passerini reactions, and it became a useful method for constructing a large library of α-hydroxy carboxamide scaffolds from a small set of substrates [24].

The lipase-catalyzed *in situ* acetaldehyde formation could also be combined with organocatalytic aldehyde–aldehyde C–C bond coupling reactions. By employing CALB as the biocatalyst and trifluoromethyl substituted diphenylprolinol as the organocatalyst, a set of different aromatic aldehydes reacted with acetaldehyde generated *in situ* from vinyl acetate in the environmentally friendly deep eutectic solvents [DESs, glycerol–choline chloride (2:1)], affording the corresponding aromatic 1,3-diols in up to 70% yields and up to 99% ee, followed by reduction of the resulting aldehydes with $NaBH_4$ (Scheme 6.26). DES and CALB were reused up to six cycles without loss of enzymatic activity, and a significant amount of organocatalyst remained in DES for use in the next cycle. When some fresh organocatalyst was added, the product was obtained with stable yield and excellent enantioselectivity in the next reaction cycles [25].

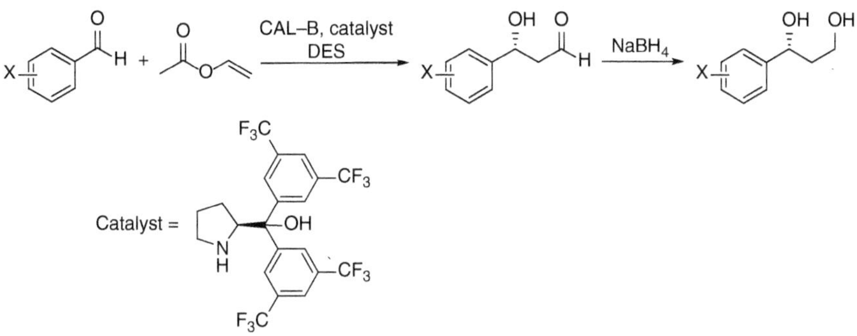

Scheme 6.26 One-pot chemoenzymatic process involves lipase-catalyzed hydrolysis and organocatalytic aldehyde–aldehyde C–C bond formation.

Hydrolases such as lipases catalyze not only hydrolysis of carboxylic esters but also the reverse reactions such as esterification, transesterification, aminolysis, and ammonolysis under suitable reaction conditions. In addition to these normal hydrolase activities, some hydrolases, mainly lipase, also catalyze the transformation of carboxylic acids into the corresponding peroxy acids in the presence of hydrogen peroxide, i.e. perhydrolysis reaction. The resulting peroxy acids can act as an oxidizing agent in the subsequent oxidation reactions. As such, the lipase-catalyzed perhydrolysis reaction has been used as a progressive source of H_2O_2 in the oxidation of various kinds of compounds, thus expanding the versatility of hydrolases in synthetic chemistry [26, 27].

The reactive peracid intermediates of hydrolase-catalyzed perhydrolysis have been demonstrated to be able to oxidize alkenes, ketones, and amines into the corresponding epoxides, esters, or oximes, respectively. In 1990, the immobilized lipase B (Novozyme 435) was shown to mediate the transformation of medium-chain saturated fatty acids into the corresponding peroxy acids in the presence of hydrogen peroxide, which was then used for the *in situ* epoxidation of alkenes (Scheme 6.27) [28].

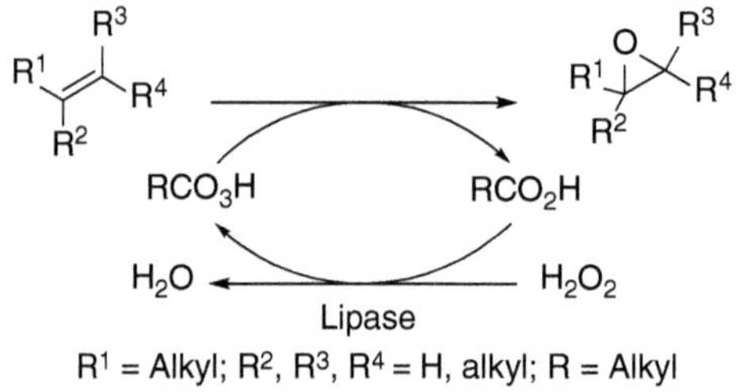

Scheme 6.27 One-pot lipase-catalyzed perhydrolysis and chemical epoxidation. Source: Based on Björkling et al. [28].

When unsaturated fatty acids were treated with hydrogen peroxide in the presence of CALB, the lipase catalyzed the conversion of the unsaturated fatty acids into the corresponding unsaturated peroxy acid, followed by the epoxidation via an uncatalyzed Prileshajev reaction to give epoxidized fatty acids (Scheme 6.28) [29]. Similarly, unsaturated fatty acid esters were perhydrolyzed under lipase catalysis to peroxy acids, and the subsequent epoxidation of themselves to afford the epoxidized fatty acids (Scheme 6.29) [30]. This chemoenzymatic reaction was also applied to plant oils (such as rapeseed oil, sunflower oil and linseed oil), which were epoxidized with good yields [31]. The epoxidized plant oils and fatty acids are mainly used as plasticizers and stabilizers for polyvinyl chloride (PVC) and other plastic materials, among other applications such as diluents for paints, corrosion protecting agents, and as additives to lubricating oils.

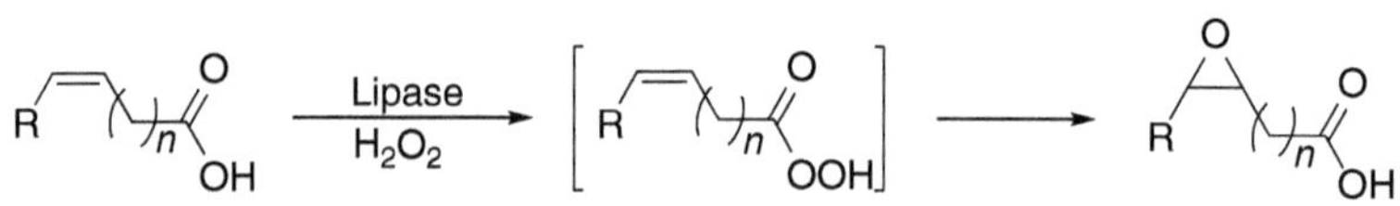

Scheme 6.28 One-pot chemoenzymatic epoxidation of unsaturated fatty acids. Source: Based on Warwel et al. [29].

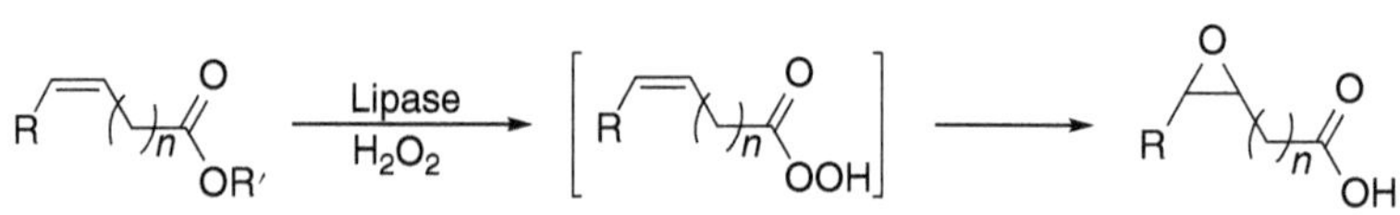

Scheme 6.29 One-pot chemoenzymatic epoxidation of unsaturated fatty acid esters. Source: Based on Klaas et al. [30].

The lipase/H_2O_2 system was also effective for the epoxidation of other alkene substrates with diverse structural features (Scheme 6.30 and Table 6.1). When

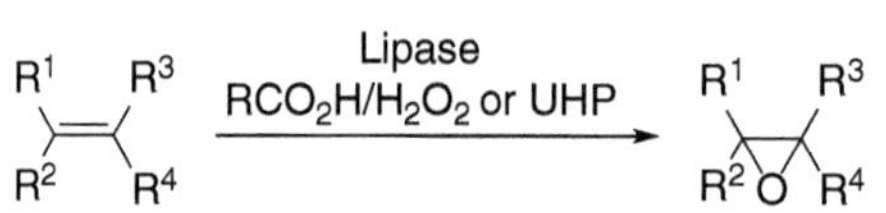

Scheme 6.30 Chemoenzymatic epoxidation of alkenes by lipase/H_2O_2 system.

Table 6.1 Chemoenzymatic epoxidation of alkenes by lipase/H_2O_2 system.

Entry	Alkene	Epoxide	
1	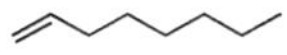	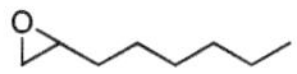	[32]
2	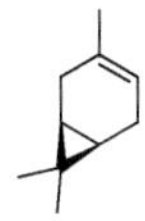 (+)-3-carene	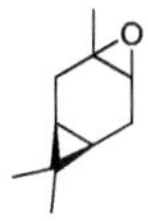	[33]
3	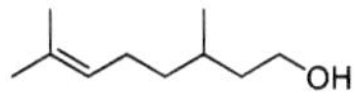 citronellol	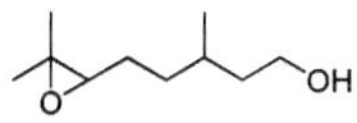	[34]
4	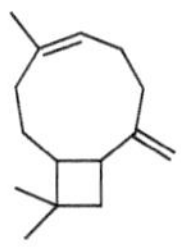 β-caryophyllene	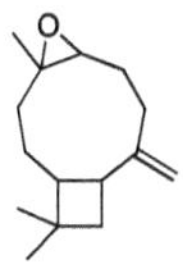	[35]
5	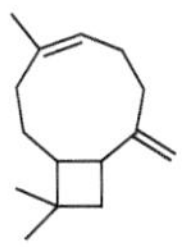 β-caryophyllene	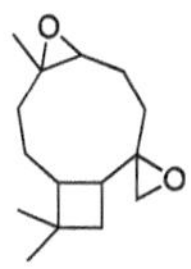	[35]
6	Cyclohexene	Cyclohexene oxide	[36]
7	2-Methylcyclohexene	2-Methylcyclohexene oxide	[36]
8	Cycloheptene	Cycloheptene oxide	[36]
9	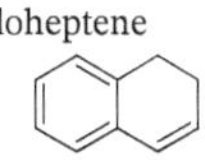 1,2-dihydronaphthalene	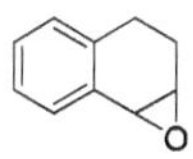	[36]
10	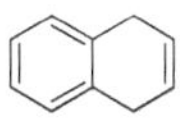 1,4-dihydronaphthalene	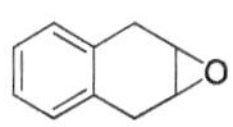	[36]
11	4-Octene	4-Octene oxide	[36]
12	Styrene	Styrene oxide	[36]
13	2-Bromostyrene	2-Bromostyrene oxide	[36]
14	3-Bromostyrene	3-Bromostyrene oxide	[36]
15	4-Bromostyrene	4-Bromostyrene oxide	[36]
16	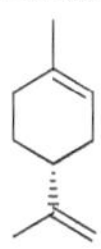 limonene	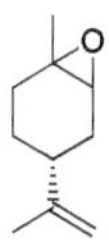	[36]
17	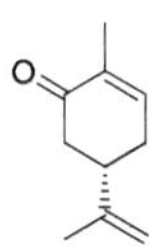 (*S*)-carvone	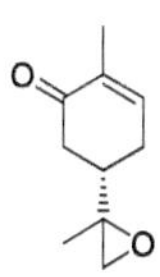	[36]
18	α-Pinene	α-Pinene oxide	[37]
19	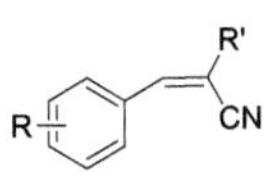	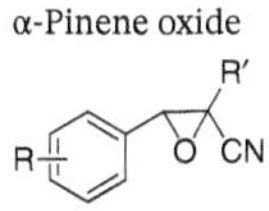	[38]

various carboxylic acids, hydrogen peroxide, and 1-octene were mixed in toluene in the presence of lipase Novozym 435, 1-octene was transformed into 1,2-epoxyoctane with up to 68% yield [32]. The chemoenzymatic epoxidation of β-caryophyllene afforded mono-epoxide and di-epoxide. The ratio of the two products depended on the experimental conditions including the source of the lipase, the oxidizing agent (H_2O_2 aq. (AHP) or urea–hydrogen peroxide [UHP]) and the carboxylic acid, and the organic medium. When CALB, the most efficient catalyst, was used, mono-epoxide was obtained with >99% conversion in hexane, while di-epoxide was produced in ethyl acetate or toluene with >99% conversion [35].

A series of cyclic and linear alkenes were tested for the chemoenzymatic epoxidation using a lipase from *Rhizomcor miehei*. The corresponding epoxides were obtained in moderate to excellent yields by choosing acetonitrile as solvent, lauric acid as peracid precursor, and the urea–hydrogen peroxide as oxidant. Interestingly, a complete preference for the oxidation of the exocyclic alkene bond of (*S*)-carvone was observed, while the internal alkene bond of limonene was epoxidized [36]. To overcome the toxicity of aqueous hydrogen peroxide to the enzyme, and for it to be delivered slowly into the reaction mixture, nontoxic and anhydrous UHP were utilized as the oxidant. The UHP complex allowed the progressive liberation of H_2O_2 in the reaction medium, avoiding the presence of an excess of free hydrogen peroxide. A range of alkenes were epoxidized using UHP instead of free H_2O_2 with yields ranging from 75% to 100% [39].

This chemoenzymatic epoxidation was also applied to the allylated gallic and vanillic acids. By treating allylated gallic acid with caprylic acid and H_2O_2 in the presence of CALB, a mixture of the tri-epoxidized and tetra-epoxidized derivatives of gallic acid were obtained in 77 : 23 ratio (Scheme 6.31). Similarly, the epoxidation of the diallylated vanillic acid gave a mixture of mono- and di-epoxidized products with the di-epoxidized derivative as the major product (Scheme 6.32) [40].

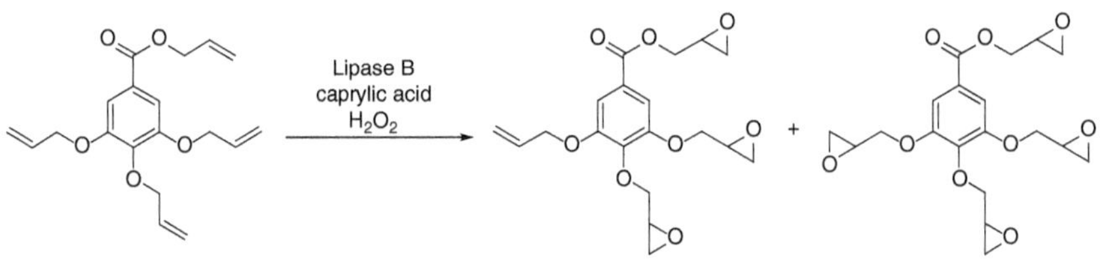

Scheme 6.31 One-pot chemoenzymatic epoxidation of allylated gallic acid.

The lipase-mediated perhydrolysis had also been incorporated with Baeyer–Villiger oxidation (BVO) of a range of ketones. The first example was reported in 1995, in which the BVO of some 2- and 3-substituted cyclopentanones and cyclohexanones were realized by using myristic acid and H_2O_2 in the presence of CALB [41]. The immobilized CALB was found to catalyze the perhydrolysis of ethyl acetate to generate peroxyacetic acid in the presence of UHP. This perhydrolysis was thus examined for the BVO of various substituted cyclohexanones, which were converted into the corresponding lactone with up to 96% yield in ethyl acetate

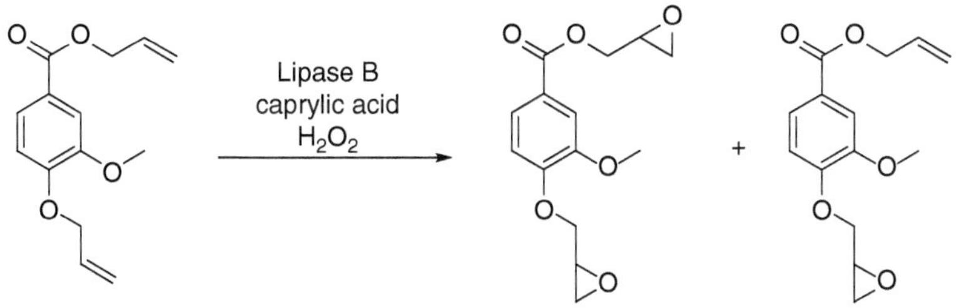

Scheme 6.32 One-pot chemoenzymatic epoxidation of diallylated vanillic acid. Source: Based on Aouf et al. [40].

(Scheme 6.33) [42]. In contrast to cyclohexanones, the oxidation of cyclopentanones also gave the corresponding δ-valerolactones, but the lactone product reacted further with ethyl acetate via the catalysis of lipase to deliver δ-hydroxy esters and/or δ-acetylated esters, depending on the structure of the cyclopentanones (Scheme 6.34) [43].

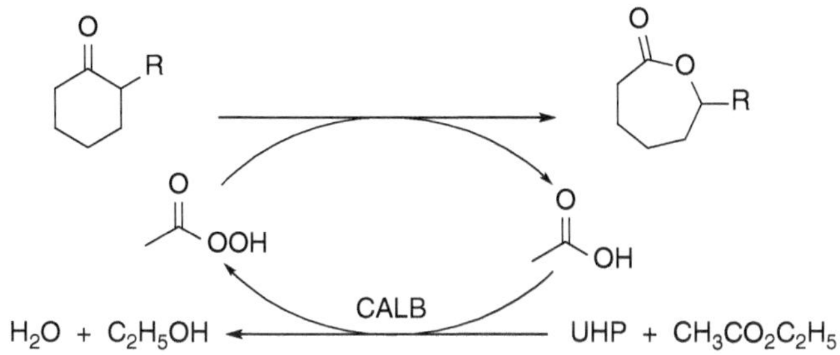

Scheme 6.33 One-pot chemoenzymatic perhydrolysis and Baeyer–Villiger oxidation of substituted cyclohexanones. Source: Based on Ríos et al. [42].

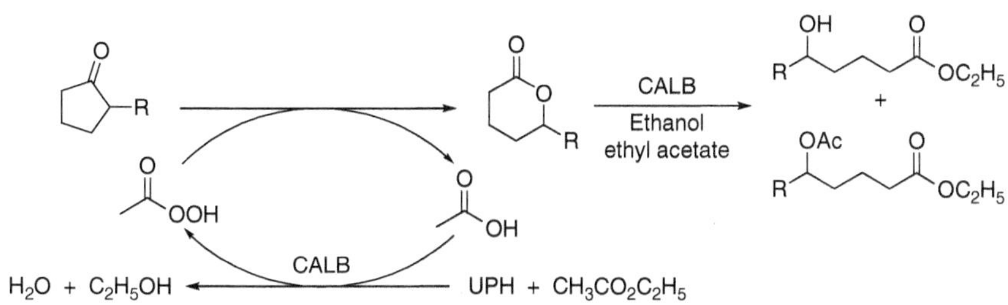

Scheme 6.34 One-pot chemoenzymatic perhydrolysis and Baeyer–Villiger oxidation of substituted cyclopentanones. Source: Based on Rios et al. [43].

After the reaction condition optimization, the one-pot chemoenzymatic BVO of several 3-substituted cyclobutanones was performed at high substrate concentration utilizing CALB, ethyl acetate, and UHP complex as the oxidizing system, affording the corresponding γ-butyrolactones with up to 99% isolated yields (Scheme 6.35) [44]. Although the chemoenzymatic BVO was a useful transformation, CALB required fairly long reaction times and sometimes showed poor stability in the presence of H_2O_2. To address these challenges, CALB was immobilized onto silica with multimodal pore structure (MH) and typical SBA-15. These supporting materials were chemically modified with organosilanes terminated with methyl, octyl, and hexadecyl groups prior to enzyme immobilization. The resulting biocatalysts showed higher activity than CALB alone and good stability in the presence of 60% aqueous H_2O_2. The immobilized lipase biocatalyst on chemically modified MH with methyl-terminated organosilanes (MH-Me-L) catalyzed the chemoenzymatic BVO of cyclohexanones and cyclobutanones, giving the corresponding lactones with 80–98% yields, but low isolated yields were obtained for cyclopentanone and cycloheptanone. Formation of δ-hydroxy acid from lactone hydrolysis reaction was detected in the chemoenzymatic BVO of cyclopentanone [45].

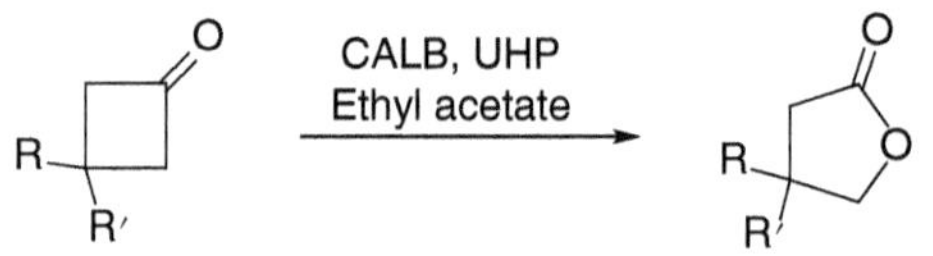

Scheme 6.35 One-pot chemoenzymatic perhydrolysis and Baeyer–Villiger oxidation of substituted cyclobutanones. Source: From González-Martínez et al. [44].

CALB was also immobilized on multiwalled carbon nanotubes using a simple physical adsorption technique. The resulting new heterogeneous nanobiocatalyst (CALB-MWCNTs) catalyzed the chemoenzymatic BVO of various cyclic ketones with diverse structural features to the corresponding lactones in 86–95% isolated yields using 30% aqueous H_2O_2 as the oxidant in the presence octanoic acid under mild conditions. The nanobiocatalyst could be reused five times without significant loss of activity at 20 °C, and three times at 30 °C [46]. Furthermore, the BVO of 2-methylcyclohexanone with this highly stable and active nanobiocatalyst was also carried out in a flow reactor, affording 6-methyl-ε-caprolactone with 87% yield and >99% selectivity. Environmentally friendly ethyl acetate was used as both solvent and precursor for *in situ* generation of peracid in this continuous flow process [47].

The substituted lactones obtained from chemoenzymatic BVO of substituted ketones has been subjected to ring-opening polymerization (ROP). Chemoenzymatic BVO of various substituted cyclohexanones was carried out in ethyl acetate using the immobilized CALB as the biocatalyst. Subsequent ROP of the resulting substituted lactones was initiated by adding a small amount of butanol after evaporation of the solvent. The one-pot sequential chemoenzymatic BVO–ROP process generated the corresponding substituted polyesters in the range of M_w 1000 to 12 800 (Scheme 6.36). The molecular weights were affected by the structural features of the substituted cyclohexanones [48].

Under similar reaction conditions, the lipase-mediated BVO of levoglucosenone (LGO), a valuable chiral platform chemical available by pyrolysis of cellulose, resulted in a mixture of (*S*)-γ-hydroxymethyl-α,β-butenolide (HBO) and its formate (FBO), followed by acid hydrolysis to give optically pure (*S*)-HBO. In the same

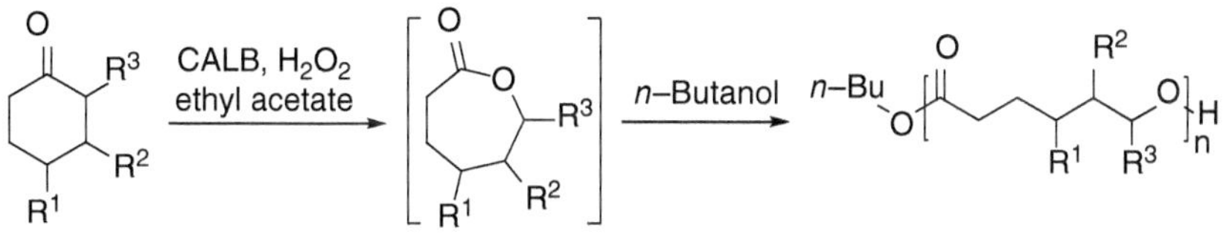

Scheme 6.36 One-pot sequential chemoenzymatic BVO-ROP of substituted cyclohexanones.

manner, 2H-LGO, which could be obtained by palladium-catalyzed hydrogenation of LGO, underwent BVO and acid hydrolysis to provide (*S*)-γ-hydroxymethyl-γ-butyrolactone (2H-HBO, Scheme 6.37). Both HBO and 2H-HBO are highly valuable intermediates for the synthesis of drugs or flavors [49].

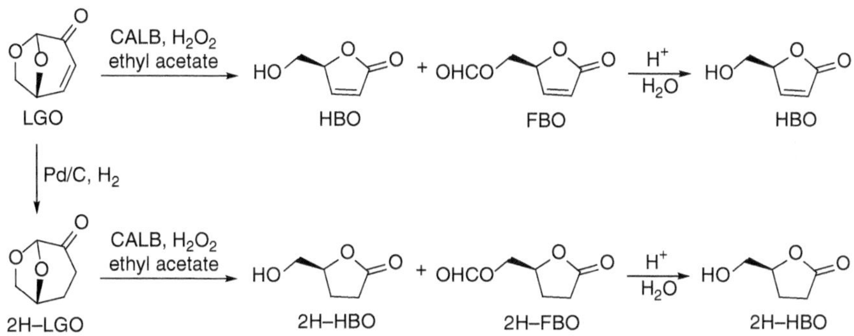

Scheme 6.37 One-pot chemoenzymatic perhydrolysis and Baeyer–Villiger oxidation of levoglucosenone (LGO) and dihydro levoglucosenone (2H-LGO).

The reactive peracid intermediate, which was generated *in situ* by lipase-catalyzed perhydrolysis reaction, could also oxidize amines to give the corresponding oximes in a one-pot chemoenzymatic process. The substituted benzylamines and α-methylbenzylamines were oxidized in short reaction times (one hour) by employing CALB as the enzyme, UHP as the chemical oxidant, and ethyl acetate as both the peracid precursor and the solvent, and the corresponding aldoximes and ketoximes were isolated in high purity with 71–82% and 90–98% yields (Scheme 6.38), respectively [50]. When the oxidation of benzylamine was studied in detail, concomitant formation of benzaldehyde (<1%), benzonitrile (<1%), and *N*-benzylidenebenzylamine (5–9%) was detected by GC analyses. The amount of *N*-benzylidenebenzylamine increased when the reaction was carried out at higher temperatures (37 and 45 °C).

A range of phenyl-substituted anilines were also oxidized using immobilized lipase Novozym 435 and UHP in ethyl acetate at room temperature to afford the corresponding azoxybenzenes in 63–94% yields (Scheme 6.39) [51]. Similarly, the chemoenzymatic oxidation of various *N*-alkylimines was conducted in acetonitrile using CALB as the biocatalyst, UHP as the peroxide donor, and octanoic acid as the acyl donor, generating a mixture of *E*- and *Z*-*N*-alkyloxaziridines in 33–99%

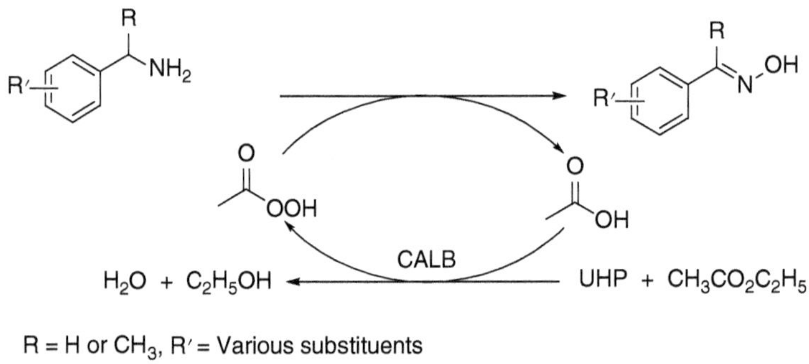

Scheme 6.38 One-pot chemoenzymatic oxidation of substituted benzylamines and α-methylbenzylamines.

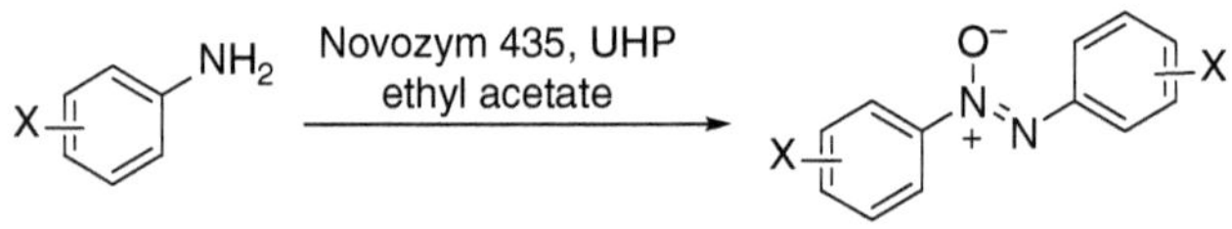

Scheme 6.39 One-pot chemoenzymatic oxidation of substituted anilines. Source: From Yang et al. [51].

conversions, with the E-isomer being the major product and the ratio of E- and Z-isomers (65 : 35–100 : 0) being dependent on the stereoelectronic nature of the substituents in the N-alkylimines (Scheme 6.40) [52]. For the substrate with R being a bulky *tert*-butyl and no substituent on the phenyl ring, only the E-isomer of oxaziridine was detected.

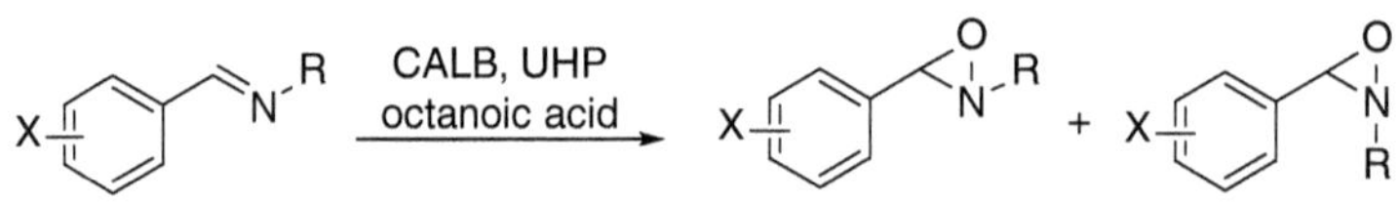

Scheme 6.40 One-pot chemoenzymatic oxidation of *N*-alkylimines. Source: Based on Yang et al. [51].

In the presence of catalytic amount (5 mol%) of 2,2,6,6-tetramethyl-1-piperidinyl-oxy (TEMPO), this system also oxidized salicyl alcohols to the corresponding salicylaldehydes in 81–95% yields under the optimal reaction conditions (Scheme 6.41) [53].

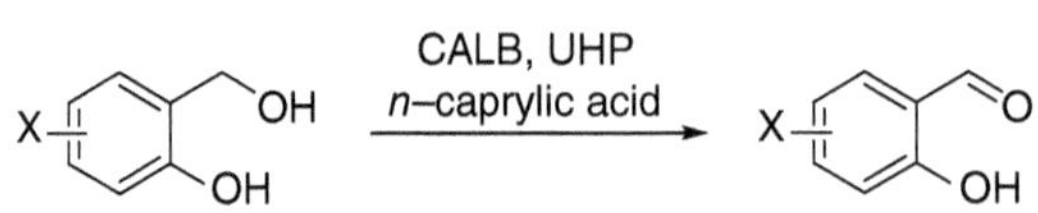

Scheme 6.41 One-pot chemoenzymatic oxidation of salicyl alcohols. Source: Based on Zhao et al. [53].

6.1.2 Carbonyl Reductases

Carbonyl reductases (or alcohol dehydrogenases) catalyze the asymmetric reduction of ketones and have been widely applied for the synthesis of optically pure alcohols in organic synthesis. These enzymes require cofactor NAD(P)H, which is usually regenerated using a co-enzyme (glucose dehydrogenase, formate dehydrogenase, or other enzymes) system or a co-substrate (such as isopropanol) system. The use of a metal-complex catalyst for the cofactor recycling is attractive, but mutual inactivation often occurs, making it impractical. Recently, a rhodium (Rh) complex was immobilized in a bipyridine-based periodic mesoporous organosilica (BPy-PMO) by mixing $[RhCp^*Cl_2]_2$ and BPy-PMO powder in MeOH, and the resulting Rh catalyst (Rh@PMO) showed high catalytic activity for the transfer hydrogenation of 2-cyclohexenone. The Rh catalyst also catalyzed the transfer hydrogenation of NAD^+ to give nicotinamide adenine dinucleotide cofactor (NADH), and thus was used in the horse liver alcohol dehydrogenase (HLADH)-catalyzed reduction of 4-phenyl-2-butanone for NADH regeneration. (*S*)-(+)-4-Phenyl-2-butanol was obtained with 91% conversion and >98% ee after 42 hours (Scheme 6.42) [54].

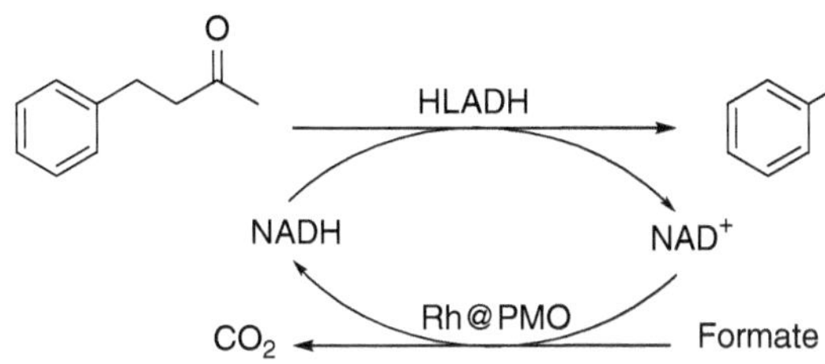

Scheme 6.42 Rh-catalyzed NADH regeneration for enzymatic ketone reduction. Source: Based on Himiyama et al. [54].

In addition to metal-catalyzed cofactor regeneration, the metal-catalyzed ketone or aldehyde formation under aqueous conditions at room temperature has also been carried out with enzymatic reduction in one-pot concurrent manner, in which the newly formed carbonyl group served as the substrate of an alcohol dehydrogenases (ADH) to produce alcohol. It was reported that $(Me_3P)CpRu(NCMe)_2{}^+$ encapsulated in a tetrahedral Ga_4L_6 (L = N,N'-bis(2,3-dihydroxybenzoyl)-1,5-diaminonaphthalene) cluster catalyzed the isomerization of allylic alcohols into the corresponding aldehydes. This transformation was then incorporated with an enzymatic reduction of the resulting aldehydes. In the reaction system containing HLADH, NADPH, yeast formate dehydrogenase (FDH), sodium formate, and the supramolecular Ru catalyst, 1-propenol was converted into propanol in 61% yield after six hours at 37 °C via concurrent metal-catalyzed isomerization and enzymatic reduction (Scheme 6.43) [23].

Similarly, the enantioselective transformation of substituted allylic alcohols into the corresponding optically active alcohols had been realized by employing a ruthenium complex and commercially available ketoreductases. In order to achieve this, the researchers first investigated the Ru-catalyzed redox isomerization of

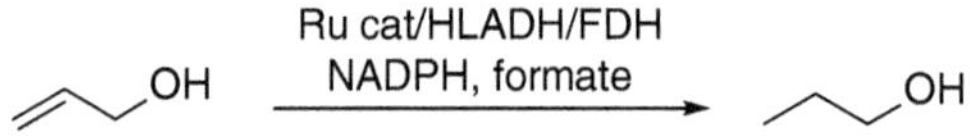

Scheme 6.43 One-pot concurrent Ru-catalyzed isomerization and enzymatic reduction. Source: Based on Wang et al. [23].

allylic alcohols in the presence of the ketoreductase under the reaction conditions required for the enzymatic reduction. It was found that a bis(allyl)ruthenium(IV) complex catalyzed effective redox isomerization of α-vinylbenzyl alcohol under these conditions. The ketone product was obtained with 99% yield in 2.5 hours and the concomitant reduction of ketone via Ru-catalyzed transfer hydrogenation was not observed. By using a commercial KRED Screening Kit, the ketoreductases were identified to catalyze the reduction of the resulting ketones without inhibition by the metal catalyst, affording both enantiomers of the corresponding chiral alcohols with 97–>99% ee. After the two reactions were optimized independently and the compatible reaction conditions established, the chemoenzymatic transformation of various substituted allylic alcohols was carried out in a one-pot process sequential manner. Once the Ru-catalyzed isomerization was finished, the selected ketoreductase and NADPH were added into the reaction mixture. The corresponding saturated alcohols were isolated with 85–90% yields and >98% ee values. The Ru-catalyzed isomerization and enzymatic reduction were then carried out in a one-pot concurrent mode with the metal catalyst and the enzyme being added at the beginning. The corresponding saturated alcohols were isolated with 60–86% yields and >99% ee values (Scheme 6.44). The lower yields were due to the incomplete bioreduction since ketone intermediates were detected in the product mixtures [55].

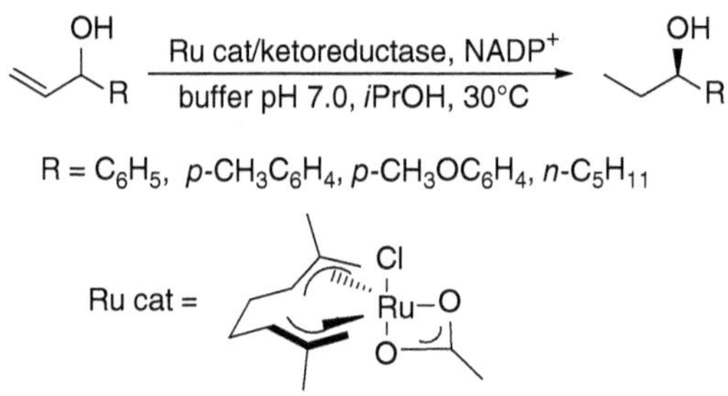

Scheme 6.44 One-pot concurrent Ru-catalyzed isomerization and enzymatic reduction.

The combination of the Ru-catalyzed isomerization of racemic allylic alcohols and the enantioselective bioreduction was also accomplished in a mixture (50/50, w/w) of phosphate buffer and DESs consisting of choline chloride (ChCl)/glycerol (Gly) (1 : 2). This chemoenzymatic cascade process was carried out in both sequential and concurrent modes, leading to formation of the corresponding chiral alcohols with high yields and excellent ee values. Ketone intermediate was also detected in the product mixture with 4–32% in the concurrent process [56].

The metal nanoparticles (Pd or Pt NPs) catalyzed the hydrogenation of the C=C bond of 3-methyl-2-cyclohexenone, affording racemic 3-methyl-2-cyclohexanone.

A NAD-dependent thermostable alcohol dehydrogenase (TADH) in combination with the metal nanoparticles (Pd or Pt NPs) was tested as the catalysts for the double reduction of 3-methyl-2-cyclohexenone in a one-flask setting. A conversion of 42% was achieved with 14% yield of 3-(1*S*,3*S*)-methylcyclohexanol by using Pd NPs (Scheme 6.45), but no reaction was observed using Pt NPs. The interactions between the metal catalyst and the enzyme were proposed to result in their mutual inhibition [57].

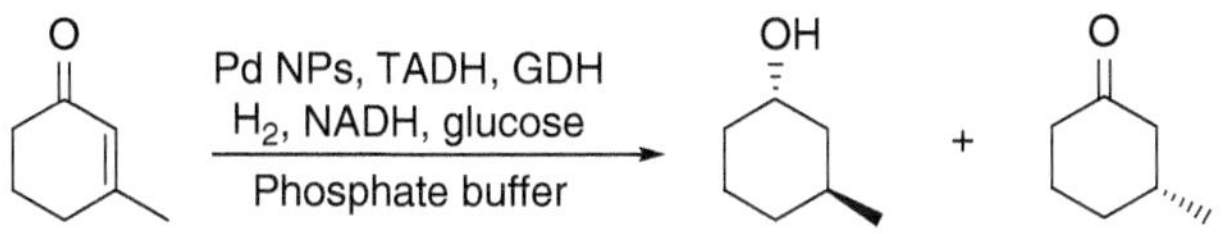

Scheme 6.45 One-pot concurrent Pd-catalyzed hydrogenation and enzymatic reduction.

Organocatalytic aldol reaction of aldehyde and ketones generates β-hydroxy ketones, which can be reduced to furnish 1,3-diols. The organocatalytic aldol reactions are usually carried out in organic solvents, but also can proceed in aqueous medium. This makes it feasible to combine an aldol reaction with an enzymatic reduction of the resulting β-hydroxy ketone in one-pot setting. A proline derivative organocatalyst was found to catalyze the aldol reaction of *m*-chlorobenzaldehyde with acetone in a phosphate buffer (pH 7.5, 50 mM) with 28% (v/v) 2-propanol to give the β-hydroxy ketone. The aldol reaction proceeded smoothly at high substrate concentration of 500 mM, but only 22% conversion was obtained at 50 mM substrate concentration, at which enzymatic reductions are normally performed. Fortunately, an ADH from *Rhodococcus* sp. was identified as a biocatalyst for stereoselective reduction of ketones at such high substrate loading in an organic water-miscible medium. The organocatalyst and the enzyme ADH were added at the same time into the reaction mixture containing 500 mM *m*-chlorobenzaldehyde, nine equivalent of acetone, 28% (v/v) 2-propanol, and cofactor NAD^+ in phosphate buffer (pH 7.5, 50 mM). The one-pot process resulted in a 50% conversion and production of the desired (1*R*,3*S*)-diol with 33% isolated yield (Scheme 6.46) [58].

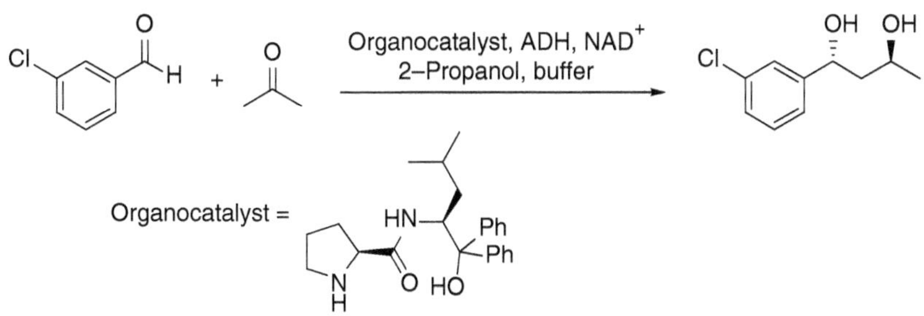

Scheme 6.46 One-pot concurrent organocatalytic aldol reaction and enzymatic reduction. Source: Based on Rulli et al. [58].

The hydroxy group resulting from enzymatic ketone or aldehyde reduction can be converted into other functional groups. Therefore, enzymatic ketone or aldehyde reduction has also been combined with chemical transformations to convert ketones or aldehydes into other compounds. For example, alcohol dehydrogenases catalyze the reduction of ketoesters to give the corresponding hydroxyesters. For the γ- and δ-ketoesters, the products γ- and δ-hydroxyesters may undergo intramolecular cyclization to form lactones under suitable reaction conditions. When the bioreduction of methyl levulinate by alcohol dehydrogenases from *Rhodococcus ruber* (ADH-A) or *Lactobacillus brevis* (LBADH) was carried out at pH 9.0 and 30 °C, the (*S*)- or (*R*)-γ-valerolactone was obtained as the only product via the spontaneous cyclization of the γ-hydroxyester (Scheme 6.47). The lactone formation was greatly affected by the structures of hydroxyesters and the reaction conditions. At lower pH, the lactone formation proceeded to a lower extent. The bioreduction of ethyl levulinate gave a mixture of the γ-hydroxyester and γ-valerolactone under the same reaction conditions. For methyl 5-oxohexanoate and ethyl 5-oxohexanoate, a mixture of 5-hydroxyhexanoate and δ-caprolactone was obtained with 5-hydroxyhexanoate as the major product. For a bulkier substrate methyl 4-oxo-4-phenylbutanoate, the reduction using alcohol dehydrogenase from *Ralstonia* sp. (RasADH) as the biocatalyst led to the formation of only methyl 4-hydroxy-4-phenylbutanoate in 97% conversion and 99% ee. For the cases of incomplete concurrent lactone formation, the corresponding lactones were isolated by treating the reaction mixture with 1 M HCl solution after the bioreduction was completed in a one-pot two-step tandem protocol [59].

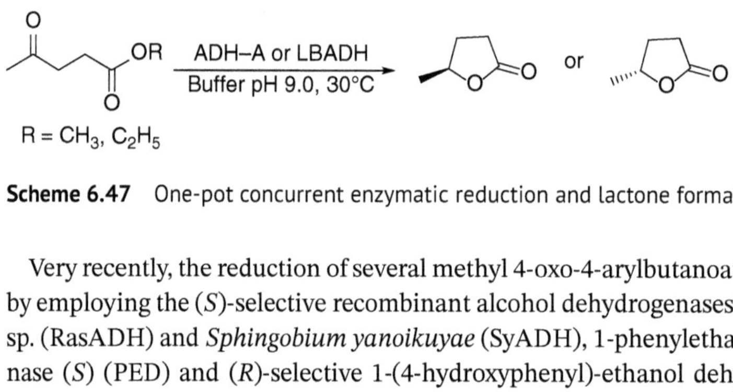

Scheme 6.47 One-pot concurrent enzymatic reduction and lactone formation.

Very recently, the reduction of several methyl 4-oxo-4-arylbutanoate was achieved by employing the (*S*)-selective recombinant alcohol dehydrogenases from *Ralstonia* sp. (RasADH) and *Sphingobium yanoikuyae* (SyADH), 1-phenylethanol dehydrogenase (*S*) (PED) and (*R*)-selective 1-(4-hydroxyphenyl)-ethanol dehydrogenase (*R*) (HPED) from *Aromatoleum aromaticum* in Tris–HCl buffer (pH 7.5) with 2-PrOH (90 : 10, v/v), affording a mixture of the corresponding (*S*)- and (*R*)-configured methyl 4-hydroxy-4-arylbutanoate and γ-aryl-γ-butyrolactone, respectively. After biotransformations, treatment with trifluoroacetic acid led to completion of lactone formation, giving the respective (*S*)- or (*R*)-γ-aryl-γ-butyrolactones with up to >99% ee (Scheme 6.48) [60].

DKR reduction of dimethyl 2-benzoylsuccinate and its derivatives by commercially available ketoreductases followed by lactonization was recently reported to afford the corresponding lactones in 52–90% yields, > 20 : 1 dr and 86–>98% ee as shown in Scheme 6.49. Dimethyl 2-benzoylsuccinate and its derivatives could be synthesized by Stetter reaction of aldehydes with fumarate utilizing Glorius's Isa-NHC catalyst. Chemical formation and enzymatic reduction of dimethyl

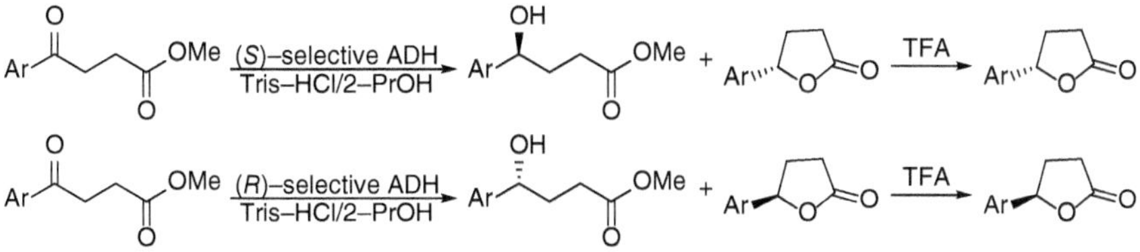

Scheme 6.48 One-pot concurrent enzymatic reduction and lactone formation. Source: Based on Borowiecki et al. [60].

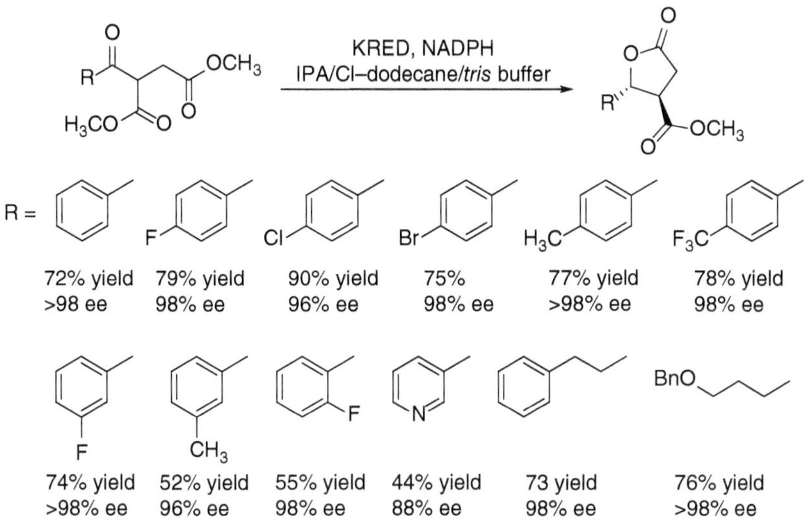

Scheme 6.49 One-pot concurrent chemoenzymatic DKR reduction and lactone formation of α-acetyl β-ketoesters.

2-(4-chlorobenzoyl)succinate were performed in a biphasic reaction medium containing isopropanol, 1-chlorododecane, and Tris buffer in one-pot one-step process, affording methyl (2*R*,3*R*)-2-(4-chlorophenyl)-5-oxotetrahydrofuran-3-carboxylate in 27% yield from 4-chlorobenzaldehyde. The product yield was improved to 75% by conducting the chemoenzymatic cascade sequentially using tetrahydrofuran (THF) as solvent for Stetter reaction and switching the reaction medium to the biphasic system (Scheme 6.50).

Dimethyl 2-benzoylsuccinate could also be prepared by photoredox-induced decarboxylation of phenylglyoxylic acid sodium salt followed by a Giese-type addition to dimethyl maleate in the presence of (Ir[dF(CF_3)ppy]$_2$(dtbpy))PF_6 catalyst. Combination of the photochemical reaction with enzymatic DKR reduction in a single one-pot operation produced methyl (2*R*,3*R*)-5-oxo-2-phenyltetrahydrofuran-3-carboxylate in 49% yield, whereas the lactone product was obtained with 73% yield in a sequential process (Scheme 6.51) [61].

In addition to lactone formation, the alcohol product from enzymatic reduction can also undergo other chemical transformations under same reaction conditions or by adding additional reagents. The resting and lyophilized cells of *Rhodotorula*

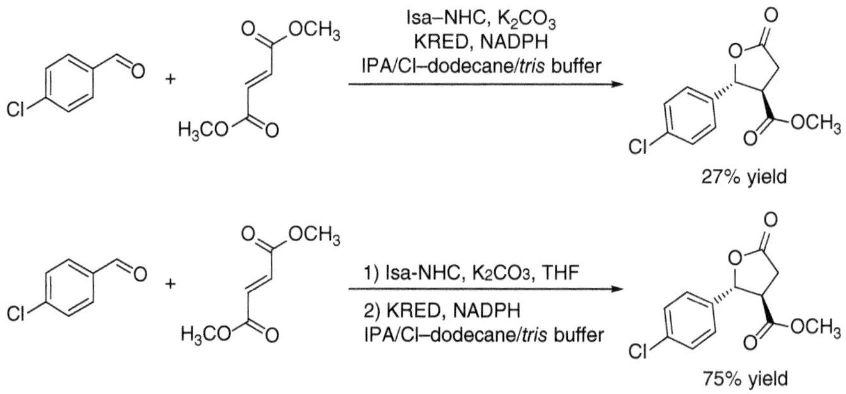

Scheme 6.50 One-pot concurrent and sequential Stetter reaction and enzymatic DKR reduction.

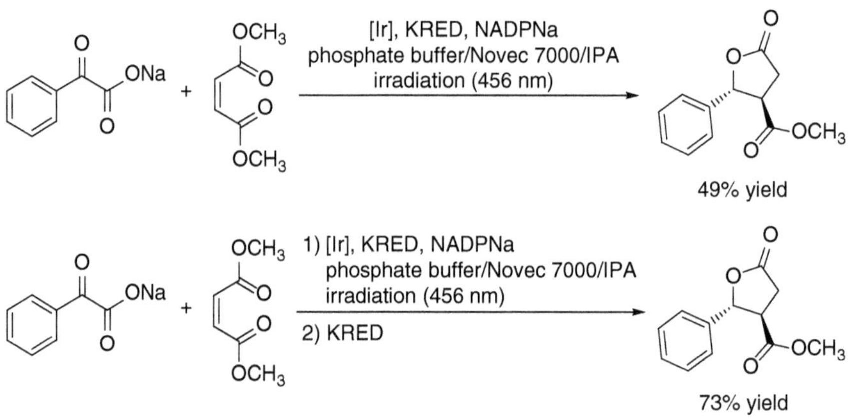

Scheme 6.51 One-pot concurrent and sequential photocatalytic reaction and enzymatic DKR reduction. Source: Based on Maskeri et al. [61].

sp. LSL, which was isolated from a local land farming, catalyzed the reduction of prochiral arylketones into the corresponding (*R*)-alcohols with > 99% ee at pH 6.5 for α-chloroacetohenones and pH 4.3 for α-bromoacetohenones. When the reduction of α-chloroacetohenone was carried out at pH 12, a mixture of (*R*)-phenyloxirane (65%) and (*R*)-1-phenylethane-1,2-diol (35%) was obtained. Further experiments suggested that the formation of (*R*)-1-phenylethane-1,2-diol resulted from the substitution of chloro group in α-chloroacetohenone by OH^- at pH 12, followed by the biocatalytic reduction of the resulting α-hydroxyacetohenone. (*R*)-Phenyloxirane was produced by the biocatalytic reduction of α-chloroacetohenone and subsequent cyclization (Scheme 6.52). When the reaction was performed at pH 10, the formation of (*R*)-1-phenylethane-1,2-diol was minimized to give (*R*)-phenyloxirane as the sole product. (*R*)-1-Phenylethane-1,2-diol was prepared when α-bromoacetohenone was reduced with 100% conversion and >99% ee by this biocatalyst at pH 12 (Scheme 6.53). The resting and lyophilized cells of *Rhodotorula* sp. LSL showed high resistance in alkaline media, enabling one-pot chemoenzymatic synthesis of enantiopure epoxides and diols from aromatic α-haloketones [62].

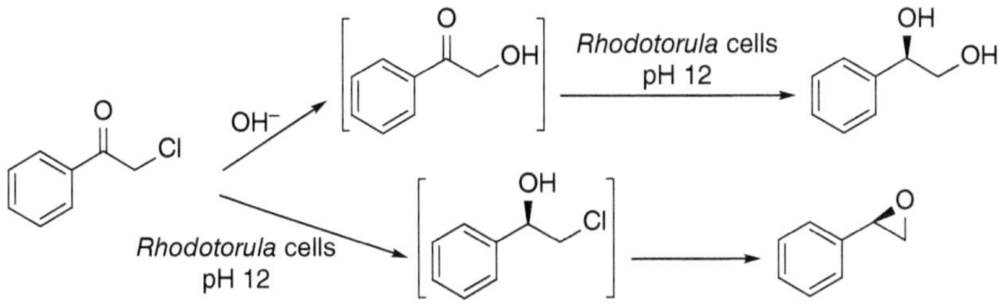

Scheme 6.52 One-pot chemoenzymatic formation of enantiopure epoxides and diols.

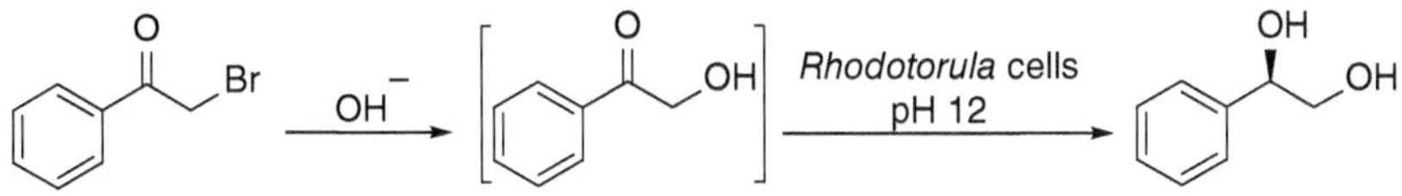

Scheme 6.53 One-pot chemoenzymatic synthesis of enantiopure diols.

Ketoreductases catalyze the reduction of β-ketonitriles into optically active β-hydroxynitriles with exquisite stereoselectivity. Ruthenium complexes can catalyze the hydration of nitriles in water to afford amides. As such, bioreduction/hydration cascade process was developed by combining the two reactions in aqueous medium to directly convert β-ketonitriles into optically active β-hydroxyamides in a concurrent mode. A commercial kit of KREDs and two enzymes, namely the (*R*)-selective ADH from *Lactobacillus kefir* and (*S*)-selective ADH from *R. ruber*, were examined toward the reduction of 3-(3′-chlorophenyl)-3-oxopropanenitrile and the corresponding amide. A commercial enzyme KRED-P2-H07 was found to have an excellent activity toward the β-ketonitrile, but very poor activity with the β-ketoamide. For the hydration of 3-(3′-chlorophenyl)-3-oxopropanenitrile and 3-(3′-chlorophenyl)-3-hydroxypropanenitrile catalyzed by Ru complex, 3-(3′-chlorophenyl)-3-hydroxypropanenitrile was completely converted into the β-hydroxyamide, but 3-(3′-chlorophenyl)-3-oxopropanenitrile was mostly recovered. KRED-P2-H07 and Ru complex were then combined to convert 3-(3′-chlorophenyl)-3-oxopropanenitrile to the corresponding β-hydroxyamide in a one-pot concurrent process. When the transformation was performed at 40 °C, a mixture of enantiomerically pure (*R*)-3-(3′-chlorophenyl)-3-hydroxypropanenitrile and (*R*)-3-(3′-chlorophenyl)-3-hydroxypropaneamide was obtained in the ratio of 79 : 21, showing a high efficiency of the enzyme but a low activity of the Ru catalyst. Another commercial enzyme KRED-P2-C11 catalyzed the reduction of β-ketonitrile and β-ketoamide to give (*R*)-β-hydroxynitrile and (*R*)-β-hydroxyamide with high conversions and >99% ee. By employing KRED-P2-C11 and Ru complex, 3-(3′-chlorophenyl)-3-oxopropanenitrile was converted into (*R*)-3-(3′-chlorophenyl)-3-hydroxypropaneamide at 60 °C and 100 mM substrate concentration, with 97% conversion and >99% ee. Some other aromatic β-ketonitriles were effectively transformed into the corresponding (*R*)- or (*S*)-β-hydroxyamides by this dual Ru/ketoreductase catalytic system of Ru complex and suitable ketoreductase. In some cases, the resulting (*R*)-β-hydroxyamides were isolated with very high yield (92–94%, Scheme 6.54). While these results have demonstrated the successful simultaneous action of enzymes and metal catalysts in one-pot, it was still challenging to find suitable ketoreductases for other β-ketonitriles and preparation of the hydroxyamide antipodes. Consequently, appropriate ketoreductases were found, which together with a metal-catalyzed hydration step allowed a sequential one-pot/two-step process for the preparation of aromatic β-hydroxyamide antipodes and the transformation of other cyclic and acyclic β-ketonitriles [63].

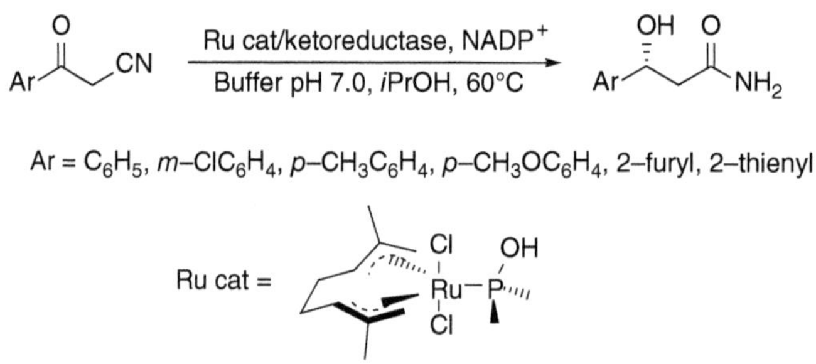

Scheme 6.54 One-pot concurrent enzymatic reduction and Ru-catalyzed hydration.

Biocatalytic oxidation of primary alcohols produced aldehydes, which could react *in situ* with hydroxylamine to give aldoximes. In the presence of hydroxylamine, primary alcohols were transformed to the corresponding aldoximes by an acetic acid bacterium (Scheme 6.55), which lacks aldehyde dehydrogenase activity. This offered a new synthetic method for the one-pot preparation of aldoximes in water [64].

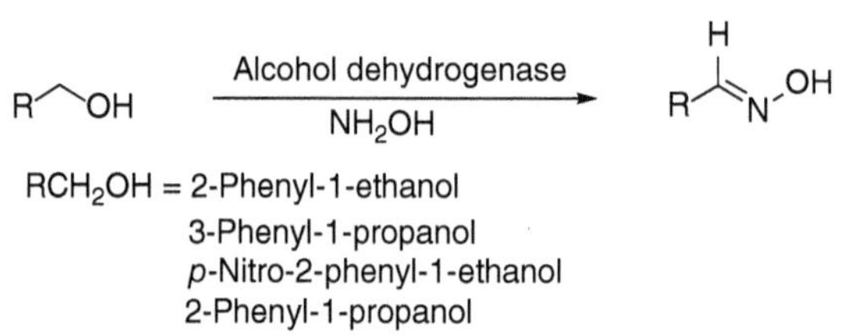

Scheme 6.55 One-pot concurrent enzymatic dehydrogenation and aldoxime formation.

Gluconobacter oxydans whole cell biocatalyst catalyzed the oxidation of *n*-aliphatic alcohols to the corresponding aldehydes, which underwent aldol coupling to give α,β-unsaturated aldehydes in the presence of lysine organocatalyst. The alcohols were oxidized into aliphatic carboxylic acids without adding lysine organocatalyst to intercept the aldehyde intermediate. A single-pot process involving biocatalytic oxidation and organocatalytic C–C coupling was established to convert a range of C2–C6 alcohols into the corresponding α,β-unsaturated aldehydes under mild aqueous conditions (Scheme 6.56) [65].

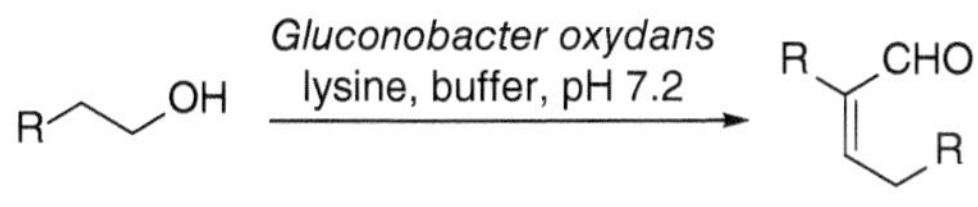

Scheme 6.56 One-pot concurrent chemoenzymatic transformation of aliphatic alcohols into α,β-unsaturated aldehydes. Source: Based on Stewart et al. [65].

6.1.3 Enoate Reductases

Enoate reductases catalyze the stereoselective hydrogenation of C=C bonds with an electron-withdrawing group, creating up to two new chiral centers. They have been increasingly recognized as useful catalysts for organic synthesis. Enoate-reductase-catalyzed reduction involves hydride transfer from the reduced flavin prosthetic group (mostly a flavin mononucleotide $FMNH_2$) to the β-carbon atom of the alkene substrate followed by protonation of the resulting enolate anion in a *trans* manner. The oxidized FMN is reduced to $FMNH_2$ by the cofactor NAD(P)H, which in turn is usually enzymatically regenerated using a glucose dehydrogenase or formate dehydrogenase. Recently, chemical and photochemical methods have been developed for the regeneration of $FMNH_2$ for enoate reductases. The transition metal complex $[Cp^*Rh(bpy)(H_2O)]^{2+}$ catalyzed the reduction of FMN with formate to give $FMNH_2$. The enzymatic regeneration system with cofactor NAD(P)H was replaced by this chemical method for the reduction of ketoisophorone catalyzed by a chromate reductase from *Thermus scotoductus* SA-01 (CrS, Scheme 6.57). This system was further applied to the reduction of various α,β-unsaturated carbonyl compounds, giving the corresponding products with higher conversion and ee values compared to the NADPH-based enzymatic regeneration [66].

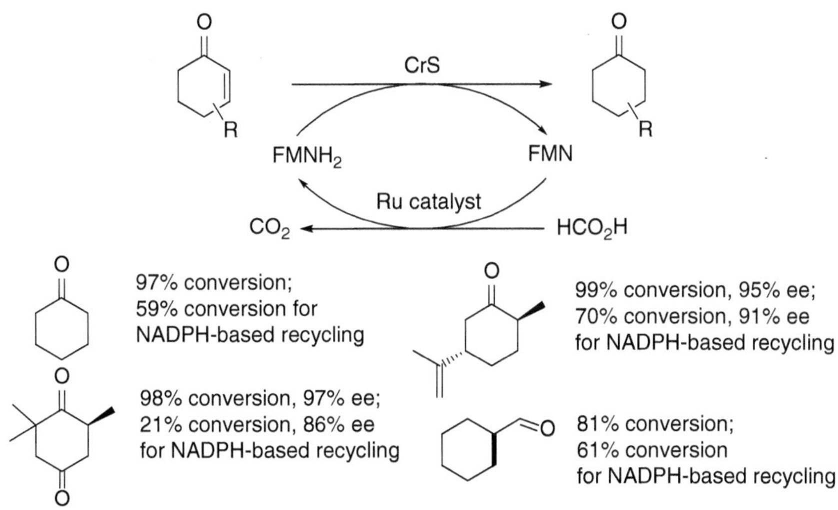

Scheme 6.57 One-pot chemoenzymatic reduction of α,β-unsaturated carbonyl compounds using Ru-catalyzed cofactor recycling.

Elemental Cr metal was also used as stoichiometric reductant to regenerate $FMNH_2$ for the stereoselective reduction of C=C bonds using old yellow enzymes (OYEs) as catalysts. With the addition of FMN as electron mediator, several α,β-unsaturated ketones or aldehydes were reduced by the OYE homolog from *Bacillus subtilis* (YqjM) in high yields and enantioselectivity (Scheme 6.58) [67].

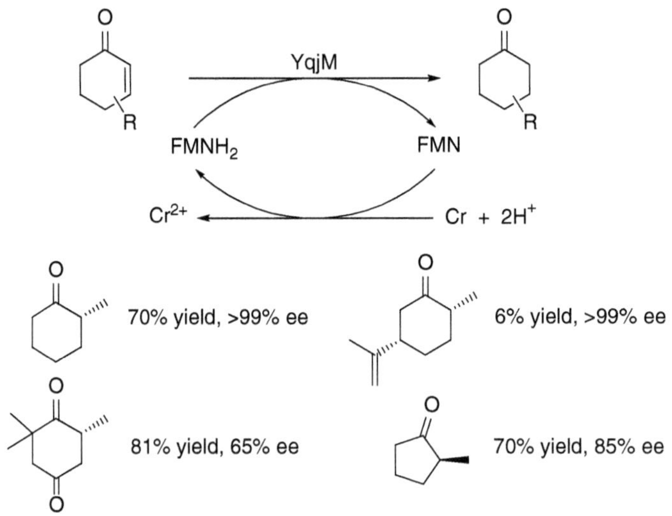

Scheme 6.58 One-pot chemoenzymatic reduction of α,β-unsaturated ketones using Cr as reductant. Source: Based on Rauch et al. [67].

Regeneration of the reduced $FMNH_2$ was also achieved by photoexciting FMN in the presence of ethylenediaminetetraacetic acid (EDTA) as sacrificial electron donor. Upon irradiating with commercial LEDs (λ = 450 nm) as light source, transfer of electrons from EDTA to the OYE YqjM's flavin prosthetic group resulted in the reduction of 2-methylcyclohexenone to give (*R*)-2-methylcyclohexanone (Scheme 6.59) [68].

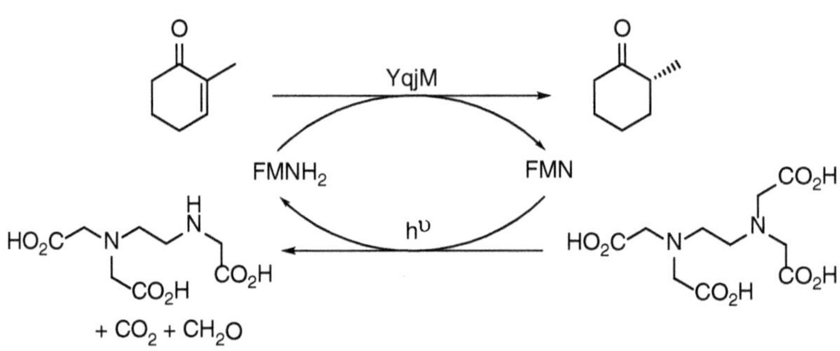

Scheme 6.59 One-pot photochemoenzymatic reduction of 2-methylcyclohexenone. Source: Based on Rauch et al. [68].

6.1.4 Transaminases

In Chapter 3, we discussed that an ADH from *L. kefir* (LKADH) catalyzed the selective reduction of methyl keto group of (3*E*)-non-3-ene-2,8-dione, affording (3*E*)-8(*R*)-hydroxy-non-3-en-2-one with concomitant formation of 1-((2*R*, 6*R*)-6-methyltetrahydro-2*H*-pyran-2yl)propan-2-one as a minor product via intramolecular oxa-Michael reaction (IMOMR). In contrast, (*R*)-selective ω-transaminase ATA-117 catalyzed the amination of (3*E*)-non-3-ene-2,8-dione followed by spontaneous intramolecular aza-Michael reaction (IMAMR), furnishing a mixture of (−)-pinidinone and its *trans*-stereoisomer with >99% conversion, which was easily epimerized by treating with methanol to give (−)-pinidinone (Scheme 6.60), a naturally occurring defense alkaloid, in 86% isolated yield with >99% de and >99% ee. By using (*S*)-selective ATA113, (*S*,*S*)-pinidinone was obtained in 91% yield with >99% de and >99% ee in the same way. The ω-transaminases (TA)-triggered aza-Michael reaction was expanded to the stereoselective synthesis of other 2,6-disubstituted piperidines from readily accessible ketoenones (Scheme 6.61) [69].

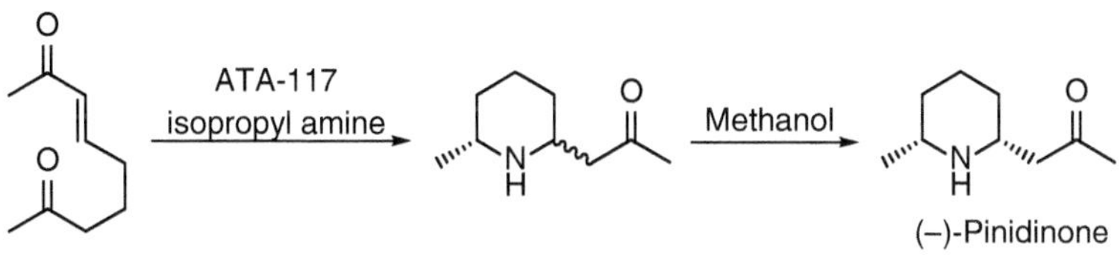

Scheme 6.60 TA-mediated IMAMR for the synthesis of (−)-pinidinone.

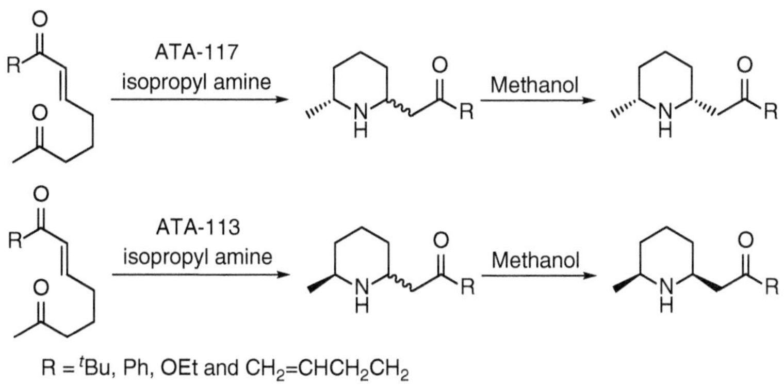

Scheme 6.61 TA-mediated IMAMR for the synthesis of 2,6-disubstituted piperidines. Source: Based on Ryan et al. [69].

TA-triggered intramolecular aza-Michael reaction was also applied to ketoynones for the synthesis of cyclic β-enaminones. By choosing suitable (*R*)- or (*S*)-selective transaminase both enantiomers of various cyclic β-enaminones were acquired in 29–95% yields and 83–>99% ee (Scheme 6.62). The spontaneous aza-Michael reaction effectively drove the thermodynamic equilibrium of enzymatic transamination toward product formation, with only two equivalents of isopropylamine being required. A powerful application of this biocatalytic aza-Michael transformation was the facile construction of stereo-defined fused alkaloid motifs, as demonstrated by one-pot synthesis of polyfunctionalized hexahydroquinolines, in which the resulting cyclic β-enaminones from biocatalytic aza-Michael reaction underwent *in situ* carbo-[3+3] annulation with an alternative ketoynone (Scheme 6.63) [70].

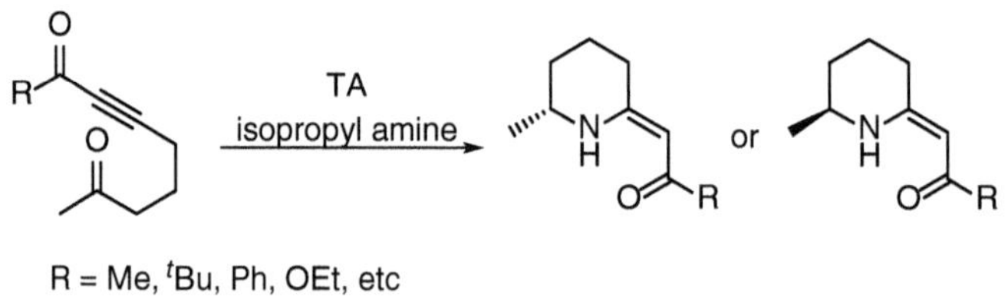

Scheme 6.62 TA-mediated IMAMR for the synthesis of cyclic β-enaminones.

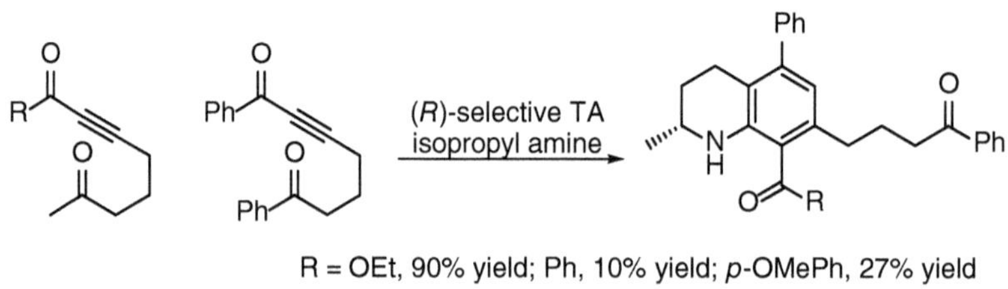

Scheme 6.63 One-pot concurrent TA-mediated IMAMR and carbo-[3+3] annulation. Source: Based on Taday et al. [70].

Transaminase ATA-113 catalyzed the regioselective amination of α-diketones in the presence of either (*S*)-aminotetralin or isopropylamine as amine donor, generating the corresponding α-amino ketones, which underwent *in situ* oxidative dimerization to give the substituted pyrazines in 32–65% yields (Scheme 6.64). α-Amino ketones are known to react with β-keto esters to afford substituted pyrroles, i.e. Knorr pyrrole synthesis. Several aryl α-diketones were tested with ethyl acetoacetate under the action of ATA-117, regioselective amination took place at the methyl keto group, and the *in situ* Knorr pyrrole synthesis resulted in the

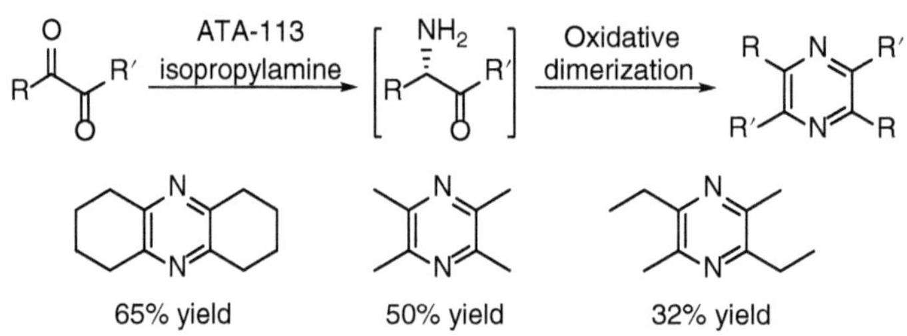

Scheme 6.64 One-pot concurrent TA-mediated regioselective amination of α-diketones and oxidative dimerization.

corresponding substituted pyrroles in 21–48% yields (Scheme 6.65). Furthermore, ethyl 3-aminobutyrate was used instead of ethyl acetoacetate, in which the β-amino acid ester served as amine donor and the precursor of β-keto ester; the product ethyl 2,5-dimethyl-4-phenyl-1*H*-pyrrole-3-carboxylate was obtained in 58% conversion as determined by NMR analysis (Scheme 6.66) [71].

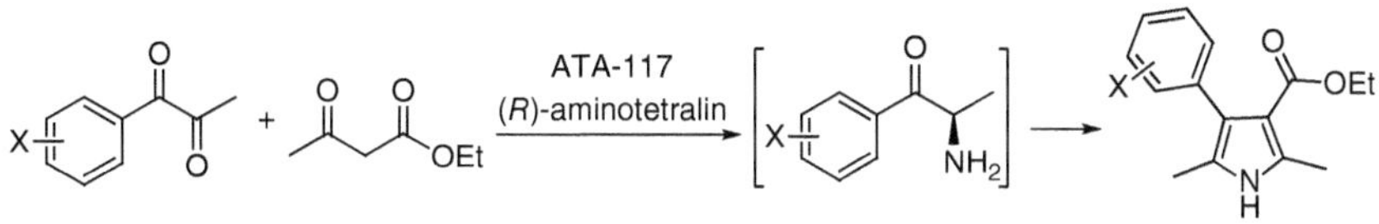

Scheme 6.65 One-pot concurrent TA-mediated regioselective amination of α-diketones and Knorr pyrrole synthesis.

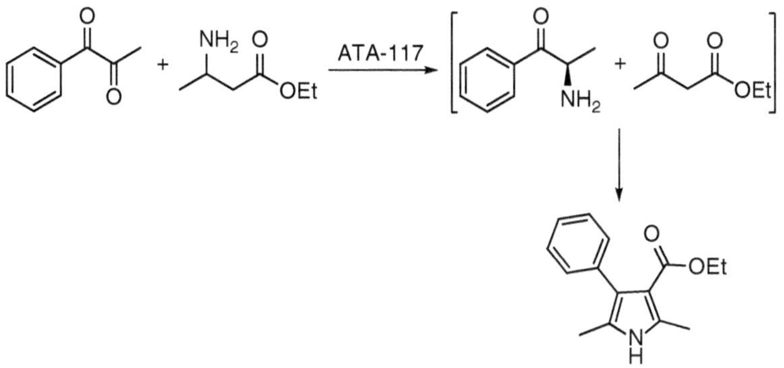

Scheme 6.66 One-pot concurrent TA-mediated Knorr pyrrole synthesis without external amine donor. Source: Based on Xu et al. [71].

6.1.5 Monoamine Oxidases

Monoamine oxidases (MAO) catalyze the stereoselective oxidation of a wide range of chiral amines to the corresponding imines. By coupling this enzymatic amine oxidation with a chemical reduction of the resulting imines in one-pot, the racemic amines can be deracemized to one enantiomer. In addition to the chemoenzymatic deracemization process discussed in detail in Chapter 5, the enzymatic amine oxidation could be combined with other chemical reactions, enabling the synthesis of diverse compounds [72]. Bechi et al. showed that combination of the MAO-N D9 biocatalyst, a mutant of the FAD-containing MAO from *Aspergillus niger*, with a catalytic amount of CuI/H_2O_2 in a one-pot reaction resulted in the conversion of 1,2,3,4-tetrahydroisoquinoline (THIQ) to 3,4-dihydroisoquinolin-1(2*H*)-one in 74% yield (Scheme 6.67) [73]. In the reaction the imine intermediate was oxidized to cyclic lactams by H_2O_2 and CuI. This sequential double oxidation when applied to the aromatization of tetrahydropyridines (THPs) generated pyridines or pyridinium ions. The first C–N oxidation of THP was catalyzed by MAO-N enzyme; the resulting dihydropyridine intermediate was aromatized via either O_2-mediated oxidation or tautomerization followed by a second MAO-N-catalyzed oxidation. The reaction was carried out in aqueous buffer with DMSO or isooctane as the cosolvent. Some aryl-substituted THPs were converted into the corresponding pyridines or pyridinium ions with yields up to 99% (Scheme 6.68) [74].

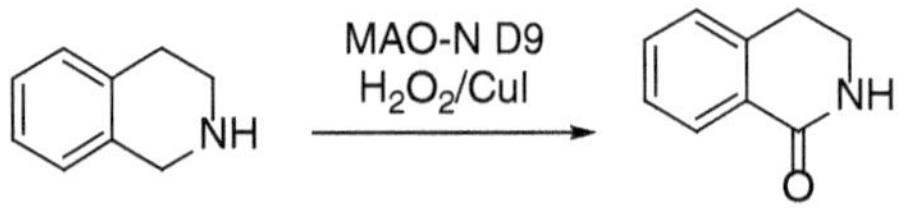

Scheme 6.67 One-pot concurrent conversion of 1,2,3,4-tetrahydroisoquinoline. Source: Based on Bechi et al. [73].

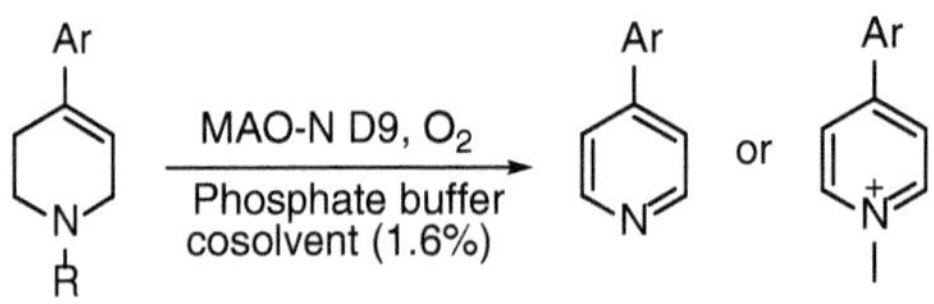

Scheme 6.68 One-pot concurrent conversion of aryl-substituted tetrahydropyridines. Source: Based on Toscani et al. [74].

MAO-N mutants also enantioselectively oxidized a variety of 3,4-substituted *meso*-pyrrolidines. The resulting Δ1-pyrroline intermediates reacted with potassium cyanide (KCN) or trimethylsilyl cyanide (TMSCN) to give the corresponding nitriles, which was further converted to the amino acids through hydrolysis. The enzymatic oxidation and the cyanation were carried out either in a sequential manner or in a one-pot concurrent process. In the step-wise procedure the concentrated methyl *tert*-butyl ether extracts of the biooxidation were treated with TMSCN/MeOH in CH_2Cl_2. In the concurrent process, KCN was used and the reaction was performed in aqueous buffer (Scheme 6.69), but the diastereoselectivity of HCN addition was usually lower than the sequential process [75].

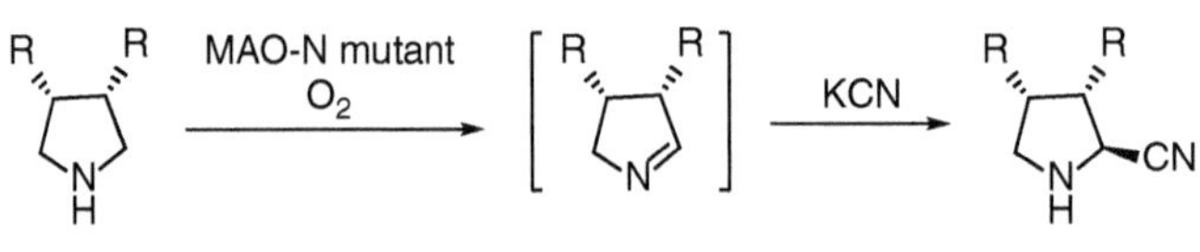

Scheme 6.69 One-pot concurrent enzymatic amine oxidation and cyanation.

In addition, the presence of cyanide in the aqueous reaction mixture inactivated the enzyme and the accumulation of product resulted in irreversible enzyme inhibition. Bisulfite was thus used to trap the imine intermediate. Upon optimization of the reaction conditions, the MAO-catalyzed oxidation of 6,6-dimethyl-3-azabicyclo[3.1.0]hexane and the subsequent sulfonation were carried out in one-pot by adding a mixture of bisulfite and 6,6-dimethyl-3-azabicyclo[3.1.0]hexane into a solution of mutant enzyme (MAON291) and catalase at a constant rate over 24 hours.

The product amino sulfonate was obtained in >99% ee with a small amount of imine (<10%), and the combined yield was >95%. This amine-oxidase-mediated desymmetrization and concurrent bisulfite adduct formation cascade was then applied to the synthesis of a key structural feature in Boceprevir (Scheme 6.70), a Merck's new drug treatment for hepatitis C [76].

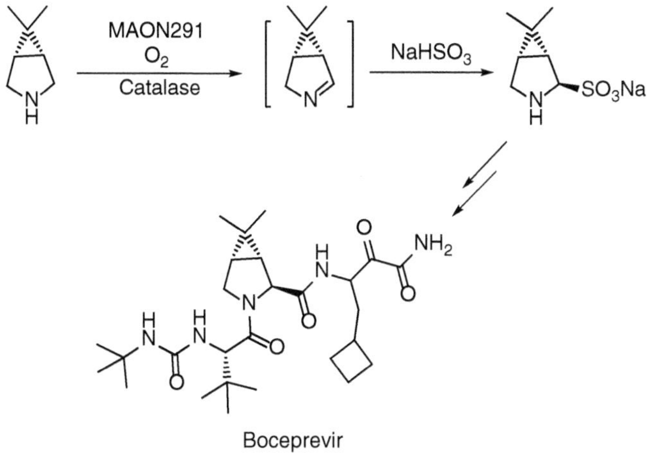

Scheme 6.70 Synthesis of Boceprevir involving one-pot concurrent enzymatic amine oxidation and sulfonation.

The imines produced by MAO-*N*-catalyzed oxidation of cyclic amines could undergo the three-component Ugi reaction, in which the imine reacted with a carboxylic acid and isocyanide to give the highly functionalized 3,4-substituted prolyl peptides in good yields and diastereomeric ratios (Scheme 6.71) [77]. Application of this sequential strategy of biooxidation and multicomponent reaction led to the synthesis of an important drug candidate telaprevir (Scheme 6.72) [78]. When the carboxylic acid was replaced with different electron-deficient phenolic compounds, the Ugi–Smiles multicomponent reaction generated a wide range of *N*-aryl proline amides with average to high yields and excellent diastereomeric ratios (Scheme 6.73) [79].

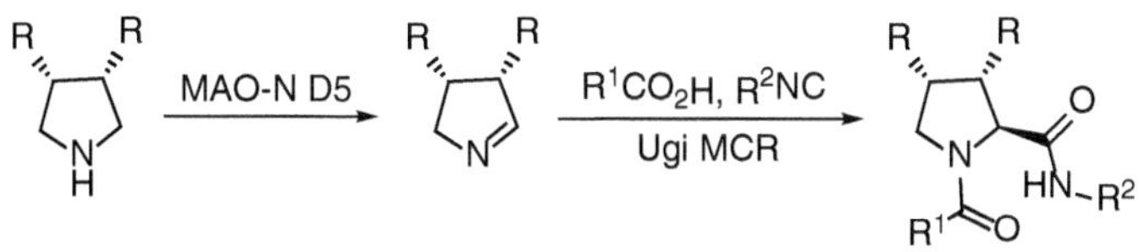

Scheme 6.71 MAO-*N*-catalyzed oxidation of cyclic amines and subsequent Ugi three-component reaction. Source: From Znabet et al. [77].

MAO-*N*-catalyzed oxidation of cyclic amines and the multicomponent reactions were usually carried out sequentially. The imines also underwent aza-Friedel–Crafts (aza-FC) reaction, in which an aromatic or a heteroaromatic compound was added

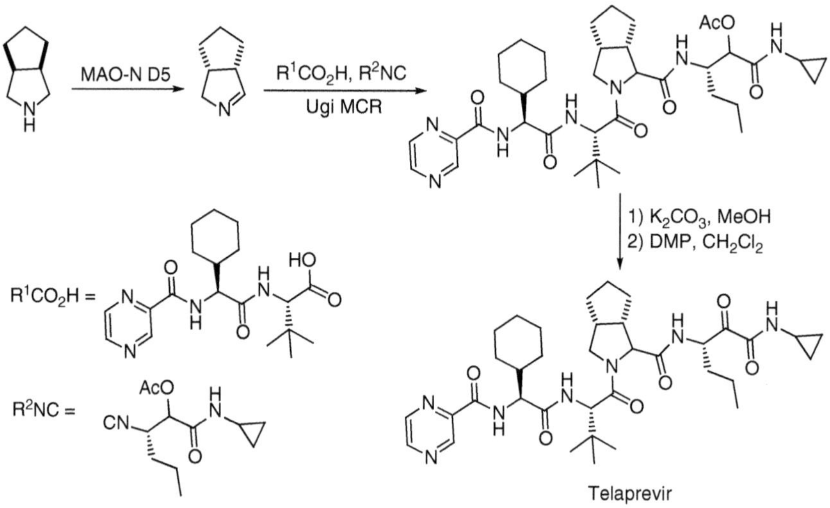

Scheme 6.72 Synthesis of telaprevir involving MAO-*N*-catalyzed oxidation and Ugi three-component reaction. Source: Based on Znabet et al. [78].

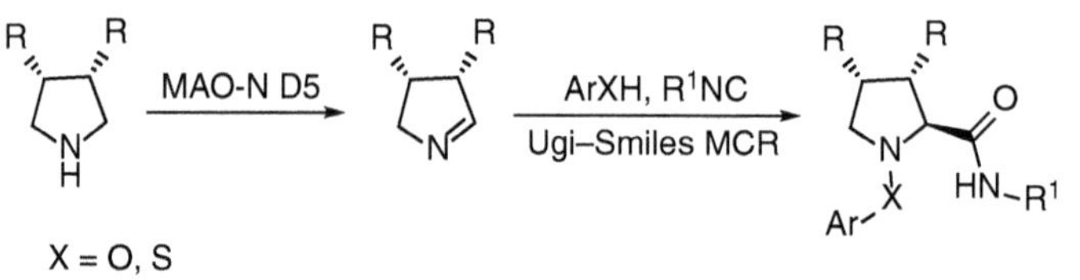

Scheme 6.73 MAO-*N*-catalyzed oxidation of cyclic amines and subsequent Ugi–Smiles three-component reaction. Source: Based on Znabet et al. [79].

unto the C=N bond. MAO-*N*-catalyzed oxidation of *meso*-pyrrolidine had been coupled with the aza-FC reaction. The chemoenzymatic oxidative aza-FC process was performed with pyrrole as the C-nucleophile in phosphate buffer in one-pot concurrent manner, and 2-pyrrolylpyrrolidine was obtained with high diastereoselectivity (Scheme 6.74). However, higher yield and ee values of the product were achieved when the biocatalytic oxidation and aza-FC reaction were carried out in a sequential manner. As such, the direct α-functionalization of a range of pyrrolidines was realized by means of biocatalytic oxidation in aqueous buffer, followed by the addition of the C-nucleophile without an additional catalyst, generating 2-substituted pyrrolidines in good yields and high enantioselectivity (Scheme 6.75) [80].

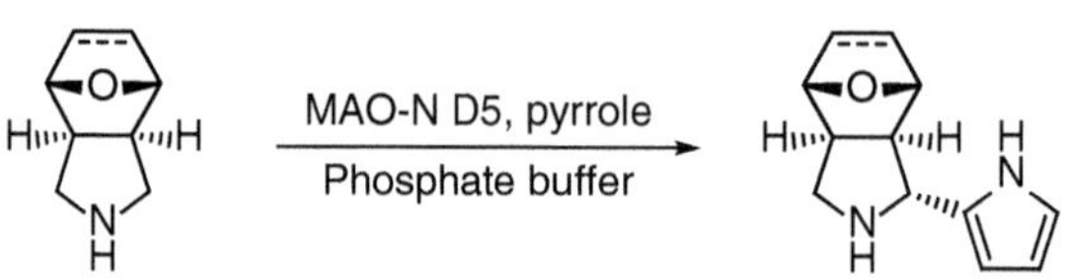

Scheme 6.74 One-pot concurrent biocatalytic oxidation and aza-FC reaction.

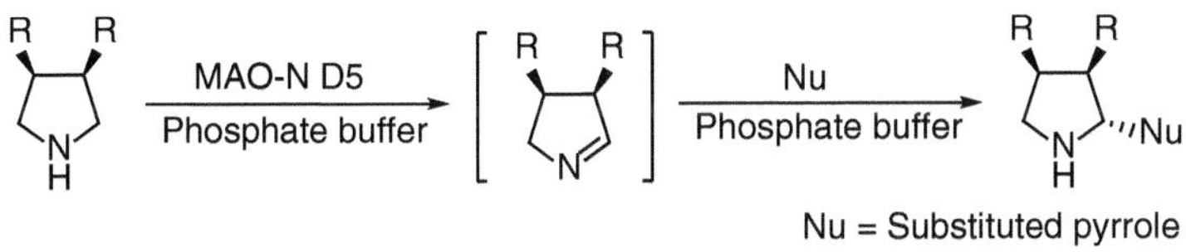

Scheme 6.75 One-pot sequential biocatalytic oxidation and aza-FC reaction. Source: Based on de Graaff et al. [80].

Recently, the gold-catalyzed C–C bond formation between imine and alkyne was combined with the MAO-*N*-mediated oxidation of THIQs in a one-pot process, leading to an effective cross-dehydrogenative coupling of THIQs with alkynes. Specifically, a series of *N*-alkyl-substituted THIQs were alkynylated at C-1 position in high yields, by employing MAO-N D5 mutant as biocatalyst and $HAuCl_4 \cdot H_2O$ as the chemical catalyst concurrently in one pot (Scheme 6.76) [81].

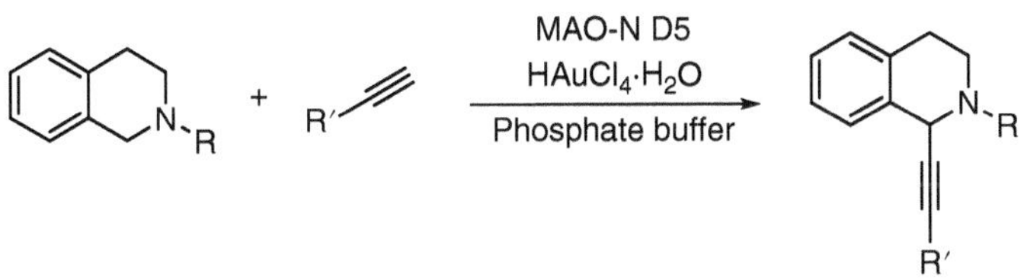

Scheme 6.76 One-pot concurrent biocatalytic oxidation and gold-catalyzed C–C bond formation. Source: Based on Odachowski et al. [81].

MAO-N variants and nicotine oxidase biocatalyst 6-HDNO catalyzed the oxidation/aromatization of 3-pyrrolines to afford the corresponding pyrroles in high yields. Furthermore, the MAO-N variants working in combination with the ruthenium Grubbs catalyst in a one-pot process led to the conversion of diallylamines and diallylanilines to the corresponding pyrroles via chemoenzymatic metathesis–aromatization sequence. A wide range of pyrroles with diverse structural features have been synthesized by exploiting this one-pot cascade (Scheme 6.77). The yields of product pyrroles were greatly affected by the size and electronic properties of the *N*-substituent, and higher yields were obtained for aryl substituents [82].

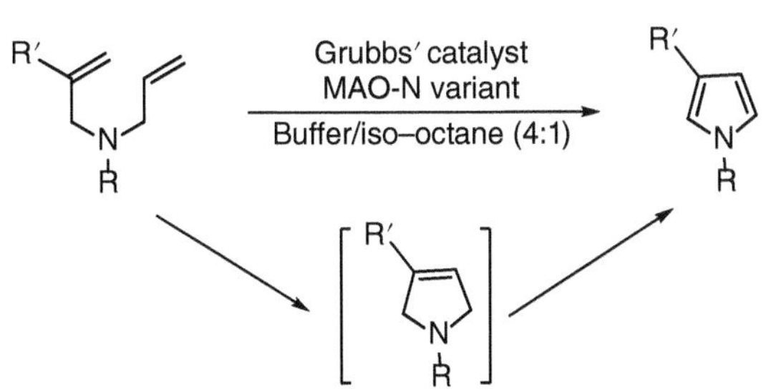

Scheme 6.77 One-pot concurrent Ru-catalyzed metathesis and MAO-*N*-mediated aromatization.

D-Amino acid oxidase from porcine kidney (pkDAO) and L-amino acid oxidase from *Crotalus atrox* catalyzed the oxidation of amino groups in amino acids to form imines, which could be intercepted by cyanide anion via nucleophilic addition. This chemoenzymatic cascade resulted in the synthesis of 2-amino-2-cyano-3-phenylpropanoic acid from phenylalanine and potassium cyanide in a single-pot procedure. Similarly, mutant pkDAO (Y228L/R283G) catalyzed the oxidation of (*R*)-α-methylbenzylamine in the presence of potassium cyanide, affording racemic-2-methyl-2-phenylglycinonitrile in high yield (Scheme 6.78). The α-aminonitrile was further transformed into the unnatural α-amino acid via nitrilase-catalyzed hydrolysis. This offered a new chemoenzymatic cascade for the synthesis of primary α-aminonitriles and unnatural α-amino acids in aqueous systems [83].

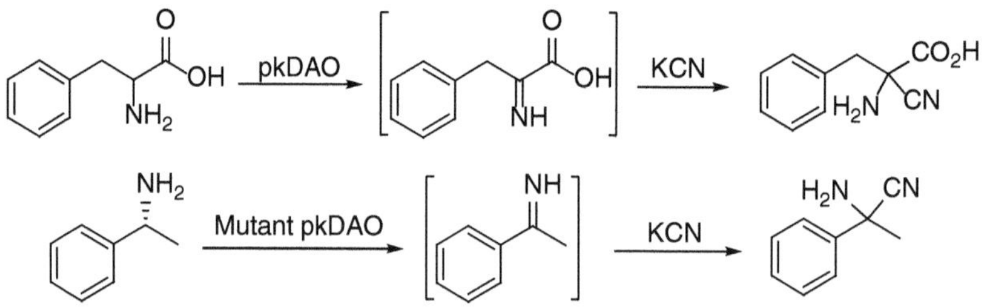

Scheme 6.78 One-pot concurrent chemoenzymatic synthesis of α-aminonitriles.

6.1.6 Cytochrome P450s

A P450 enzyme from *Bacillus megaterium* (P450 BM3) catalyzed the epoxidation of unsaturated fatty acids and the activity was found to be dependent on the chain length of the substrate. It showed a preference for 12-tridecenoic acid over 10-undecenoic acid and the symmetrical alkenes such as *trans*-3-hexene. The biocatalytic epoxidation was successfully combined with the cross-metathesis of 10-undecenoic acid and *trans*-3-hexene catalyzed by Hoveyda–Grubbs second-generation ruthenium carbine catalysts. The one-pot process was carried out in the isooctane/buffer (1 : 4, v/v) biphasic medium, and the cross-metathesis products, *cis*- and *trans*-10-tridecenoic acids, were converted into the corresponding epoxides (Scheme 6.79). In the absence of P450 BM3, the yield of the self-metathesis product of 10-undecenoic acid was higher than that of the cross-metathesis product. Compared to the metathesis reaction alone, the self-metathesis of 10-undecenoic acid was suppressed in the chemoenzymatic transformation and the epoxide was obtained in up to 90% yield. This was even higher than that of sequential

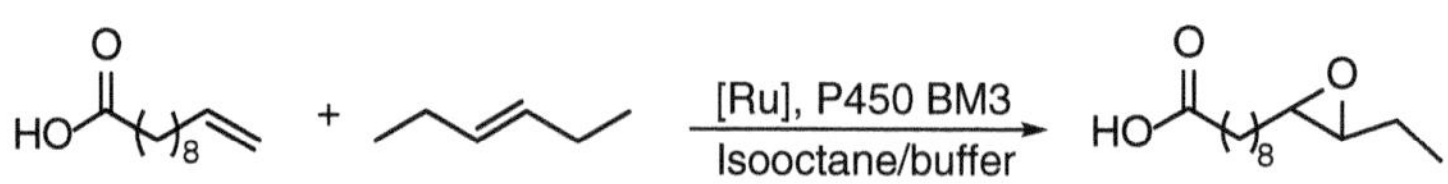

Scheme 6.79 One-pot ruthenium-catalyzed cross-metathesis and P450-catalyzed epoxidation.

metathesis and epoxidation reaction, in which the yield of epoxide was limited by the concentration of the cross-metathesis product C13 alkene that was determined by the equilibrium ratio of the different alkenes. These results suggested that a dynamic equilibration of alkenes existed in the system and a selective epoxidation of the cross-metathesis product could be achieved [84].

This cooperative catalysis has been shown to be applicable to other alkenes depending on the substrate specificity of the P450 enzymes. For example, P450 BM3 enzyme was shown to catalyze the epoxidation of (*E*)-4-(hex-3-en-1-yloxy)benzoic acid, with low activity toward 4-(but-3-en-1-yloxy)benzoic acid. The Ru-catalyzed cross-metathesis of 4-(but-3-en-1-yloxy)benzoic acid with *trans*-3-hexene proceeded smoothly in the biphasic system of dioctyl phthalate and buffer, generating (*E*)-4-(hex-3-en-1-yloxy)benzoic acid in reasonable yield. The tandem metathesis/epoxidation reaction with 4-(but-3-en-1-yloxy)benzoic acid and *trans*-3-hexene, conducted with a Hoveyda–Grubbs second-generation Ru catalyst and P450 BM3 in a one-pot manner, gave rise to 4-(3-(3-methyloxiran-2-yl)propoxyl)benzoic acid methyl ester as the major epoxide product in 75% yield (Scheme 6.80). However, the yield of the cross-metathesis product was only 58% when the two reactions were performed sequentially.

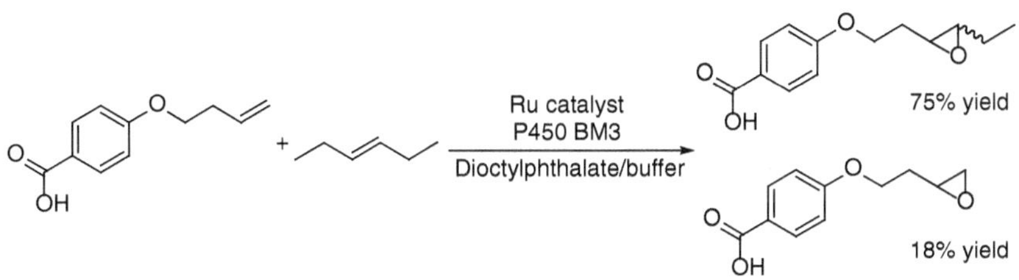

Scheme 6.80 One-pot Ru-catalyzed cross-metathesis and P450-catalyzed epoxidation.

Three P450 BM3 mutants RLYF, KT2 and RH47 catalyzing the epoxidation of vinylarenes were evaluated for the epoxidation of β-methylstyrene and β-ethylstyrene in the presence of (*Z*)-stilbene in potassium phosphate buffer using a glucose dehydrogenase (GDH) system for the regeneration of NADPH. These enzymes showed selectivity for the epoxidation of β-methylstyrene and β-ethylstyrene over (*Z*)-stilbene. For example, P450 KT2 catalyzed the selective epoxidation of (*E*)-β-methylstyrene to give (2*R*,3*R*)-2-methyl-3-phenyloxirane in 41% yield with less than 3% yield of (*Z*)-stilbene oxide. The Ru-catalyzed alkene metathesis of (*Z*)-stilbene and (*Z*)-2-butene or (*E*)-3-hexene when coupled with enzymatic epoxidation in a one-pot process generated up to 50% of the corresponding aryl epoxides of the cross-metathesis products. However, isomerization of (*Z*)-stilbene into (*E*)-stilbene and the epoxidation of (*Z*)-stilbene occurred, resulting in a mixture of the desired epoxide, (*E*)-stilbene and (*Z*)-stilbene oxide (Scheme 6.81). In order to develop this tandem catalysis into an efficient one-pot process substantial catalyst (both metal catalyst and enzyme) and reaction engineering are required to improve both selectivity and catalytic activity [85].

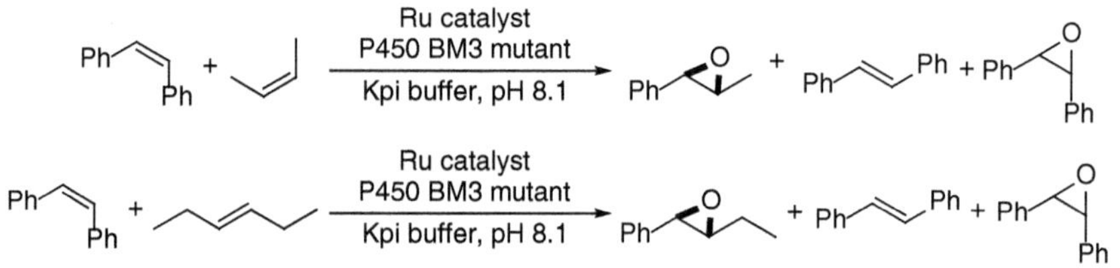

Scheme 6.81 One-pot concurrent cross-metathesis and enzymatic epoxidation of mixed alkenes.

6.1.7 Halohydrin Dehalogenases

Halohydrin dehalogenase (HheC) catalyzed the enantioselective azidolysis of epoxides to give 1,2-azido alcohols. The copper-catalyzed 1,3-dipolar cycloaddition of azides and alkynes generated 1,4-disubstituted triazoles that have wide applications in drug discovery and polymer chemistry, among others. The combination of enzymatic azidolysis of epoxides with click reaction in a one-pot process would offer an attractive approach to access these important chiral hydroxyl triazoles. HheC from *Agrobacterium radiobacter* catalyzed the azidolysis of substituted styrene oxides at β-position with high enantioselectivity. Ligation of the resulting chiral 1,2-azido alcohols to phenylacetylene by click reaction in a one-pot procedure gave rise to the corresponding chiral hydroxy triazoles with excellent ee (Scheme 6.82) [86].

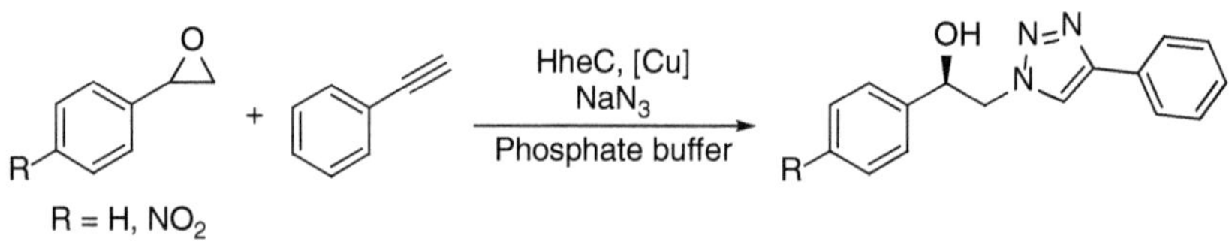

Scheme 6.82 One-pot concurrent enzymatic azidolysis of epoxides and click reaction with phenylacetylene. Source: Based on Campbell-Verduyn et al. [86].

HheC also catalyzed the dehalogenation and ring closure of β-halo alcohol to form epoxide and the subsequent ring opening of the resulting epoxide by azide anion, producing β-azido alcohol. Wiktor et al. extended the above one-pot enzymatic azidolysis and click reaction cascade one step further, by using β-halo alcohol as the starting material. β-Halo alcohol in turn could be prepared by the enzymatic reduction of α-haloketone. The authors constructed two whole cell biocatalysts: one (CT cells) expressed AdhT, an *R*-selective ADH from *Thermoanaerobacter* sp., and HheC, an *R*-selective HheC from *Agrobacterium radiobacter* AD1, while the other (BL cells) expressed AdhL, the *S*-selective ADH from *L. brevis*, and HheBGP1, the HheC of low enantioselectivity from *Mycobacterium* sp. GP1. The whole cell biocatalysts were first applied to convert α-haloketone to the enantiopure epoxide without addition of NaN_3. The enzymatic azidolysis and click reaction cascade were then performed in a one-pot mode by adding more whole cell biocatalyst together with NaN_3, alkyne, and the click reaction catalyst into the reaction mixture. As a result, both enantiomers of β-hydroxytriazoles were prepared with high ee from the readily available α-haloketones (Scheme 6.83) [87].

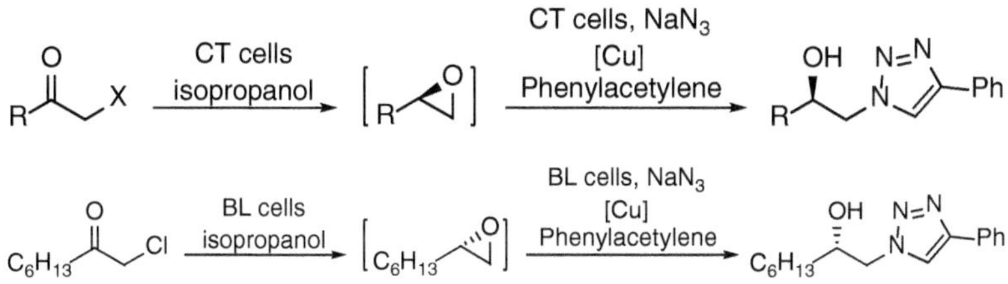

Scheme 6.83 One-pot chemoenzymatic synthesis of optically pure β-hydroxytriazoles involving multiple enzymatic reactions. Source: Based on Wiktor et al. [87].

The (*R*)-β-hydroxytriazoles were also synthesized in about 80% isolated yields by mixing *Rhodotorula mucilaginosa* LSL cells, α-bromoacetophenone, sucrose, NaN_3, phenylacethylene or propargyl alcohol, $CuSO_4 \cdot 5H_2O$, and sodium ascorbate in a phosphate buffer (Scheme 6.84). This one-pot chemoenzymatic cascade involved the biocatalytic reduction of ketone, the nucleophilic substitution of bromide by azide anion, and the Cu-catalyzed [2+3] cycloaddition of alkyne and azide [88].

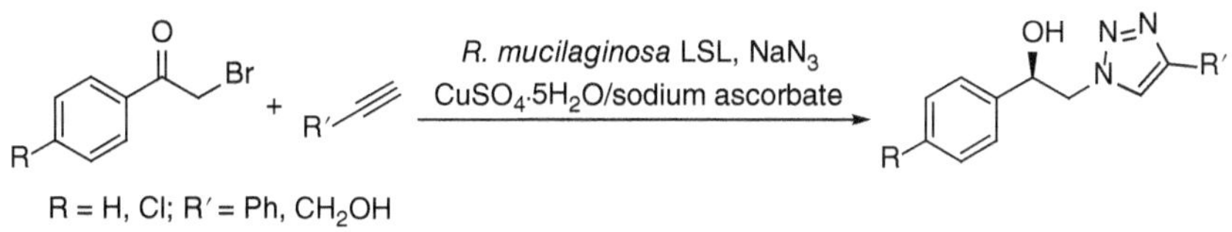

Scheme 6.84 One-pot chemoenzymatic synthesis of (*R*)-β-hydroxytriazoles involving enzymatic reduction of α-bromoketones.

6.1.8 Vanadium Haloperoxidases

Vanadium-dependent haloperoxidases catalyze the formation of hypohalous acids from the corresponding halide anions using H_2O_2 as the oxidant. It is well known that hypochlorite and hypobromite readily undergo electrophilic substitution at the phenol moiety, resulting in halogenation of phenols. In this context, vanadium-dependent chloroperoxidase from *Curvularia inaequalis* (CiVCPO) was applied for the *in situ* generation of hypobromite (BrO^-), and subsequent electrophilic bromination of thymol. A mixture of monobrominated and dibrominated thymol was obtained with the turnover number of CiVCPO exceeding 2 000 000 (Scheme 6.85). This vanadium chloroperoxidase is a useful robust enzyme for the chemoenzymatic halogenation of phenols than vanadium bromoperoxidase and heme-containing peroxidases in terms of operational stability toward H_2O_2 [89].

The reactive hypohalites also undergo hydroxyhalogenation with alkenes, affording *vic*-halohydrins, a family of valuable building blocks in organic synthesis. Vanadium chloroperoxidase CiVCPO was also evaluated for the hydroxyhalogenation

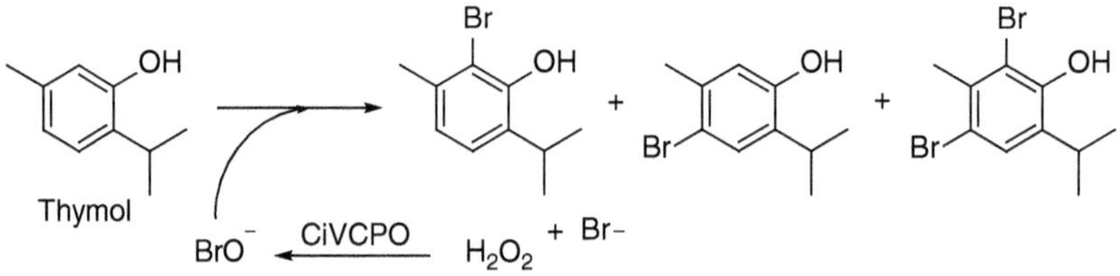

Scheme 6.85 Chemoenzymatic bromination of thymol.

of styrene. In citrate buffer (pH 5), 2-bromo-1-phenylethanol was obtained in 90% yield with only 2% styrene oxide (Scheme 6.86). The hydroxybromination reaction followed Markovnikov's rule in regioselectivity and showed no enantioselectivity, suggesting a chemoenzymatic mechanism. A broad range of alkenes were transformed into the corresponding halohydrins in 51–91% yields using this chemoenzymatic process with KBr or KCl [90].

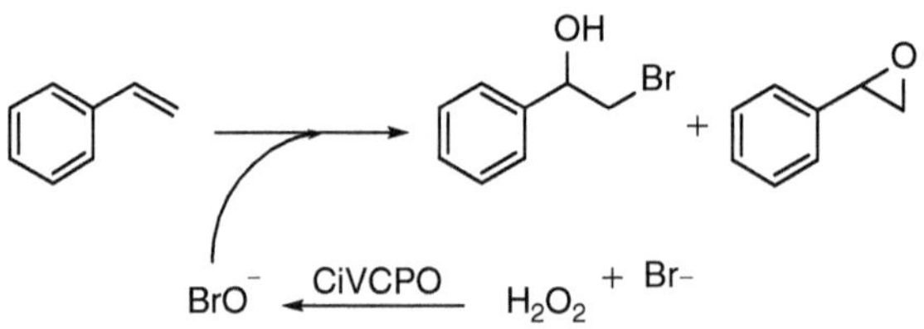

Scheme 6.86 Chemoenzymatic hydroxybromination of styrene.

Hypohalites (XO^-) reacted with γ,δ-unsaturated carboxylic acids, resulting in halolactonization. This reaction was well established for the synthesis of halolactones. The hypohalites could be generated by different methods. Recently, the abovementioned CiVCPO-catalyzed *in situ* generation of hypohalites from H_2O_2 and halides was applied to the halocyclization of various unsaturated carboxylic acids. The reaction proceeded up to >99% conversion and the corresponding halolactones were produced in 56–87% selectivity with hydroxylactone being the only by-product (Scheme 6.87) [91]. The chemoenzymatic bromolactonization of 4-pentenoic acid was scaled up to 100 g scale, and a mixture of 5-(bromomethyl)dihydrofuran-2(3*H*)-one and 5-(hydroxymethyl)dihydrofuran-2(3*H*)-one in about 2 : 1 molar ratio was isolated in 90% yield [92].

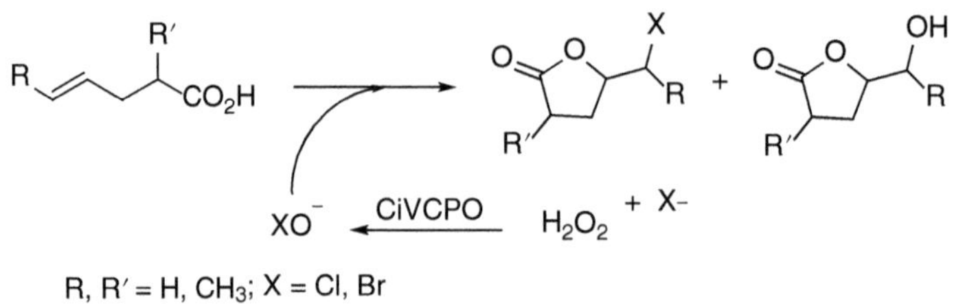

Scheme 6.87 Chemoenzymatic halolactonization of unsaturated carboxylic acids. Source: Based on Younes et al. [91].

A chemoenzymatic halocyclization of unsaturated alcohols was also achieved using this robust vanadium chloroperoxidase, and a broad range of cyclic haloethers were formed in 52–99% conversion with 52–85% selectivity (Scheme 6.88). Similar to the lactonization, the hydroxyethers (X = OH) were detected as the sole byproduct [91]. (−)-Carveol and (+)-citronellol were converted almost quantitatively, and the haloetherificated products were isolated in 60 and 50% yield (Scheme 6.89), respectively.

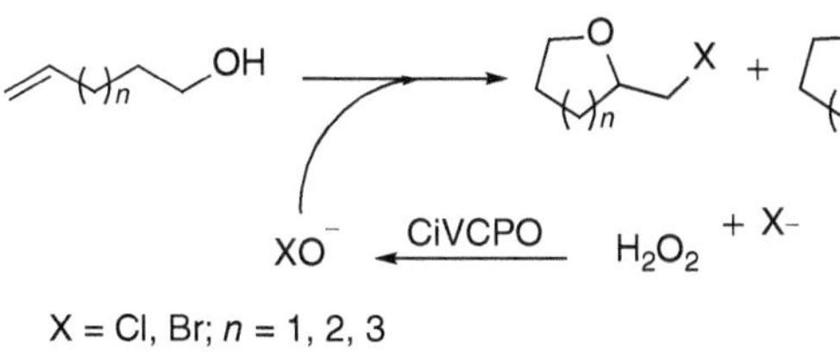

Scheme 6.88 Chemoenzymatic haloetherification of alkenols.

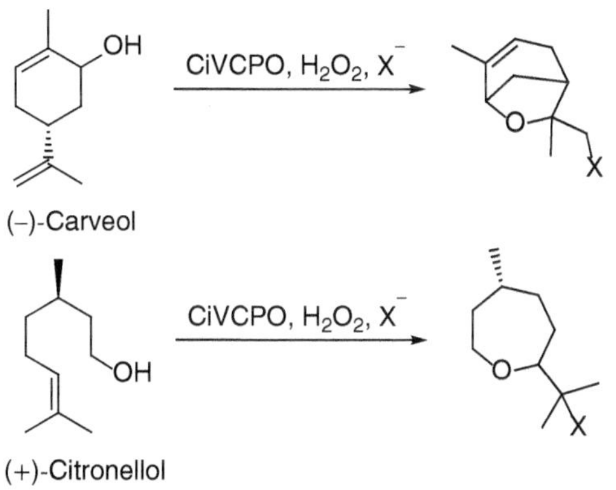

Scheme 6.89 Chemoenzymatic haloetherification of (−)-carveol and (+)-citronellol.

Hypohalogenites (XO^-) also initiated the (aza-)Achmatowicz reaction by oxidatively cleaving the α-heterofunctionalized furan ring to generate a reactive dicarbonyl intermediate, which cyclized to form the corresponding pyranone or piperidinone moiety. A chemoenzymatic (aza-)Achmatowicz reaction was developed by employing the vanadium chloroperoxidase from *C. inaequalis* in combination with H_2O_2 and a catalytic amount of bromide. A few α-heterofunctionalized furans were thus transformed into the corresponding pyranones or piperidinones in 50–82% isolated yields with the diastereomeric ratio being 65 : 35 to 80 : 20 (Scheme 6.90) [93].

Chloroperoxidase from *Caldariomyces fumago* (CfCPO) had been found to catalyze the cleavage of α-hydroxylated furan ring with H_2O_2 to generate a reactive dicarbonyl intermediate, through oxygen transfer from an oxoferryl porphyrin species of CfCPO enzyme to furan core and subsequent ring fission of the oxygenated furan ring. The dicarbonyl intermediate was then recyclized to form

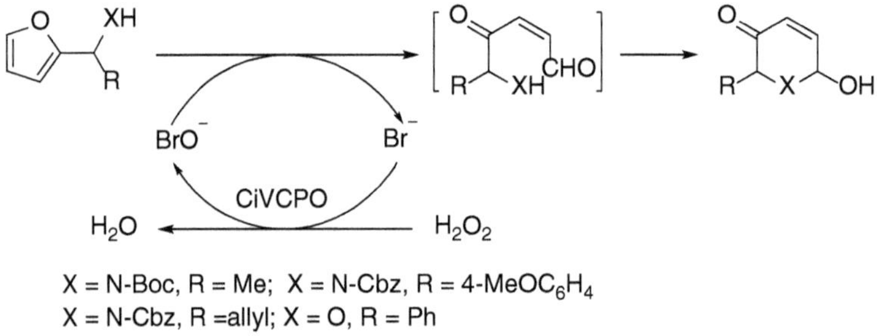

Scheme 6.90 Chemoenzymatic (aza-) Achmatowicz reaction. Source: From Fernández-Fueyo et al. [93].

the pyranone product. Owing to the sensitivity of CfCPO to high concentration of hydrogen peroxide, glucose oxidase (*A. niger*, GOx) was used for the *in situ* generation of H_2O_2 via reduction of aerial oxygen. The ring expansion of various optically active furylcarbinols was achieved with these enzymatic transformations of glucose oxidase and chloroperoxidase, generating the corresponding pyranones in up to 78% isolated yields through the subsequent recyclization of the dicarbonyl intermediates (Scheme 6.91) [94].

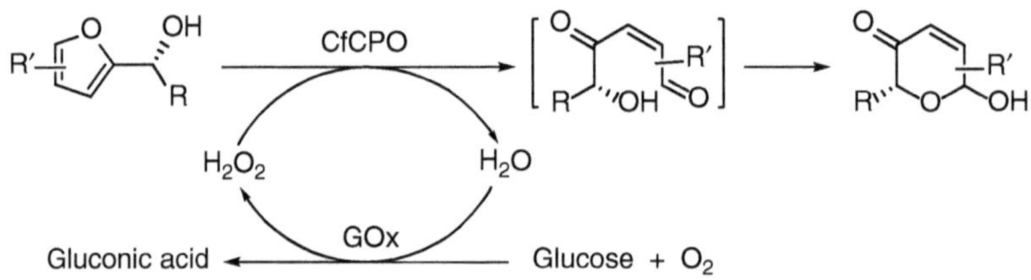

Scheme 6.91 Chloroperoxidase-initiated Achmatowicz reaction. Source: Based on Thiel et al. [94].

Halonium ions are known to trigger allenols to undergo oxidative cyclization, furnishing heterocyclic products with vinyl halide moiety, which can serve as a versatile functional handle for further transformations. Chloroperoxidase CfCPO in combination with glucose oxidase catalyzed the *in situ* generation of reactive halonium species from halide salts in the presence of air and glucose. The oxidase-mediated redox cascade was applied to the oxidative bromocyclization of allenic alcohols, and the corresponding functionalized furan heterocycles were isolated in up to 93% yields (Scheme 6.92). Furthermore, the stereochemical integrity of stereochemically defined allenols could remain uncompromised, as shown in Scheme 6.93. Similarly, bromolactonization of 3,4-dienoate was also realized to give γ-bromo-δ-lactone as the major product in 34% isolated yield (Scheme 6.94) [95].

6.1.9 Laccases

Resin-supported peptide (resin-(Leu)$_{27}$-Trp-Trp-Aib-D-Pro-Pro) is an effective catalyst for asymmetric α-oxyamination of aldehydes in aqueous media in the presence

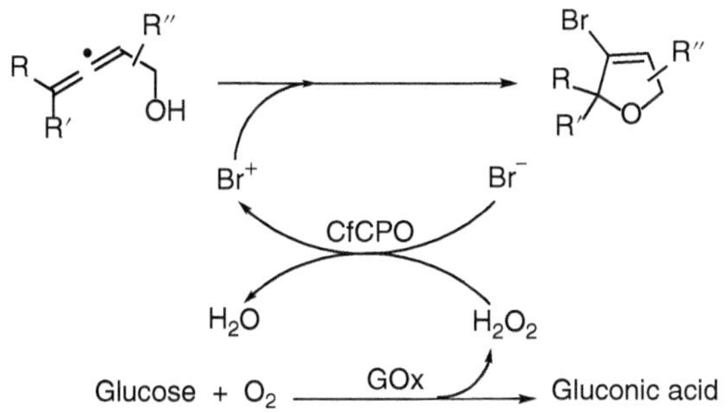

Scheme 6.92 Chloroperoxidase-triggered bromocyclization of α-allenols.

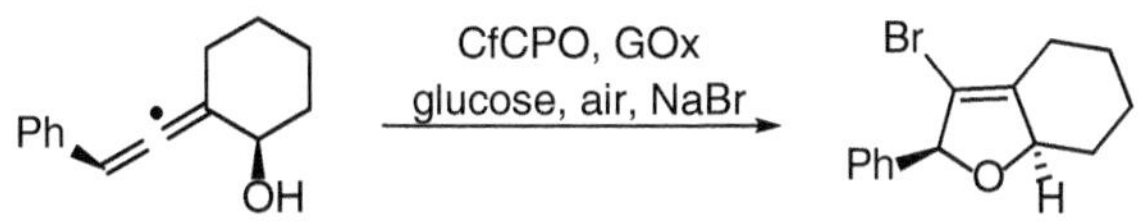

Scheme 6.93 Chloroperoxidase-mediated synthesis of a stereochemically defined bicyclic vinyl bromide.

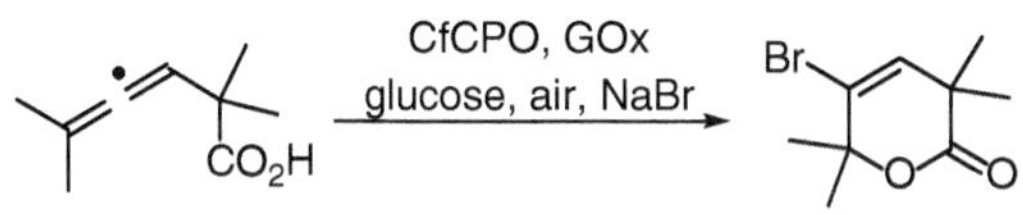

Scheme 6.94 Chloroperoxidase-triggered bromolactonization of 3,4-dienoate. Source: Based on Naapuri et al. [95].

of Fe(II) or Cu(I) co-catalyst. Laccase has been recognized as a versatile oxidative enzyme. The laccase-mediated air oxidation was successfully combined with the peptide-catalyzed α-oxyamination of aldehydes without using a metal co-catalyst. The asymmetric α-oxyamination of aldehydes resulted in the corresponding oxyaminated aldehydes in 65–80% isolated yields and 86–90% ee. When the reaction was carried out in the presence of a surfactant Tween 80 for a longer time, the oxyaminated carboxylic acids were isolated in 60–78% yields and 89–92% ee (Scheme 6.95). Laccase oxidized TEMPO to give the oxoammonium ion, which underwent addition

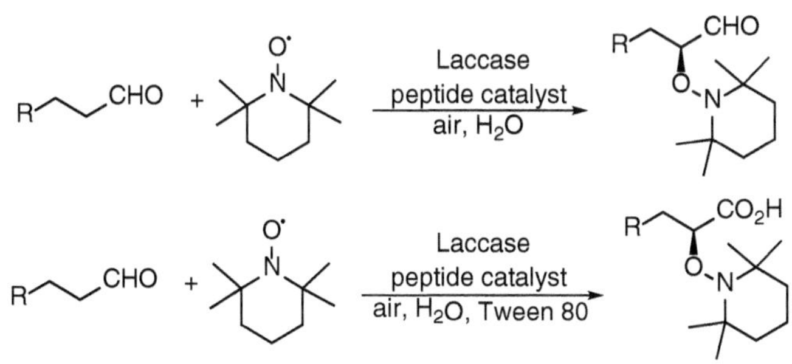

Scheme 6.95 Chemoenzymatic α-oxyamination of aldehydes.

to the enamine formed between the peptide catalyst and the aldehyde, resulting in α-oxyamination of the aldehydes. Indeed, the preformed oxoammonium ion proceeded with the enantioselective α-oxyamination of 5-phenylpental to furnish the product with 60% yield and 86% ee. Since an enamine was reported to be more readily oxidized than TEMPO, the oxidation of the enamine by laccase followed by the coupling of the oxidized enamine intermediate with TEMPO was investigated [96].

The laccase/TEMPO system was recently found to catalyze the aromatization of 2,5-dihydrofurans to give the corresponding furans. Metal-catalyzed ring-closing metathesis (RCM) reaction of diallyl ethers offered a straightforward synthesis of substituted 2,5-dihydrofurans. These two reactions were combined in one-pot processes to prepare the furans from the diallyl ethers. The metathesis/aromatization cascade of (1-(allyloxy)allyl)benzene was first carried out in a one-pot two-step manner using Grubbs second-generation catalyst and *Tinea versicolor* laccase lyophilized powder/TEMPO. Under the optimized conditions in the medium of sodium phosphate buffer/isooctane (1 : 2) (1-(allyloxy)allyl)benzene was transformed into 2-phenylfuran with 81% conversion. The chemoenzymatic transformation was then conducted in a one-pot concurrent mode with conversion reaching 93%, and 2-phenylfuran was isolated in 62% yield. A set of furan derivatives were synthesized in 21–99% conversions and 20–76% isolated yields from the corresponding substituted diallyl ethers through this chemoenzymatic concurrent cascade of Grubbs-catalyzed RCM and laccase/TEMPO-catalyzed aromatization (Scheme 6.96) [97].

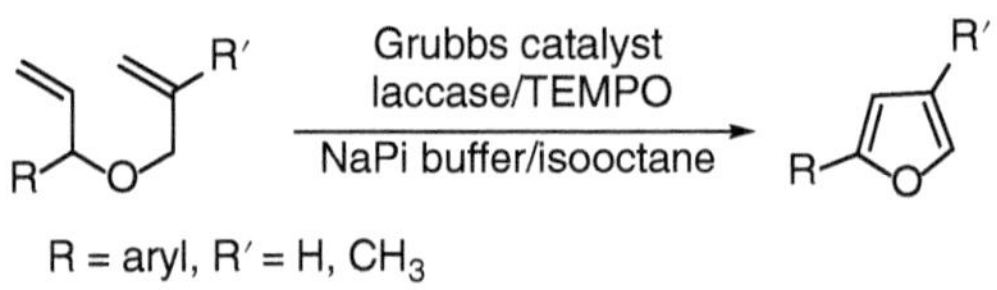

Scheme 6.96 Chemoenzymatic metathesis/aromatization cascade for the synthesis of substituted furans. Source: Based on Risi et al. [97].

Organocatalysts such as (*S*)-2-[diphenyl(trimethylsilyl- oxy)methyl]pyrrolidine mediated the reaction of quinones with aldehydes, leading to the enantioselective formation of α-arylated aldehydes. The quinones were usually generated *in situ* from 1,4-dihydroxybenzenes by electrochemical method or with chemical oxidizing agents such as $PhI(OAc)_2$. Laccases also catalyzed the two-electron oxidation of electron-rich (poly-)phenols and aromatic amines with aerial oxygen as the terminal oxidant. As such, laccase-catalyzed oxidation of 1,4-dihydroxybenzenes was applied to the umpolung of their intrinsic nucleophilic reactivity to electrophiles for the organocatalytic reaction with aldehydes. By combining a laccase from *Agaricus bisporus* with (*S*)-2-[diphenyl(trimethylsilyloxy)methyl]pyrrolidine, an asymmetric bio- and organocatalytic cascade was developed for the α-arylation of aldehydes, which were *in situ* transformed into the corresponding cyclic hemiacetals (Scheme 6.97). A series of 3-substituted 2,3-dihydrobenzofuran-2,5-diols were prepared in up to 98% yields and up to 97%

ee from substituted 1,4-dihydroxybenzenes and aldehydes in a simple one-pot concurrent chemoenzymatic process [98].

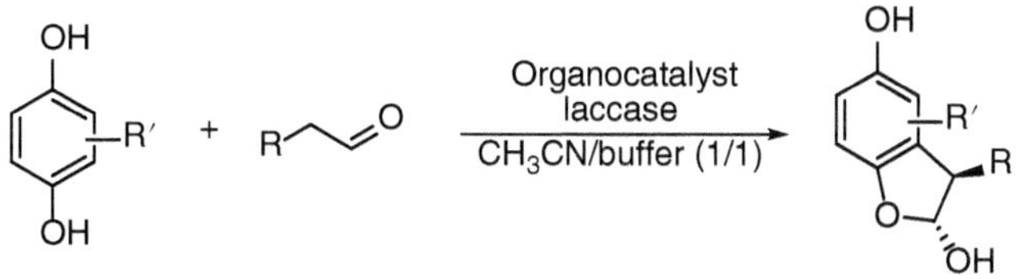

Scheme 6.97 One-pot concurrent chemoenzymatic synthesis of 3-substituted 2,3-dihydrobenzofuran-2,5-diols.

6.2 Integration of Chemical Reaction with Metabolism of Living Organisms

In the above discussion, the biotransformation usually involved single reactions. Recently, advances in synthetic biology have enabled the utilization of engineered microbes for chemical production from renewable feedstocks. However, many important chemicals such as pharmaceuticals and other fine chemicals cannot be directly accessed using this approach. In this context, efforts have also been made to combine non-enzymatic reactions with engineered microbial metabolism for synthesis of complex molecules [99]. The metabolites generated *in situ* by living engineered microbes can be further transformed into value-added chemicals using traditional chemical synthetic reactions or used as reagents for the chemical synthesis. For example, bacterially produced hydrogen gas was utilized as hydrogen source for transition-metal-catalyzed hydrogenation of alkenes. By using water-soluble alkene caffeic acid as the substrate, biocompatible Royer palladium catalyst (2.44% palladium on polyethy leneimine (PEI)/silica gel) was identified as an effective hydrogenation catalyst and combined with *in situ* hydrogen production by *Escherichia coli* DD-2 in minimal media containing additives $Fe(NH_4)_2(SO_4)_2$ and casamino acids. Caffeic acid was hydrogenated with 100% conversion and 87% isolated yield. This integrated system of metal-catalyzed hydrogenation with metabolism of living organisms showed excellent substrate tolerance and a diversity of alkenes and 3-phenylpropiolic acid were hydrogenated to afford the products in 63–95% yields (Scheme 6.98) [100].

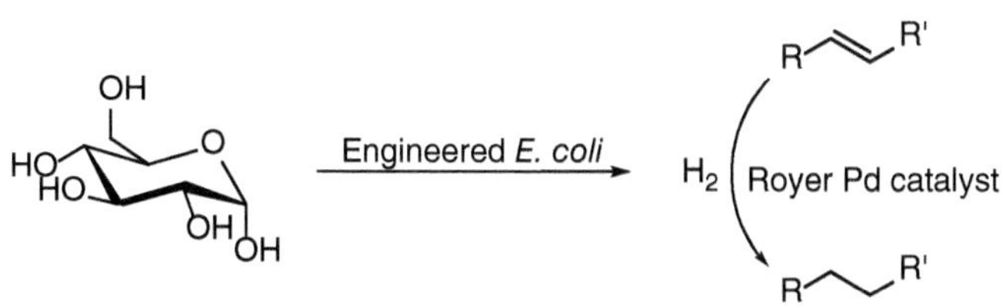

Scheme 6.98 Integration of Pd-catalyzed hydrogenation of alkenes with metabolism of engineered *E. coli*. Source: Based on Sirasani et al. [100].

Various transition metal catalysts (Cu, Rh, Pd, Fe, etc.) catalyzed cyclopropanation of olefins with ethyl diazoacetate (EDA). Among them, iron(III) phthalocyanine catalyst (FePcCl) was identified as an efficient biocompatible catalyst for olefin cyclopropanation in aqueous media with growing *E. coli*. By integrating this catalyst with engineered *E. coli* capable of producing styrene, a set of nonnatural phenyl cyclopropanes were synthesized directly from D-glucose through microbial metabolism followed by non-biological carbene-transfer reaction in a one-pot process (Scheme 6.99) [101]. Vitamin E-derived designer micelles were found to not only increase the membrane permeability of *E. coli* by interacting with the bacterial outer membrane and thus facilitating styrene production, but also enhanced the accommodation of transition metal catalysts for olefin cyclopropanation. Addition of the surfactant TPGS-750-M (2% wt/v) resulted in two-fold increase in cyclopropane production titer (from 282 to 560 mg/l) [102].

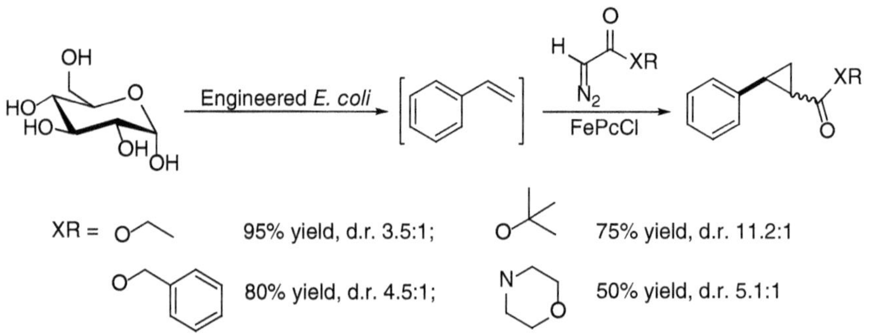

Scheme 6.99 Integration of microbial production of styrene with Fe-catalyzed olefin cyclopropanation. Source: Based on Wallace and Balskus [101].

An oxidative decarboxylase, the membrane-bound desaturase-like enzyme UndB from *Pseudomonas* was found to catalyze two decarboxylation steps of α,ω-dicarboxylic acids, generating the corresponding α,ω-dienes. This enzymatic transformation was combined with Ru-catalyzed RCM to access cycloalkenes. By employing *E. coli* whole cells expressing *und*B gene as the biocatalyst and a commercially available Hoveyda–Grubbs ruthenium(II) catalyst for the metathesis, cyclopentene, cyclohexene, and cycloheptene were synthesized from the corresponding α,ω-dicarboxylic acids with 83%, 80%, and 77% conversions, respectively, in a one-pot concurrent process in a biphasic medium of phosphate buffer and *n*-dodecane (90/10, v/v, Scheme 6.100). The cascade was also performed in a one-pot sequential mode, and similar results were obtained. This whole-cell decarboxylation–metathesis cascade was also applied to transform oleic acid into cycloheptene in 44% conversion (Scheme 6.101).

Oleic acid could be transformed into sebacic acid via enzymatic cascades of hydration, oxidation, BVO, and hydrolysis, or azelaic acid by the same enzymatic transformations followed by further double oxidation. A single *E. coli* strain (C10) was constructed by co-expressing PfBVMO, MlADH, OhyA2, and TLL with different expression levels to implement the multienzyme cascade for the

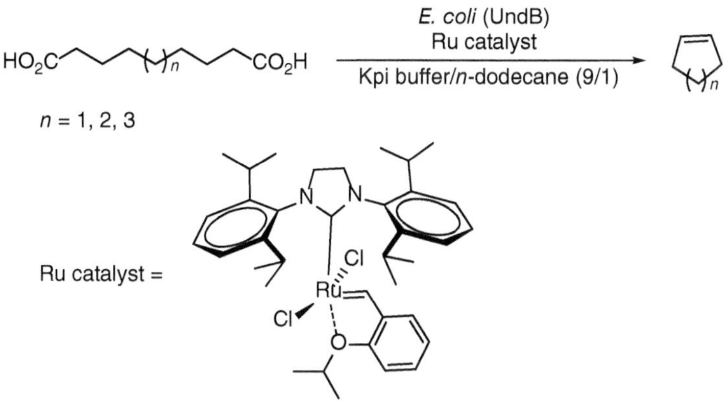

Scheme 6.100 One-pot concurrent chemoenzymatic transformation of α,ω-dicarboxylic acids to cycloalkenes.

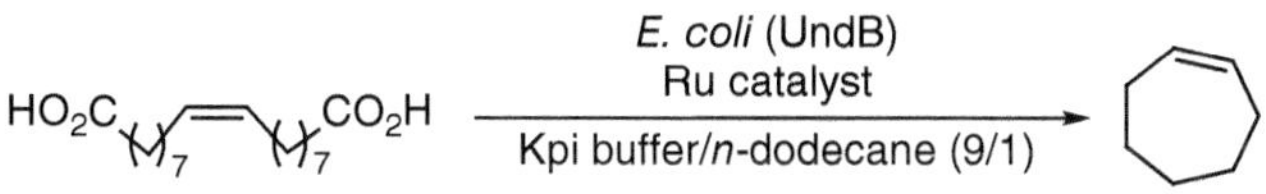

Scheme 6.101 One-pot concurrent chemoenzymatic transformation of oleic acid to cycloheptene.

production of sebacic acid from oleic acid. Similarly, a single *E. coli* strain (C9) for the multienzymatic transformation of oleic acid to azelaic acid was obtained by first constructing an *E. coli* strain for the production of 9-hydroxynonanoic acid and then incorporating ChnD, ChnE, and FadL genes into the same strain. By combination of *E. coli* strain C9 with the above whole-cell decarboxylation–metathesis cascade, direct transformation of oleic acid into cyclopentene was effected with 14% conversion in a one-pot concurrent process (Scheme 6.102). Higher conversion of 65% was achieved by adding *E. coli* (UndB) and Ru catalyst after oleic acid was converted into azelaic acid. The lower conversion in the concurrent mode was due to the cross-reactivity of *E. coli* (UndB) toward decarboxylation of oleic acid. Oleic acid was converted into cyclohexene in 6% conversion when the *E. coli* (C10), *E. coli* (UndB) strains, and Ru catalyst were added simultaneously in one pot (Scheme 6.103), while 22% was realized by adding *E. coli* (UndB) strains and Ru catalyst at a late stage. The low activity of PfBVMO possibly contributed to the modest conversion of cyclohexene, which should be improved by engineering a more active BVMO.

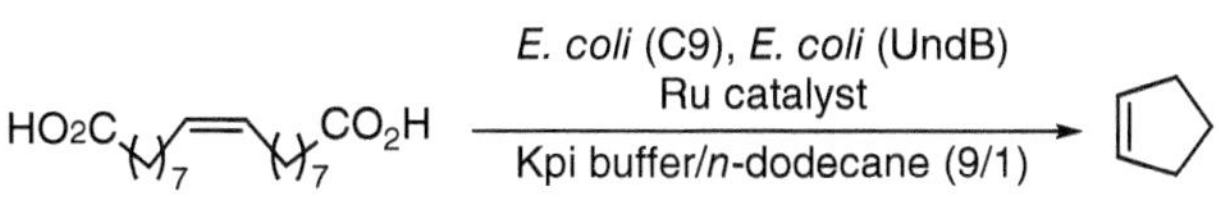

Scheme 6.102 One-pot concurrent chemoenzymatic transformation of oleic acid to cyclopentene.

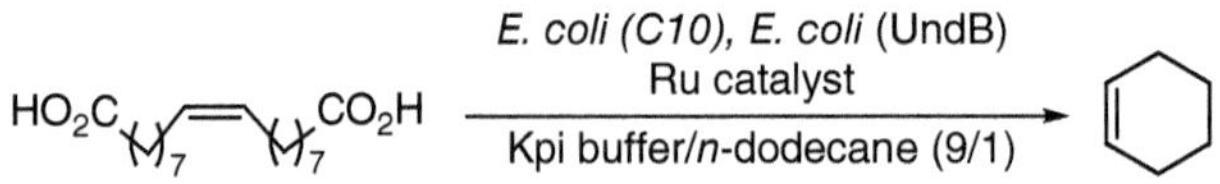

Scheme 6.103 One-pot concurrent chemoenzymatic transformation of oleic acid to cyclohexene.

Oleic acid is industrially produced from plant oils such as canola oil and olive oil. Lipase catalyzes the hydrolysis of olive oil to give oleic acid. An *E. coli* (TLL) strain was constructed by expressing the lipase from *Thermomyces lanuginosus* and the lyophilized cells showed good activity for the hydrolysis of olive oil to oleic acid. The enzymatic hydrolysis of olive oil to oleic acid was integrated into the above chemoenzymatic cascades for the synthesis of cyclopentene, cyclohexene, and cycloheptene from olive oil, an attractive bio-based feedstock. When lyophilized *E. coli* (TLL), *E. coli* (UndB), and Ru catalyst were added simultaneously in one-pot, 610–670 μM of cycloheptene was produced from 1 g/l olive oil in a concurrent process (Scheme 6.104). For the production of cyclopentene from olive oil, the lyophilized *E. coli* (TLL), *E. coli* (C9), *E. coli* (UndB), and Ru catalyst were added sequentially in one pot, and 1 g/L olive oil was converted into 760 μM of cyclopentene (Scheme 6.105). Similarly, by adding lyophilized *E. coli* (TLL), *E. coli* (C10), *E. coli* (UndB), and Ru catalyst sequentially in one-pot, 710 μM cyclohexene was generated from 1 g/l olive oil (Scheme 6.106). Cyclopentene and cyclohexene were produced from olive oil via chemoenzymatic cascades consisting of nine or seven enzymatic transformations and a metal-catalyzed reaction. These results demonstrated that engineered multistep chemoenzymatic cascades are feasible and offer powerful tools to convert renewable resources into value-added products [103].

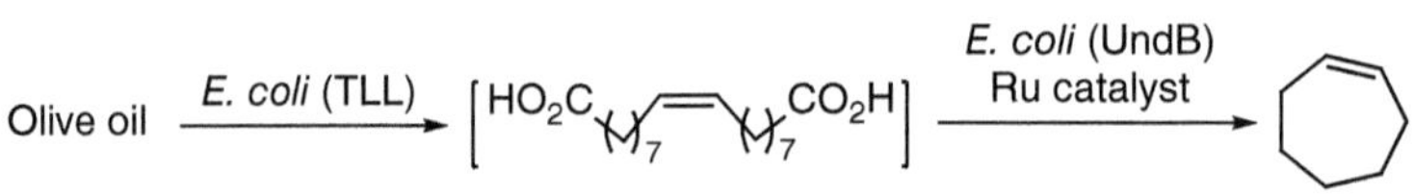

Scheme 6.104 One-pot concurrent chemoenzymatic transformation of olive oil to cycloheptene.

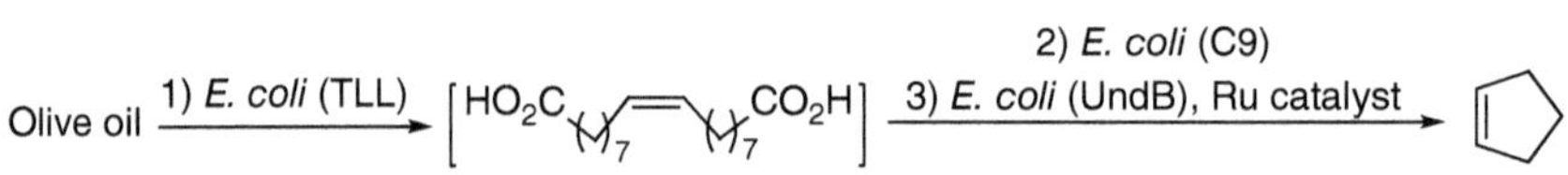

Scheme 6.105 One-pot sequential chemoenzymatic transformation of olive oil to cyclopentene.

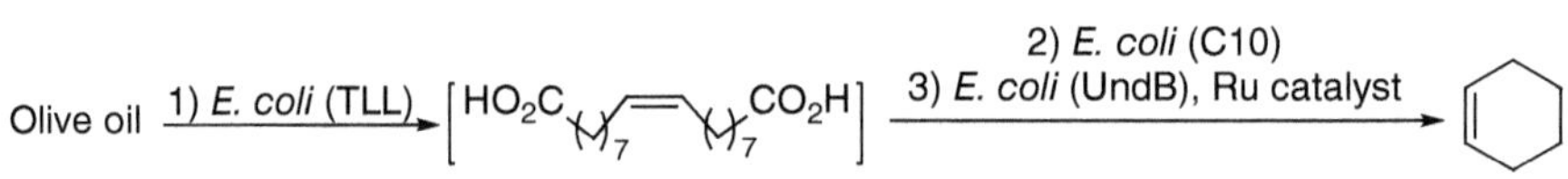

Scheme 6.106 One-pot sequential chemoenzymatic transformation of olive oil to cyclohexene.

6.3 One-pot Concurrent Chemoenzymatic Cascades via Compartmentalization

As discussed in the previous sections, the compatibility of chemo- and biocatalytic steps has been achieved in many cases. However, due to the intrinsic complexity of one-pot concurrent cascades, this is usually not the case and the requirements are not easy to be met in many situations, especially in the combination of biocatalysts with transition metal catalysts. The different catalysts should have shared substrate profile and compatible reaction conditions without cross-inhibitions of metal ions, substrates, and intermediates. Mutual inactivation of biocatalysts and chemical catalysts are often encountered. Also, their reaction conditions are usually quite different in the reaction medium, reaction temperature, and pH. In order to address these incompatibilities, a spatial separation strategy has been developed, in which the chemical reaction and the biotransformation were carried out in different compartments [104].

It has been reported that Grubbs second-generation catalyst was combined with pig liver esterase to convert diethyl diallylmalonate into cyclic malonic acid monoester. The metal catalyst and biocatalyst were encapsulated in octyl-grafted alginate amide and chitosan-coated alginate, respectively, and the RCM and hydrolysis were conducted in water and water-*t*-butanol (4 : 1), respectively, in a one-pot sequential mode. When the two reactions were carried out in a concurrent manner, side product formation increased, leading to 50% conversion with only 1% of the desired product. The core–shell beads were created to locate Grubbs catalyst and esterase PLE in different compartments, by coating chitosan-coated alginate beads of PLE with octyl-grafted alginate amide for Grubbs catalyst encapsulation multiple times. The resulting core–shell beads with Grubbs catalyst located between the alginate amide layers enabled the RCM to occur first before the substrate diffused into the PLE-containing core for hydrolysis. This significantly enhanced the selectivity toward the desired product to 75% at 84% conversion by using a large bead of 13.5 mm diameter allowing enough diffusion route of the Grubbs catalyst-containing shell (Scheme 6.107) [105].

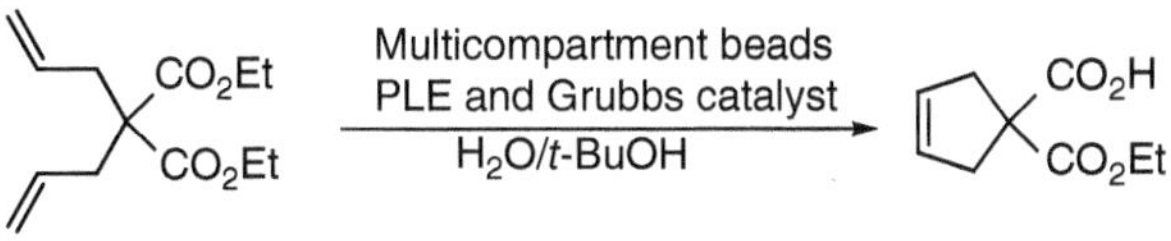

Scheme 6.107 One-pot concurrent conversion of diethyl diallylmalonate into cyclic malonic acid monoester by compartmentation of catalysts. Source: Based on Pauly et al. [105].

The metal catalyst (Pd/Cu) promoted Liebeskind–Srogl (L–S) coupling of thioesters with boronic acids at neutral pH and room temperature in THF with a phosphite additive, affording the corresponding ketones. Further studies showed that acetic phenyl thioester was hydrolyzed in a mixture of water and isopropanol (90/10, v/v) under otherwise identical conditions. Fortunately, the more stable *t*-butyl-thioester avoided hydrolysis and was coupled with phenylboronic acid to furnish the desired ketone in 78% yield. The L–S reaction was combined with an enantioselective enzymatic reduction for the production of chiral alcohols. The *R*-selective ADH from *L. kefir* (LKADH) and *S*-selective one from *R. ruber* (ADH-A) were used for the reduction of the *in situ*-formed ketone. Compatibility issues were solved by employing a membrane reactor with two chambers divided by a polydimethylsiloxane (PDMS) membrane to separate enzyme-deactivating components. The concurrent catalytic process was applied to acetic *t*-butyl-thioester with different substituted phenylboronic acid, and the corresponding antipodal chiral alcohols were obtained in high yields and 99% ee (Scheme 6.108) [106].

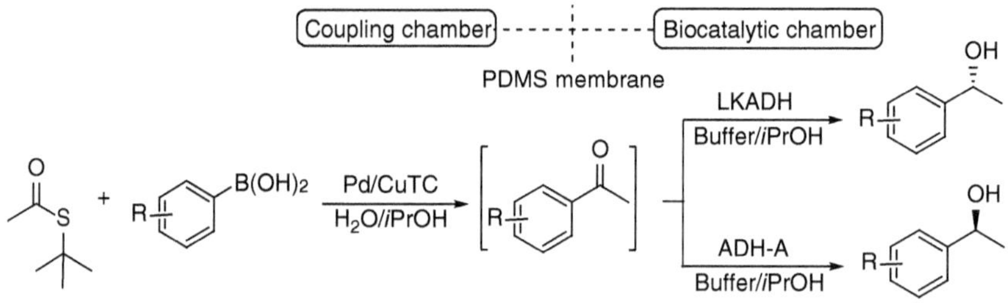

Scheme 6.108 One-pot Pd/Cu-promoted Liebeskind–Srogl (L–S) coupling with enzymatic reduction. Source: Based on Schaaf et al. [106].

Wacker oxidation of styrene into acetophenone using CuCl/$PdCl_2$ as a catalyst system could be carried out under aqueous condition that is compatible with the enzymatic reduction of ketones. However, the Cu salt strongly inhibited the enzyme activity of the ADH in the subsequent ketone reduction, preventing the incorporation of these two reactions into a one-pot concurrent process. Compartmentalization of the reactions was then applied to successfully combine Wacker oxidation of styrene with an enzymatic ketone reduction to convert styrene enantioselectively into 1-phenyl-ethanol in a one-pot process. The Wacker oxidation was performed in a PDMS thimble, which was placed in a flask where the formed acetophenone diffused into and was reduced by an ADH from *L. kefir* (Scheme 6.109). This site-isolation strategy enabled the biotransformation in a one-pot process, which formally corresponded to an asymmetric hydration of alkenes, affording various (*R*)-1-arylethanols with high conversions and 98–99% ee [107].

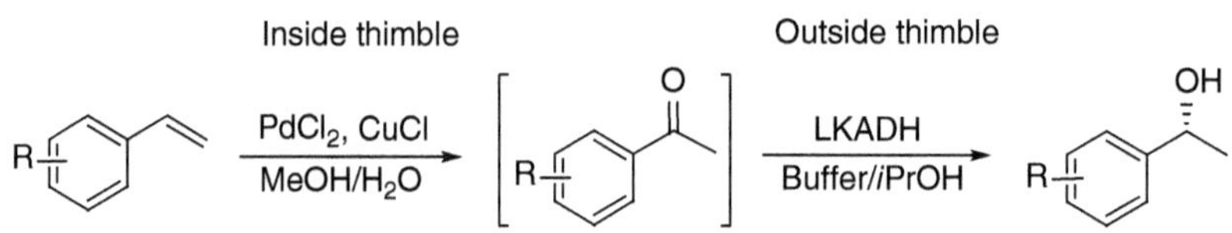

Scheme 6.109 One-pot Pd/Cu-promoted Wacker oxidation with enzymatic reduction.

Pd/Cu-catalyzed Wacker oxidation was also coupled with an enzymatic transamination of the resulting ketone intermediate. The transaminase from *Vibrio fluvialis*

(VF-TA) was used for the transamination reaction, but Cu^+ ion deactivated this enzyme. As such, similar compartmentation of the catalytic systems was adopted to achieve one-pot direct transformation of styrenes into 1-phenylethylamines through Wacker oxidation and enzymatic transamination (Scheme 6.110). The desired (*S*)-configurated chiral amines were obtained with72–93% conversions and 97–99 % ee [108]. The transaminase could be replaced with an amine dehydrogenase, in which ammonia instead of L-alanine was used as the amino donor. Styrene was converted into (*R*)-1-phenylethylamine in 96% conversion (76% isolated yield) and 99% ee using a double mutant of the leucine dehydrogenase from *Exigobacterium sibiricum* (EsLeuDH-DM) as the biocatalyst (Scheme 6.111) [109].

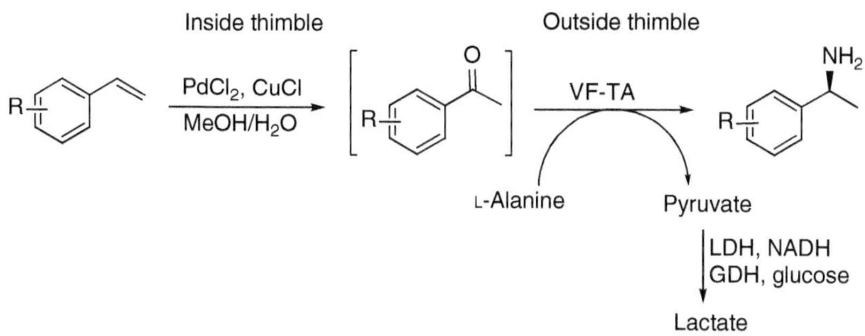

Scheme 6.110 One-pot Pd/Cu-promoted Wacker oxidation with enzymatic transamination.

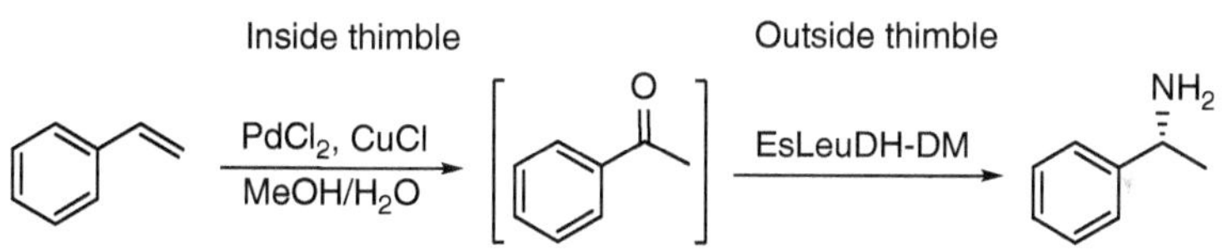

Scheme 6.111 One-pot Pd/Cu-promoted Wacker oxidation with enzymatic reductive amination. Source: Based on Uthoff and Gröger [109].

The regioselective flavin-dependent tryptophan halogenases catalyzed the chlorination and bromination of indole derivatives under mild conditions. By using a similar membrane compartmentalization approach, the halogenase-catalyzed regioselective bromination was integrated with Pd-assisted Suzuki–Miyaura coupling between the brominated intermediate and phenyl boronic acid, leading to the formation of arylated products in a one-pot process. A major challenge in the combination of halogenase enzymes with Pd-catalyst was that the halogenase protein resulted in significant reduction in the cross-coupling yields. As such, the biotransformation mixture by tryptophan halogenase was passed through a 10-kDa molecular weight cut-off (MWCO) cellulose membrane to remove the protein prior to the Pd-catalyzed cross-coupling reaction, affording the corresponding arylated product in good conversion. Although the membrane filtration worked well, a solution deoxygenation step was necessitated before the coupling reaction and high Pd-loading was required. In this context, the individual subunits of halogenase were cross-linked with each other, as well as the partner reductase and dehydrogenase proteins, via surface lysines by glutaraldehyde, to form the halogenase–reductase–dehydrogenase cross-linked enzyme aggregates (HRD-CLEAs). Using CLEAs as biocatalyst, the Pd-loading was reduced to 10%. A further improvement was achieved by employing PDMS thimbles as the compartmentalization tool with the cross-coupling reagents within the PDMS thimble. The Pd-loading could be reduced to 2% and the solution deoxygenation step was avoided. A range of functionally diverse arylated products could be synthesized via this one-pot chemoenzymatic process using regioselective and regiodivergent halogenases with different substrates and varied boronic acids (Scheme 6.112). This offered a new way to achieve discrimination between two similar, unactivated C–H positions, which is beyond the scope of current chemocatalytic methods [110].

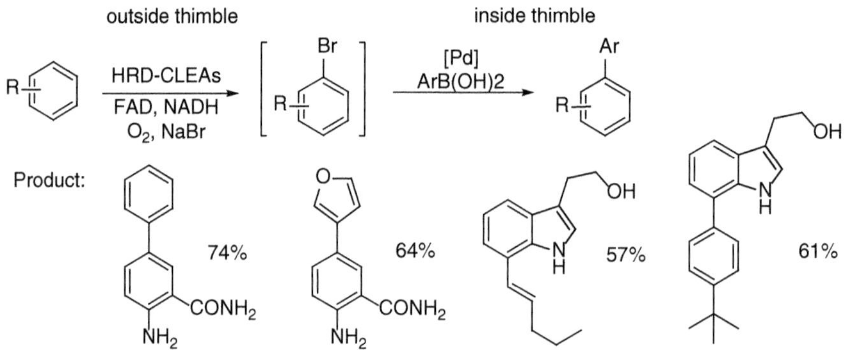

Scheme 6.112 One-pot enzymatic bromination and Pd-catalyzed Suzuki–Miyaura coupling.

In addition to the spatial separation strategy of conducting chemical reaction and biotransformation in different compartments of one reactor, the different reactions could also be carried out in different reactors with a continuous flow setting. Although this is not a one-pot setting, its efficiency is close to that of the one-pot process.

It was described earlier that a one-pot concurrent process of the proline-derivative-catalyzed aldol reaction of *m*-chloroacetophenone with acetone and the subsequent reduction of the resultant β-hydroxyketone catalyzed by an ADH could be performed in a buffer with isopropanol, but the conversion was low. In order to improve the transformation efficiency, immobilized organo- and biocatalysts were utilized, and the asymmetric organocatalytic aldol reaction and the bioreduction were carried in different fixed-bed reactors. In this setup, the aldol reaction of *m*-chloroacetophenone with acetone and the subsequent reduction of the resultant β-hydroxyketone proceeded in cyclohexane to give (*R*,*S*)-(1-(3-chlorophenyl)-1,3-butandiol in 89% yield with excellent diastereo- and enantioselectivity (d.r. >35 : 1, >99% ee, Scheme 6.113) [111].

2-Keto-3-deoxy sugar acids (KDs) represent the first partially defunctionalized intermediate in the production of bio-based building blocks from carbohydrates.

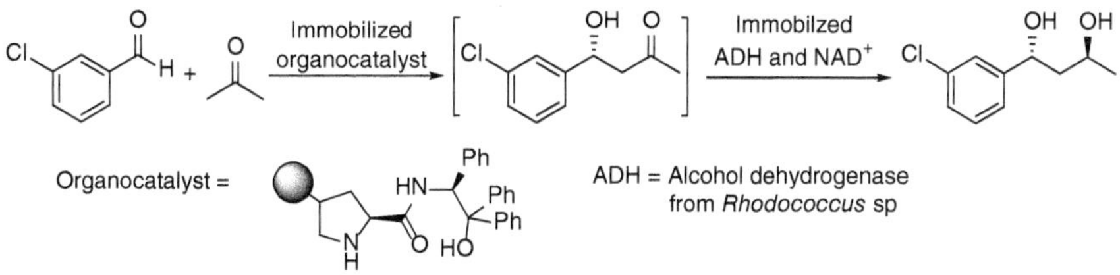

Scheme 6.113 Continuous flow chemoenzymatic synthesis of 1,3-diols. Source: Based on Heidlindemann et al. [111].

The synthesis of KDs could be achieved by an oxidation of the sugars followed by a dehydration of the sugar acids formed. It was found that different aldoses such as D-glucose, D-galactose, L-arabinose, and D-xylose could be selectively oxidized to the corresponding aldonic acids by a 0.5% Au/Al_2O_3 catalyst under medium alkaline aqueous conditions (pH 9–10). Several classes of dehydratases acted on different α,β-dihydroxy acids to give the dehydrated products. Among them, dihydroxyacid dehydratases (DHAD) from *Sulfolobus solfataricus* dihydroxyacid dehydratases (SsDHAD) exhibited high activity for a broad range of sugar acids in the pH range from 6 to 7.5 and temperature range from 50 to 80 °C. The gold-catalyzed oxidation and enzymatic dehydration were combined in an attempt to convert sugars to 2-keto-3-deoxy sugar acids in a one-pot setting. However, addition of SsDHAD directly into the reaction mixture of D-glucose oxidation resulted in inactivation of the gold catalyst. Further studies showed that the H_2O_2 generated in the gold-catalyzed oxidation inhibited the enzyme SsDHAD. Therefore, a continuous flow setting with the one-pot-incompatible catalysts in different compartments was developed to realize this chemoenzymatic cascade. In the gold-catalyzed oxidation of L-arabinose in a tank reactor at pH 9, the reaction mixture was pumped through an automatic titrator to set the pH as 6.5 and then into a second reaction vessel with catalase to remove H_2O_2. The H_2O_2-free solution was pumped over a Ni-sepharose column loaded with SsDHAD. By employing this continuous flow system, not only were the incompatibility issues solved but also each individual reaction step was performed under the optimized conditions. L-Arabinose and other sugars were effectively transformed into the corresponding 2-keto-3-deoxy sugar acids (Scheme 6.114). This represents an efficient synthesis route of 2-keto-3-deoxy sugar acids from the corresponding sugars and serves as an example for combination of heterogeneous inorganic catalysis with enzyme catalysis in a continuous flow system [112].

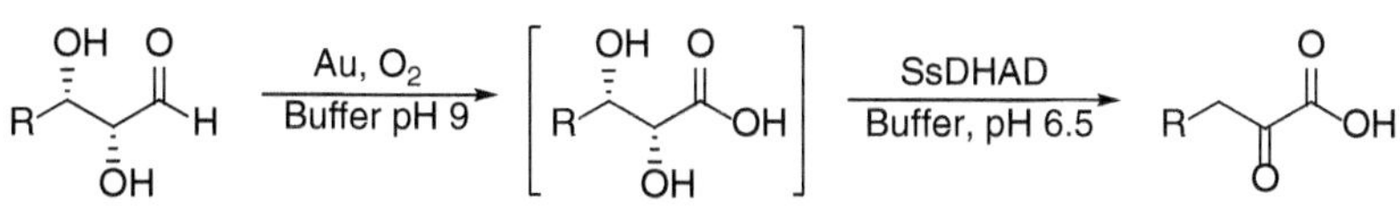

Scheme 6.114 Continuous flow chemoenzymatic synthesis of 2-keto-3-deoxy sugar acids.

The aza-Michael addition of benzylamine onto ethyl *trans*-crotonate produced racemic ethyl 3-(benzylamino)butanoate as the major product. The aminolysis of ethyl 3-(benzylamino)butanoate also occurred, resulting in the formation of benzyl 3-(benzylamino)butanamide. The side reaction could be reduced by using a plug-flow reactor (PFR). Lipase (Novozym 435) catalyzed the kinetic resolution of this chiral β-amino acid ester via aminolysis with benzylamine to give ethyl (*S*)-3-(benzylamino)butanoate and benzyl (*R*)-3-(benzylamino)butanamide. This enzymatic reaction could be carried out in a continuous stirred tank reactor (CSTR) or a packed-bed reactor (PBR). The thermal aza-Michael addition and lipase-catalyzed kinetic resolution were coupled by conducting them in a PFR and PBR, respectively, in a continuous flow setting. A chemoenzymatic continuously operating process was developed for the synthesis of ethyl (*S*)-3-(benzylamino)butanoate (Scheme 6.115), which was obtained in >98% ee with a space-time yield of 0.4 kg/l/d for the total reactor system and 1.8 kg/l/d for the PBR [113].

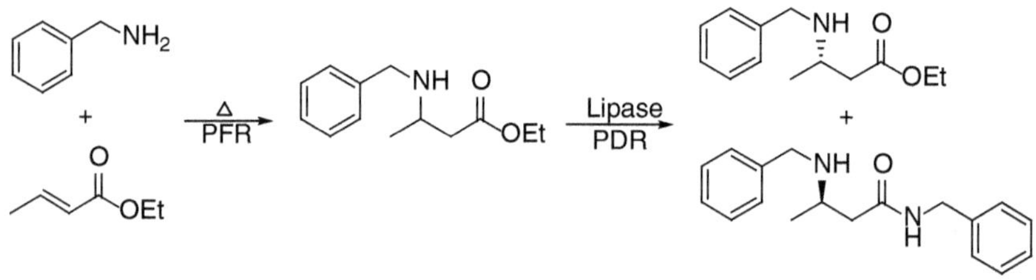

Scheme 6.115 Continuous flow chemoenzymatic synthesis of ethyl (*S*)-3-(benzylamino)butanoate.

6.4 Summary and Outlook

Although one-pot concurrent chemoenzymatic reaction, the most efficient mode of chemoenzymatic cascades, presents the highest compatibility requirements in reaction kinetics and reaction conditions of chemical and biocatalytic transformations, there are increasing number of such cascade processes reported in the last few years. While lipase is the most used enzyme in this endeavor due to its availability, stability, and versatility, we have seen that other enzymes have attracted more and more attention and been combined with various chemical reactions to enable various transformations that are useful in organic synthesis. Especially, multienzyme cascades and metabolically engineered bio-pathway, which transform readily available or bio-based cheap feedstocks such as glucose into valuable building blocks, have been integrated with chemical reactions, allowing the sustainable production of the desired products. In the process engineering aspect, delicately designed compartmentalization strategies make the non-compatible chemical and biochemical reactions implementable in the one-pot-like single-operation process, which also tremendously improves the production efficiency. Advances in (bio)catalysts design and preparation together with the reaction and process engineering are expected to further the development of state-of-art one-pot concurrent chemoenzymatic cascades and their applications in chemical production.

References

1 Makkee, M., Kieboom, A.P.G., Van Bekkum, H., and Roels, J.A. (1980). Combined action of enzyme and metal catalyst, applied to the preparation of D-mannitol. *Journal of the Chemical Society, Chemical Communications* (19): 930–931.

2 Makkee, M., Kieboom, A.P.G., and van Bekkum, H. (1985). Combined action of an enzyme and a metal catalyst on the conversion of d-glucose/d-fructose mixtures into d-mannitol. *Carbohydrate Research* 138 (2): 237–245.

3 Jung, H.M., Koh, J.H., Kim, M.-J., and Park, J. (2000). Concerted catalytic reactions for conversion of ketones or enol acetates to chiral acetates. *Organic Letters* 2 (3): 409–411.

4 Jung, H.M., Koh, J.H., Kim, M.-J., and Park, J. (2000). Practical ruthenium/lipase-catalyzed asymmetric transformations of ketones and enol acetates to chiral acetates. *Organic Letters* 2 (16): 2487–2490.

5 Kim, M.-J., Choi, M.Y., Han, M.Y. et al. (2002). Asymmetric transformations of acyloxyphenyl ketones by enzyme–metal multicatalysis. *The Journal of Organic Chemistry* 67 (26): 9481–9483.

6 Millet, R., Träff, A.M., Petrus, M.L., and Bäckvall, J.-E. (2010). Enantioselective synthesis of syn- and anti-1,3-aminoalcohols via β-aminoketones and subsequent reduction/dynamic kinetic asymmetric transformation. *Journal of the American Chemical Society* 132 (43): 15182–15184.

7 El-Sepelgy, O., Brzozowska, A., and Rueping, M. (2017). Asymmetric chemoenzymatic reductive acylation of ketones by a combined iron-catalyzed hydrogenation–racemization and enzymatic resolution cascade. *ChemSusChem* 10 (8): 1664–1668.

8 Langvik, O., Sandberg, T., Warna, J. et al. (2015). One-pot synthesis of (R)-2-acetoxy-1-indanone from 1,2-indanedione combining metal catalyzed hydrogenation and chemoenzymatic dynamic kinetic resolution. *Catalysis Science & Technology* 5 (1): 150–160.

9 Mäki-Arvela, P., Sahin, S., Kumar, N. et al. (2008). Cascade approach for synthesis of R-1-phenyl ethyl acetate from acetophenone: effect of support. *Journal of Molecular Catalysis A: Chemical* 285 (1): 132–141.

10 Mäki-Arvela, P., Sahin, S., Kumar, N. et al. (2008). One-pot chemo-biocatalytic synthesis of R-1-phenylethyl acetate from acetophenone hydrogenation over Pd/Al_2O_3 catalyst. *Applied Catalysis A: General* 350 (1): 24–29.

11 Vongvilai, P., Angelin, M., Larsson, R., and Ramström, O. (2007). Dynamic combinatorial resolution: direct asymmetric lipase-mediated screening of a dynamic nitroaldol library. *Angewandte Chemie International Edition* 46 (6): 948–950.

12 Vongvilai, P., Larsson, R., and Ramström, O. (2008). Direct asymmetric dynamic kinetic resolution by combined lipase catalysis and nitroaldol (henry) reaction. *Advanced Synthesis & Catalysis* 350 (3): 448–452.

13 Foley, A.M., Gavin, D.P., Deasy, R.E. et al. (2018). Dynamic kinetic resolution of 2-methyl-2-nitrocyclohexanol: combining the intramolecular

nitroaldol (henry) reaction & lipase-catalysed resolution. *Tetrahedron* 74 (13): 1435–1443.

14 Zhang, Y., Hu, L., and Ramström, O. (2013). Double parallel dynamic resolution through lipase-catalyzed asymmetric transformation. *Chemical Communications* 49 (18): 1805–1807.

15 Zhang, Y., Vongvilai, P., Sakulsombat, M. et al. (2014). Asymmetric synthesis of substituted thiolanes through domino thia-michael–henry dynamic covalent systemic resolution using lipase catalysis. *Advanced Synthesis & Catalysis* 356 (5): 987–992.

16 Hu, L., Schaufelberger, F., Zhang, Y., and Ramström, O. (2013). Efficient asymmetric synthesis of lamivudine via enzymatic dynamic kinetic resolution. *Chemical Communications* 49 (88): 10376–10378.

17 Sakulsombat, M., Zhang, Y., and Ramström, O. (2012). Dynamic asymmetric hemithioacetal transformation by lipase-catalyzed γ-lactonization: in situ tandem formation of 1,3-oxathiolan-5-one derivatives. *Chemistry – A European Journal* 18 (20): 6129–6132.

18 Zhang, Y., Schaufelberger, F., Sakulsombat, M. et al. (2014). Asymmetric synthesis of 1,3-oxathiolan-5-one derivatives through dynamic covalent kinetic resolution. *Tetrahedron* 70 (24): 3826–3831.

19 Hu, L., Zhang, Y., and Ramström, O. (2014). Lipase-catalyzed asymmetric synthesis of oxathiazinanones through dynamic covalent kinetic resolution. *Organic & Biomolecular Chemistry* 12 (22): 3572–3575.

20 Nunes, J.P.M., Veiros, L.F., Vaz, P.D. et al. (2011). Asymmetric synthesis of trans-4,5-dioxygenated cyclopentenone derivatives by organocatalyzed rearrangement of pyranones and enzymatic dynamic kinetic resolution. *Tetrahedron* 67 (15): 2779–2787.

21 Han, K., Kim, Y., Park, J., and Kim, M.-J. (2010). Chemoenzymatic synthesis of the calcimimetics (+)-NPS R-568 via asymmetric reductive acylation of ketoxime intermediate. *Tetrahedron Letters* 51 (27): 3536–3537.

22 Wen, Y.-Q., Hertzberg, R., Gonzalez, I., and Moberg, C. (2014). Minor enantiomer recycling: application to enantioselective syntheses of beta blockers. *Chemistry – A European Journal* 20 (13): 3806–3812.

23 Wang, Z.J., Clary, K.N., Bergman, R.G. et al. (2013). A supramolecular approach to combining enzymatic and transition metal catalysis. *Nature Chemistry* 5: 100.

24 Żądło-Dobrowolska, A., Koszelewski, D., Paprocki, D. et al. (2017). Enzyme-promoted asymmetric tandem passerini reaction. *ChemCatChem* 9 (15): 3047–3053.

25 Muller, C.R. and Meiners, I. (2014). Dominguez de Maria, P., Highly enantioselective tandem enzyme-organocatalyst crossed aldol reactions with acetaldehyde in deep-eutectic-solvents. *RSC Advances* 4 (86): 46097–46101.

26 Kapoor, M. and Gupta, M.N. (2012). Lipase promiscuity and its biochemical applications. *Process Biochemistry* 47 (4): 555–569.

27 Tang, Q., Popowicz, G.M., Wang, X. et al. (2016). Lipase-driven epoxidation is a two-stage synergistic process. *ChemistrySelect* 1 (4): 836–839.

28 Björkling, F., Godtfredsen, S.E., and Kirk, O. (1990). Lipase-mediated formation of peroxycarboxylic acids used in catalytic epoxidation of alkenes. *Journal of the Chemical Society, Chemical Communications* (19): 1301–1303.

29 Warwel, S. and Rüsch gen. Klaas, M. (1995). Chemo-enzymatic epoxidation of unsaturated carboxylic acids. *Journal of Molecular Catalysis B: Enzymatic* 1 (1): 29–35.

30 Klaas, M.R.G. and Warwel, S. (1996). Chemoenzymatic epoxidation of unsaturated fatty acid esters and plant oils. *Journal of the American Oil Chemists' Society* 73 (11): 1453–1457.

31 Aouf, C., Durand, E., Lecomte, J. et al. (2014). The use of lipases as biocatalysts for the epoxidation of fatty acids and phenolic compounds. *Green Chemistry* 16 (4): 1740–1754.

32 Klaas, M.R.G. and Warwel, S. (1997). Lipase-catalyzed preparation of peroxy acids and their use for epoxidation. *Journal of Molecular Catalysis A: Chemical* 117 (1): 311–319.

33 Moreira, M.A. and Nascimento, M.G. (2007). Chemo-enzymatic epoxidation of (+)-3-carene. *Catalysis Communications* 8 (12): 2043–2047.

34 da Silva, J.M.R. and Nascimento, M.D.G. (2012). Chemoenzymatic epoxidation of citronellol catalyzed by lipases. *Process Biochemistry* 47 (3): 517–522.

35 da Silva, J.M.R., Bitencourt, T.B., Moreira, M.A., and Nascimento, M.D.G. (2013). Enzymatic epoxidation of β-caryophyllene using free or immobilized lipases or mycelia from the Amazon region. *Journal of Molecular Catalysis B: Enzymatic* 95: 48–54.

36 Méndez-Sánchez, D., Ríos-Lombardía, N., Gotor, V., and Gotor-Fernández, V. (2014). Chemoenzymatic epoxidation of alkenes based on peracid formation by a Rhizomucor miehei lipase-catalyzed perhydrolysis reaction. *Tetrahedron* 70 (6): 1144–1148.

37 Meyer-Waßewitz, J., Holtmann, D., Ansorge-Schumacher, M.B. et al. (2017). An organic-single-phase CSTR process for the chemo-enzymatic epoxidation of α-pinene enables high selectivity and productivity. *Biochemical Engineering Journal* 126: 68–77.

38 Yang, F., Zhang, X., Li, F. et al. (2016). Chemoenzymatic synthesis of α--cyano epoxides by a Tandem–Knoevenagel–epoxidation reaction. *European Journal of Organic Chemistry* 2016 (7): 1251–1254.

39 Ankudey, E.G., Olivo, H.F., and Peeples, T.L. (2006). Lipase-mediated epoxidation utilizing urea–hydrogen peroxide in ethyl acetate. *Green Chemistry* 8 (10): 923–926.

40 Aouf, C., Lecomte, J., Villeneuve, P. et al. (2012). Chemo-enzymatic functionalization of gallic and vanillic acids: synthesis of bio-based epoxy resins prepolymers. *Green Chemistry* 14 (8): 2328–2336.

41 Lemoult, S.C., Richardson, P.F., and Roberts, S.M. (1995). Lipase-catalysed Baeyer–Villiger reactions. *Journal of the Chemical Society, Perkin Transactions* 1 (2): 89–91.

42 Ríos, M.Y., Salazar, E., and Olivo, H.F. (2007). Baeyer–Villiger oxidation of substituted cyclohexanones via lipase-mediated perhydrolysis utilizing urea–hydrogen peroxide in ethyl acetate. *Green Chemistry* 9 (5): 459–462.

43 Rios, M.Y., Salazar, E., and Olivo, H.F. (2008). Chemo-enzymatic Baeyer–Villiger oxidation of cyclopentanone and substituted cyclopentanones. *Journal of Molecular Catalysis B: Enzymatic* 54 (3): 61–66.

44 González-Martínez, D., Rodríguez-Mata, M., Méndez-Sánchez, D. et al. (2015). Lactonization reactions through hydrolase-catalyzed peracid formation Use of lipases for chemoenzymatic Baeyer–Villiger oxidations of cyclobutanones. *Journal of Molecular Catalysis B: Enzymatic* 114: 31–36.

45 Drożdż, A., Chrobok, A., Baj, S. et al. (2013). The chemo-enzymatic Baeyer–Villiger oxidation of cyclic ketones with an efficient silica-supported lipase as a biocatalyst. *Applied Catalysis A: General* 467: 163–170.

46 Markiton, M., Boncel, S., Janas, D., and Chrobok, A. (2017). Highly active nanobiocatalyst from lipase noncovalently immobilized on multiwalled carbon nanotubes for Baeyer–Villiger synthesis of lactones. *ACS Sustainable Chemistry & Engineering* 5 (2): 1685–1691.

47 Szelwicka, A., Zawadzki, P., Sitko, M. et al. (2019). Continuous flow chemo-enzymatic Baeyer–Villiger oxidation with superactive and extra-stable enzyme/carbon nanotube catalyst: an efficient upgrade from batch to flow. *Organic Process Research & Development* 23 (7): 1386–1395.

48 Zhong, J., Xu, F., Wang, J. et al. (2014). Candida antarctica lipase B-catalyzed synthesis of polyesters: starting from ketones via a tandem BVO/ROP process. *RSC Advances* 4 (17): 8533–8540.

49 Flourat, A.L., Peru, A.A.M., Teixeira, A.R.S. et al. (2015). Chemo-enzymatic synthesis of key intermediates (S)-γ-hydroxymethyl-α,β-butenolide and (S)-γ-hydroxymethyl-γ-butyrolactone via lipase-mediated Baeyer–Villiger oxidation of levoglucosenone. *Green Chemistry* 17 (1): 404–412.

50 Mendez-Sanchez, D., Lavandera, I., Gotor, V., and Gotor-Fernandez, V. (2017). Novel chemoenzymatic oxidation of amines into oximes based on hydrolase-catalysed peracid formation. *Organic & Biomolecular Chemistry* 15 (15): 3196–3201.

51 Yang, F., Wang, Z., Zhang, X. et al. (2015). A green chemoenzymatic process for the synthesis of azoxybenzenes. *ChemCatChem* 7 (21): 3450–3453.

52 Bitencourt, T.B. and da Graça Nascimento, M. (2009). Chemo-enzymatic synthesis of N-alkyloxaziridines mediated by lipases and urea-hydrogen peroxide. *Green Chemistry* 11 (2): 209–214.

53 Zhao, Z., Zhang, L., Li, F. et al. (2017). A novel oxidation of salicyl alcohols catalyzed by lipase. *Catalysts* 7 (12): 354.

54 Himiyama, T., Waki, M., Maegawa, Y., and Inagaki, S. (2019). Cooperative catalysis of an alcohol dehydrogenase and rhodium-modified periodic mesoporous organosilica. *Angewandte Chemie International Edition* 58 (27): 9150–9154.

55 Ríos-Lombardía, N., Vidal, C., Liardo, E. et al. (2016). From a sequential to a concurrent reaction in aqueous medium: ruthenium-catalyzed allylic alcohol isomerization and asymmetric bioreduction. *Angewandte Chemie International Edition* 55 (30): 8691–8695.

56 Cicco, L., Ríos-Lombardía, N., Rodríguez-Álvarez, M.J. et al. (2018). Programming cascade reactions interfacing biocatalysis with transition-metal catalysis in deep eutectic solvents as biorenewable reaction media. *Green Chemistry* 20 (15): 3468–3475.

57 Coccia, F., Tonucci, L., Del Boccio, P. et al. (2018). Stereoselective double reduction of 3-methyl-2-cyclohexenone, by use of palladium and platinum nanoparticles, in tandem with alcohol dehydrogenase. *Nanomaterials* 8 (10): 853.

58 Rulli, G., Duangdee, N., Hummel, W. et al. (2017). First tandem-type one-pot process combining asymmetric organo- and biocatalytic reactions in aqueous media exemplified for the enantioselective and diastereoselective synthesis of 1,3-diols. *European Journal of Organic Chemistry* 2017 (4): 812–817.

59 Díaz-Rodríguez, A., Borzęcka, W., Lavandera, I., and Gotor, V. (2014). Stereodivergent preparation of valuable γ- or δ-hydroxy esters and lactones through one-pot cascade or tandem chemoenzymatic protocols. *ACS Catalysis* 4 (2): 386–393.

60 Borowiecki, P., Telatycka, N., Tataruch, M. et al. (2020). Biocatalytic asymmetric reduction of γ-keto esters to access optically active γ-aryl-γ-butyrolactones. *Advanced Synthesis & Catalysis* 362 (10): 2012–2029.

61 Maskeri, M.A., Schrader, M.L., and Scheidt, K.A. (2020). A sequential umpolung/enzymatic dynamic kinetic resolution strategy for the synthesis of γ-lactones. *Chemistry – A European Journal* 26 (26): 5794–5798.

62 Aguirre-Pranzoni, C., Bisogno, F.R., Orden, A.A., and Kurina-Sanz, M. (2015). Lyophilized *Rhodotorula* yeast as all-in-one redox biocatalyst: access to enantiopure building blocks by simple chemoenzymatic one-pot procedures. *Journal of Molecular Catalysis B: Enzymatic* 114: 19–24.

63 Liardo, E., González-Fernández, R., Ríos-Lombardía, N. et al. (2018). Strengthening the combination between enzymes and metals in aqueous medium: concurrent ruthenium-catalyzed nitrile hydration – asymmetric ketone bioreduction. *ChemCatChem* 10 (20): 4676–4682.

64 Zambelli, P., Pinto, A., Romano, D. et al. (2012). One-pot chemoenzymatic synthesis of aldoximes from primary alcohols in water. *Green Chemistry* 14 (8): 2158–2161.

65 Stewart, K.N., Hicks, E.G., and Domaille, D.W. (2020). Merger of whole cell biocatalysis with organocatalysis upgrades alcohol feedstocks in a mild, aqueous, one-pot process. *ACS Sustainable Chemistry & Engineering* 8 (10): 4114–4119.

66 Bernard, J., van Heerden, E., Arends, I.W.C.E. et al. (2012). Chemoenzymatic reduction of conjugated C=C double bonds. *ChemCatChem* 4 (2): 196–199.

67 Rauch, M.C.R., Gallou, Y., Delorme, L. et al. (2019). Metals in biotechnology: Cr-driven stereoselective reduction of conjugated C=C double bonds. *ChemBioChem.* 21: 1112–1115.

68 Rauch, M.C.R., Huijbers, M.M.E., Pabst, M. et al. (1868). Photochemical regeneration of flavoenzymes – an old yellow enzyme case-study. *Biochimica et Biophysica Acta (BBA) – Proteins and Proteomics* 2020 (1): 140303.

69 Ryan, J., Šiaučiulis, M., Gomm, A. et al. (2016). Transaminase triggered aza-michael approach for the enantioselective synthesis of piperidine scaffolds. *Journal of the American Chemical Society* 138 (49): 15798–15800.

70 Taday, F., Ryan, J., Argent, S.P. et al. (2020). Asymmetric construction of alkaloids by employing a key ω-transaminase cascade. *Chemistry – A European Journal* 26 (17): 3729–3732.

71 Xu, J., Green, A.P., and Turner, N.J. (2018). Chemo-enzymatic synthesis of pyrazines and pyrroles. *Angewandte Chemie International Edition* 57 (51): 16760–16763.

72 Batista, V.F., Galman, J.L., Pinto, D.C.G.A. et al. (2018). Monoamine oxidase: tunable activity for amine resolution and functionalization. *ACS Catalysis* 8 (12): 11889–11907.

73 Bechi, B., Herter, S., McKenna, S. et al. (2014). Catalytic bio-chemo and bio-bio tandem oxidation reactions for amide and carboxylic acid synthesis. *Green Chemistry* 16 (10): 4524–4529.

74 Toscani, A., Risi, C., Black, G.W. et al. (2018). Monoamine oxidase (MAO-N) whole cell biocatalyzed aromatization of 1,2,5,6-tetrahydropyridines into pyridines. *ACS Catalysis* 8 (9): 8781–8787.

75 Köhler, V., Bailey, K.R., Znabet, A. et al. (2010). Enantioselective biocatalytic oxidative desymmetrization of substituted pyrrolidines. *Angewandte Chemie International Edition* 49 (12): 2182–2184.

76 Li, T., Liang, J., Ambrogelly, A. et al. (2012). Efficient, chemoenzymatic process for manufacture of the boceprevir bicyclic [3.1.0]proline intermediate based on amine oxidase-catalyzed desymmetrization. *Journal of the American Chemical Society* 134 (14): 6467–6472.

77 Znabet, A., Ruijter, E., de Kanter, F.J.J. et al. (2010). Highly stereoselective synthesis of substituted prolyl peptides using a combination of biocatalytic desymmetrization and multicomponent reactions. *Angewandte Chemie* 122 (31): 5417–5420.

78 Znabet, A., Polak, M.M., Janssen, E. et al. (2010). A highly efficient synthesis of telaprevir by strategic use of biocatalysis and multicomponent reactions. *Chemical Communications* 46 (42): 7918–7920.

79 Znabet, A., Blanken, S., Janssen, E. et al. (2012). Stereoselective synthesis of N-aryl proline amides by biotransformation–Ugi-Smiles sequence. *Organic & Biomolecular Chemistry* 10 (5): 941–944.

80 de Graaff, C., Oppelaar, B., Péruch, O. et al. (2016). Stereoselective monoamine oxidase-catalyzed oxidative aza-Friedel–Crafts reactions of meso-pyrrolidines in aqueous buffer. *Advanced Synthesis & Catalysis* 358 (10): 1555–1560.

81 Odachowski, M., Greaney, M.F., and Turner, N.J. (2018). Concurrent biocatalytic oxidation and C–C bond formation via gold catalysis: one-pot alkynylation of N-alkyl tetrahydroisoquinolines. *ACS Catalysis* 8 (11): 10032–10035.

82 Scalacci, N., Black, G.W., Mattedi, G. et al. (2017). Unveiling the biocatalytic aromatizing activity of monoamine oxidases MAO-N and 6-HDNO: development of chemoenzymatic cascades for the synthesis of pyrroles. *ACS Catalysis* 7 (2): 1295–1300.

83 Kawahara, N., Yasukawa, K., and Asano, Y. (2017). New enzymatic methods for the synthesis of primary α-aminonitriles and unnatural α-amino acids by oxidative cyanation of primary amines with d-amino acid oxidase from porcine kidney. *Green Chemistry* 19 (2): 418–424.

84 Denard, C.A., Huang, H., Bartlett, M.J. et al. (2014). Cooperative tandem catalysis by an organometallic complex and a metalloenzyme. *Angewandte Chemie* 126 (2): 475–479.

85 Denard, C.A., Bartlett, M.J., Wang, Y. et al. (2015). Development of a one-pot tandem reaction combining ruthenium-catalyzed alkene metathesis and enantioselective enzymatic oxidation to produce aryl epoxides. *ACS Catalysis* 5 (6): 3817–3822.

86 Campbell-Verduyn, L.S., Szymanski, W., Postema, C.P. et al. (2010). One pot 'click' reactions: tandem enantioselective biocatalytic epoxide ring opening and [3+2] azide alkyne cycloaddition. *Chemical Communications* 46 (6): 898–900.

87 Szymanski, W., Postema, C.P., Tarabiono, C. et al. (2010). Combining designer cells and click chemistry for a one-pot four-step preparation of enantiopure β-hydroxytriazoles. *Advanced Synthesis & Catalysis* 352 (13): 2111–2115.

88 Aguirre-Pranzoni, C., Tosso, R.D., Bisogno, F.R. et al. (2018). Preparation of chiral β-hydroxytriazoles in one-pot chemoenzymatic bioprocesses catalyzed by *Rhodotorula mucilaginosa*. *Process Biochemistry*.

89 Fernández-Fueyo, E., van Wingerden, M., Renirie, R. et al. (2015). Chemoenzymatic halogenation of phenols by using the haloperoxidase from *Curvularia inaequalis*. *ChemCatChem* 7 (24): 4035–4038.

90 Dong, J.J., Fernández-Fueyo, E., Li, J. et al. (2017). Halofunctionalization of alkenes by vanadium chloroperoxidase from *Curvularia inaequalis*. *Chemical Communications* 53 (46): 6207–6210.

91 Younes, S.H.H., Tieves, F., Lan, D. et al. (2020). Chemoenzymatic halocyclization of γ,δ-unsaturated carboxylic acids and alcohols. *ChemSusChem* 13 (1): 97–101.

92 Höfler, G.T., But, A., Younes, S.H.H. et al. (2020). Chemoenzymatic halocyclization of 4-pentenoic acid at preparative scale. *ACS Sustainable Chemistry & Engineering* 8 (7): 2602–2607.

93 Fernández-Fueyo, E., Younes, S.H.H., Rootselaar, S.V. et al. (2016). A biocatalytic aza-achmatowicz reaction. *ACS Catalysis* 6 (9): 5904–5907.

94 Thiel, D., Doknić, D., and Deska, J. (2014). Enzymatic aerobic ring rearrangement of optically active furylcarbinols. *Nature Communications* 5 (1): 5278.

95 Naapuri, J., Rolfes, J.D., Keil, J. et al. (2017). Enzymatic halocyclization of allenic alcohols and carboxylates: a biocatalytic entry to functionalized O-heterocycles. *Green Chemistry* 19 (2): 447–452.

96 Akagawa, K. and Kudo, K. (2011). Peptide/laccase cocatalyzed asymmetric α-oxyamination of aldehydes. *Organic Letters* 13 (13): 3498–3501.

97 Risi, C., Zhao, F., and Castagnolo, D. (2019). Chemo-enzymatic metathesis/aromatization cascades for the synthesis of furans: disclosing the aromatizing activity of laccase/TEMPO in oxygen-containing heterocycles. *ACS Catalysis* 9 (8): 7264–7269.

98 Suljić, S., Pietruszka, J., and Worgull, D. (2015). Asymmetric bio- and organocatalytic cascade reaction – laccase and secondary amine-catalyzed α-arylation of aldehydes. *Advanced Synthesis & Catalysis* 357 (8): 1822–1830.

99 Wallace, S., Schultz, E.E., and Balskus, E.P. (2015). Using non-enzymatic chemistry to influence microbial metabolism. *Current Opinion in Chemical Biology* 25: 71–79.

100 Sirasani, G., Tong, L., and Balskus, E.P. (2014). A biocompatible alkene hydrogenation merges organic synthesis with microbial metabolism. *Angewandte Chemie International Edition* 53 (30): 7785–7788.

101 Wallace, S. and Balskus, E.P. (2015). Interfacing microbial styrene production with a biocompatible cyclopropanation reaction. *Angewandte Chemie International Edition* 54 (24): 7106–7109.

102 Wallace, S. and Balskus, E.P. (2016). Designer micelles accelerate flux through engineered metabolism in *E. coli* and support biocompatible chemistry. *Angewandte Chemie International Edition* 55 (20): 6023–6027.

103 Wu, S., Zhou, Y., Gerngross, D. et al. (2019). Chemo-enzymatic cascades to produce cycloalkenes from bio-based resources. *Nature Communications* 10 (1): 5060.

104 Schmidt, S., Castiglione, K., and Kourist, R. (2018). Overcoming the incompatibility challenge in chemoenzymatic and multi-catalytic cascade reactions. *Chemistry – A European Journal* 24 (8): 1755–1768.

105 Pauly, J., Gröger, H., and Patel, A.V. (2019). Developing multicompartment biopolymer hydrogel beads for tandem chemoenzymatic one-pot process. *Catalysts* 9 (6): 547.

106 Schaaf, P., Bayer, T., Koley, M. et al. (2018). Biocompatible metal-assisted C–C cross-coupling combined with biocatalytic chiral reductions in a concurrent tandem cascade. *Chemical Communications* 54 (92): 12978–12981.

107 Sato, H., Hummel, W., and Gröger, H. (2015). Cooperative catalysis of noncompatible catalysts through compartmentalization: wacker oxidation and enzymatic reduction in a one-pot process in aqueous media. *Angewandte Chemie International Edition* 54 (15): 4488–4492.

108 Uthoff, F., Sato, H., and Gröger, H. (2017). Formal enantioselective hydroamination of non-activated alkenes: transformation of styrenes into enantiomerically pure 1-phenylethylamines in chemoenzymatic one-pot synthesis. *ChemCatChem* 9 (4): 555–558.

109 Uthoff, F. and Gröger, H. (2018). Asymmetric synthesis of 1-phenylethylamine from styrene via combined wacker oxidation and enzymatic reductive amination. *The Journal of Organic Chemistry* 83 (16): 9517–9521.

110 Latham, J., Henry, J.-M., Sharif, H.H. et al. (2016). Integrated catalysis opens new arylation pathways via regiodivergent enzymatic C–H activation. *Nature Communications* 7: 11873.

111 Heidlindemann, M., Rulli, G., Berkessel, A. et al. (2014). Combination of asymmetric organo- and biocatalytic reactions in organic media using immobilized catalysts in different compartments. *ACS Catalysis* 4 (4): 1099–1103.

112 Sperl, J.M., Carsten, J.M., Guterl, J.-K. et al. (2016). Reaction design for the compartmented combination of heterogeneous and enzyme catalysis. *ACS Catalysis* 6 (10): 6329–6334.

113 Strompen, S., Weiß, M., Gröger, H. et al. (2013). Development of a continuously operating process for the enantioselective synthesis of a β-amino acid ester via a solvent-free chemoenzymatic reaction sequence. *Advanced Synthesis & Catalysis* 355 (11-12): 2391–2399.

7

Photocatalytic and Biocatalytic Cascade Transformations

In the previous chapters, chemoenzymatic cascades in which chemical and biocatalytic transformations proceed in a one-pot sequential or concurrent manner have been discussed. The chemical reactions are usually catalyzed by a metal catalyst or organocatalyst. In recent years, the emergence of light-activated catalysts has led to the exploration of cascade transformations that combine a photochemical reaction with a biocatalytic transformation. This presents opportunities for the development of new synthetic routes for constructing molecules that may not be otherwise accomplishable. Photocatalysis with light as the energy source can proceed under milder conditions than its "dark" counterparts. It is known that enzymes catalyze highly specific transformations under mild and environmentally benign conditions. A light-activated catalyst working with an enzyme in pairs can promote more effective reactions. Combining light with enzymes brings the liaison of two fields of "green chemistry," which would enable new synthetic processes with higher efficiency and less environmental impacts [1, 2].

When light and enzymes are brought together, a few individual areas of photobiocatalysis have been developed in recent years [3–5]. These include the fields of light-driven enzymes, light-activated cofactor regeneration, light-triggered enzyme promiscuity, and photochemical and biocatalytic cascade reactions. In the first three cases, light serves as an energy source to assist the enzymatic reactions. While light-driven enzymes will be briefly presented here, the focus of this chapter will be on light-activated cofactor recycling, light-triggered nonnaturally occurring reactions, and the combination of a photochemical reaction with a biocatalytic transformation, and their potential applications in synthetic organic chemistry.

7.1 Photoenzymes

Light-driven enzymes or photoenzymes are enzymes that are not active in the absence of light, and require light to catalyze the biochemical reaction, i.e. the transformation of a substrate to a product. To date, four types of photoenzymes, i.e. the protochlorophyllide-reductases, photolyases, photosystems, and photodecarboxylases, have been discovered [6].

Chemo-Enzymatic Cascade Reactions, First Edition. Dunming Zhu.

Light-dependent NADPH:protochlorophyllide oxidoreductase (LPOR) catalyzes the reduction of protochlorophyllide to chlorophyllide (Scheme 7.1). The substrate protochlorophyllide is excited to a higher energy level by the absorption of a photon that induces the conformational changes in the enzyme for the reduction reaction. LPOR was thought to be evolved for oxygenic photosynthesis in response to the oxygenated atmosphere. However, recently this enzyme has also been found in the anoxygenic phototrophic α-proteobacterium *Dinoroseobacter shibae* DFL12T [7].

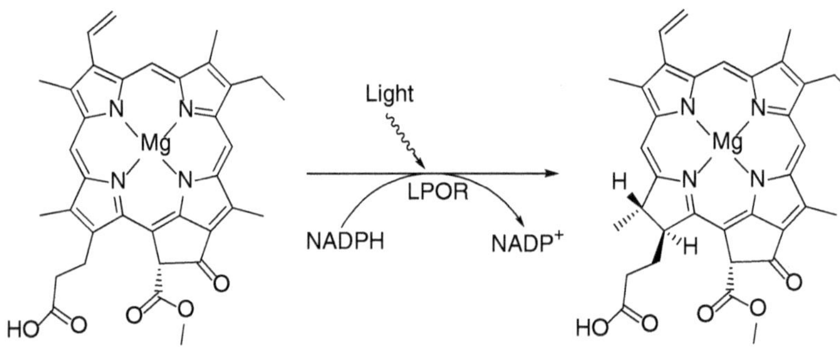

Scheme 7.1 LPOR-catalyzed reduction of protochlorophyllide to chlorophyllide

Photolyases are a class of flavoproteins requiring a flavin adenine dinucleotide (FAD) molecule as the key cofactor. These enzymes utilize UV–vis or blue light to repair the sun-induced DNA damage in forms of cyclobutane pyrimidine dimers (CPD photolyase) and pyrimidine (6-4) pyrimidone photoproducts (6-4 photolyase). CPD photolyases are divided into three classes (I–III) and single-stranded DNA (ssDNA)-specific one. The crystal structures of photolyases from different subfamilies showed overall similarities with the anionic hydroquinone ($FADH^-$) cofactor in a folded configuration but differences in $FADH^-$ and DNA recognition and electron transfer pathways. The different binding modes of $FADH^-$ and substrate in the divergent photolyase family may result in different reaction dynamics and the wide distribution of repair quantum yields (QYs) across the different classes of photolyases [8].

Photosystems are functional protein complexes that absorb light and transfer energy and electrons, resulting in the primary photochemistry of photosynthesis. There are two types of photosystems: plastocyanin: ferredoxin oxidoreductase (photosystem I, PSI) and water: plastoquinone oxidoreductase (photosystem II, PSII), and they are known to play important roles in photosynthesis. The protein complex includes a reaction center, which is surrounded by light-harvesting complexes for the absorption of light. The reaction center is an enzyme that uses light to initiate the electron transfer. In type I photosystems, ferredoxin-like iron–sulfur cluster proteins serve as terminal electron acceptors, whereas in type II photosystems electrons are ultimately shuttled to a quinone terminal electron acceptor [6, 9].

Recently, a photoenzyme that catalyzed the decarboxylation of fatty acids in the presence of blue light to give *n*-alkanes or alkenes (Scheme 7.2) was discovered in the microalga *Chlorella variabilis* NC64A. This fatty acid photodecarboxylase (CvFAP)

was active under illumination with blue light (400–520 nm), which is correlated with the absorbance of FAD. This implied that FAD might be the light-capturing part of the protein. The electron transfer from the fatty acid to the photoexcited FAD initiates the decarboxylation of the fatty acids. Time-resolved and cryogenic trapping UV–visible absorption spectroscopy studies characterized a red-shifted flavin intermediate involved in the formation of the final alkane product via hydrogen atom transfer (HAT) from a nearby active site cysteine residue (Scheme 7.3). This enzyme belongs to the glucose-methanol-choline oxidoreductase family and has been evaluated in the light-driven production of various chain-length hydrocarbons from the corresponding carboxylic acids. It showed highest conversion for the decarboxylation of palmitic acid, margaric acid, stearic acid, and arachidic acid [10–13].

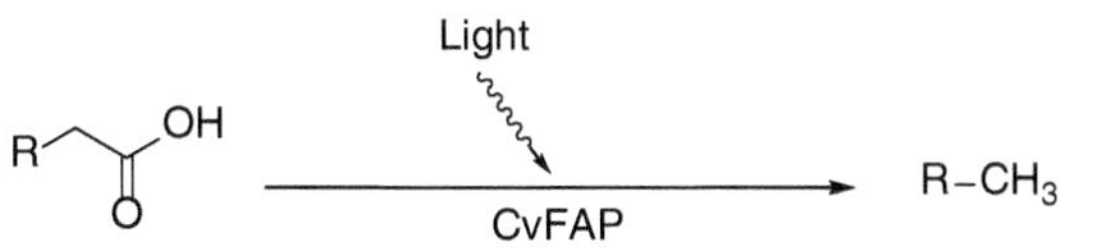

Scheme 7.2 Photoinduced CvFAP-catalyzed decarboxylation of fatty acids

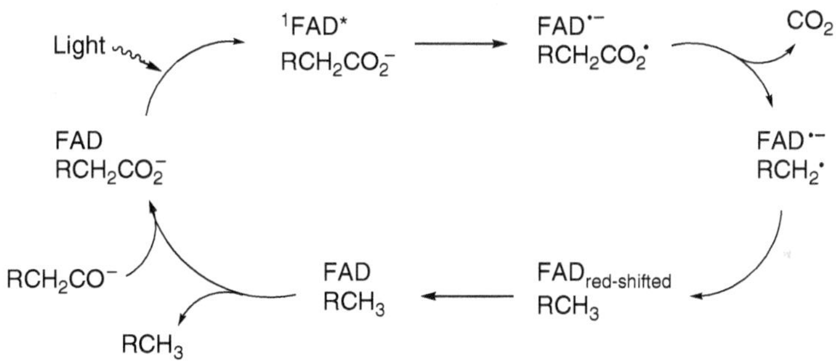

Scheme 7.3 Proposed reaction mechanism of photoinduced CvFAP-catalyzed decarboxylation of fatty acids

In addition to showing different activity of *cis*- and *trans*-oleic acid, CvFAP could differentiate the chiral configuration of α-carbon although the stereoselectivity of the wild-type enzyme was low for substrate 2-hydroxyoctanoic acid. Thankfully, variant G462Y, which was obtained by scanning of large amino acids at hotspots to narrow the substrate binding tunnel, catalyzed the decarboxylation of 2-hydroxyoctanoic acid under the irradiation with blue light to give the unreacted (*R*)-2-hydroxyoctanoic acid in 51% yield and 99% ee. The catalytic efficiency (k_{cat}/K_m) of the mutant G462Y (6.99/s/mM) was also improved 30 times compared to the wild-type enzyme (0.23/s/mM). This engineered fatty acid photodecarboxylase thus catalyzed the kinetic resolution of various α-hydroxy acids and α-amino acids by decarboxylating the (*S*)-enantiomer, providing the unreacted *R*-configured α-functionalized carboxylic acids with high yields and enantioselectivity (up to 99 %

ee, Scheme 7.4). Unlike the usual biocatalytic reactions, NADPH regeneration for the redox enzymes or prior preparation of esters for the hydrolytic reactions is not required for this light-driven kinetic resolution process [14].

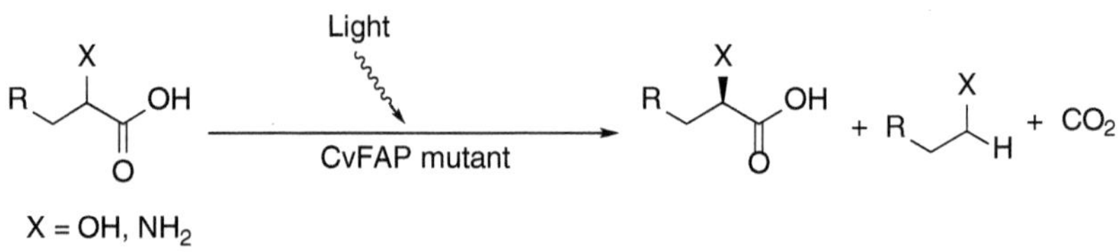

Scheme 7.4 CvFAP-mutant-catalyzed kinetic resolution of α-functionalized carboxylic acids

Recently, this fatty acid photodecarboxylase was also found to catalyze the decarboxylation of hydroxy fatty acids to yield the corresponding secondary fatty alcohols. Ricinoleic acid (C18), 12-hydroxy stearic acid (C18), and 16-hydroxy palmitic acid (C16) were decarboxylated by CvFAP in Tris–HCl buffer (pH 8.5, 100 mM), giving the corresponding fatty alcohols in about 80%, 60%, and 30% yields, respectively. By combination with the immobilized lipase from *Rhizopus oryzae* (ROL), castor oil was converted into (*R,Z*)-octadec-9-en-7-ol in up to 41.7% conversion via a one-pot cascade of lipase-catalyzed hydrolysis of castor oil and photoenzymatic decarboxylation of ricinoleic acid (Scheme 7.5) [15].

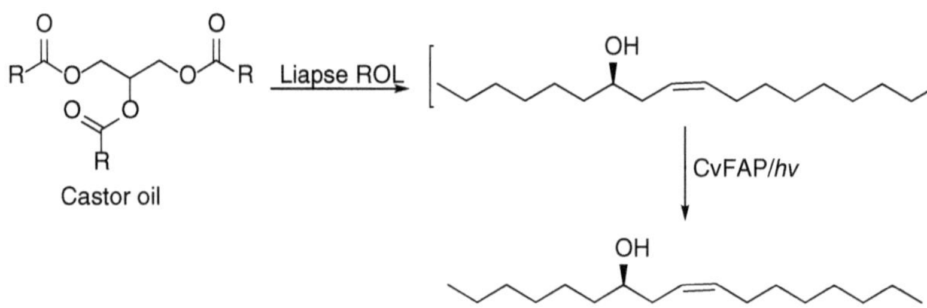

Scheme 7.5 Photoenzymatic cascade transformation of castor oil into (*R,Z*)-octadec-9-en-7-ol. Source: Based on Ma et al. [15].

The photoenzymatic decarboxylation was combined with the hydration of (poly)unsaturated fatty acids catalyzed by the oleate hydratase (LrOhyA) from *Lactobacillus reuteri*. The photoenzymatic cascade enabled the transformation of a broad range of (poly)unsaturated fatty acids into the corresponding enantiomerically enriched secondary fatty alcohols with up to 92% yields and up to 99% ee in a one-pot two-step procedure (Scheme 7.6) [16]. 5,8-Diol synthase (AnDS) from *Aspergillus*

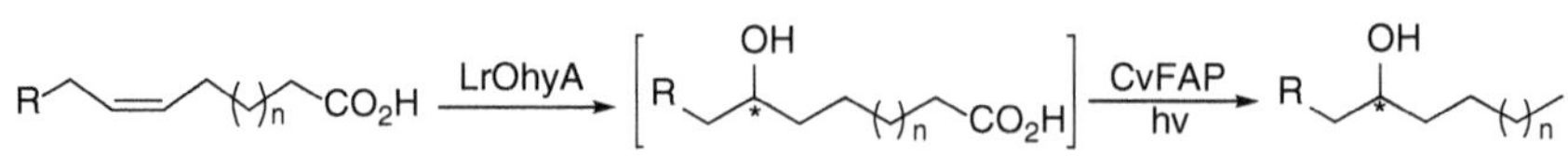

Scheme 7.6 Photoenzymatic cascade transformation of (poly)unsaturated fatty acids to chiral secondary fatty alcohols. Source: Based on Zhang et al. [16].

nidulans catalyzed the dihydroxylation of oleic acid. A photoenzymatic cascade was also established by coupling this synthase and the fatty acid photodecarboxylase to convert oleic acid into the chiral diol in 95% conversion (Scheme 7.7).

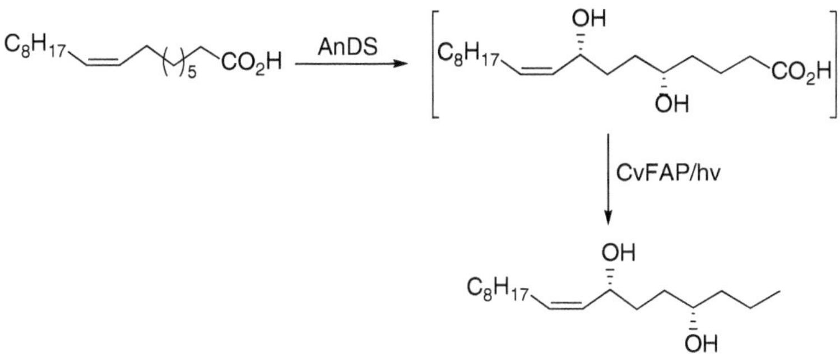

Scheme 7.7 Photoenzymatic cascade transformation of oleic acid into the chiral diol.

CvFAP was further applied to the multienzymatic cascades for the synthesis of long-chain aliphatic amines and esters from renewable fatty acids in a one-pot process. Combination of an ADH from *Micrococcus luteus* (MlADH), a mutant transaminase from *Vibrio fluvialis* (Vf-ATA), and a mutant CvFAP in a whole-cell cascade resulted in the transformation of ricinoleic acid and oleic acid into (*S*,*Z*)-heptadec-9-en-7-amine and 9-aminoheptadecane in 78% and 58% yields, respectively. Conversions of ricinoleic acid and oleic acid to 10-(heptanoyloxy)dec-8-ene and octylnonanoate were achieved in 65% and 69% yields, respectively, by integrating the ADH, an engineered Baeyer–Villiger monooxygenase from *Pseudomonas putida* (PpBVMO) and the Cv-FAP in a one-pot cascade reaction (Scheme 7.8). For the cascade reactions with oleic acid as starting material, a fatty acid double-bond hydratase from *Stenotrophomonas maltophilia* (OhyA) was included [17].

7.2 Light-Activation of Redox Enzymes Without Cofactor Regeneration

Oxidoreductases catalyze various redox reactions, such as asymmetric reduction, C–H hydroxylation, epoxidation, and Baeyer–Villiger oxidation, and have been increasingly applicable to the synthesis of pharmaceuticals, food additives, and fuels. An oxidoreductase usually requires a redox equivalent (cofactor) for electron transfer that drives the redox biocatalytic reaction in the active site of an enzyme. The most common cofactors are nicotinamide adenine dinucleotide, NAD(P)H, which mediate the electron transfer for various biocatalytic redox transformations. Since the cofactors are quite expensive, it is not cost-effective to supply a stoichiometric amount of cofactor for synthetic purposes, and thus much effort has been made to develop efficient methods for the regeneration of cofactors. A range of enzymatic methods have been developed for *in situ* regeneration of cofactor in the presence of a co-substrate and applied in the synthetic processes. Recently,

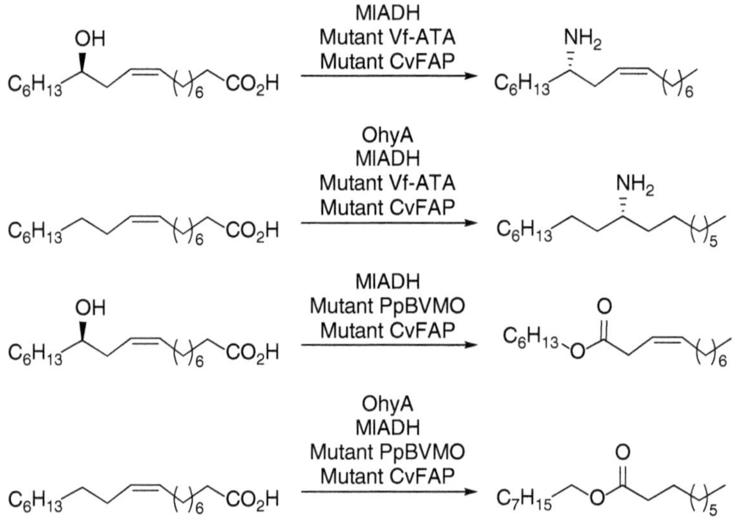

Scheme 7.8 Photoenzymatic cascade transformation of long-chain fatty acids into aliphatic amines and esters.

electrochemical and photochemical cofactor regenerations have also been explored. Electrochemical cofactor regeneration utilizes electrons generated by electrodes, and thus is considered as a clean approach. The problems of direct electrochemical reduction of nicotinamide cofactors are low regioselectivity, high over-potential, and electrode fouling. Photochemical cofactor regeneration harnesses the light energy for the electron transfer and has also gained increasing attention. In addition, some redox enzymes can function by photo-metiated electron transfer from electron donor to the enzyme without cofactor regeneration. This will be discussed in detail here and light-activated cofactor regeneration will be presented in the next section.

Photosensitizers absorb ultraviolet (or visible) light, which promotes electrons to higher energy states. While the photoexcited electrons occupy the lowest unoccupied molecular orbitals (LUMO), the oxidation of electron donors, such as water, triethanolamine (TEOA), or ethylenediaminetetraacetic acid (EDTA), provides electrons to fill the holes vacated by excited electrons. The photoexcited electrons can be directly transferred to the active site of oxidoreductases with prosthetic groups of metal clusters (e.g., Fe, Mo, Ni) or organic moieties (e.g., heme, flavins, Scheme 7.9a), or indirectly transferred to the redox enzyme by an electron mediator that shuttles electrons between photosensitizers and enzymes (Scheme 7.9b), thus driving redox biotransformations. In the natural reaction mechanisms of the redox enzymes, it is the cofactors such as NAD(P)H that provide the electrons for the catalytic transformation (Scheme 7.9c). The photoactivation of redox enzymes, such as dehydrogenases, hydrogenases, nitrogenases, cytochrome P450s, and flavoenzymes, allows for the development of photobiocatalytic cofactor-free reactions for the production of value-added chemicals, thus avoiding the requirement of expensive cofactors [18].

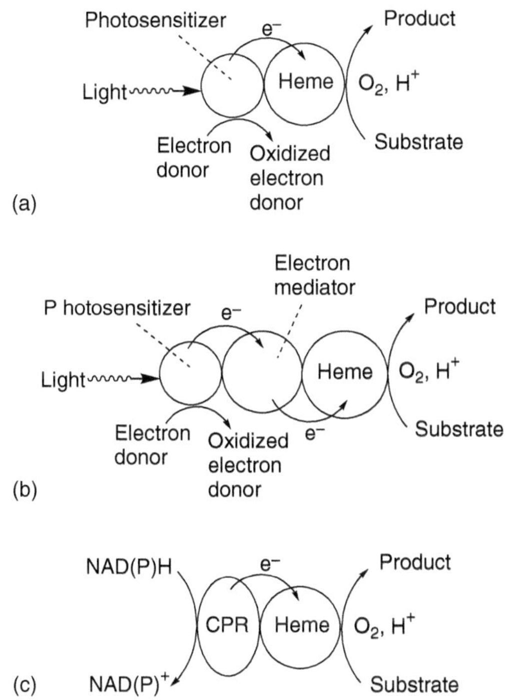

Scheme 7.9 Photoactivation of redox enzymes (P450 enzyme is used as the example). (a) Direct electron transfer from photosensitizer to enzyme. (b) Indirect electron transfer from photosensitizer to enzyme. (c) Natural electron transfer pathway.

The heme-containing cytochrome P450s catalyze hydroxylation, oxidation, sulfoxidation, decarboxylation, or dealkylation of a variety of substrates such as alkanes, fatty acids, terpenes, steroids, and xenobiotics. The P450 enzymes are typically composed of a flavin-containing reductase domain that draws electrons from cofactor NAD(P)H, and a heme domain that receives electrons from the reductase domain for the bioconversion of substrates. The efficient supply of electrons to the heme moiety is essential to initiate catalytic biotransformation by P450s. The direct transfer of photoexcited electrons from the photosensitizer to the P450 heme domain can successfully activate the enzyme in the absence of NAD(P)H. Recently, efficient reduction of the heme center by photoexcited eosin Y (EY) or Ru(II)-diimine complexes in the absence of NAD(P)H has been reported for different P450 enzymes including P450 BM3 variants. The cofactor-free, whole-cell P450 photobiocatalytic system was applied to the catalytic biotransformation of various substrates including simvastatin, lovastatin, omeprazole, steroids, and lauric acid (Scheme 7.10) [19, 20]. The direct electron transfer to the P450 heme iron can also be realized by using flavin mononucleotide (FMN) as the photosensitizer in the presence of EDTA as an electron donor, enabling the bacterial CYP102A1 to catalyze the photobiocatalytic C-hydroxylation reactions of 4-nitrophenol and lauric acid in the absence of NAD(P)H and NAD(P)H-P450 reductase (CPR) [21].

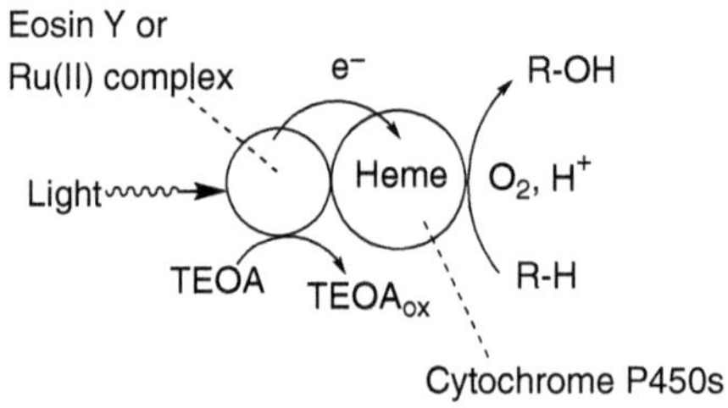

Scheme 7.10 Photoactivation of P450 enzyme for the cofactor-free hydroxylation. Source: Based on Park et al. [19]; Tran et al. [20].

Rieske non-heme iron oxygenases (ROs) activate molecular oxygen to generate reactive oxygen species that hydroxylate alkyl substrates and promote other oxidative transformations. Similar to P450s, these enzymes are also promising biocatalysts for the functionalization of C–H bonds, but are dependent on a complex electron transfer system and require in situ cofactor regeneration. An *Escherichia coli* whole-cell system was developed by using light-harvesting complexes to mediate the electron transfer from sacrificial electron donors to RO, thus catalyzing the hydroxylation of C–H bonds. When EY, 5(6)-carboxyeosin, or rose bengal (RB) was used as photosensitizer, and EDTA, 3-(*N*-morpholino)propanesulfonic acid (MOPS), or MES served as sacrificial electron donor, (*R*)-limonene, toluene, and indene were transformed into (1*R*,5*S*)-carveol, benzyl alcohol, and 1-indenol and *cis*- or *trans*-1,2-indanediol, respectively, by the mutant cumene dioxygenase (mutCDO) from *Pseudomonas fluorescens* IP01 under visible light irradiation (Scheme 7.11). Cofactor NAD(P)H was not needed, and the product yield was up to 1.3 g/l and reaction rate reached up to 1.6 mM/h [22].

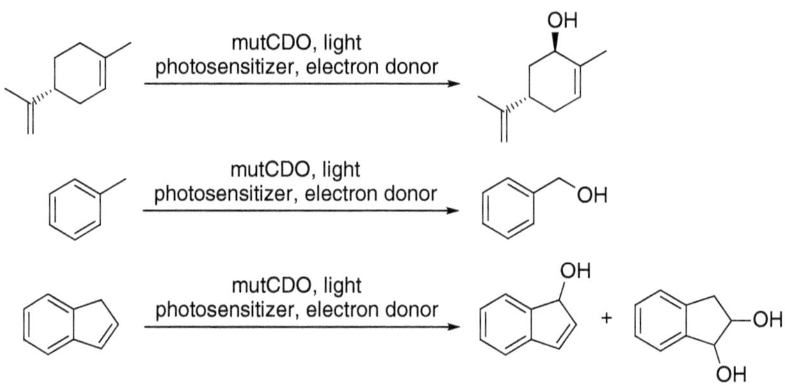

Scheme 7.11 Photoactivation of Rieske non-heme iron oxygenases for the cofactor-free hydroxylation.

Carbon dots (CDs) are attractive photosensitizers for biological systems because of their water solubility, photostability, and readily modified surface chemistry. As such, CDs were exploited as light absorbers in the solar-driven hydrogenation of fumarate to succinate catalyzed by a fumarate reductase (FccA), a flavoenzyme

c_3 from *Shewanella oneidensis* MR-1. Fumarate is reduced to succinate at the flavin active site by accepting one hydride from the bound flavin and one proton from a nearby arginine. By using the positively charged ammonium-terminated CDs (CD-$NHMe_2^+$) as photosensitizer, a high enzyme-based turnover number (TON) of $6.0 \pm 0.6 \times 10^3$ mol succinate (mol FccA)$^{-1}$ was achieved within 24 hours (Scheme 7.12). In contrast, the negatively charged carboxylate-terminated CDs (CD-CO_2^-) displayed little or no activity. The electrostatic interactions of the positively charged ammonium with the negatively charged enzyme at the CD–enzyme interface facilitated the transfer of photoexcited electrons directly to the enzyme, leading to the high photocatalytic activity observed with CD-$NHMe_2^{+23}$.

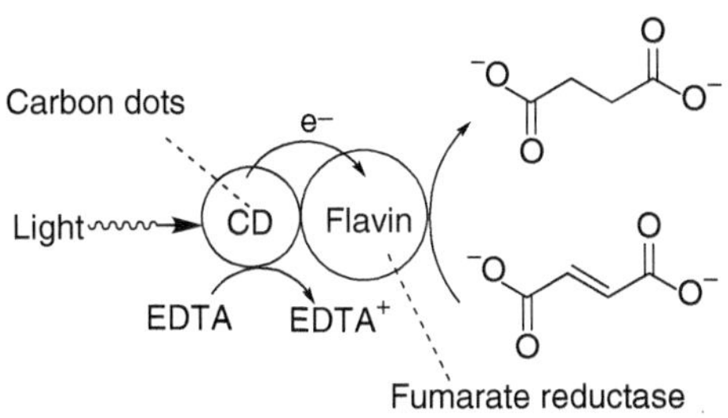

Scheme 7.12 Photoactivation of fumarate reductase using carbon dots as photosensitizer.

The modular surface chemistry of CDs also significantly affected the photoinduced reduction of protons to H_2 catalyzed by a [NiFeSe]-hydrogenase (H_2ase) from *Desulfomicrobium baculatum*. Negligible H_2 generation was observed with CD-CO_2^-, while CD-$NHMe_2^+$ exhibited high photoactivity when EDTA was used as the sacrificial electron donor and the reaction was performed at pH 6, reaching a TON for H_2ase of $52 \pm 8 \times 10^3$ mol H_2 (mol H_2ase)$^{-1}$ after 48 hours (Scheme 7.13) [23]. On the other hand, a TON of >50 000 mol H_2 (mol H_2ase)$^{-1}$ was reported for the light-driven H_2 production in water with [NiFeSe]-H_2ase in combination with a heptazine carbon nitride polymer (melon, CNx) as the photosensitizer [24].

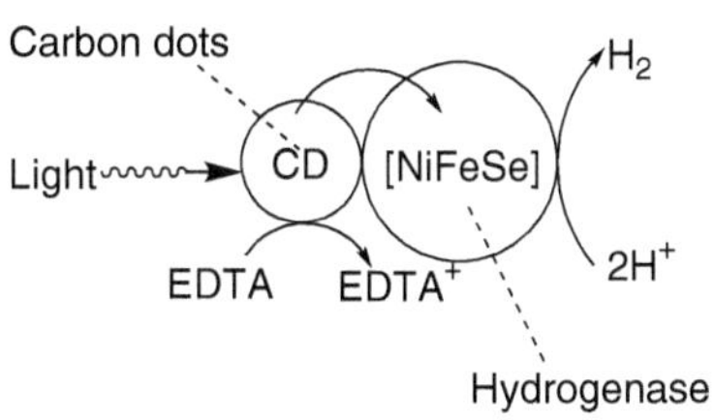

Scheme 7.13 Photoactivation of hydrogenase using carbon dots as photosensitizer. Source: Based on Hutton et al. [23].

The co-immobilization of a visible-light-absorbing chromophore and a redox enzyme onto the supporting materials such as metal oxide nanoparticles could also facilitate the transfer of the photoexcited electrons from the chromophore to the enzyme, thus promoting the photochemical redox reactions. TiO_2 nanoparticles (TiO_2 NPs) were modified with enzyme flavocytochrome c3 (fcc3) and the visible-light-absorbing chromophore [Ru(bpy)$_2$(4,4′-(PO$_3$H$_2$)$_2$bpy)]Br$_2$ (RuP) (bpy = 2,2′-bipyridine), and then suspended in 2-(*N*-morpholino)ethanesulfonic acid (MES) buffer containing disodium fumarate. Irradiation of the stirred suspension by visible light resulted in the hydrogenation of fumarate to succinate with a TON of 5800 and an average turnover frequency (TOF) of 0.4 s^{-1} after 4 hours (Scheme 7.14) [25].

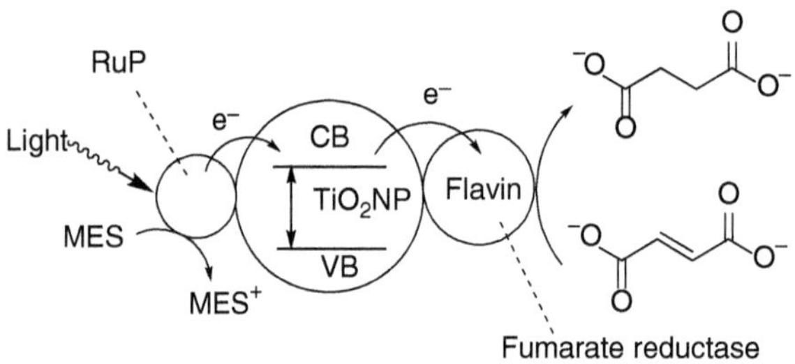

Scheme 7.14 TiO_2 nanoparticles-mediated photobiocatalytic reduction of fumarate. Source: Based on Bachmeier et al. [25].

Similarly, the formate dehydrogenase (FDH) from *Desulfovibrio vulgaris* Hildenborough (DvH) was immobilized onto dye-sensitized TiO_2 nanoparticles with a ruthenium *tris*-2,2′-bipyridine (RuP) complex or a diketopyrrolopyrrole (DPP) serving as a photosensitizer. Upon UV irradiation, the photoexcited electrons were transferred from dye to the conduction band (CB) of TiO_2 and conveyed to the catalytic W-center of FDH for the transformation of CO_2 into HCO_2^-. TEOA served as a sacrificial electron donor for the reduction of the oxidized dye. TOFs of 11 ± 1.0 and 5 ± 0.6 s^{-1} (after six hours) were observed by using RuP and DPP-sensitized TiO_2 (Scheme 7.15), respectively. A comparison of enzymatic and synthetic catalysts in combination with dye-sensitized TiO_2 nanoparticles without diffusional mediators for CO_2 reduction and H_2 evolution showed that enzymes outperform the synthetic systems in terms of TOF. Among the systems compared, the present RuP|TiO_2|FDH system exhibited the highest TOF for CO_2 reduction. The strong interfacial interactions facilitated rapid electron transfer from TiO_2 to the enzyme, thus playing an important role in the high activity and stability of the dye|TiO_2|FDH [26].

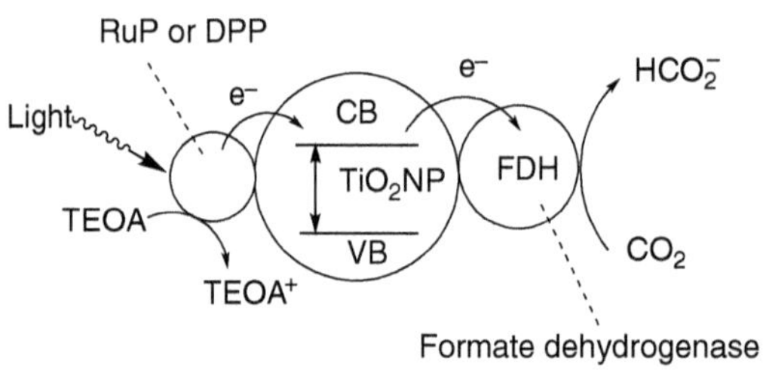

Scheme 7.15 TiO_2 nanoparticles-mediated photobiocatalytic reduction of CO_2.

Flavoenzymes such as Baeyer–Villiger monooxygenase (BVMO) and old yellow enzymes (OYEs) contain a tightly bound flavin molecule (e.g., FAD, FMN), which mediates hydrogen transfer with two electrons (H^-), one electron (H), or no electrons (H^+), resulting in various chemical transformations. The reduced nicotinamide cofactors NAD(P)H usually serve as a redox equivalent for the regeneration of the prosthetic flavins. Reetz et al. developed a photoactivation method for BVMO by adding free FAD as reductants rather than NAD(P)H. Photochemical reduction of FAD using EDTA as the sacrificial electron donor resulted in enantioselective Baeyer–Villiger oxidations of cyclic ketones with a high enantioselectivity of 97% (Scheme 7.16). However, the photoreduced FAD reacted rapidly with oxygen, the co-substrate of BVMO catalysis, in a diffusion-controlled manner, yielding FAD with concomitant formation of H2O2. This caused approximately 95 % of the reducing equivalents supplied by EDTA to not be coupled with enzymatic reactions, leading to the significantly lower apparent TOF than using NADPH as reductant [27].

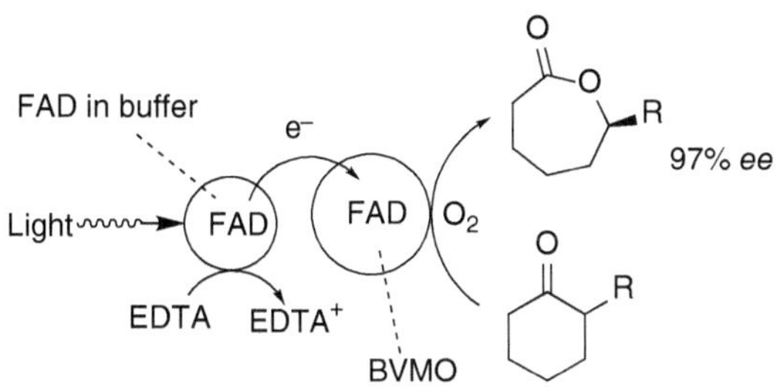

Scheme 7.16 Light-driven enzymatic enantioselective Baeyer–Villiger oxidation of cyclic ketones.

This decoupling problem was avoided in the light-driven OYE-catalyzed asymmetric reduction of C=C bonds. OYEs are oxygen-independent flavoenzymes that catalyze the reactions under anaerobic conditions. When FMN was used as a photosensitizer and EDTA as an electron donor, the direct photoactivation of an OYE from *Bacillus subtilis* (YqjM) catalyzed the reduction of ketoisophorone with TTN and TOF being 1.09×10^4 and 1.26×10^4/h, respectively. Similarly, UV and visible light promoted the electron transfer from titanium dioxide-based photosensitizer to the prosthetic FMN group of an OYE from *Thermus scotoductus* SA-01 (TsOYE) with water as the sacrificial electron donor, resulting in the asymmetric reduction of ketoisophorone into (*R*)-levodione in 66% conversion and 86% ee (Scheme 7.17) [28].

Lee et al. reported that xanthene dyes such as RB were able to transfer photoinduced electrons directly to the prosthetic FMN moiety of OYEs from an electron donor, TEOA, facilitating the conversion of 2-methylcyclohexenone to (*R*)-2-methylcyclohexanone with 80–90% conversion and >99% ee (Scheme 7.18) [29]. The ene-reductase (ER) TsOYE and RB were co-immobilized in an alginate hydrogel. The resulting RB-TsOYE-loaded alginate capsules also facilitated the asymmetric reduction of 2-methylcyclohexenone into enantiopure (*R*)-2-methylcyclohexanone with >99% ee and up to 70% conversion under irradiation with light without NAD(P)H [30]. Therefore, light could activate OYEs through

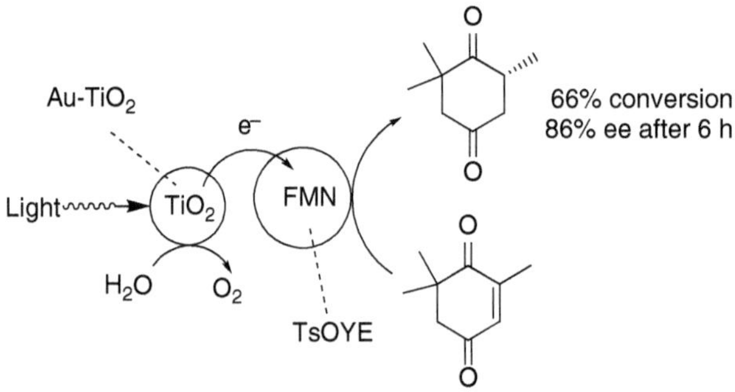

Scheme 7.17 TiO_2-based photosensitizer-mediated photobiocatalytic enantioselective reduction of ketoisophorone. Source: Based on Mifsud et al. [28].

direct transfer of photoexcited electrons from a photosensitizer to the prosthetic flavin moiety, triggering the NAD(P)H-free, enantioselective reduction of C=C bonds, a greener and sustainable approach for producing value-added chemicals using light energy.

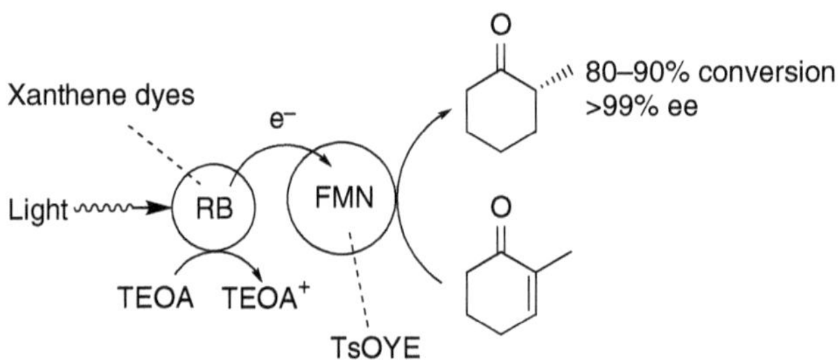

Scheme 7.18 Rose Bengal-mediated photobiocatalytic enantioselective reduction of 2-methylcyclohexenone.

The spontaneous aerobic oxidation of $FADH_2$ in the light-driven regeneration of flavin cofactor was shown to be inhibited by adding catalase into the reaction system. In the presence of catalase and free added FAD, photoinduced regeneration of the catalytically active $FADH_2$ prosthetic group of styrene monooxygenase from *Pseudomonas* sp. VLB120 (StyA) resulted in epoxidation of various styrenes, affording the corresponding styrene oxides with >95% ee [31].

Recently, it was reported that in the flavin-dependent tryptophan halogenase-catalyzed chlorination reaction, photoinduced reduction of enzyme-bound flavin also suppressed the reaction of free flavin radical anion with oxygen. By using blue light and EDTA as sacrificial electron donor for the regeneration of $FADH_2$, the halogenase PyrH from *Streptomyces rugosporus* effectively catalyzed the regioselective chlorination of tryptophan (0.5 mM) to 5-chlorotryptophan in 70% conversion without adding free FAD and catalase, although the conversion was lower than the reaction with free FAD and catalase. This simplified the regeneration of $FADH_2$ for FAD-dependent enzymes (Scheme 7.19) [32].

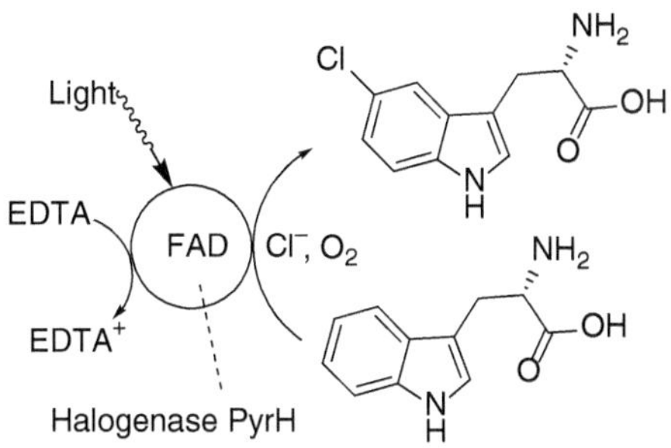

Scheme 7.19 Photobiocatalytic regioselective chlorination of tryptophan. Source: Based on Lee et al. [29].

In morpholine-based buffers, such as MOPS, light irradiation of FMN promoted cyclohexanone monooxygenase (CHMO)-catalyzed oxidation of cyclohexanone to ε-caprolactone under aerated conditions with comparable conversion and enantioselectivity with the reaction using the enzymatic recycling system glutamate dehydrogenase (GDH)/glucose/$NADP^+$ (Scheme 7.20). The flavin cofactor was photochemically regenerated by using MOPS as the sacrificial electron donor, and the triplet states of the excited flavin and the reactive flavin semiquinone (FMNsq) radical anion were stabilized by MOPS via formation of the spin-correlated ion pair ^{3}[flavin$^-$–MOPS$\cdot^+$], circumventing the oxygen dilemma in enzymatic photoreactions that require oxygen for the catalysis [33]. Similarly, morpholines, including MOPS, enabled the enoate reductase XenB from *Pseudomonas* sp. to effectively catalyze the asymmetric reduction of ketoisophorone under aerobic conditions, affording (*R*)-levodione in 91% GC-yield and >94% ee [34].

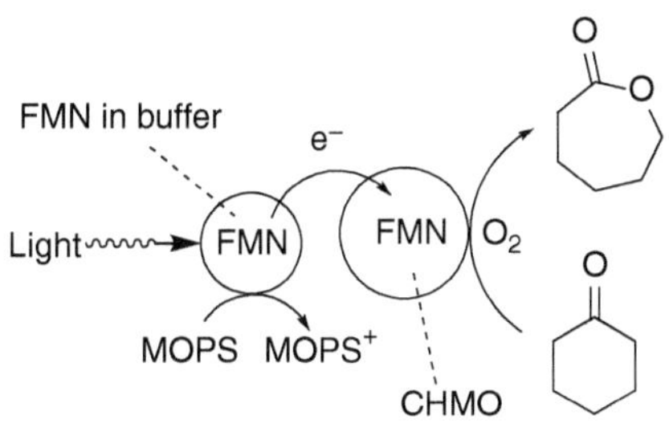

Scheme 7.20 Light-driven CHMO-catalyzed Baeyer–Villiger oxidation of cyclic ketones.

7.3 Light-Activated Cofactor Regeneration for Redox Enzymes

The photoexcited electrons can also be indirectly transferred to the redox enzymes through a natural cofactor (e.g., NAD(P)H, $FADH_2$) or artificial mediator (e.g., methyl viologen, MV). In other words, the regeneration of natural cofactors such as NAD(P)H and $FADH_2$ can be accomplished by photochemical methods. Efficient regeneration of nicotinamide cofactors plays an essential role in redox biocatalysis, and many reports have appeared for their photochemical regeneration.

The oxidoreductases catalyze synthetically useful reactions that require NAD(P)H as the cofactor. As such, an electron mediator is usually employed to drive the regioselective hydride transfer reaction in light-driven NAD(P)H regeneration. The electron flows from electron donor to photosensitizer, then to electron mediator, followed by regioselective hydride transfer to $NAD(P)^+$, affording enzymatically active NAD(P)H. In this context, $[Cp^*Rh(bpy)H_2O]^{2+}$ has been widely used as an artificial electron mediator for the photochemical regeneration of NAD(P)H, with TEOA, EDTA, and water being the sacrificial electron donors (Scheme 7.21) [35].

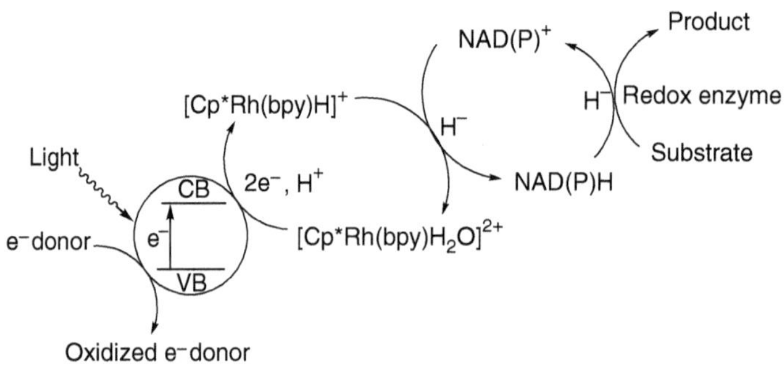

Scheme 7.21 Light-driven NAD(P)H regeneration for redox enzymes. Source: Based on Lee et al. [29].

Upon the bandgap excitation of photocatalyst $W_2Fe_4Ta_2O_{17}$ by visible light an electron was transferred to $[Cp^*Rh(bpy)H_2O]^{2+}$. The reduced Rh complex abstracted a proton from the aqueous solution, and subsequently transferred hydride to NAD^+ to give nicotinamide adenine dinucleotide cofactor (NADH), which served as cofactor for the GDH-catalyzed conversion of α-ketoglutarate into L-glutamate (Scheme 7.22) [36]. CdS nanocrystals grown on SiO_2 beads also served as a visible-light-absorbing photocatalyst to achieve similar photochemical NADH regeneration, which was successfully coupled with the redox-enzymatic synthesis of L-glutamate [37].

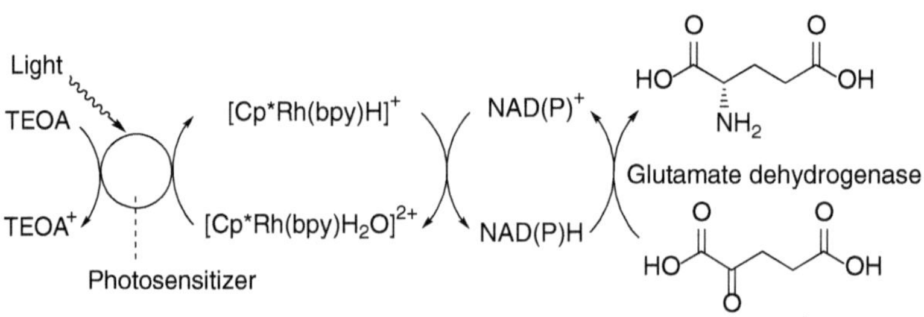

Scheme 7.22 Light-driven NAD(P)H regeneration for the enzymatic synthesis of L-glutamate. Source: Based on Park et al. [36].

Various xanthene dyes were tested as photosensitizers for the visible-light-driven, nonenzymatic regeneration of NADH in this redox reaction catalyzed by L-glutamate dehydrogenase. Among them, phloxine B, erythrosine B, eosin Y, and RB showed superior performance in the light-induced electron transfer from the donor molecule to the mediator when the Rh-based organometallic compound was employed as the hydride-transfer mediator, generating NADH in an "enzymatically active" form with high conversion yields and TONs of NAD^+ [38]. Zn–porphyrin complexes were also promising light-harvesting molecules for the regeneration of NADH utilizing solar energy. The Zn complexes formed noncovalent interactions with electron mediator $[Cp^*Rh(bpy)H_2O]^{2+}$, and the photoexcited electron could be efficiently transferred from Zn–porphyrin to $[Cp^*Rh(bpy)H_2O]^{2+}$, resulting in effective light-driven NADH regeneration for the redox enzymatic reaction to convert α-ketoglutarate to L-glutamate [39]. Similarly, proflavine could serve as an efficient photosensitizer for the light-induced NADH regeneration for the synthesis of L-glutamate in the presence of $[Cp^*Rh(bpy)H_2O]^{2+}$ as the electron mediator. However, when proflavine was replaced with flavin derivatives (FAD, FMN, lumichrome, and riboflavin) with a similar tricyclic ring, no NADH regeneration was observed with light irradiation. In contrast to proflavine, flavin derivatives did not promote the NAD^+ reduction but promoted the NADH oxidation [40].

$[Ru(bpy)_3]^{2+}$ dye has strong absorbance and photoluminescence (PL) in the visible light region (at around 450 and 600 nm, respectively) and acts as a light-harvesting photosensitizer in photocatalytic water splitting. The electrostatic interactions between light-harvesting dyes and $[Cp^*Rh(bpy)(H_2O)]^{2+}$ facilitated the electron flow, enabling simultaneous photocatalytic oxidation of water and reduction of cofactor NAD^+. The photocatalytic cofactor regeneration was efficiently coupled with a glutamate dehydrogenase-catalyzed redox reaction to synthesize L-glutamate [41].

Deposition of cadmium sulfide quantum dots (CdS QDs) on the inner wall of protamine–titania (PTi) microcapsules generated an artificial thylakoid. The CdS QDs in the PTi microcapsules were able to harvest the visible light, resulting in photocatalytic NADH regeneration in the presence of TEOA as the sacrificial electron donor and $[Cp^*Rh(bpy)H_2O]^{2+}$ as the electron mediator. The photocatalytic NADH regeneration was coupled with a FDH for CO_2 reduction. By compartmentalizing the FDH in the same PTi microcapsules with CdS QDs, the resulting artificial thylakoid–enzyme coupled system effectively promoted the photocatalytic reduction of carbon dioxide, producing formate at a rate of 1500 μM/h (Scheme 7.23) [42].

OYEs that catalyze the asymmetric reduction of activated C=C bonds contain an FMN prosthetic group. A hydride is transferred from a reduced cofactor NAD(P)H to FMN, and subsequently delivered to the substrate, resulting in the hydrogenation of the substrate by accepting a proton from a Tyr-residue of the OYE. As discussed above, UV and visible light induced the direct electron transfer from photosensitizers such as RB to the prosthetic FMN group of the OYEs, leading to the NAD(P)H-free, asymmetric reduction of C=C bonds. However, such direct electron transfer was not effective to provide the reduced FMN for the enzymatic reduction

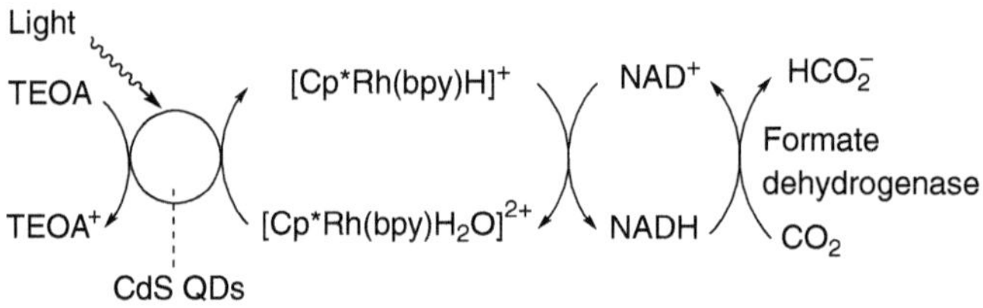

Scheme 7.23 Light-driven NADH regeneration for the enzymatic reduction of CO_2. Source: Based on Zhang et al. [42].

in some cases. For example, the use of CdSe quantum dots (QDs) as photosensitizer did not result in the accumulation of reduced FMN. Methyl viologen (MV^{2+}) was then employed as an electron relay system to facilitate the formation of reduced FMN. Upon exposure to visible light of an aqueous solution containing CdSe-QD, MV^{2+}, and TEOA, $MV^{+\cdot}$ radical was generated and subsequently reduced FMN, thus driving the enantioselective reduction of ketoisophorone by the OYE from *B. subtilis* (YqjM) to give (*R*)-levodione with 64% ee (Scheme 7.24) [43].

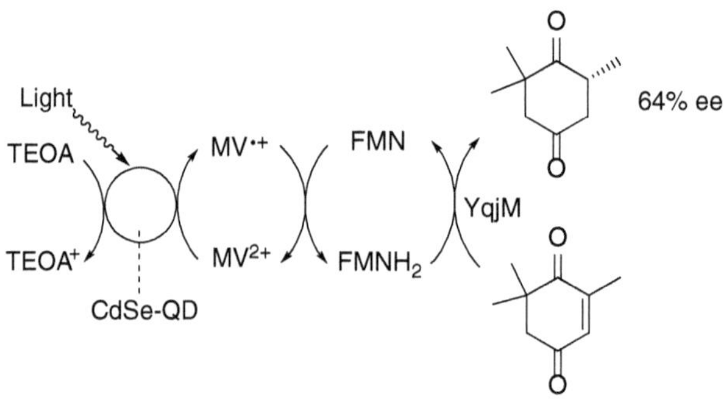

Scheme 7.24 MV^{2+}-mediated light-driven enzymatic reduction of ketoisophorone. Source: Based on Burai et al. [43].

In the OYE-catalyzed stereoselective hydrogenation of C=C bonds, light-driven regeneration of the NADH could also be accomplished. Under illumination with visible light in the presence of an organometallic electron mediator $[Cp^*Rh(bpy)H_2O]^{2+}$ and sacrificial electron donor TEOA, photoinduced electron transfer from N-doped carbon nanodots (N-CDs) to $[Cp^*Rh(bpy)H_2O]^{2+}$ triggered highly regioselective reduction of NAD^+ and its analogs $mNAD^+$s [1-benzyl-3-carbamoylpyridinium (mNH_2^+), 1-butyl-3-carbamoylpyridinium (mBu^+), and 1-benzyl-3-carboxypyridinium ion ($mCOOH^+$)] to afford the reduced cofactors. The regenerated enzymatically active NADH and its analogs subsequently delivered hydride to the FMN prosthetic group of an OYE from *Thermus scotoductus* (TsOYE) for effective stereoselective hydrogenation of α,β-unsaturated ketones or aldehydes such as 2-methyl-2-cyclohexen-1-one and *trans*-cinnamaldehyde (Scheme 7.25) [44].

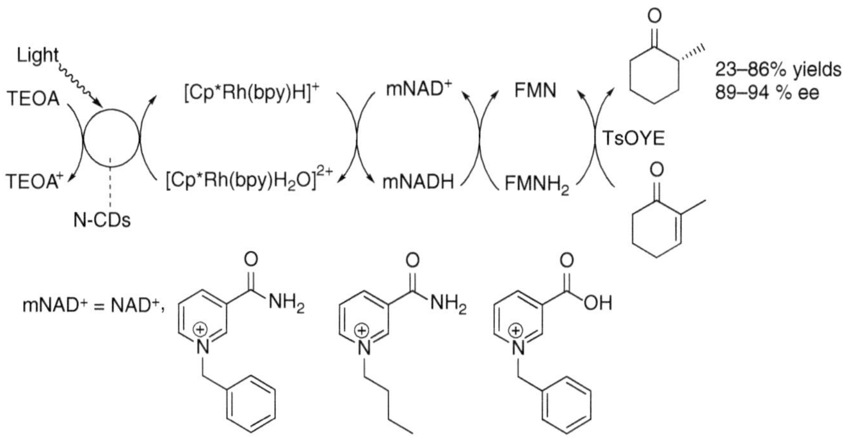

Scheme 7.25 Light-driven cofactor regeneration for the enzymatic reduction of 2-methyl-2-cyclohexen-1-one. Source: Based on Kim et al. [44].

A hetero-binuclear dyad $[(tbbpy)_2\ Ru(tpphz)Rh(Cp^*)Cl]Cl(PF_6)_2$ (Rutpphz-RhCp*, tbbpy = 4,4′-di-*tert*-butyl-2,2′-bipyridine, tpphz = tetrapyrido[3,2-a:2′,3′-c:3″,2″-h:2‴,3‴-j]phenazine) together with TEOA as the sacrificial electron donor was shown to enable the light-driven regeneration of the *N*-benzyl-3-carbamoyl-pyridinium cation, NAD^+, and $NADP^+$. This cofactor recycling system was coupled with the lactate dehydrogenase-catalyzed transformation of pyruvate to lactate. However, a low TON of 350 per photocatalyst was achieved (Scheme 7.26). This could be explained by the observation that a one-electron reduction of the oxidized nicotinamide cofactors also occurred during the reaction, resulting in the continuous accumulation of inactive dimeric cofactor species (Scheme 7.27) [45].

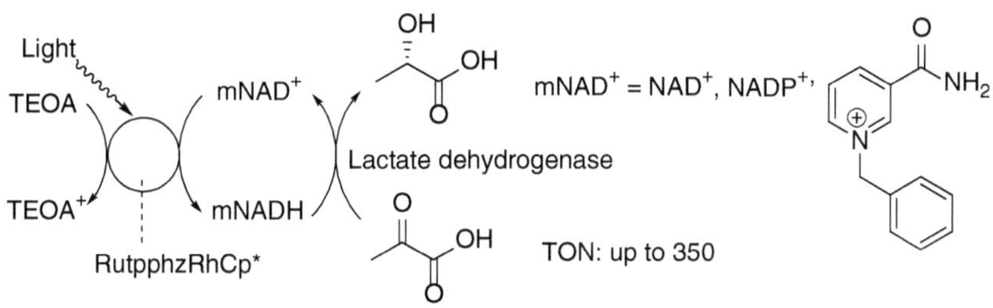

Scheme 7.26 Light-driven cofactor regeneration for the enzymatic reduction of pyruvate.

Cyanobacteria have been known to undergo photosynthetic water splitting, which supplies electrons for reductive enzymatic reactions. This light-driven biotransformation in cyanobacterium *Synechocystis* sp. PCC 6803 has been employed for cofactor regeneration of imine reductase-catalyzed reactions. Three recombinant imine reductases (IREDs) in the cyanobacterium *Synechocystis* sp. PCC 6803 were tested with eight cyclic imine substrates in a photo-bioreactor at 30 °C. Under the optimized reaction conditions, the recombinant IRED-A from *Streptomyces* sp.

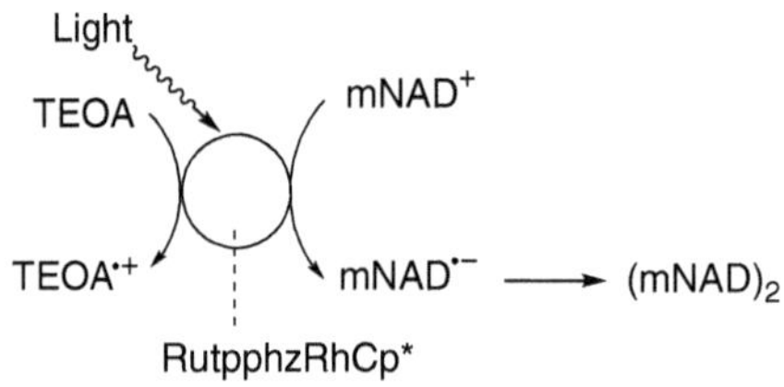

Scheme 7.27 Dimer formation during light-driven cofactor regeneration. Source: Based on Mengele et al. [45].

GF3587 catalyzed the reduction of prochiral imines to give the amine products with up to >99% ee (Scheme 7.28) [46].

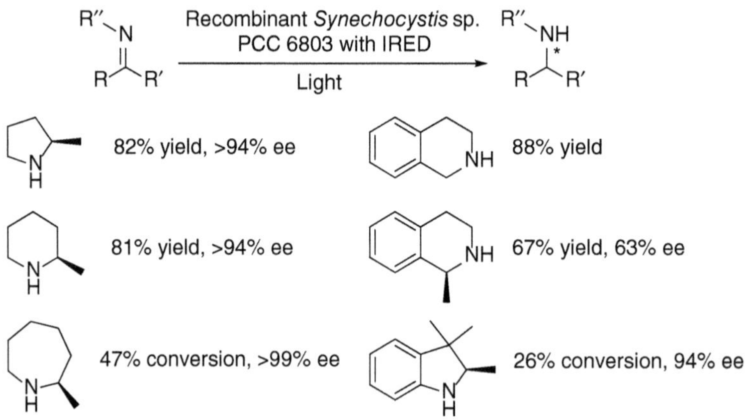

Scheme 7.28 Photosynthetic cofactor regeneration for imine reductase-catalyzed reactions. Source: Based on Büchsenschütz et al. [46].

7.4 Photoinduced Catalytic Promiscuity of Redox Enzymes

In photoredox catalysis, photoexcited molecules are commonly used to generate organic radicals from the small organic molecules. The redox enzymes usually require a cofactor for the catalysis. Under photoexcitation, the photoactive cofactors could generate radical intermediates that catalyze radical-driven reactions. For example, NADH (or NADPH) serves as a hydride (H^-) source and a weak single-electron reductant in its ground state. Upon photoexcitation it becomes a potent single-electron reductant that can reduce an array of functional groups. The resulting cofactor cation radical has a low homolytic bond dissociation free energy for the C4–H bond, thus enabling transfer of hydrogen atoms (H·). This could result in new reactions other than those catalyzed by the enzymes, achieving enzyme catalytic promiscuity. As such, under irradiation with visible light, the nicotinamide-dependent ketoreductase catalyzed a highly enantioselective radical dehalogenation of α-halolactones. A diversity of halolactones were converted into the (*R*)-enantiomer of the dehalogenated lactones when using a variant of the short-chain dehydrogenase of *Lactobacillus kefiri* (LKADH), while the alcohol dehydrogenase from *Ralstonia* species (RasADH) gave the (*S*)-enantiomer as the major product (Scheme 7.29) [47]. In this example, the charge-transfer complex of either enantiomer of halolactone and NAD(P)H within the active site of KRED was formed. Photoexcitation of the charge-transfer band enabled an electron transfer from the cofactor to the substrate, generating NAD(P)H cation radical and halolactone anion radical. The C–Br bond of halolactone anion radical was cleaved to give Br^- and lactone radical, which then abstracted a hydrogen atom (H·) from NAD(P)H cation radical affording the dehalogenated lactone and $NAD(P)^+$. The conformational restriction within the enzyme's active site enabled the hydrogen-atom transfer in an enantioselective manner to give the (*R*) or (*S*)-configurated product.

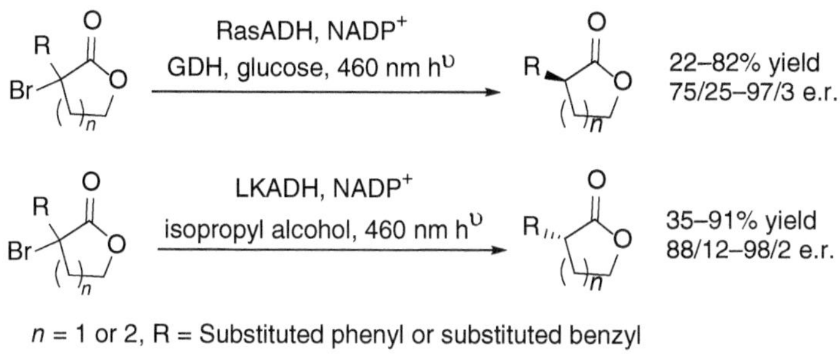

Scheme 7.29 Light-driven ketoreductase-catalyzed enantioselective radical dehalogenation of α-bromolactones. Source: Based on Emmanuel et al. [47].

However, α-bromoamides and α-bromo-α-allyl-γ-lactone could not be dehalogenated under the above reaction conditions, since they were not able to form charge-transfer complexes in the enzyme's active site. This was overcome by employing a photoredox catalyst such as Eosin Y. The photoredox catalyst mediated the single electron transfer from NAD(P)H to the substrate. When various α-bromoamides and α-bromo-α-allyl-γ-lactone were subjected to a LKADH variant under irradiation with 530 nm LEDs in the presence of Eosin Y, they were dehalogenated in 58–71% yields and 80/20 to 97/3 enantiomeric ratio (Scheme 7.30) [48].

Similarly, under visible light irradiation, xanthene-based photocatalysts enabled a double-bond reductase to catalyze an enantioselective deacetoxylation. With RB as the photocatalyst, various racemic α-substituted α-acetoxytetralones were deacetoxylated by the double bond reductase from *Nicotiana tabacum* (NtDBR) under the irradiation with 530 nm LEDs, affording the corresponding α-substituted tetralones in 38–85% yields and 70/30–97/3 enantiomeric ratio (Scheme 7.31) [48].

Another photoinduced nonnatural reaction was observed for flavin-dependent ERs. This type of enzymes are known to stereoselectively reduce activated alkenes, which occurs by stereoselective delivery of hydride from flavin hydroquinone (FMNhq). With visible light photoexcitation the photoredox catalyst $Ru(bpy)_3Cl_2$

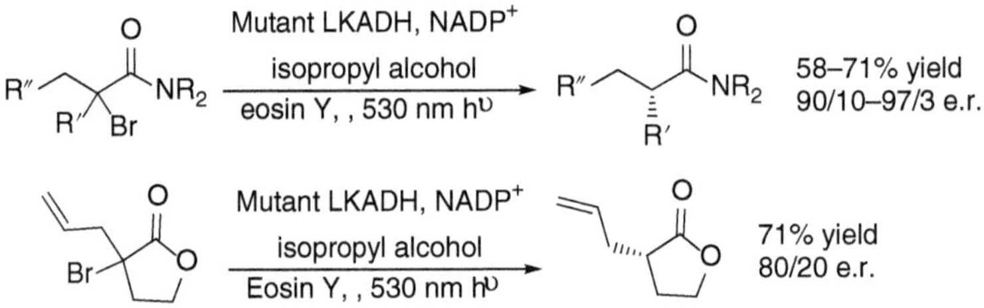

Scheme 7.30 Light-driven ketoreductase-catalyzed enantioselective radical dehalogenation of α-bromoamides and α-bromolactones. Source: Based on Biegasiewicz et al. [48].

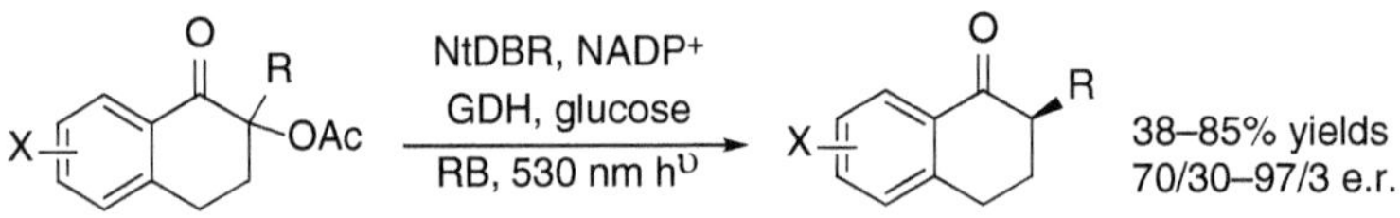

Scheme 7.31 Light-driven double-bond reductase-catalyzed enantioselective deacetoxylation of α-substituted α-acetoxytetralones. Source: Based on Biegasiewicz et al. [48].

mediated single-electron transfer from FMNhq to the substrate ketone, generating a ketyl radical, which was quenched by a HAT, thus resulting in the reduction of the ketone substrate. A series of aromatic ketones with *para* and *meta* substituents were reduced with morphinone reductase (MorB) from *P. putida* and $Ru(bpy)_3Cl_2$ under irradiation with 460 nm LEDs, giving the corresponding chiral alcohols with good to high yields and moderate enantioselectivity (Scheme 7.32) [49].

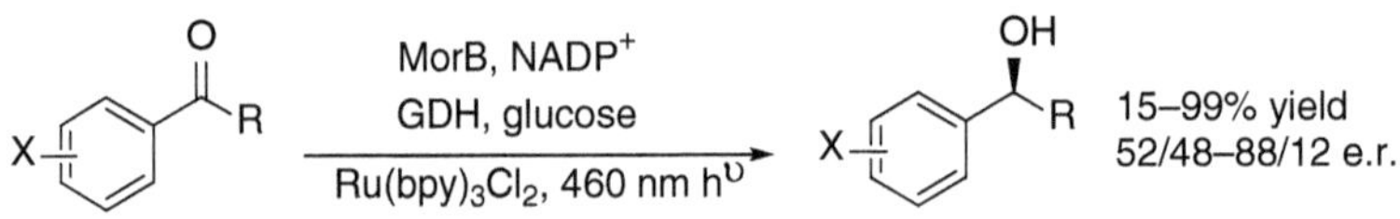

R = CH_3, X = *p*-OCH_3, *p*-CF_3, *p*-F, *m*-CH_3, *m*-Cl, *o*-CH_3
X = H, R = 2-Thiophenyl, *n*-C_5H_{11}

Scheme 7.32 Light-driven ene reductase-catalyzed enantioselective reduction of ketones. Source: From Sandoval et al. [49].

ERs selectively catalyze the asymmetric reduction of alkenes with at least one electron-withdrawing group such as aldehyde, ketone, carboxylic ester, or nitro group, and are not active toward unactivated alkenes. Under blue light irradiation an ER from *Nostoc punctiforme* (NostocER) was found to promote the asymmetric hydrogenation of 2-(1-phenylvinyl)pyridine in the presence of $Ru(bpy)_3Cl_2$ (bpy = 2,2′-bipyridine), furnishing (*S*)-2-(1-phenylethyl)pyridine in 96% yield and 84% ee. Upon irradiation, an electron was transferred from the photoredox catalyst to vinyl pyridine generating an anion radical that was converted to a neutral benzylic radical via protonation in solution. This radical was sufficiently stable to diffuse into the active site of ER, where stereoselective HAT to the radical occurred to give the hydrogenated product. This photoinduced enzymatic reduction was expanded to a diversity of vinyl pyridines (Scheme 7.33) [50].

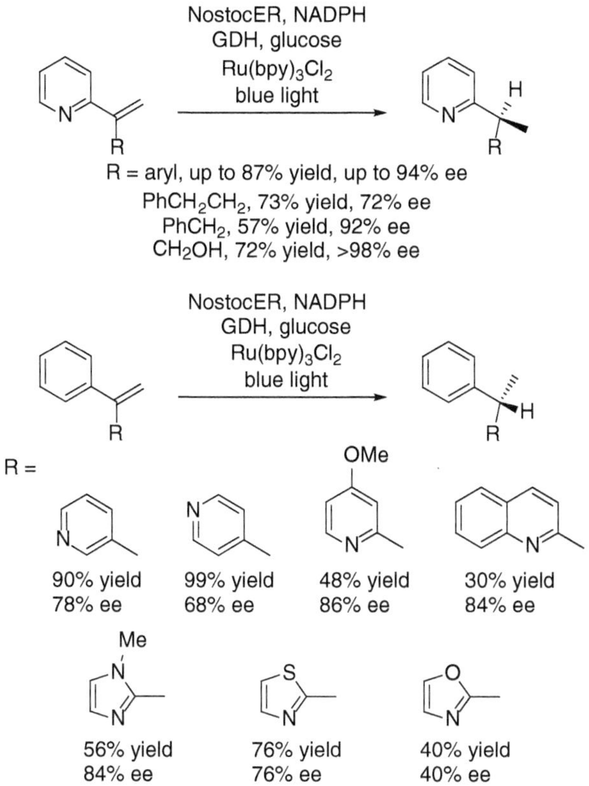

Scheme 7.33 Photoinduced ene reductase-catalyzed enantioselective hydrogenation of vinyl pyridines. Source: Based on Nakano et al. [50].

ERs catalyze the stereoselective reduction of activated alkenes via hydride transfer from the FMNhq to the electrophilic β-position of alkene. FMNhq in ER was also found to function as a single-electron reductant, enabling ER-catalyzed enantioselective radical dehalogenation of α-bromoesters [51]. However, ERs are not be able to catalyze radical dehalogenation of α-chloroamides since electron transfer from FMNhq [$E_{1/2}$ = −0.45 V versus saturated calomel electrode (SCE)] to α-chloroamides ($E_{p/2}^{red}$ = −1.65 V versus SCE) is thermodynamically challenging. But visible light excitation was able to enable single-electron transfer from FMNhq* ($E_{1/2}$* = −2.26 V versus SCE) to α-chloroamide in an ER enzyme. In a recent endeavor, photoexcitation of ERs with cyan LED light (497 nm) promoted the dehalogenation of α-chloroamides leading to the formation of a prochiral α-acyl radical, which underwent an asymmetric radical cyclization to generate five-, six-, seven-, and eight-membered lactams with high enantioselectivity (Scheme 7.34). The single electron transfer was proposed to occur through direct excitation of an electron donor–acceptor complex of FMNhq and the substrate within the enzyme active site, and the stereochemical preference and stereoselectivity were conferred by the enzyme active site [52].

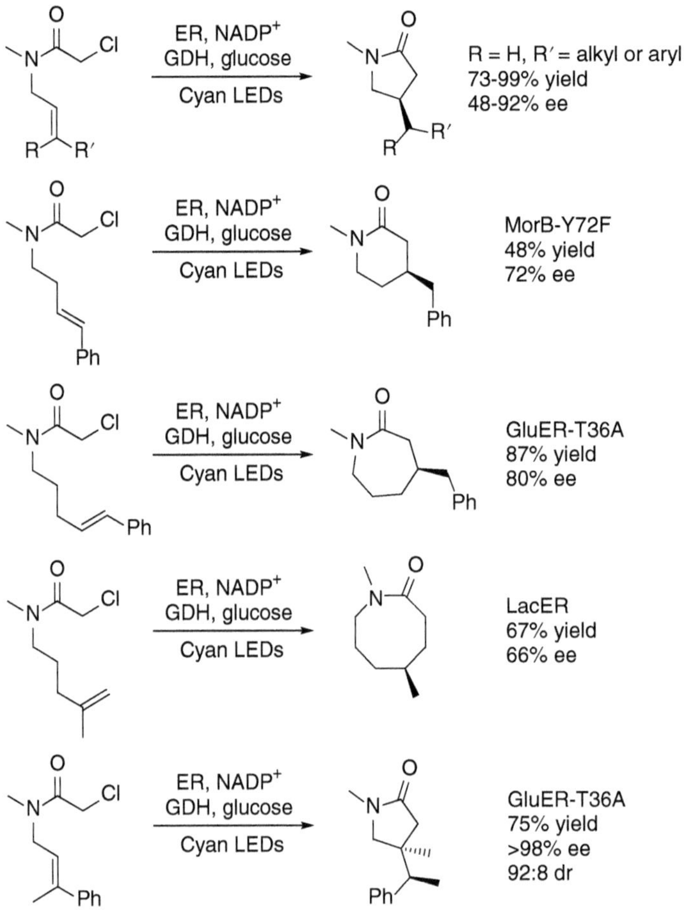

Scheme 7.34 Photoinduced ER-catalyzed asymmetric radical cyclization of α-chloroamides.

The photoexcitation-generated α-acyl radical from α-haloamides also could couple with phenyl C–H bond, leading to redox-neutral radical cyclization to give enantioenriched oxindoles. Using MorB with a large active site, methyl 2-bromo-2-(methyl(phenyl)carbamoyl)butanoate was transformed to racemic methyl 3-ethyl-1-methyl-2-oxoindoline-3-carboxylate in good yield under cyan LED light in tricine buffer, while 12-oxophytodienoate reductase (OPR1) catalyzed the radical dehalogenative cyclization with improved stereoselectivity, affording methyl (*S*)-3-ethyl-1-methyl-2-oxoindoline-3-carboxylate in 78% yield with 60% ee. A mechanism was proposed that photoinduced electron transfer generating FMNsq reduced α-chloroamide to an α-acyl radical, which underwent stereoselective cyclization to form a vinylogous α-amido radical, subsequent oxidation of which by FMN generated the product oxindole (Scheme 7.35). The substrate scope of this photoinduced biocatalytic radical cyclization was evaluated and a series of oxindoles with variant substituent at α-carbon and on the phenyl ring were prepared in high yields and ee (Scheme 7.36) [53].

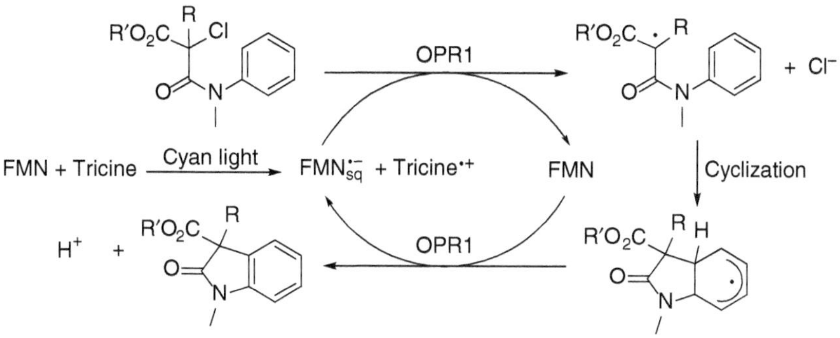

Scheme 7.35 Proposed reaction mechanism of photoinduced ER-catalyzed radical cyclization for the synthesis of oxindoles.

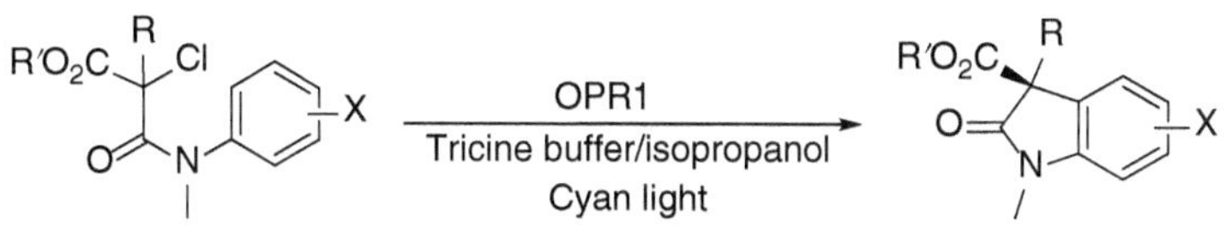

R′ = Me, X = H, R = Me, Et, *n*-Pr, *n*-Bu, *i*-Pr and allyl, 39–97% yields, 56–95% ee
R′ = Et, X = H, R = Et, 90% yield, 92% ee
R′ = *n*-Pr, X =H, R = Et, 80% yield, 88% ee
R′ = Me, R = Et, X = *p*-Cl, *p*-Br, *p*-Me, *p*-Et, *p*-OMe, 37–96% yields, 68–96% ee
R′ = Me, R = Et, X = *m*-Me, 86% yield, 4-Me/6-Me = 2.1(82% ee)/1(72% ee)

Scheme 7.36 Photoinduced ER-catalyzed radical cyclization for the synthesis of oxindoles. Source: Based on Black et al. [53].

Very recently, the light-induced ER-catalyzed intramolecular radical dehalogenative cyclization was extended into intermolecular radical hydroalkylation of terminal alkenes with α-halo carbonyl compounds. With blue LED irradiation Ers promoted hydroalkylation of α-methyl styrene with α-bromo acetophenone in Tris buffer under anaerobic conditions affording (*S*)-1,4-diphenylpentan-1-one in up to 88% yield and 96% ee. By choosing a suitable ER, different α-methyl aryl substituted ethylenes were hydroalkylated with a diversity of α-halo ketones, carboxylic esters, and amides, affording the corresponding carbonyl compounds bearing a γ-stereocenter with up to 99% yield and 99% ee (Scheme 7.37) [54].

7.5 Photocatalysis and Biocatalysis Cascades

From the previous sections, we can see that similar to Nature's photosynthesis, light energy is used to feed electrons directly into the enzyme, or to regenerate the

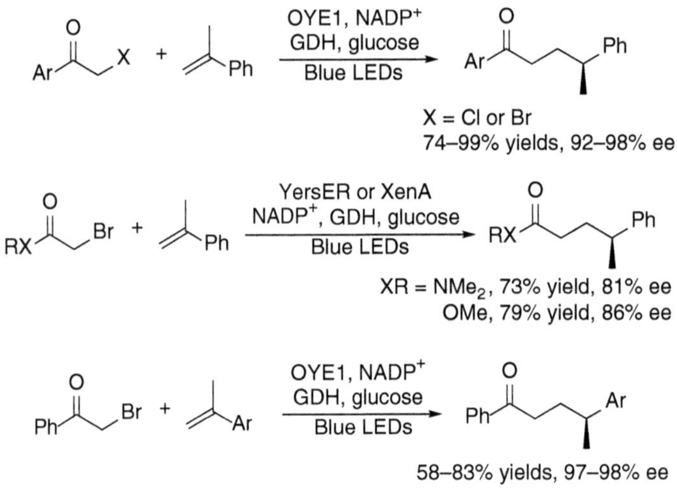

Scheme 7.37 Photoinduced ER-catalyzed intermolecular radical coupling of terminal alkenes with α-halo carbonyl compounds. Source: Based on Huang et al. [54].

cofactors such as NAD(P)H, thus facilitating the enzymatic reaction. In the latter case, photoinduced single electron transfer also resulted in HAT, leading to different reactions other than the native reaction of the enzyme, i.e. catalytic promiscuity. In the "true" cascade of photocatalysis and biocatalysis, a photochemical reaction generates an intermediate that serves as the substrate of the biotransformation (or vice versa). Both photochemical and biocatalytic reactions can be performed at or near room temperature, offering an advantage for the integration of enzyme catalysis with photoreaction. Recently, a linear cascade in which a light-induced chemical transformation was combined with a biocatalytic reaction in a sequential or concurrent one-pot process has been developed, offering a new strategy for a sustainable chemical synthesis [55]. Recent advances in this emerging area will be discussed in this section.

Peroxidases (EC 1.11.1x) catalyze various interesting transformations such as C–H oxidation and halogenation, by making use of hydrogen peroxide. But they often suffer from irreversible oxidative inactivation by high level of H_2O_2 in the reaction mixture. As such, keeping the H_2O_2 level at a certain value during the whole reaction is crucial for the biocatalytic processes, in particular on a large scale. The photoexcited FMN* by blue light can be reduced to $FMNH_2$ at the expense of a sacrificial electron donor. FMNH2 is then reoxidized by atmospheric O_2 to form H_2O_2. This continuous, light-driven process guarantees a constant, but low H_2O_2 concentration in solution for the biocatalytic processes. The photocatalyzed H_2O_2 generation was combined with peroxidase-catalyzed halogenation. A number of different electron-rich (hetero) aromatic compounds were effectively brominated by using the vanadium-dependent haloperoxidase (AmVHPO) from the cyanobacterium *Acaryochloris marina* under irradiation (455 nm) in the presence of KBr in MES buffer. Similar chlorination of aromatic compounds was achieved by using the fungal VHPO from *Curvularia inaequalis* (CiVHPO) in the presence of KCl under the same reaction conditions (Scheme 7.38). It is worth noting that the redox-active MES buffer served as an efficient sacrificial electron donor [56].

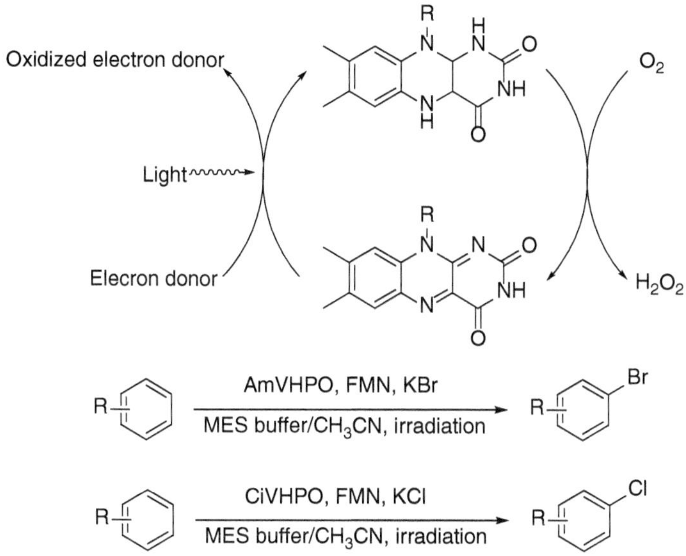

Scheme 7.38 FMN-mediated light-driven halogenation of aromatic compounds catalyzed by vanadium-dependent haloperoxidases.

TiO_2 doped with gold nanoparticles (Au-TiO_2) has been reported as a heterogeneous photocatalyst for the oxidation of methanol and water in the presence of air, generating H_2O_2. The Au-TiO_2 nanoparticles were evaluated for water oxidation under irradiation with different narrow-emission-band light sources; evolution of H_2O_2 occurred at 375 nm UV irradiation, reaching ca 600 μM. This was then successfully applied as an *in situ* H_2O_2 source for the bromination of 2,6-dimethoxypyridine catalyzed by AmVHPO in phosphate buffer, in which water served as the source of electrons (Scheme 7.39) [56].

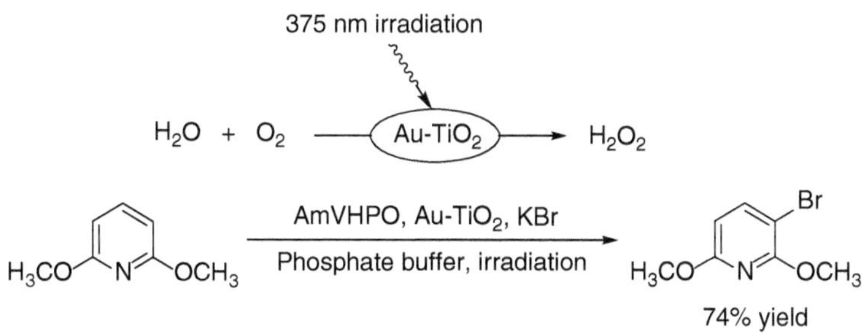

Scheme 7.39 Au-TiO_2-mediated light-driven halogenation of aromatic compounds catalyzed by vanadium-dependent haloperoxidase. Source: Based on Seel et al. [56].

Photocatalyst sodium anthraquinone sulfonate (SAS) was known to reductively activate O_2 to generate H_2O_2 with methanol as the sacrificial electron donor.

This light-driven H_2O_2 generation was combined with the vanadium-dependent chloroperoxidase from *Curvularia inaequalis* (CiVCPO) to enable the halogenation of thymol, yielding 4-bromothymol in 91% yield with traces of 2-bromo isomer and 2,4-dibromo thymol in the presence of NaBr (Scheme 7.40). Furthermore, the photochemical-enzymatic system also promoted hydroxybromination of styrene to give a mixture of hydroxybromide (61% yield) and epoxide (17% yield), and bromolactonization of 4-pentenoic acid to afford the bromolactone in 80% yield (Scheme 7.41) [57].

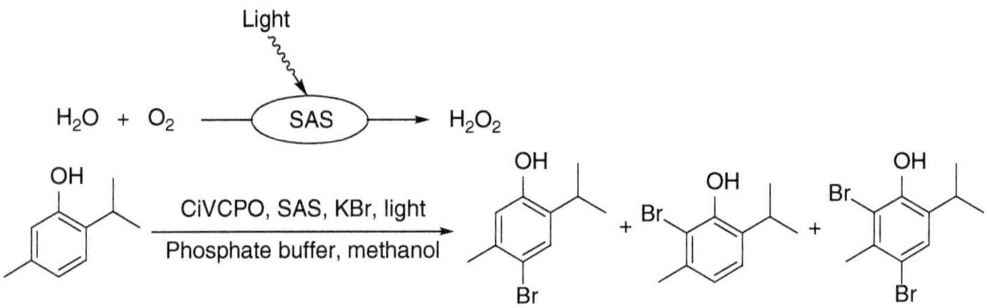

Scheme 7.40 SAS-mediated light-driven bromination of thymol catalyzed by vanadium-dependent chloroperoxidase.

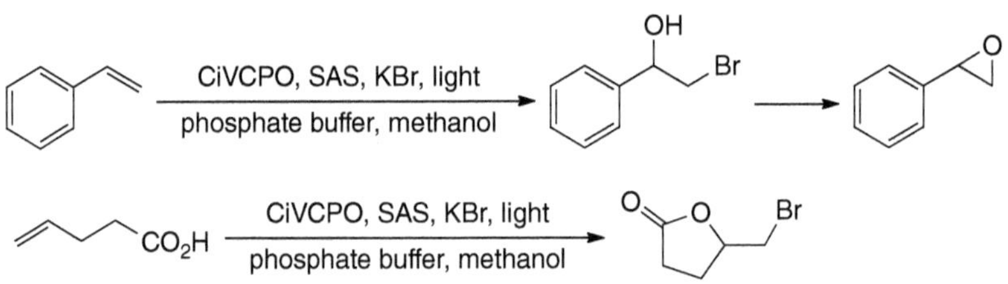

Scheme 7.41 SAS-mediated light-driven hydroxybromination of styrene and 4-pentenoic acid catalyzed by vanadium-dependent chloroperoxidase. Source: From Yuan et al. [57].

Peroxygenases also use simple H_2O_2 as the stoichiometric oxidant and are attractive catalysts for selective oxyfunctionalization chemistry. However, their application in synthetic chemistry remains challenging because of their poor robustness against H_2O_2. A straightforward approach to overcome this problem is to keep H_2O_2 at low concentration in the reaction mixture via *in situ* generation of the oxidant. In this context, a few visible-light-driven methods for *in situ* generation of H_2O_2 have been developed, rendering the peroxygenase catalytically active. In the presence of visible-light-active Au-loaded TiO_2, water was partially oxidized to H_2O_2. This *in situ* H_2O_2 generation was combined with the unspecific peroxygenase from *Agrocybe aegerita* (AaeUPO) to effect the stereoselective oxyfunctionalization of hydrocarbons. Several hydrocarbons such as cyclic alkanes and phenylalkanes were hydroxylated with moderate conversion and high stereoselectivity by simply using this catalytic system, water, and visible light (Scheme 7.42) [58].

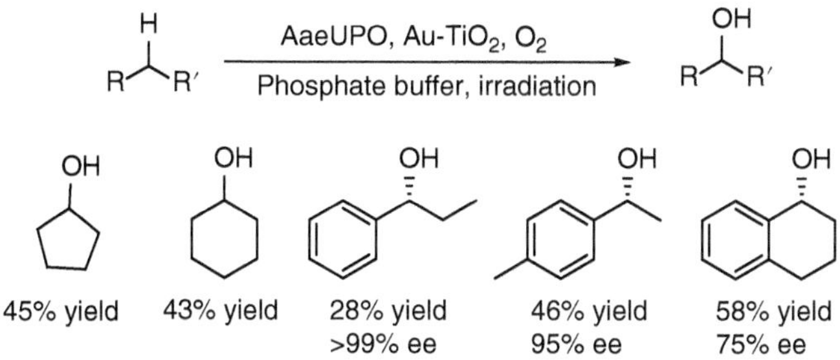

Scheme 7.42 Au-TiO_2-mediated light-driven hydroxylation of hydrocarbons catalyzed by peroxygenase. Source: Based on Zhang et al. [58].

Recently, simple graphitic carbon nitride (g-C_3N_4) was demonstrated to be a promising photocatalyst for the O_2 reduction with formate or methanol as the sacrificial electron donor to generate H_2O_2. This heterogeneous photocatalytic generation of H_2O_2 was coupled with peroxygenase-catalyzed hydroxylation reaction. Ethyl benzene was effectively hydroxylated by the peroxygenase from *Agrocybe aegerita* (AaeUPO), affording (*R*)-1-phenyl ethanol with a TON of the biocatalyst of more than 60 000 (Scheme 7.43) [59].

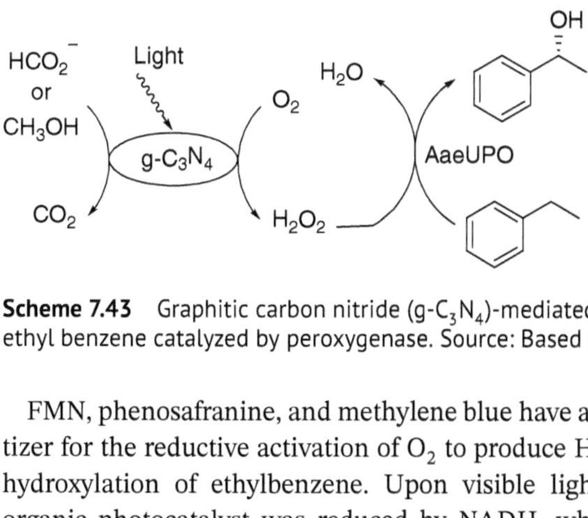

Scheme 7.43 Graphitic carbon nitride (g-C_3N_4)-mediated light-driven hydroxylation of ethyl benzene catalyzed by peroxygenase. Source: Based on van Schie et al. [59].

FMN, phenosafranine, and methylene blue have also been utilized as photosensitizer for the reductive activation of O_2 to produce H_2O_2 in peroxygenase-catalyzed hydroxylation of ethylbenzene. Upon visible light irradiation the flavin-based organic photocatalyst was reduced by NADH, which was regenerated from the sacrificial electron donor formate catalyzed by the FDH from *Candida boidinii* (CbFDH). The reduced photocatalyst was oxidized by O_2, generating a controlled supply of H_2O_2 for enzymatic hydroxylation of ethylbenzene catalyzed by a recombinant peroxygenase from *Agrocybe aegerita* (AaeUPO, Scheme 7.44) [60].

Combination of SAS-mediated light-driven generation of H_2O_2 with vanadium-dependent chloroperoxidase CiVCPO also enabled the hydroxylation of ethylbenzene or cyclohexane, yielding a mixture of the corresponding alcohol and ketone that was formed by alcohol oxidation (Scheme 7.45) [57].

ERs catalyze the stereoselective reduction of carbon–carbon double bonds to generate enantioenriched products. It was often observed that only one isomer of the *E*/*Z* mixtures of alkenes could be reduced by this enzymatic reaction, due to

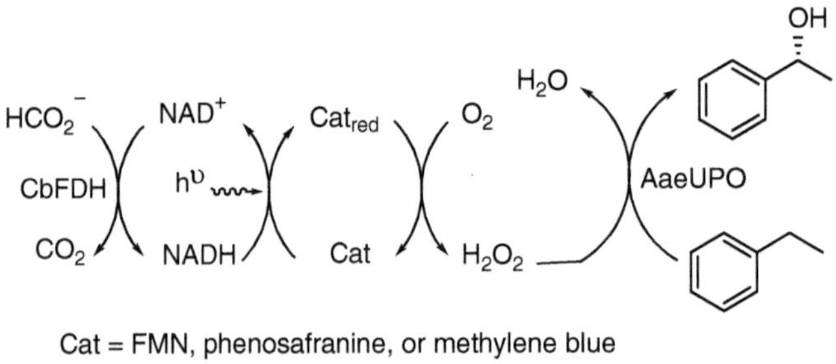

Scheme 7.44 Flavin-based organic-photocatalyst-mediated light-driven hydroxylation of ethyl benzene catalyzed by peroxygenase. Source: Based on Willot et al. [60].

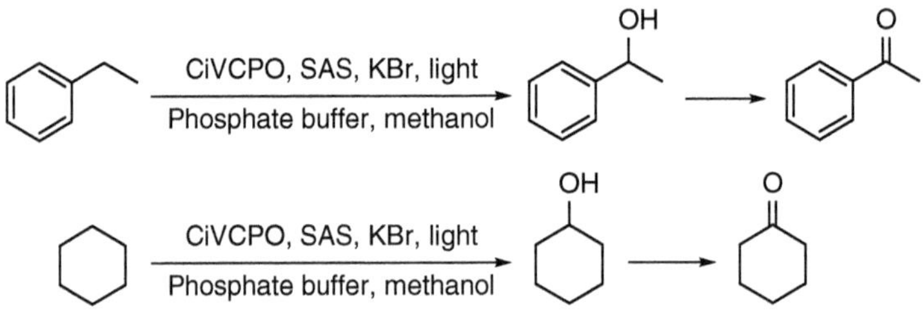

Scheme 7.45 SAS-mediated light-driven hydroxylation of hydrocarbons catalyzed by vanadium-dependent chloroperoxidase. Source: Based on Willot et al. [60].

the different interactions between the *E* and *Z* isomers of alkene substrate and the active site of the enzyme. For example, an ER from *Yersinia bercovieri* (YersER) exclusively catalyzed the reduction of (*E*)-2-phenylbut-2-enedioic acid dimethyl ester to give dimethyl (*R*)-2-phenylsuccinate in high yield and excellent enantioselectivity by using a glucose dehydrogenase for cofactor regeneration. In order to enable the stereoconvergent reduction of *E*/*Z* mixtures of alkenes or reduction of the unreactive stereoisomers of alkenes, the photocatalyzed isomerization of alkenes was incorporated with enzymatic reduction of alkenes. A range of photocatalysts were evaluated for the simultaneous, cooperative photoisomerization and enzymatic reduction of (*Z*)-2-phenylbut-2-enedioic acid dimethyl ester; dimethyl (*R*)-2-phenylsuccinate was obtained in moderate to high yields, with the highest conversions and yields being achieved by using 5% of FMN or cationic iridium(III) complex [Ir(dmppy)$_2$(dtbbpy)]PF$_6$ (dmppy, 4-methyl-2-(4-methylphenyl)pyridine; dtbbpy, 4,4′-di-tert-butyl-2,2′-bipyridine; Scheme 7.46). The *E*/*Z* mixtures of a diversity of aryl alkenes containing diverse combinations of functional groups were effectively reduced to give the corresponding aryl alkenes with high yields and high stereoselectivity by employing the cationic iridium(III) complexes or FMN as the photocatalyst and various ERs as the enzyme catalyst in the presence of blue light (Scheme 7.47) [61].

The concurrent one-pot process showed benefits in yield and enantioselectivity over the one-pot sequential reactions. For example, for the (*Z*)-α-cyano-α,β-unsaturated esters, when the photoisomerization with Ir(III) or FMN and the

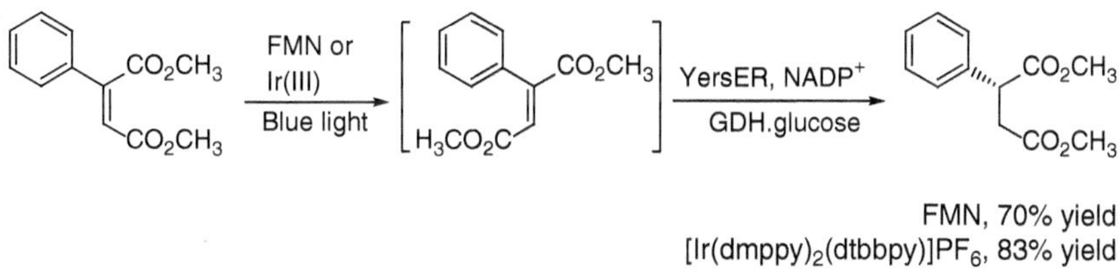

Scheme 7.46 Simultaneous photocatalytic isomerization and enzymatic reduction of (*Z*)-2-phenylbut-2-enedioic acid dimethyl ester.

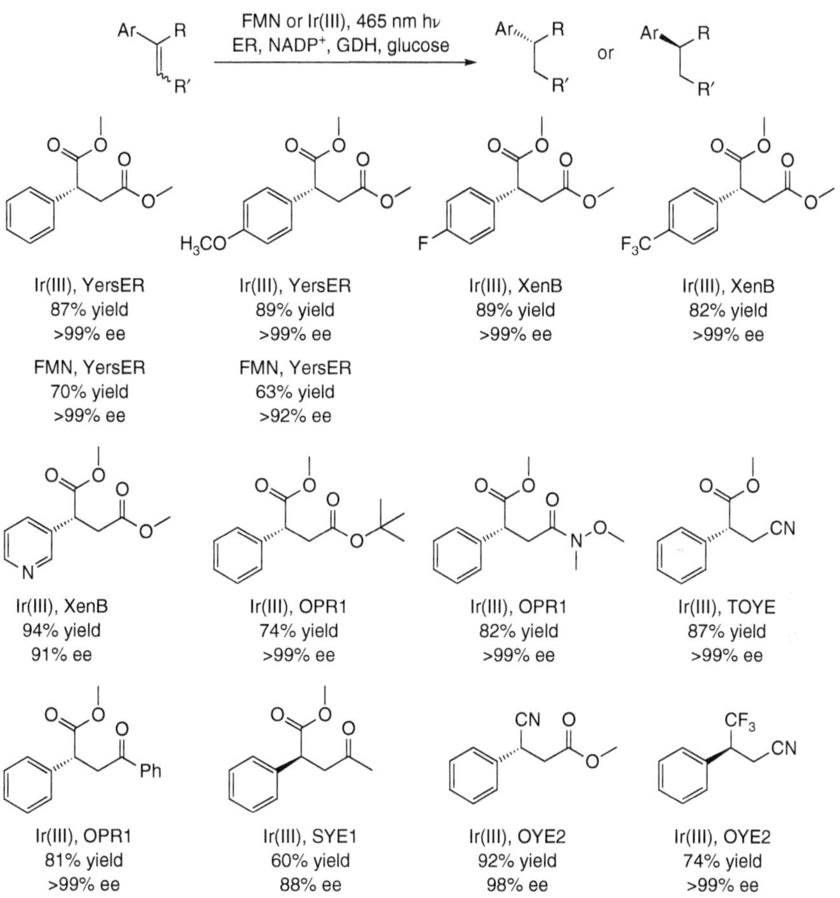

Scheme 7.47 Simultaneous photocatalytic isomerization and enzymatic reduction of aryl alkenes. Source: Based on Litman et al. [61].

reduction with ER OYE2 were performed in a sequential manner, lower yields and ee values of the products were obtained compared to the one-pot concurrent procedure (Scheme 7.48).

Mercaptoketones are the precursors for the preparation of optically pure 1,3-mercaptoalkanol, a class of volatile sulfur compounds with aroma activity

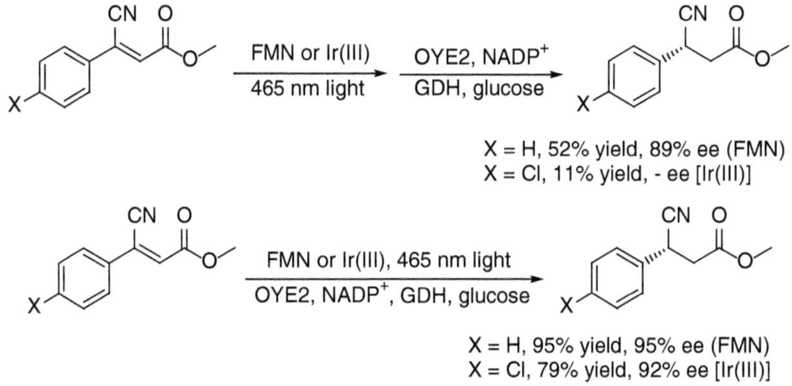

Scheme 7.48 Sequential and concurrent photocatalytic isomerization and enzymatic reduction of (*Z*)-α-cyano-α,β-unsaturated esters.

that widely exist in foods and beverages, by chemical and biocatalytic reductions. Mercaptoketones could be synthesized by Michael addition of thiophenol to vinyl ketones with borax under basic conditions. However, these conditions were incompatible with the biocatalytic reaction conditions preventing development of a concurrent chemoenzymatic one-pot process for the production of 1,3-mercaptoalkanol directly from thiophenol and vinyl ketones. As such, an alternative photocatalytic Michael addition of thiophenol to vinyl ketones in a neutral aqueous medium was developed. Under visible light irradiation with $Ru(bpy)_3Cl_2$ as the photocatalyst, Michael reaction of thiols to vinyl ketones proceeded smoothly in potassium phosphate buffer at pH 7.0, giving mercaptoketones in quantitative conversion. Photocatalytic thio-Michael addition was combined with the biocatalytic ketone reduction in a one-pot process. By employing KRED 311 or 349 with opposite enantioselectivity, two of the 384 KREDs identified and isolated through a metagenomics approach from the Prozomix library, enantiomerically pure (*R*)- or (*S*)-1,3-mercaptoalkanols, were synthesized in high yields with excellent enantioselectivity through this one-pot photobiocatalytic cascade reaction (Scheme 7.49) [62].

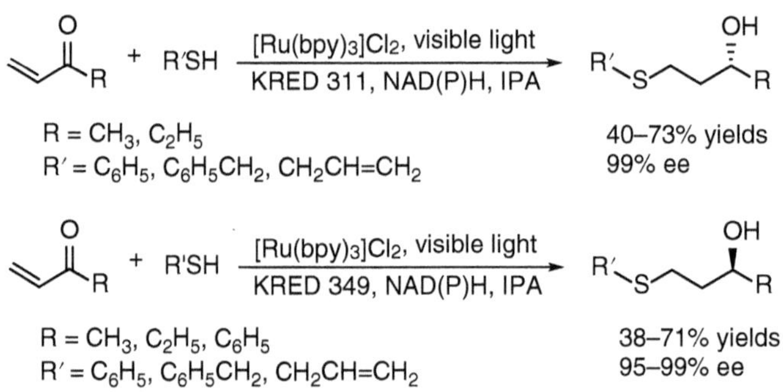

Scheme 7.49 Photocatalytic thio-Michael addition and enzymatic reduction cascade for enantiocomplementary synthesis of (*R*)- or (*S*)-1,3-mercaptoalkanols. Source: Based on Lauder et al. [62].

It has been reported that the lipase from wheat germ (WGL) type I showed catalytic promiscuity to catalyze the highly enantioselective Mannich reaction between 3-substituted-2*H*-1,4-benzoxazines and acetone, affording the corresponding β-amino ketone derivatives in 15–54% yields and up to 95% ee. On the other hand, 2-arylindoles could be oxidized to 2-arylindol-3-one under visible light irradiation with $Ru(bpy)_3Cl_2$ (bpy = 2,2′-bipyridine) as the photoredox catalyst. The oxidation products have a ketimine structure meaning that they could undergo Mannich reaction. This was found to be the case when a combination of photocatalytic oxidation of 2-phenylindole and Mannich reaction of the resulting 2-phenylindol-3-one with acetone by using $Ru(bpy)_3Cl_2$ as photocatalyst and a lipase from WGL as the biocatalyst led to the desired product in 62% yield and 89:11 enantiomeric ratio (e.r.) (Scheme 7.50) [63].

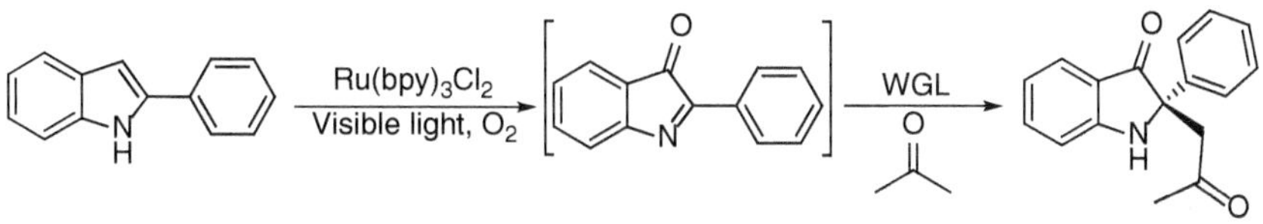

Scheme 7.50 One-pot concurrent photocatalytic oxidation and lipase-catalyzed Mannich reaction. Source: Based on Ding et al. [63].

The scope of this one-pot concurrent process was then evaluated, and a series of 2-arylindoles were photocatalytically oxidized to the corresponding 2-arylindol-3-ones, which were subsequently alkylated with ketones via WGL-catalyzed Mannich reaction, generating 2,2-disubstituted indol-3-ones with a chiral quaternary carbon center at C2 of the indoles in 10–70% yields and moderate stereoselectivity. This showed the feasibility of combining visible light catalysis with the nonnatural catalytic activity of hydrolases, offering a mild and useful method for the one-pot stereoselective synthesis of complex compounds (Scheme 7.51) [63].

SAS could serve as the photo-organo catalyst for the oxyfunctionalization of alkanes, furnishing aldehydes or ketones as the products. This light-driven alkane functionalization was combined with enzymatic transformations in a one-pot concurrent process to yield various kinds of products by employing different enzymes. Benzyl alcohol was readily photooxidized by SAS, and then converted to phenyl formate via Baeyer–Villiger reaction catalyzed by 4-hydroxyacetophenone monooxygenase (HAPMO). Similarly, cyclohexanol was converted into ε-caprolactone with CHMO as the biocatalyst. The coupling of photooxidation of toluene with hydroxynitrile lyase (HNL)-catalyzed reaction gave chiral mandelonitrile, while benzoin was obtained when the photooxidation was carried out in the presence of benzaldehyde lyase (BAL). Ethylbenzene was transformed into chiral 1-phenylethylamine or 1-phenylethanol by combining the light-driven oxidation with transaminase-catalyzed amino transfer reaction or ketoreductase-catalyzed ketone reduction, respectively. 1-Methylcyclohexene was

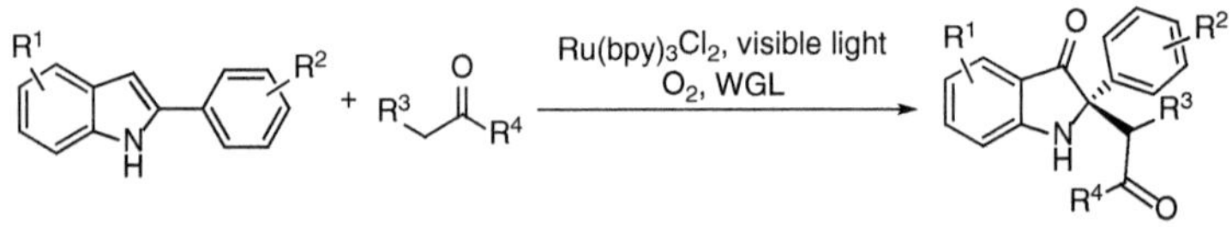

R1	R2	R3	R4	Yield (%)	e.r.
H	H	H	H	62	89:11
H	*p*-CF_3	H	H	37	82:18
H	*p*-F	H	H	54	86:14
H	*p*-Cl	H	H	47	87:13
H	*p*-Br	H	H	40	87:13
H	*p*-CH_3	H	H	63	81:19
H	*p*-OCH_3	H	H	70	71:29
H	*m*-F	H	H	33	80:20
H	*m*-CH_3	H	H	59	83:17
H	*m*-OCH_3	H	H	53	80:20
5-F	H	H	H	36	86:14
5-Br	H	H	H	32	85:15
5-CH_3	H	H	H	30	87:13
5-OCH_3	H	H	H	27	89:11
6-CH_3	H	H	H	30	84:16

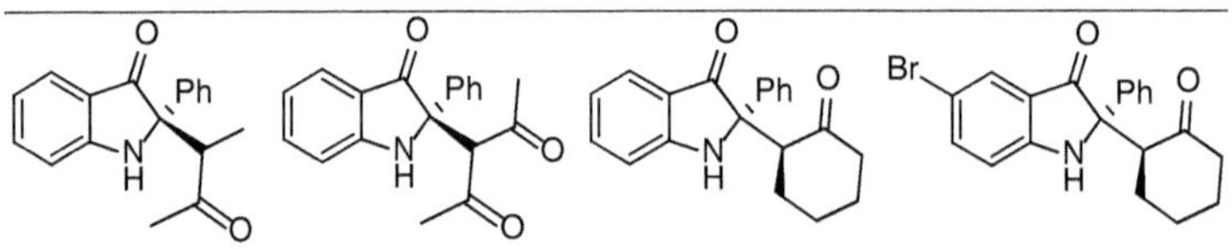

55% yield
4:1 d,r., 93:7 e.r.

58% yield
66:34 e.r.

59% yield
5:1 d.r., 82;18 e.r.

10% yield
3:1 d.r., 83:17 e.r.

Scheme 7.51 One-pot synthesis of 2,2-disubstituted indol-3-ones via photocatalytic oxidation and lipase-catalyzed Mannich reaction. Source: Based on Ding et al. [63].

first photocatalytically oxidized to 2-methylcyclohexenone, which was reduced by an ER to give 2-methylcyclohexanone in 99 % ee (Scheme 7.52). When the transformation was performed in a concurrent one-pot process, the incompatibilities of the photooxidation and the enzymatic reactions often limited the total conversions. As such, the photocatalytic oxidation and biocatalytic reactions were also performed in sequential one-pot mode, resulting in a few-fold increased formation of the final products. The asymmetric C–H bond functionalization of simple alkanes was thus achieved by the combination of photo-organo redox catalysis and biocatalysis [64].

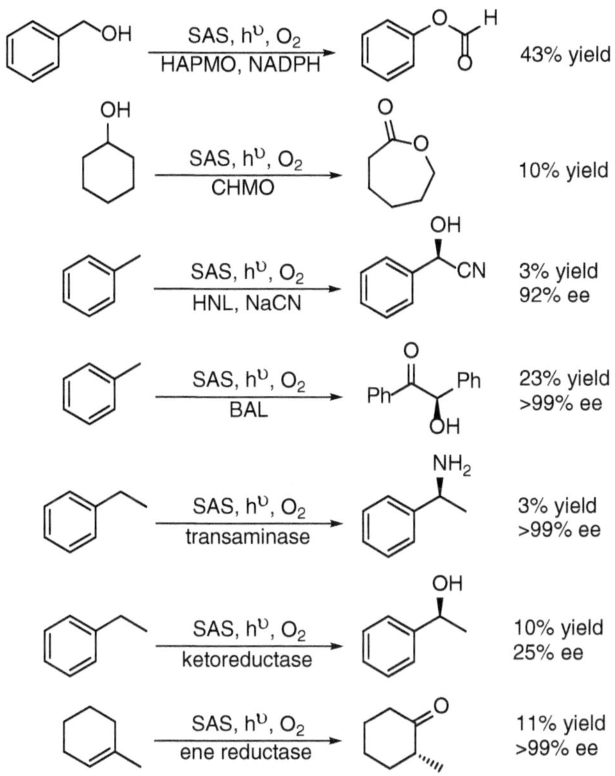

Scheme 7.52 One-pot concurrent cascades of photocatalytic oxidation and enzymatic functionalization.

Chemical transformations that install heteroatoms into C–H bonds are of significant interest because they streamline the construction of value-added small molecules. The above C–H oxyfunctionalization of ethyl benzene by photocatalytic oxidation and enzymatic reduction afforded 1-phenylethanol in low yield and ee. This was a limiting factor in the application of this highly enabling C–H oxyfunctionalization in the construction of chiral alcohol motifs widely existing in pharmaceuticals and natural products. 9-Mesityl-10-methylacridinium ion (Acr^+-Mes ClO_4^-) was found as a photocatalyst for the C–H bond oxygenation of (*p*-methoxphenyl)ethane, providing the acetophenone product in 89 % isolated yield after four hours. When this photocatalytic oxidation was combined with enzymatic ketone reduction in a one-pot concurrent manner, the co-substrate isopropanol for NAD(P)H regeneration was deleterious to the reactions under irradiation. As such, isopropanol was added into the reaction mixture after the photooxidation was completed; the resulting ketone was then reduced to (*R*)-(4′-methoxyphenyl)ethanol in 85% yield and >99:1 e.r. The single-flask photooxidation/enzymatic reduction process was successfully applied to the C–H hydroxylation of a broad range of substrates, offering a powerful strategy for the synthesis of chiral alcohols with high regioselectivity and enantioselectivity (Scheme 7.53) [65]. For ethyl 4-aryl-butanoate and ethyl 5-aryl-pentanoate, the sequential photooxidation and enzymatic reduction resulted in the direct hydroxylation of the benzyl C–H bond, and the hydroxyl esters formed were spontaneously cyclized to furnish the corresponding lactones with good yields and excellent enantioselectivity (Scheme 7.54).

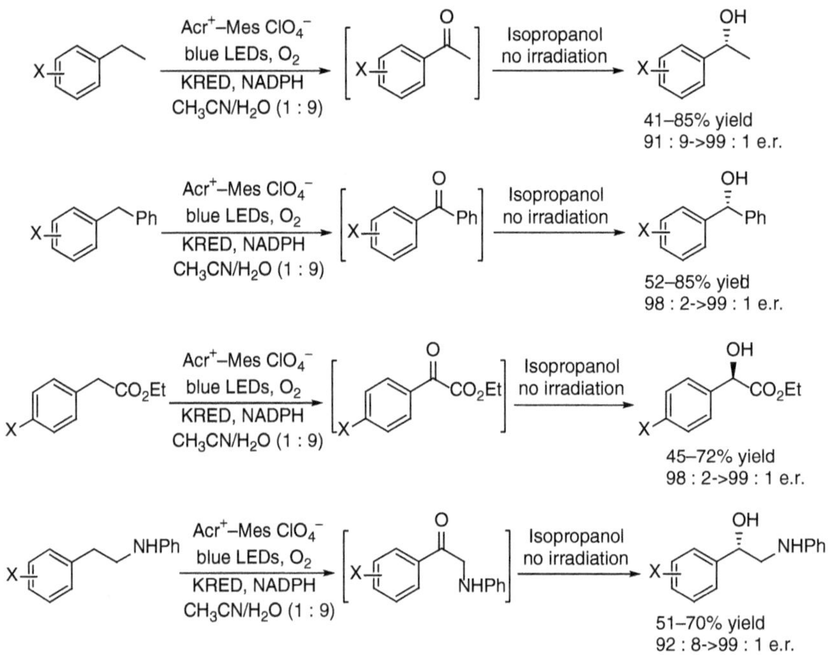

Scheme 7.53 One-pot cascades of photocatalytic oxidation and enzymatic reduction leading to C–H hydroxylation. Source: Based on Betori et al. [65].

Heterogeneous photocatalysts graphitic carbonitride (g-C_3N_4) and Au-doped TiO_2 (Au-TiO_2) also catalyzed the oxidation of cyclohexane under visible light irradiation to give cyclohexanone, which could serve as the substrate of CHMO. When the light-driven oxyfunctionalization of cyclohexane and the enzymatic Baeyer–Villiger oxidation were performed in a one-pot sequential procedure by using Au-TiO_2 or g-C_3N_4 as photocatalyst and *E. coli* whole cells expressing CHMO from *Acinetobacter calcoaceticus* (AcCHMO) as biocatalyst (Scheme 7.55), ε-caprolactone was obtained in about 360 μM concentration from 50 mM cyclohexane or 420 μM from 200 mM substrate, respectively. In the case of Au-TiO_2, the product concentration sharply decreased to about 100 μM when the concentration of cyclohexane was increased to 100 mM, possibly due to the poisoning of the photocatalyst at higher substrate concentration [66].

In the presence of the photocatalyst SAS, primary and secondary alcohols were oxidized to aldehydes and ketones under atmospheric conditions with visible light

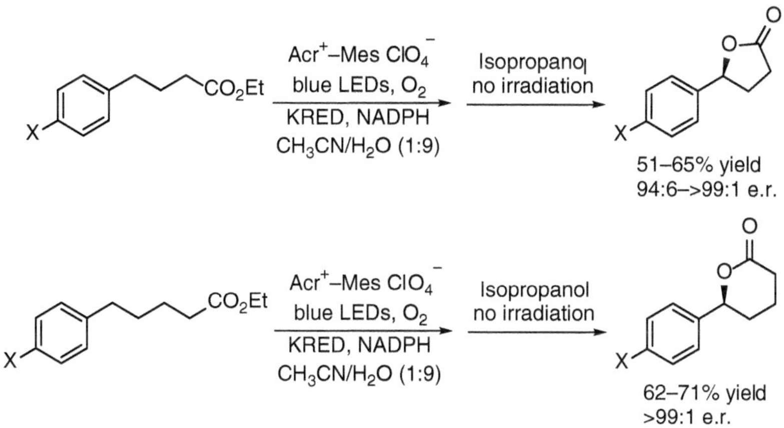

Scheme 7.54 Lactone synthesis via one-pot cascades of photocatalytic oxidation and enzymatic reduction.

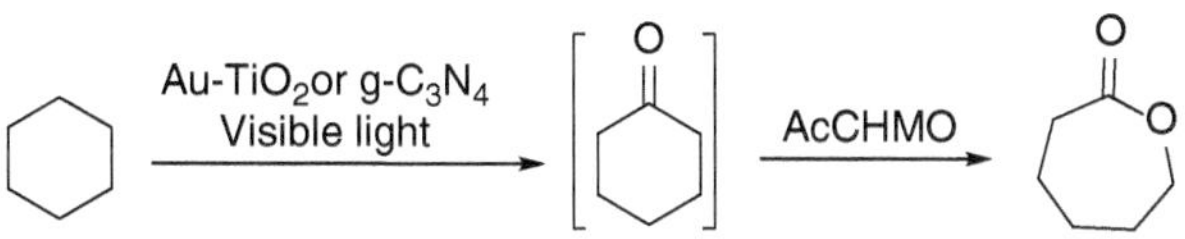

Scheme 7.55 One-pot transformation of cyclohexane to ε-caprolactone by photocatalytic oxidation and enzymatic reduction.

irradiation ($\lambda > 400$ nm). The ω-transaminases are known to catalyze the enantioselective reductive amination of the intermediate aldehydes and ketones. When the photooxidation of various alcohols and enzymatic reductive amination were carried out in a one-pot one-step procedure, only low conversion of alcohol to amine was achieved. The conversion was significantly improved by performing the reactions in a one-pot two-step procedure. A series of aliphatic and aromatic alcohols were transformed into the corresponding (*R*)- or (*S*)-amines by this photooxidation and reductive amination cascade using either (*R*)- or (*S*)-selective transaminases (Scheme 7.56) [67].

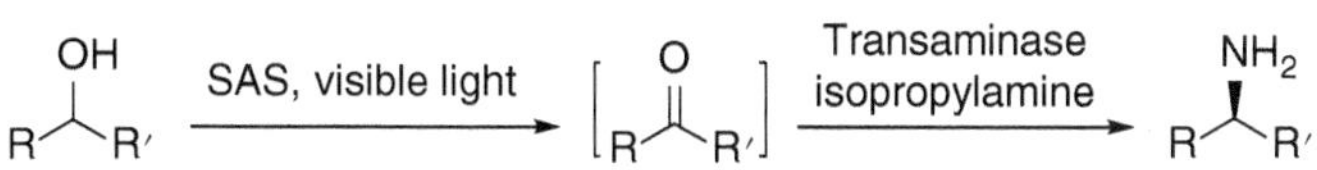

Scheme 7.56 Synthesis of chiral amines via sequential photooxidation of alcohols and enzymatic amination. Source: Based on Gacs et al. [67].

Tremendous efforts and progress have been made for the dynamic kinetic resolution (DKR) of chiral amines over the past decade. A major effort for the DKR of amines has been the search for efficient amine racemization catalysts, and transition metals, such as Ru, Pd, Co, and Ni, have been used as the catalysts for amine racemization. This was discussed in detail in Chapter 4. Recently, a photoredox-mediated

HAT protocol was developed for the racemization of amines upon coupling with lipase-catalyzed kinetic resolution of amines leading to photoenzymatic DKR of racemic amines under mild conditions. After a variety of photocatalysts and HAT catalyst thiols were evaluated for the racemization of (*S*)-4-phenyl-2-butanamine under white LED irradiation (32 W), Ir(ppy)$_2$(dtb-bpy)PF$_6$ and *n*-octanethiol were identified as an effective pair of photocatalyst and HAT catalyst for the racemization. The use of 2 mol% of Ir(ppy)$_2$(dtb-bpy)PF$_6$ [Ir(III)] and 50 mol% of *n*-octanethiol (*n*-OctSH) in the presence of 4 Å molecular sieves (4 Å-MS) resulted in complete racemization of (*S*)-4-phenyl-2-butanamine. When *Candida antarctica* lipase B (Novozym 435) and methyl β-methoxypropionate were added into the racemization system, racemic 4-phenyl-2-butanamine was converted into the corresponding (*R*)-amide in 98% yield and 99% ee. By employing this photoredox-mediated racemization of amines in conjunction with lipase-catalyzed kinetic resolution the DKR of a variety of primary amines was achieved, affording a single enantiomer of the corresponding (*R*)-amides in 61–97% yield and 99% ee (Scheme 7.57). Notably, DKR of hexane-2,5-diamine (*dl*:*meso* = 1:1) was also realized by using this photoenzymatic protocol, and the corresponding diamide was isolated in 64% yield as single diastereomer with 99% ee (Scheme 7.58) [68].

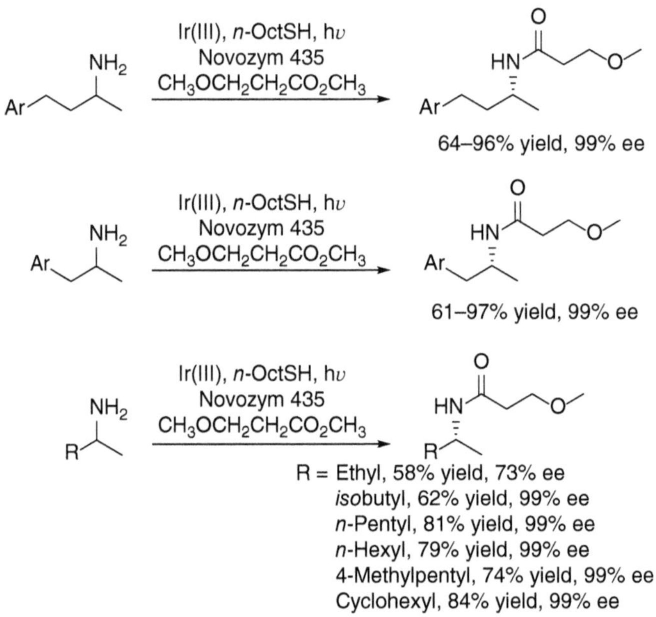

Scheme 7.57 Photoenzymatic dynamic kinetic resolution of racemic amines.

In addition to the DKR, cyclic deracemization had emerged as an attractive strategy for conversion of a racemic mixture into a single enantiomer. In a typical cyclic deracemization, one enantiomer of the racemic compound (A) is enantioselectively converted to another compound (B), while the other enantiomer accumulates; a subsequent reaction converts compound (B) back to compound (A)

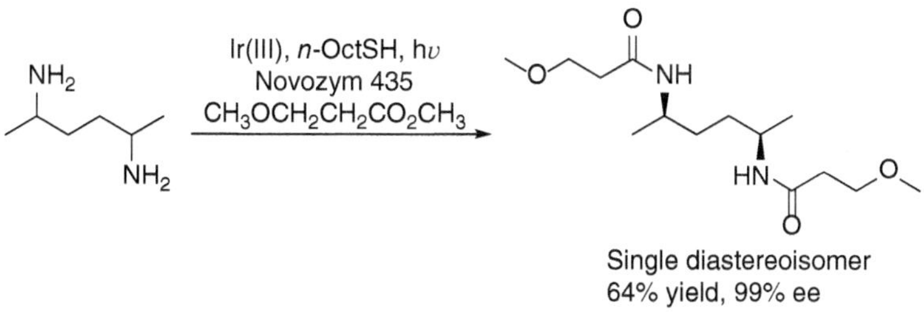

Scheme 7.58 Photoenzymatic DKR of hexane-2,5-diamine. Source: Based on Yang et al. [68].

in racemic form. Over time, such selective recycling leads to the enrichment of the unreactive enantiomer of compound (A). This strategy had been applied for the chemoenzymatic deracemization of various racemic amines by using an enantioselective amine oxidase and a non-enantioselective reducing reagent, as discussed in Chapter 5. More recently, the visible-light-driven reduction of imines to enantioenriched amines was achieved by a similar strategy of combining photocatalytic reduction of cyclic amines with monoamine oxidase-catalyzed enantioselective oxidation of the resulting amines. By using $Na_3[Ru(sppy)_3]$ as a photocatalyst and ascorbic acid as a reductant, 2-cyclohexyl-1-pyrroline was reduced to racemic 2-cyclohexyl-1-pyrrolidine in aqueous phosphate buffer at pH 8.0 (Scheme 7.59) under irradiation with blue LED (405 nm). Photoexcitation of $Na_3[Ir(sppy)_3]$ photosensitizer in the presence of the cyclic imine resulted in a highly reactive α-amino alkyl radical that was intercepted by HAT from ascorbate or thiol donors to afford the corresponding amine in racemic form. This imine photoreduction was combined with the enantioselective amine oxidation to the corresponding imine catalyzed by a monoamine oxidase (MAO-N-9), affording (*R*)-2-cyclohexyl-1-pyrrolidine in high yield and ee value. The concurrent photoredox- and enzymatic catalysis protocol were applied to the asymmetric synthesis of aliphatic and aromatic cyclic amines (Scheme 7.60) [69].

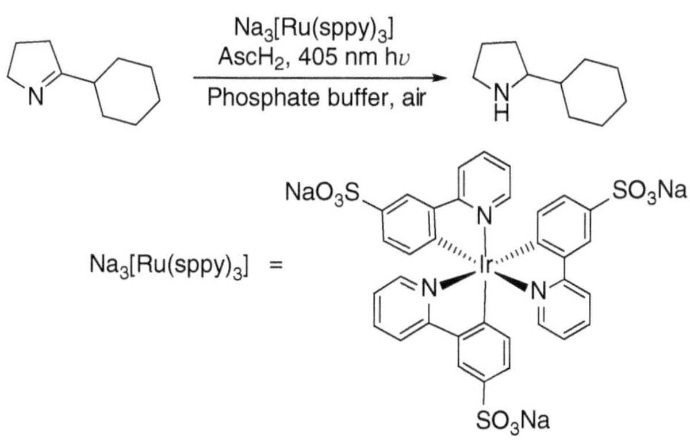

Scheme 7.59 Photocatalytic reduction of imines.

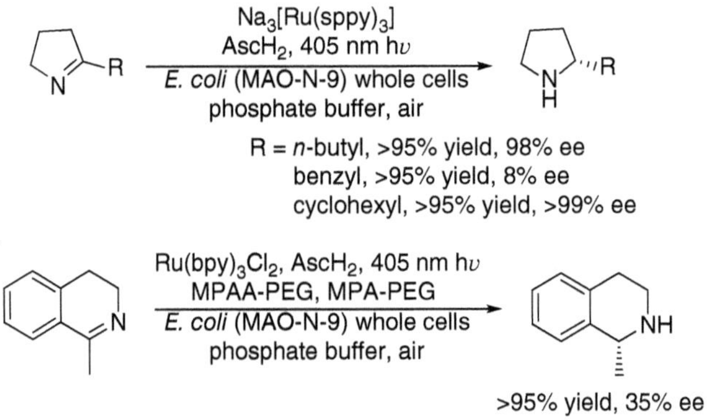

Scheme 7.60 Synthesis of chiral amines via concurrent photocatalytic reduction of imines and stereoselective enzymatic oxidation. Source: Based on Guo et al. [69].

In the presence of photocatalyst SAS, 2-phenylpropanoic acid was transformed to a benzylic radical by decarboxylation under irradiation with blue LEDs. The radical intermediate captured O_2 and was converted into acetophenone with 98% yield. The photoinduced decarboxylative carbonylation worked well with a variety of carboxylic acids, generating the corresponding acetophenone derivatives in up to 99% yields. The resulting ketones were subsequently reduced by using (*R*)-selective carbonyl reductase (KtCR) from *Kluyveromyces thermotolerans* or (*S*)-selective carbonyl reductase (YueD) from *Bacillus*, providing the corresponding (*R*)- or (*S*)-alcohols, respectively. However, when the light-driven decarboxylative carbonylation of carboxylic acids and the bioreduction were carried out in the concurrent one-pot mode, the alcohol products were not obtained. It might be due to the capability of the light-catalytic system to oxidize the alcohol. When the cascade reactions were performed in a one-pot sequential procedure using the *E. coli* whole cells harboring the carbonyl reductases as the biocatalyst, a diversity of chiral alcohols with complementary (*R*)- or (*S*)-configurations were prepared with up to 93% yields and up to 99% ee (Scheme 7.61). This photochemoenzymatic one-pot sequential process offered a mild, effective and highly stereoselective method for the transformation of α-substituted carboxylic acids into the corresponding enantioenriched chiral alcohols [70].

With $Ru(bpy)_3Cl_2$ (bpy = 2,2′-bipyridine) as the photocatalyst, *p*-tolylglycine was decarboxylated under visible light irradiation generating α-amino radicals, which were further oxidized to the iminium ions. The Friedel–Crafts reaction of the iminium ion with indoles proceeded subsequently affording 3-substituted indoles (Scheme 7.62). *p*-Tolylglycine could be obtained by the hydrolysis of ethyl *p*-tolylglycinate catalyzed by lipase CALB. The enzyme-catalyzed hydrolysis was combined with visible-light-driven decarboxylation, followed by the oxidization of α-amino radicals and the Friedel–Crafts reaction of the iminium ion with indole (Scheme 7.63). The cascade reactions proceeded smoothly when they were performed in a one-pot procedure with a wide range of substituted phenylglycinates and indoles and gave rise to α-aminoalkylation products, such as 3-substituted indoles, in up to 96% yields (Scheme 7.64) [71].

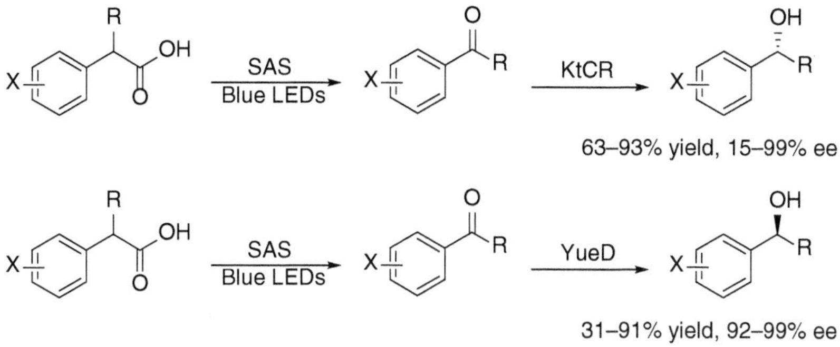

Scheme 7.61 Transformation of α-substituted carboxylic acids into chiral alcohols by sequential one-pot photoinduced decarboxylative carbonylation and enzymatic reduction.

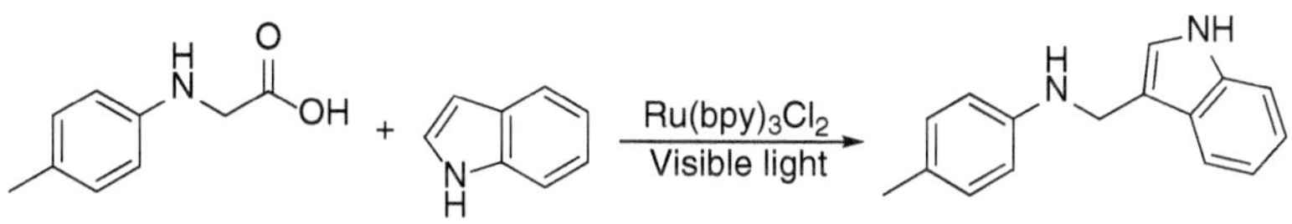

Scheme 7.62 Photoredox catalytic aminoalkylation of indoles.

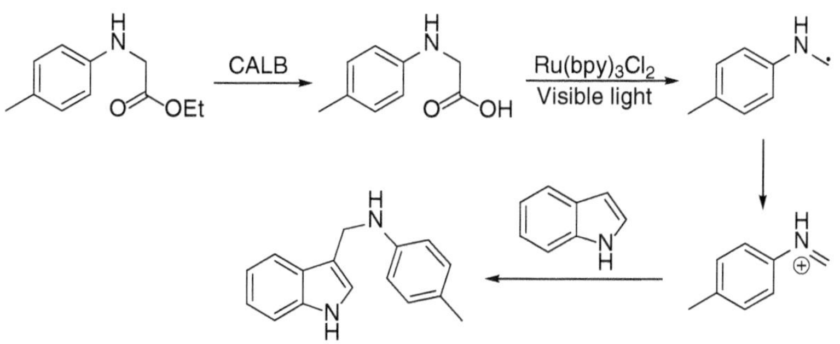

Scheme 7.63 Reaction sequence of one-pot enzymatic hydrolysis and photoredox catalytic aminoalkylation of indoles.

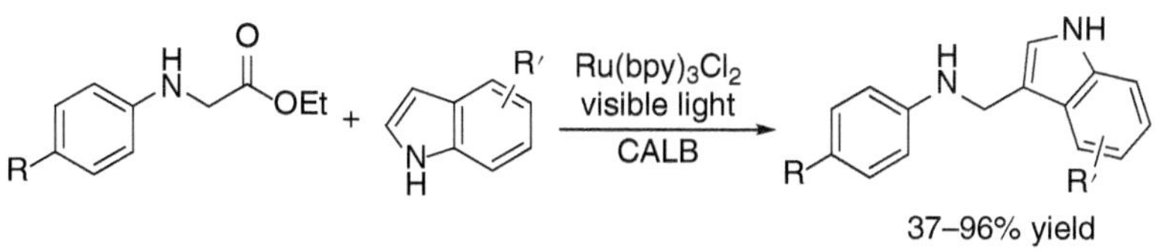

Scheme 7.64 One-pot synthesis of 3-substituted indoles by enzymatic hydrolysis and photoredox aminoalkylation of indoles. Source: From He et al. [71].

The above iminium ion intermediate also could undergo Mannich reaction with β-naphthol, followed by transimination with iminium ion and subsequent intramolecular cyclization to furnish a 1,3-oxazine derivative (Scheme 7.65). Various α- or β-naphthols and ethyl *N*-aryl glycinates were tested and the corresponding 1,3-oxazine derivatives were prepared with up to 69% yields by this one-pot photobiocatalytic process (Scheme 7.66) [72].

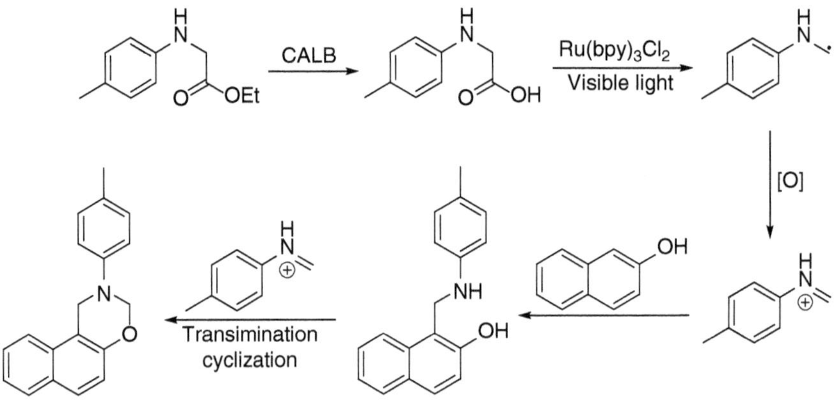

Scheme 7.65 Reaction sequence of one-pot enzymatic hydrolysis and photoredox catalytic aminoalkylation of β-naphthol.

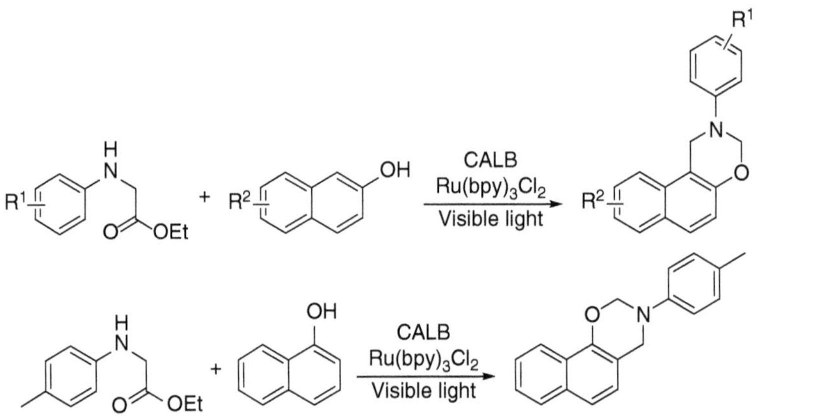

Scheme 7.66 One-pot enzymatic hydrolysis and photoredox catalytic aminoalkylation of naphthols. Source: Based on Zhang et al. [72].

Acridines could also serve as photocatalyst for the decarboxylation of carboxylic acids, via HAT within the acridine-carboxylic acid hydrogen bond complex followed by loss of CO_2, generating alkyl radicals. The alkyl radicals then underwent α-C–H-HAT with a cobaloxime catalyst, resulting in the formation of alkenes. The dual photoinduced acridine-catalyzed O–H-HAT and cobaloxime-catalyzed C–H-HAT led to the dehydrodecarboxylation of carboxylic acids into alkenes, as shown in Scheme 7.67.

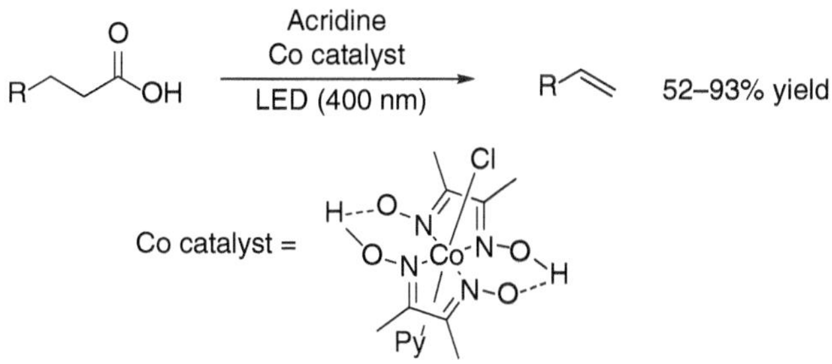

Scheme 7.67 One-pot photoinduced dual catalytic dehydrodecarboxylation of carboxylic acids.

Triglycerides are the main components of vegetable and algal oils, which are renewable feedstock materials. Lipases catalyze the hydrolysis of triglycerides under mild reaction conditions, generating long-chain fatty acids. It can be envisioned that combination of lipase-catalyzed hydrolysis of triglycerides with dual catalytic dehydrodecarboxylation of carboxylic acids would enable the direct conversion of renewable oils into valuable alkenes. Indeed, Amano lipase PS from *Burkholderia cepacia* catalyzed the hydrolysis of triglycerides from different resources. This enzymatic reaction was compatible with the dual catalytic dehydrodecarboxylation process, leading to the development of a scalable triple-catalytic cooperative chemoenzymatic lipase–acridine–cobaloxime process that enabled direct conversion of a variety of plant oils to the corresponding long-chain terminal alkenes (Scheme 7.68) serving as precursors to bio-derived polymers [73].

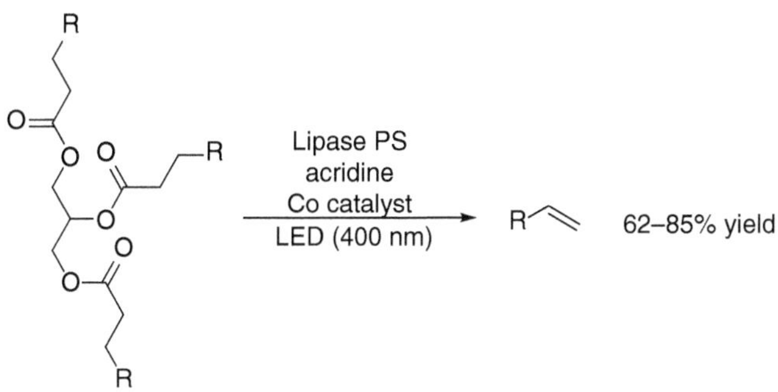

Scheme 7.68 One-pot photoinduced triple-catalytic chemoenzymatic conversion of plant oils to alkenes.

7.6 Summary and Outlook

Light usually induces the formation of reactive species in a radical form through energy or electron transfer, leading to novel chemical transformations. Although photochemical reactions have gradually emerged as practical methods for organic synthesis, their stereocontrol remains a challenging task [74]. On the other hand, enzymes provide unmatched stereocontrol over the catalyzed reactions by their unique substrate-binding scaffolds. Photo-triggered formation of radical species in enzyme active sites offers attractive strategies to modulate the electron transfer process and to control the stereochemical outcome of the reaction, thus enabling a wide array of novel nonnatural enzymatic reactions with high enantio- and stereoselectivity. As discussed above, this has been successfully implemented with redox enzymes such as ketoreductases and ERs. In recent years, photochemical and biocatalytic cascade reactions, in which the two types of reaction are performed sequentially or concurrently, have been increasingly investigated, and have shown synergetic effects in the sustainable synthesis of the desired chemicals. These new developments at the intersection of photochemistry and biocatalysis together with light-driven enzymes will prosper, and offer new methodologies for addressing challenges in chemical synthesis.

References

1 Seel, C.J. and Gulder, T. (2019). Biocatalysis fueled by light: on the versatile combination of photocatalysis and enzymes. *ChemBioChem* 20 (15): 1871–1897.

2 Schmermund, L., Jurkaš, V., Özgen, F.F. et al. (2019). Photo-biocatalysis: biotransformations in the presence of light. *ACS Catalysis* 9 (5): 4115–4144.

3 Kim, J. and Park, C.B. (2019). Shedding light on biocatalysis: photoelectrochemical platforms for solar-driven biotransformation. *Current Opinion in Chemical Biology* 49: 122–129.

4 Maciá-Agulló, J.A., Corma, A., and Garcia, H. (2015). Photobiocatalysis: the power of combining photocatalysis and enzymes. *Chemistry – A European Journal* 21 (31): 10940–10959.

5 Guo, X., Okamoto, Y., Schreier, M.R. et al. (2020). Reductive amination and enantioselective amine synthesis by photoredox catalysis. *European Journal of Organic Chemistry* 2020 (10): 1288–1293.

6 Björn, L.O. (2018). Photoenzymes and related topics: an update. *Photochemistry and Photobiology* 94 (3): 459–465.

7 Gabruk, M. and Mysliwa-Kurdziel, B. (2015). Light-dependent protochlorophyllide oxidoreductase: phylogeny, regulation. *and Catalytic Properties. Biochemistry* 54 (34): 5255–5262.

8 Zhang, M., Wang, L., and Zhong, D. (2017). Photolyase: dynamics and mechanisms of repair of sun-induced DNA damage. *Photochemistry and Photobiology* 93 (1): 78–92.

9 Bao, H. and Burnap, R.L. (2016). Photoactivation: the light-driven assembly of the water oxidation complex of photosystem II. *Frontiers in Plant Science* 7: 578.

10 Sorigué, D., Légeret, B., Cuiné, S. et al. (2017). An algal photoenzyme converts fatty acids to hydrocarbons. *Science* 357 (6354): 903–907.

11 Huijbers, M.M.E., Zhang, W., Tonin, F., and Hollmann, F. (2018). Light-driven enzymatic decarboxylation of fatty acids. *Angewandte Chemie International Edition* 57 (41): 13648–13651.

12 Zhang, W., Ma, M., Huijbers, M.M.E. et al. (2019). Hydrocarbon synthesis via photoenzymatic decarboxylation of carboxylic acids. *Journal of the American Chemical Society* 141 (7): 3116–3120.

13 Heyes, D.J., Lakavath, B., Hardman, S.J.O. et al. (2020). Photochemical mechanism of light-driven fatty acid photodecarboxylase. *ACS Catalysis* 10 (12): 6691–6696.

14 Xu, J., Hu, Y., Fan, J. et al. (2019). Light-driven kinetic resolution of α-functionalized carboxylic acids enabled by an engineered fatty acid photodecarboxylase. *Angewandte Chemie International Edition* 58 (25): 8474–8478.

15 Ma, Y., Zhang, X., Li, Y. et al. (2020). Production of fatty alcohols from non-edible oils by enzymatic cascade reactions. *Sustainable Energy & Fuels* 4 (8): 4232–4237.

16 Zhang, W., Lee, J.-H., Younes, S.H.H. et al. (2020). Photobiocatalytic synthesis of chiral secondary fatty alcohols from renewable unsaturated fatty acids. *Nature Communications* 11 (1): 2258.

17 Cha, H.-J., Hwang, S.-Y., Lee, D.-S. et al. (2020). Whole-cell photoenzymatic cascades to synthesize long-chain aliphatic amines and esters from renewable fatty acids. *Angewandte Chemie International Edition* 59 (18): 7024–7028.

18 Lee, S.H., Choi, D.S., Kuk, S.K., and Park, C.B. (2018). Photobiocatalysis: activating redox enzymes by direct or indirect transfer of photoinduced electrons. *Angewandte Chemie International Edition* 57 (27): 7958–7985.

19 Park, J.H., Lee, S.H., Cha, G.S. et al. (2015). Cofactor-free light-driven whole-cell cytochrome P450 catalysis. *Angewandte Chemie International Edition* 54 (3): 969–973.

20 Tran, N.-H., Nguyen, D., Dwaraknath, S. et al. (2013). An efficient light-driven P450 BM3 biocatalyst. *Journal of the American Chemical Society* 135 (39): 14484–14487.

21 Le, T.-K., Park, J.H., Choi, D.S. et al. (2019). Solar-driven biocatalytic C-hydroxylation through direct transfer of photoinduced electrons. *Green Chemistry* 21 (3): 515–525.

22 Özgen, F.F., Runda, M.E., Burek, B.O. et al. (2020). Artificial light-harvesting complexes enable rieske oxygenase catalyzed hydroxylations in non-photosynthetic cells. *Angewandte Chemie International Edition* 59 (10): 3982–3987.

23 Hutton, G.A.M., Reuillard, B., Martindale, B.C.M. et al. (2016). Carbon dots as versatile photosensitizers for solar-driven catalysis with redox enzymes. *Journal of the American Chemical Society* 138 (51): 16722–16730.

24 Caputo, C.A., Gross, M.A., Lau, V.W. et al. (2014). Photocatalytic hydrogen production using polymeric carbon nitride with a hydrogenase and a bioinspired synthetic Ni catalyst. *Angewandte Chemie International Edition* 53 (43): 11538–11542.

25 Bachmeier, A., Murphy, B.J., and Armstrong, F.A. (2014). A multi-heme flavoenzyme as a solar conversion catalyst. *Journal of the American Chemical Society* 136 (37): 12876–12879.

26 Miller, M., Robinson, W.E., Oliveira, A.R. et al. (2019). Interfacing formate dehydrogenase with metal oxides for the reversible electrocatalysis and solar-driven reduction of carbon dioxide. *Angewandte Chemie International Edition* 58 (14): 4601–4605.

27 Hollmann, F., Taglieber, A., Schulz, F., and Reetz, M.T. (2007). A light-driven stereoselective biocatalytic oxidation. *Angewandte Chemie International Edition* 46 (16): 2903–2906.

28 Mifsud, M., Gargiulo, S., Iborra, S. et al. (2014). Photobiocatalytic chemistry of oxidoreductases using water as the electron donor. *Nature Communications* 5: 3145.

29 Lee, S.H., Choi, D.S., Pesic, M. et al. (2017). Cofactor-free, direct photoactivation of enoate reductases for the asymmetric reduction of C=C bonds. *Angewandte Chemie International Edition* 56 (30): 8681–8685.

30 Yoon, J., Lee, S.H., Tieves, F. et al. (2019). Light-harvesting dye–alginate hydrogel for solar-driven. Sustainable biocatalysis of asymmetric hydrogenation. *ACS Sustainable Chemistry & Engineering* 7 (6): 5632–5637.

31 van Schie, M.M.C.H., Paul, C.E., Arends, I.W.C.E., and Hollmann, F. (2019). Photoenzymatic epoxidation of styrenes. *Chemical Communications* 55 (12): 1790–1792.

32 Schroeder, L., Frese, M., Müller, C. et al. (2018). Photochemically driven biocatalysis of halogenases for the green production of chlorinated compounds. *ChemCatChem* 10 (15): 3336–3341.

33 Gonçalves, L.C.P., Mansouri, H.R., Bastos, E.L. et al. (2019). Morpholine-based buffers activate aerobic photobiocatalysis via spin correlated ion pair formation. *Catalysis Science & Technology* 9 (6): 1365–1371.

34 Gonçalves, L.C.P., Mansouri, H.R., PourMehdi, S. et al. (2019). Boosting photobioredox catalysis by morpholine electron donors under aerobic conditions. *Catalysis Science & Technology* 9 (10): 2682–2688.

35 Lee, S.H., Kim, J.H., and Park, C.B. (2013). Coupling photocatalysis and redox biocatalysis toward biocatalyzed artificial photosynthesis. *Chemistry – A European Journal* 19 (14): 4392–4406.

36 Park, C.B., Lee, S.H., Subramanian, E. et al. (2008). Solar energy in production of l-glutamate through visible light active photocatalyst—redox enzyme coupled bioreactor. *Chemical Communications* 42: 5423–5425.

37 Lee, S.H., Ryu, J., Nam, D.H., and Park, C.B. (2011). Photoenzymatic synthesis through sustainable NADH regeneration by SiO2-supported quantum dots. *Chemical Communications* 47 (16): 4643–4645.

38 Lee, S.H., Nam, D.H., and Park, C.B. (2009). Screening xanthene dyes for visible light-driven nicotinamide adenine dinucleotide regeneration and photoenzymatic synthesis. *Advanced Synthesis & Catalysis* 351 (16): 2589–2594.

39 Kim, J.H., Lee, S.H., Lee, J.S. et al. (2011). Zn-containing porphyrin as a biomimetic light-harvesting molecule for biocatalyzed artificial photosynthesis. *Chemical Communications* 47 (37): 10227–10229.

40 Nam, D.H. and Park, C.B. (2012). Visible light-driven NADH regeneration sensitized by proflavine for biocatalysis. *ChemBioChem* 13 (9): 1278–1282.

41 Ryu, J., Nam, D.H., Lee, S.H., and Park, C.B. (2014). Biocatalytic photosynthesis with water as an electron donor. *Chemistry – A European Journal* 20 (38): 12020–12025.

42 Zhang, S., Shi, J., Sun, Y. et al. (2019). Artificial thylakoid for the coordinated photoenzymatic reduction of carbon dioxide. *ACS Catalysis* 9 (5): 3913–3925.

43 Burai, T.N., Panay, A.J., Zhu, H. et al. (2012). Light-driven, quantum dot-mediated regeneration of FMN to drive reduction of ketoisophorone by old yellow enzyme. *ACS Catalysis* 2 (4): 667–670.

44 Kim, J., Lee, S.H., Tieves, F. et al. (2018). Biocatalytic C=C bond reduction through carbon nanodot-sensitized regeneration of NADH analogues. *Angewandte Chemie International Edition* 57 (42): 13825–13828.

45 Mengele, A.K., Seibold, G.M., Eikmanns, B.J., and Rau, S. (2017). Coupling molecular photocatalysis to enzymatic conversion. *ChemCatChem* 9 (23): 4369–4376.

46 Büchsenschütz, H.C., Vidimce-Risteski, V., Eggbauer, B. et al. (2020). Stereoselective biotransformations of cyclic imines in recombinant cells of synechocystis sp. PCC 6803. *ChemCatChem* 12 (3): 726–730.

47 Emmanuel, M.A., Greenberg, N.R., Oblinsky, D.G., and Hyster, T.K. (2016). Accessing non-natural reactivity by irradiating nicotinamide-dependent enzymes with light. *Nature* 540: 414.

48 Biegasiewicz, K.F., Cooper, S.J., Emmanuel, M.A. et al. (2018). Catalytic promiscuity enabled by photoredox catalysis in nicotinamide-dependent oxidoreductases. *Nature Chemistry* 10 (7): 770–775.

49 Sandoval, B.A., Kurtoic, S.I., Chung, M.M. et al. (2019). Photoenzymatic catalysis enables radical-mediated ketone reduction in ene-reductases. *Angewandte Chemie International Edition* 58 (26): 8714–8718.

50 Nakano, Y., Black, M.J., Meichan, A.J. et al. (2020). Photoenzymatic hydrogenation of heteroaromatic olefins using 'Ene'-reductases with photoredox catalysts. *Angewandte Chemie International Edition* 59 (26): 10484–10488.

51 Sandoval, B.A., Meichan, A.J., and Hyster, T.K. (2017). Enantioselective hydrogen atom transfer: discovery of catalytic promiscuity in flavin-dependent 'ene'-reductases. *Journal of the American Chemical Society* 139 (33): 11313–11316.

52 Biegasiewicz, K.F., Cooper, S.J., Gao, X. et al. (2019). Photoexcitation of flavoenzymes enables a stereoselective radical cyclization. *Science* 364 (6446): 1166–1169.

53 Black, M.J., Biegasiewicz, K.F., Meichan, A.J. et al. (2020). Asymmetric redox-neutral radical cyclization catalysed by flavin-dependent 'ene'-reductases. *Nature Chemistry* 12 (1): 71–75.

54 Huang, X., Wang, B., Wang, Y. et al. (2020). Photoenzymatic enantioselective intermolecular radical hydroalkylation. *Nature*.

55 Chen, J., Guan, Z., and He, Y.-H. (2019). Photoenzymatic approaches in organic synthesis. *Asian Journal of Organic Chemistry* 8 (10): 1775–1790.

56 Seel, C.J., Králík, A., Hacker, M. et al. (2018). Atom-economic electron donors for photobiocatalytic halogenations. *ChemCatChem* 10 (18): 3960–3963.

57 Yuan, B., Mahor, D., Fei, Q. et al. (2020). A water-soluble anthraquinone photocatalyst enables methanol driven enzymatic halogenation and hydroxylation reactions. *ACS Catalysis*.

58 Zhang, W., Fernández-Fueyo, E., Ni, Y. et al. (2018). Selective aerobic oxidation reactions using a combination of photocatalytic water oxidation and enzymatic oxyfunctionalizations. *Nature Catalysis* 1 (1): 55–62.

59 van Schie, M.M.C.H., Zhang, W., Tieves, F. et al. (2019). Cascading g-C3N4 and peroxygenases for selective oxyfunctionalization reactions. *ACS Catalysis* 9 (8): 7409–7417.

60 Willot, S.J.P., Fernández-Fueyo, E., Tieves, F. et al. (2019). Expanding the spectrum of light-driven peroxygenase reactions. *ACS Catalysis* 9 (2): 890–894.

61 Litman, Z.C., Wang, Y., Zhao, H., and Hartwig, J.F. (2018). Cooperative asymmetric reactions combining photocatalysis and enzymatic catalysis. *Nature* 560 (7718): 355–359.

62 Lauder, K., Toscani, A., Qi, Y. et al. (2018). Photo-biocatalytic one-pot cascades for the enantioselective synthesis of 1,3-mercaptoalkanol volatile sulfur compounds. *Angewandte Chemie International Edition* 57 (20): 5803–5807.

63 Ding, X., Dong, C.-L., Guan, Z., and He, Y.-H. (2019). Concurrent asymmetric reactions combining photocatalysis and enzyme catalysis: direct enantioselective synthesis of 2,2-disubstituted indol-3-ones from 2-arylindoles. *Angewandte Chemie International Edition* 58 (1): 118–124.

64 Zhang, W., Fueyo, E.F., Hollmann, F. et al. (2019). Combining photo-organo redox- and enzyme catalysis facilitates asymmetric C-H bond functionalization. *European Journal of Organic Chemistry* 2019 (1): 80–84.

65 Betori, R.C., May, C.M., and Scheidt, K.A. (2019). Combined photoredox/enzymatic C—H benzylic hydroxylations. *Angewandte Chemie International Edition* 58 (46): 16490–16494.

66 Li, P., Ma, Y., Li, Y. et al. (2020). Cascade synthesis from cyclohexane to ε-caprolactone by visible-light-driven photocatalysis combined with whole-cell biological oxidation. *ChemBioChem* 21 (13): 1852–1855.

67 Gacs, J., Zhang, W., Knaus, T. et al. (2019). A photo-enzymatic cascade to transform racemic alcohols into enantiomerically pure amines. *Catalysts* 9 (4): 305.

68 Yang, Q., Zhao, F., Zhang, N. et al. (2018). Mild dynamic kinetic resolution of amines by coupled visible-light photoredox and enzyme catalysis. *Chemical Communications* 54 (100): 14065–14068.

69 Guo, X., Okamoto, Y., Schreier, M.R. et al. (2018). Enantioselective synthesis of amines by combining photoredox and enzymatic catalysis in a cyclic reaction network. *Chemical Science* 9 (22): 5052–5056.

70 Xu, J., Arkin, M., Peng, Y. et al. (2019). Enantiocomplementary decarboxylative hydroxylation combining photocatalysis and whole-cell biocatalysis in a one-pot cascade process. *Green Chemistry* 21 (8): 1907–1911.

71 He, Y.-H., Xiang, Y., Yang, D.-C., and Guan, Z. (2016). Combining enzyme and photoredox catalysis for aminoalkylation of indoles via a relay catalysis strategy in one pot. *Green Chemistry* 18 (19): 5325–5330.

72 Zhang, G.-Y., Xiang, Y., Guan, Z., and He, Y.-H. (2017). Enzyme and photoredox sequential catalysis for the synthesis of 1,3-oxazine derivatives in one pot. *Catalysis Science & Technology* 7 (9): 1937–1942.

73 Nguyen, V.T., Nguyen, V.D., Haug, G.C. et al. (2019). Alkene synthesis by photocatalytic chemoenzymatically compatible dehydrodecarboxylation of carboxylic acids and biomass. *ACS Catalysis* 9 (10): 9485–9498.

74 Wang, C. and Lu, Z. (2015). Catalytic enantioselective organic transformations via visible light photocatalysis. *Organic Chemistry Frontiers* 2 (2): 179–190.

8

Perspectives

Chemical synthetic processes have made extraordinary contributions to the production of synthetic chemicals, which are essential and conducive to our daily lives; however, they have often exerted adverse consequences on the environment to certain extent, because these processes have usually been performed based on the narrow considerations of product performance and production cost. Given the fact that numerous essential products in our daily lives are produced by chemical processes, when looking to the future, we must answer the question: How do we limit or eliminate the detrimental impacts that threaten the sustainability of our planet while meeting the expanding needs of our society? This requires that the chemical products and processes are designed by following green chemistry principles [1]. Therefore, the inherent properties of the products and their production processes have to be considered from the design stage to address whether they are persistent or readily degradable, toxic or benign, and depleting or renewable. Integration of enzyme catalysis into chemical synthesis offers a green approach to the manufacturing processes of desired chemical products by combining the advantages of both chemical reaction and biotransformation. This has been illustrated by various industrial examples such as the production of active pharmaceutical ingredients (API), cortisone [2], and sitagliptin [3].

Retrosynthesis is based on the principle that a target molecule is disconnected into precursors that can be connected using a synthetic reaction. This process is recursively applied to lead to simpler precursors until reaching at readily available starting materials. Retrosynthetic analysis involving both biotransformation and traditional chemical reaction enables design of a shorter and more effective synthetic route to a target chemical, resulting in reductions in the production cost and environmental impact [4, 5]. The ongoing advances in the discovery and design of biocatalysts, metal catalysts, and organocatalysts continuously augment the synthesis toolbox for successful implementation of this strategic approach in the preparation of complex target compounds. Additionally, computer-aided synthesis planning (CASP) facilitates the implementation of retrosynthesis concepts in synthesis design in both synthetic chemistry and synthetic biology. In organic chemistry, computer-aided design tools such as Synthia or automated system for knowledge-based continuous organic synthesis (ASKCOS) have been increasingly used to plan synthesis routes for a number of target molecules [6, 7]. Retrosynthesis

Chemo-Enzymatic Cascade Reactions, First Edition. Dunming Zhu.

workflows such as RetroPath have been deployed to streamline retrosynthesis pathway design for the bioproduction of natural and nonnatural molecules in synthetic biology [8]. Recently, Flitsch and Turner et al. developed RetroBioCat, a computer-aided planning tool for design of biocatalytic cascades [9]. In this computational tool, a set of expertly encoded reaction rules encompassing the enzyme toolbox for biocatalysis and a system for identifying literature precedent for enzymes with the correct substrate specificity were applied to identify promising biocatalytic pathways to a target molecule. In spite of these advances, CASP involving both traditional chemical reactions and biocatalytic transformations remains underdeveloped. It is desired to develop CASP tools encompassing both enzymatic and chemical steps in their synthesis toolboxes. Consequently, it is expected that design of new chemoenzymatic processes and their applications to the sustainable manufacture of a range of important target molecules will be rapidly expanded.

One-pot chemoenzymatic cascade transformations, offering simplified operation and high efficiency, play an important role in the design and implementation of sustainable synthesis. However, some key issues including the reaction condition compatibility of chemical reaction with biotransformation and the inhibition or inactivation of (bio)catalysts among other considerations have to be overcome for successful realization of these attractive synthetic processes [10]. Since green chemistry was recognized as a principle offering solutions to environmental problems in the 1990s, chemists have endeavored to design and create more small-molecule catalysts that can perform under mild aqueous conditions [11, 12]. At the same time, advances in methodologies for enzyme discovery, protein engineering, and design have continuously enlarged the toolbox of enzymes with enhanced performance in catalytic activity, selectivity, and stability, which are well beyond the commercial lipases of yore to include ketoreductases, imine reductases, amino acid dehydrogenases, transaminases, oxidases, oxygenases, halogenases, aldolases, and ammonia lyases [13]. In addition, advances in modern biotechnology enable the development of novel enzymes tailor-made for the desired organic transformations. In recent years, such a concerted effort in synthetic chemistry and biocatalysis has led to impressive development in one-pot chemoenzymatic cascade transformations, which have rapidly expanded in chemical reaction types and enzymes involved in biotransformations, and have shown their application potential in the green synthesis of industrially relevant chemicals [14].

One-pot chemoenzymatic cascade reactions can occur sequentially or concurrently. For one-pot sequential chemoenzymatic reactions, there is certain flexibility in the process implementation because the chemical and biological transformations can be operated under different conditions and additional solvents and/or reagents can be added between the reaction steps. In some cases, the components that inhibit the second reaction can be removed from the reaction mixture by simple operation. In contrast, one-pot concurrent chemoenzymatic reactions require that the reaction conditions of chemical reaction and biotransformation must be completely compatible. However, the metal- and organocatalyzed reactions and biotransformations are traditionally carried out in different reaction media, and inhibition or inactivation of (bio)catalysts often occurs due to the large number of components in the complicated reaction system. Despite these challenges, the last two decades have seen significant advances in this research area, and the chemoenzymatic dynamic kinetic resolution and deracemization of chiral molecules represent two paradigms of performing chemical and biocatalytic reactions in a one-pot concurrent process.

Although an increasing number of true one-pot concurrent chemoenzymatic cascades other than chemoenzymatic dynamic kinetic resolution and deracemization of racemates have appeared in the literature, the scope of this elegant one-pot one-step process is still limited, largely due to the incompatibility of the different catalysts and their respective reaction conditions. In this context, the spatial separation of the chemical reaction and biotransformation has been adopted to overcome these incompatibilities. Different compartmentalization approaches have been deployed to control the microenvironments, thus eliminating the inactivation of biocatalysts and/or chemical catalysts and the mutual interference of these reactions. One protocol is encapsulation of metal catalyst and biocatalyst in different compartments. A hybrid catalyst was created by coating chitosan-coated alginate beads of pig liver esterases (PLE) with octyl-grafted alginate amide for Grubbs catalyst encapsulation a few times. The resulting core–shell beads with Grubbs catalyst and esterase PLE located in different compartments enabled effective ring-closing metathesis of diethyl diallylmalonate and subsequent hydrolysis to give cyclic malonic acid monoester with high conversion and selectivity [15]. In addition to increased activity, selectivity, stability, and recyclability compared to the individual components, chemoenzymatic hybrid catalysts provide unique microenvironments for each reaction, thus promoting cascade transformations that are challenging to proceed in a one-pot setting [16].

Polydimethylsiloxane (PDMS) membrane or thimble was utilized to divide a reactor into two chambers, in which the chemical transformation and enzymatic reaction proceed separately at the same time. A few metal-catalyzed and biocatalytic reactions were successfully carried out in one-pot concurrent mode by employing this space separation methodology [17].

Recently, continuous flow technologies have attracted increasing attentions for their applications in the preparation of fine chemicals [18]. Continuous flow setting offers easy operation for carrying out each reaction in different flow reactors, enabling efficient multistep synthesis of complex molecules such as APIs [19]. As such, chemoenzymatic cascade reactions can be conducted in a continuous flow setting with the enzymatic and chemical reactions being carried out in different flow reactors to address their one-pot incompatibility. For example, gold-catalyzed direct oxidation of sugar aldehyde and the subsequent enzymatic conversion of sugar acid were carried out under continuous flow operation, efficiently delivering four different 2-keto-3-deoxy sugar acids [20]. We expect that future advances in compartmentalization methodologies will greatly expand the potential and scope of chemoenzymatic cascade reactions for industrial applications.

Advances in enzyme engineering and microbial engineering have enabled the construction of biocatalytic cascades for *in vivo* or *in vitro* multistep transformations. The combination of biocatalytic cascades with chemical transformations offers a tremendous opportunity for the synthesis of target chemicals from readily available starting materials. Particularly, some biocompatible chemical reactions have been developed and integrated with microbial transformations for the synthesis of some value-added chemicals from renewable feedstocks such as glucose and fatty acids [21]. Further development in this endeavor will result in highly effective synthetic pathways for complex unnatural compounds of industrial importance from bio-based starting materials, which would not have been expected before.

Visible light photocatalysis has emerged as one of the eco-friendly and energy sustainable transformations showing great promise for organic synthesis. Combination of photocatalysis with other catalytic modes such as biocatalysis presents tremendous opportunities for the development of new synthetic routes to the target molecules that may not be otherwise accomplishable. In addition, combining lights with enzymes brings the liaison of two fields of "green chemistry," which would greatly benefit both the chemical industry and the environment. In this context, photocatalytic enzymatic reactions have recently drawn considerable attention, and some excellent contributions have been reported [22, 23]. In addition to photocatalytic cofactor regeneration of redox enzymes, one exciting research area is that photoexcitation often induces the redox enzymes to perform nonnatural reactions, leading to catalytic promiscuity. This is because photoexcited molecules commonly generate organic radical intermediates, and the radical-driven reactions thus dominate over the natural reaction of the enzyme. Under visible light, carbonyl reductases and ene reductases catalyze completely different reactions such as asymmetric dehalogenation and C—C bond forming reactions, which are well beyond the native repertoire of these enzymes. The new chemistry renders these enzymes versatile catalysts, allowing access to new reaction products that would be inaccessible by either photocatalyst or enzyme alone [24].

Photochemical reaction has also been combined with biotransformation in a sequential or concurrent one-pot process, in which photochemical reaction generates an intermediate that serves as the substrate or reagent of the biotransformation (or vice versa). This linear cascade of photocatalysis and biocatalysis opens a new avenue for sustainable chemical synthesis [25]. In recent years, photoinduced oxidations have been coupled with various enzymes such as ketoreductase, ene reductase, transaminase cyclohexanone monooxygenase, and hydroxynitrile lyase to effect the oxidation of various substrates and further transformation to other functional groups, providing a new way for C–H activation and functionalization. Photocatalytic chiral amine racemization and imine reduction have been integrated with enzymatic reactions to achieve dynamic kinetic resolution or deracemization of chiral amines, showing attractiveness in the synthesis of enantiopure amines. Chemoenzymatic conversion of plant oils to alkenes has been realized by employing lipase-catalyzed hydrolysis and photoinduced dehydrodecarboxylation, offering a sustainable preparation of bio-based olefins. It can be expected that the scope and applications of photocatalysis and biocatalysis cascades will be rapidly expanded in the coming years.

Electrochemical method has been utilized for cofactor regeneration of redox enzymes, resulting in the so-called bioelectrocatalysis. Electrochemical cofactor regeneration requiring no addition of a second reductant but the electron from the electrode can be easily deployed for all kinds of cofactors, including natural ones and artificial mediators, showing great potential in the bioelectrosytheses of useful chemicals. Recently, bioelectrocatalysis has been widely explored in N_2 and CO_2 reduction and the production of chemicals, biofuels, and materials [26]. Redox enzymes catalyze asymmetric transformation with high enantioselectivity; electrocatalytic coenzyme regeneration has thus been applied in combination with redox enzymes to the synthesis of chiral chemicals such as optically active alcohols, amines, amino acids, and derivatives. A prominent example is the enzymatic reaction cascade that started from a bioelectrocatalytic N_2 fixation system, in which the ammonia product of N_2 reduction was utilized as the amino donor of L-alanine dehydrogenase-catalyzed amination of pyruvate to generate L-alanine, which in turn served as the amino donor of ω-transaminase-catalyzed amination of ketone, affording a chiral amine as the end-product. The reduced $MV^{\cdot +}$ for the reduction of N_2 catalyzed by a nitrogenase and NADH for L-alanine dehydrogenase were synchronously electrochemically regenerated by using a cathode as electron supply, respectively. A variety of high-value chiral amines were produced with >99% ee by this bioelectrocatalytic cascade process [27]. These bioelectrocatalytic processes utilize electricity as a clean power source, thus avoiding the formation of by-products in chemical and enzymatic coenzyme regeneration, and minimizing pollutant generation. It offers a promising strategy to address the challenges that the chemical industry will face in resources and environment in the coming generations.

In summary, the last two decades have evidenced rapid growth in the scope and application of chemoenzymatic cascade reactions. The contributions of photocatalysis and electrocatalysis have shown and will greatly enhance the development of unprecedented new synthetic methodologies for the green production of chemicals. Surely endeavors in chemoenzymatic cascade reactions will continuously expand and play a critical role in the sustainable development of chemical industry in the years to come.

References

1. Zimmerman, J.B., Anastas, P.T., Erythropel, H.C., and Leitner, W. (2020). Designing for a green chemistry future. *Science* 367 (6476): 397–400.
2. Donova, M.V. and Egorova, O.V. (2012). Microbial steroid transformations: current state and prospects. *Applied Microbiology and Biotechnology* 94 (6): 1423–1447.
3. Savile, C.K., Janey, J.M., Mundorff, E.C. et al. (2010). Biocatalytic asymmetric synthesis of chiral amines from ketones applied to sitagliptin manufacture. *Science* 329 (5989): 305–309.
4. Turner, N.J. and O'Reilly, E. (2013). Biocatalytic retrosynthesis. *Nature Chemical Biology* 9 (5): 285–288.
5. Hönig, M., Sondermann, P., Turner, N.J., and Carreira, E.M. (2017). Enantioselective chemo- and biocatalysis: partners in retrosynthesis. *Angewandte Chemie International Edition* 56 (31): 8942–8973.

6 Klucznik, T., Mikulak-Klucznik, B., McCormack, M.P. et al. (2018). Efficient syntheses of diverse, medicinally relevant targets planned by computer and executed in the laboratory. *Chem* 4 (3): 522–532.

7 Struble, T.J., Alvarez, J.C., Brown, S.P. et al. (2020). Current and future roles of artificial intelligence in medicinal chemistry synthesis. *Journal of Medicinal Chemistry* 63 (16): 8667–8682.

8 Delépine, B., Duigou, T., Carbonell, P., and Faulon, J.-L. (2018). RetroPath2.0: a retrosynthesis workflow for metabolic engineers. *Metabolic Engineering* 45: 158–170.

9 William Finnigan, Lorna J. Hepworth, Nicholas J. Turner, Sabine, Flitsch. *RetroBioCat: computer-aided synthesis planning for biocatalytic reactions and cascades.* 2020, Preprint from ChemRxiv, doi: https://doi.org/10.26434/chemrxiv.12571235.v1.

10 Rudroff, F., Mihovilovic, M.D., Gröger, H. et al. (2018). Opportunities and challenges for combining chemo- and biocatalysis. *Nature Catalysis* 1 (1): 12–22.

11 Simon, M.-O. and Li, C.-J. (2012). Green chemistry oriented organic synthesis in water. *Chemical Society Reviews* 41 (4): 1415–1427.

12 Kitanosono, T. and Kobayashi, S. (2020). Reactions in water involving the "on-water" mechanism. *Chemistry – A European Journal* 26 (43): 9408–9429.

13 Markel, U., Essani, K.D., Besirlioglu, V. et al. (2020). Advances in ultrahigh-throughput screening for directed enzyme evolution. *Chemical Society Reviews* 49 (1): 233–262.

14 Dumeignil, F., Guehl, M., Gimbernat, A. et al. (2018). From sequential chemoenzymatic synthesis to integrated hybrid catalysis: taking the best of both worlds to open up the scope of possibilities for a sustainable future. *Catalysis Science & Technology* 8 (22): 5708–5734.

15 Pauly, J., Gröger, H., and Patel, A.V. (2019). Developing multicompartment biopolymer hydrogel beads for tandem chemoenzymatic one-pot process. *Catalysts* 9 (6): 547.

16 Ye, R., Zhao, J., Wickemeyer, B.B. et al. (2018). Foundations and strategies of the construction of hybrid catalysts for optimized performances. *Nature Catalysis* 1 (5): 318–325.

17 Latham, J., Henry, J.-M., Sharif, H.H. et al. (2016). Integrated catalysis opens new arylation pathways via regiodivergent enzymatic C–H activation. *Nature Communications* 7: 11873.

18 Porta, R., Benaglia, M., and Puglisi, A. (2016). Flow chemistry: recent developments in the synthesis of pharmaceutical products. *Organic Process Research & Development* 20 (1): 2–25.

19 Pieber, B., Gilmore, K., and Seeberger, P.H. (2017). Integrated flow processing — challenges in continuous multistep synthesis. *Journal of Flow Chemistry* 7 (3): 129–136.

20 Sperl, J.M., Carsten, J.M., Guterl, J.-K. et al. (2016). Reaction design for the compartmented combination of heterogeneous and enzyme catalysis. *ACS Catalysis* 6 (10): 6329–6334.

21 Wallace, S., Schultz, E.E., and Balskus, E.P. (2015). Using non-enzymatic chemistry to influence microbial metabolism. *Current Opinion in Chemical Biology* 25: 71–79.

22 Schmermund, L., Jurkaš, V., Özgen, F.F. et al. (2019). Photo-biocatalysis: biotransformations in the presence of light. *ACS Catalysis* 9 (5): 4115–4144.

23 Seel, C.J. and Gulder, T. (2019). Biocatalysis fueled by light: on the versatile combination of photocatalysis and enzymes. *ChemBioChem* 20 (15): 1871–1897.

24 Sandoval, B.A. and Hyster, T.K. (2020). Emerging strategies for expanding the toolbox of enzymes in biocatalysis. *Current Opinion in Chemical Biology* 55: 45–51.

25 Chen, J., Guan, Z., and He, Y.-H. (2019). Photoenzymatic approaches in organic synthesis. *Asian Journal of Organic Chemistry* 8 (10): 1775–1790.

26 Chen, H., Dong, F., and Minteer, S.D. (2020). The progress and outlook of bioelectrocatalysis for the production of chemicals, fuels and materials. *Nature Catalysis* 3 (3): 225–244.

27 Chen, H., Cai, R., Patel, J. et al. (2019). Upgraded bioelectrocatalytic N2 fixation: from N2 to chiral amine intermediates. *Journal of the American Chemical Society* 141 (12): 4963–4971.

Index

Chemo-Enzymatic Cascade Reactions, First Edition. Dunming Zhu.

b

d

e

i

m

n

p

u

v